Lecture Notes in Networks and Systems

Volume 1128

The series "Lecture Notes in Networks and Systems" publishes the latest developments in Networks and Systems—quickly, informally and with high quality. Original research reported in proceedings and post-proceedings represents the core of LNNS.

Volumes published in LNNS embrace all aspects and subfields of, as well as new challenges in, Networks and Systems.

The series contains proceedings and edited volumes in systems and networks, spanning the areas of Cyber-Physical Systems, Autonomous Systems, Sensor Networks, Control Systems, Energy Systems, Automotive Systems, Biological Systems, Vehicular Networking and Connected Vehicles, Aerospace Systems, Automation, Manufacturing, Smart Grids, Nonlinear Systems, Power Systems, Robotics, Social Systems, Economic Systems and other. Of particular value to both the contributors and the readership are the short publication timeframe and the world-wide distribution and exposure which enable both a wide and rapid dissemination of research output.

The series covers the theory, applications, and perspectives on the state of the art and future developments relevant to systems and networks, decision making, control, complex processes and related areas, as embedded in the fields of interdisciplinary and applied sciences, engineering, computer science, physics, economics, social, and life sciences, as well as the paradigms and methodologies behind them.

Indexed by SCOPUS, EI Compendex, INSPEC, WTI Frankfurt eG, zbMATH, SCImago.

All books published in the series are submitted for consideration in Web of Science.

For proposals from Asia please contact Aninda Bose (aninda.bose@springer.com).

Paulo J. Sequeira Gonçalves ·
Pradeep Kumar Singh · Sudeep Tanwar ·
Gregory Epiphaniou
Editors

Proceedings of Fifth International Conference on Computing, Communications, and Cyber-Security

IC4S'05 Volume 2

Springer

Editors
Paulo J. Sequeira Gonçalves
IDMEC
Polytechnic Institute of Castelo Branco
Castelo Branco, Portugal

Pradeep Kumar Singh
Department of Computer Engineering
NMIMS Chandigarh
Chandigarh, India

Sudeep Tanwar
Department of Computer Engineering
Institute of Technology
Nirma University
Ahmedabad, Gujarat, India

Gregory Epiphaniou
Department of Security Engineering
University of Warwick
Coventry, UK

ISSN 2367-3370 ISSN 2367-3389 (electronic)
Lecture Notes in Networks and Systems
ISBN 978-981-97-7370-1 ISBN 978-981-97-7371-8 (eBook)
https://doi.org/10.1007/978-981-97-7371-8

This Springer imprint is published by the registered company Springer Nature Singapore Pte Ltd.
The registered company address is: 152 Beach Road, #21-01/04 Gateway East, Singapore 189721, Singapore

Preface

The 5th International Conference on Computing, Communications, and Cyber-Security (IC4S'05) Volume 2 held on 8th–9th April, 2024 and venue for the conference was Graphic Era Hill University (GEHU), Bhimtal Campus, Uttarakhand India. The IC4S'05 Volume 2 inaugural speech was given by Director Prof. (Col) A. K. Nair (Retd.), GEHU-Bhimtal Campus. Brief about the conference was delivered by Dr. Vrince Vimal, Dean, Graphic Era (Deemed to be University). Conference overview and details were shared by Dr. Anupam Singh, Associate Professor, Graphic Era (Deemed to be University), India.

Total 44 papers were considered for publication in (IC4S'05) Volume 2 and these were presented during two days of schedule. The conference found to be fruitful for the Scholars, researchers, and professionals from academia and industry to discuss developments and share ideas their ideas at the International Conference on Computing, Communication, and Cyber Security (IC4S).

The primary goal of the International Conference on Computing, Communications, and Cyber-Security conference is to provide a forum where experts from academia, government, and industry can debate the most recent innovations, trends, and issues in the realms of information technology, communications, and security. Conference-attendees are able to hear about the most recent findings in the field and make connections with like-minded professionals at the same time.

There were five different technical tracks presented at the conference, it includes; (i) Communication and Networks Technologies, (ii) Advanced Computing Technologies, (iii) Data Analytics and Intelligent Learning, (iv) Latest Electrical and Electronics Trend, and (v) Security and Privacy Issues.

The conference was hosted by the Graphic Era Hill University (GEHU), Bhimtal Campus, Uttarakhand India in technical association Erbil Polytechnic University, Erbil, Iraq, CSRL, India, WSG University in Bydgoszcz, Poland and many more, and Microsoft CMT portal was used as a submission system. Professionals from these organizations contributed to the technical aspects of the conference, including paper review, suggestions to improve the quality of the papers, program schedule compilation, technical sessions support, and other support activities.

The welcome address was given by the Prof. (Col) A. K. Nair (Retd.), followed by the Address of Dr. Vrince Vimal, Dean, Graphic Era (Deemed to be University). The talk was given by Prof. Vivek Sehgal, Professor and Head, Jaypee University of IT, Waknaghat, HP, India. The Address were given by Prof. Pradeep Kumar Singh, Dr. Anupam Singh, and by many other experts during this event. The short technical speech were delivered during the technical sessions by Prof. Abhinav Juneja, Professor and Head, CSIT, KIET Group of Institutions, Ghaziabad, India, Dr. Harikesh Singh, Senior Member IEEE and Associate Professor, CSE, JSSATEN, Noida, Prof. Gaurav Dubey, Professor, CSE, KIET Group of Institutions, Ghaziabad, Dr. Yugal Kumar from Jaypee University of IT, Waknaghat, HP and Dr. Jagpreet, Associate Professor, NMIMS Deemed to be University, India.

The organising committee is thankful to anonymous reviewers, TPC members and session chairs who have extended their technical support at various stages. We would like to give a special thanks to Prof. (Dr.) Kamal Ghanshala, President, Graphic Era Group for valuable guidance, constant encouragement and guidance. We extend our thanks to Prof. (Dr.) Sanjay Jasola, Vice Chancellor of the university for his motivation and positive feedback from time to time.

Our special mention to following members for their tireless support for the event in various roles; Prof. Wei-Chiang Hong, Prof. Pao Ann Hsuing, Dr. Farkhana Muchtar, Dr. Pushan Kumar Dutta, Dr. Priyanka Chawla, Dr. Sachin Kumar, Dr. Anand Srivastava, Dr. Amit Sharma, Dr. Brahm Hazela, Dr. Punit Gupta, Dr. Rajesh Gupta, Dr. Rohit Tanwar, Dr. Hari Mohan Rai, Dr. Pramod Yadav, Dr. Smita Agrawal, Dr. Parita Oza, Dr. Yugal Kumar, Dr. Samayveer Singh, Dr. Jitendra Bhatia, Dr. Sapna Juneja, Prof. Malaram Kumhar, Dr. Pronaya Bhattacharya, Dr. Jagpreet, Dr. Rajesh Gupta, Dr. Ashutosh Sharma, Dr. Anupam Singh, Dr. Shabana Ziyad, Dr. Aruna Malik, Dr. Amit Sharma, Dr. Harsh Khatter and many more.

For their time and effort, the authors of these articles have our deepest gratitude. The Technical Program Committee's help throughout the evaluation process is much appreciated by the event's organizers. We appreciate the hard work that went into hosting this event and running the programmes over the course of two days. The organizers of IC4S'05 Volume 2 would like to express their gratitude to the session chairs who presided over the many technical sessions and offered insightful feedback to the writers. During the conference's paper presentation sessions, the session chairs have imparted their technical knowledge to the attendees. The organizing team is grateful to Springer, LNNS Series for their support.

Castelo Branco, Portugal	Paulo J. Sequeira Gonçalves
Chandigarh, India	Pradeep Kumar Singh
Ahmedabad, India	Sudeep Tanwar
Coventry, UK	Gregory Epiphaniou
April 2024	

Contents

Editors and Contributors

About the Editors

Dr. Paulo J. Sequeira Gonçalves is a Professor of Electrotechnical and Industrial Engineering with the School of Technology in the Polytechnic Institute of Castelo Branco, Portugal. He is also a senior researcher at IDMEC, Instituto Superior Técnico, University of Lisbon, Portugal. Dr. Gonçalves was trained in Mechanical Engineering, achieving the M.Sc. and Ph.D. degrees from Instituto Superior Técnico, University of Lisbon, 1998 and 2005, respectively. Dr. Paulo J. Sequeira Gonçalves (P.J.S. Gonçalves) presented more than 100 paper in renowned journal and conferences in robotics, vision and computational intelligence. He served as Chair of the IEEE-CIS (computational Intelligence) Portuguese chapter, was officer of the IEEE RAS working group who developed the first RAS standard and one of the first ontology-based IEEE standards. He received several scientific awards both national and international. He organized and chaired several international conferences, and edited its proceedings. The organized several workshops in top-tier robotic IEEE conferences, e.g., ICRA, IROS and ROMAN. He server as a regular reviewer in top level journals. He was invited visiting professor in European Universities, in the field of robotics. Currently is Co-chair of the IEEE P1872.2—Standard for Autonomous Robotics (AuR) Ontology, and actively works in three other IEEE standards working groups within robotics and automation. He received funding for his research projects, in Principal Investigator and/or Researcher, from the EU, IEEE, private companies and the Portuguese National Research Funding Agency. He works as a regular expert for the European Commission and for the National Accreditation of University Study Programmes. Dr. Paulo J. Sequeira Gonçalves research keywords include Computational Intelligence, Ontologies, Robotics, Computer Vision, Industrial Automation.

Dr. Pradeep Kumar Singh is currently working as a Professor and Associate Dean, Computer Engineering, NMIMS Deemed to be University, India. He is Associate Editor of the IJISMD, [IJISMD is indexed by Scopus and Web of Science], IJAEC,

IGI Global USA, SPY, Wiley and IJISC from Romania. He is recently appointed as a Section Editor, Discover IoT, Springer Journal. He has published nearly 200 research papers. He has received three sponsored research projects grant worth Rs. 25 Lakhs. He has edited a total 16 books from Springer and Elsevier and also edited several special issues for SCI and SCIE Journals from Elsevier and IGI Global. He has Google scholar citations 3250, H-index 27 and i-10 Index 62.

Dr. Sudeep Tanwar is a Professor in the Computer Engineering Department at the Institute of Technology of Nirma University, Ahmedabad, India. He received his Ph.D. in 2016 from the Faculty of Engineering and Technology, Mewar University, India, with a specialization in Wireless Sensor Networks. His current interests include routing issues in WSN, integration of sensors in the cloud, computational aspects of smart grids, and Blockchain Technology. He has authored or co-authored more than 150+ technical research papers published in leading Peer-reviewed International journals and International Conferences from the IEEE, Elsevier, Springer, John Wiley and has authored five books: He is recipient of the Best Research Paper awards from IEEE GLOBECOM-2018 and Springer ICRIC-2019. He is a TPC member and reviewer of many international conferences across the globe. He is an associate editor of the Security and Privacy Journal, and is a member of the IAENG, ISTE, and CSTA. He has Google scholar citations 8405, H-index 52 and i-10 Index 140.

Dr. Gregory Epiphaniou currently holds a position as an Associate Professor of security engineering at the University of Warwick. His role involves bid support, applied research and publications. Part of his current research activities is formalised around a research group in wireless communications with the main focus on crypto-key generation, exploiting the time-domain physical attributes of V-V channels. He led and contributed to several research projects funded by EPSRC, IUK and local authorities totalling over £8M. He is also the main inventor of a patented-pending technology on a distributed ledger system (GB2576160A/US200042497A1). He was previously holding a position as a Reader in Cybersecurity and acted as deputy director of the Wolverhampton Cybersecurity Research Institute (WCRI). He has taught in many universities both nationally and internationally in a variety of areas related to proactive network defence with over 120 international publications in journals and conference proceedings and author in several books and chapters. He holds several industry certifications in Information Security and worked with several government agencies, including the UK MoD, in Cybersecurity related projects. He currently holds a subject matter expert panel position in the Chartered Institute for Securities and Investments. He acts as a technical committee member for several scientific conferences in Information and network security. He serves as a key member in the development of WS5 for the formation of the UK Cybersecurity Council.

Contributors

S. Abhinandhan Department of Computer Science and Engineering, Amrita School of Computing, Amrita Vishwa Vidyapeetham, Chennai, India

Sultan Ahmad Department of Computer Science, College of Computer Engineering and Sciences, Prince Sattam Bin Abdulaziz University, Alkharj, Saudi Arabia;
Department of Computer Science and Engineering, University Center for Research and Development (UCRD), Chandigarh University, Gharuan, Mohali, Punjab, India

Fatma Ahsan Department of Computer Science, Aligarh Muslim University, Aligarh, India

Abdulaziz Fae Alasiri Department of Information Systems, College of Computer Engineering and Sciences, Prince Sattam Bin Abdulaziz University, Alkharj, Saudi Arabia

Abdulaziz Abdulrhman Alhabib Department of Information Systems, College of Computer Engineering and Sciences, Prince Sattam Bin Abdulaziz University, Alkharj, Saudi Arabia

Mazen Bunayan Alharbi Department of Information Systems, College of Computer Engineering and Sciences, Prince Sattam Bin Abdulaziz University, Alkharj, Saudi Arabia

Saule Amanzholova Department of Cyber Security, International IT University, Almaty, Kazakhstan

Fareen Ansari Department of Information Technology, KIET Group of Institutions, Delhi-NCR, UttarPradesh, Muradnagar, Ghaziabad, India

Jeba Ansari Department of Information Technology, KIET Group of Institutions, Delhi-NCR, UttarPradesh, Muradnagar, Ghaziabad, India

Faisal Anwer Department of Computer Science, Aligarh Muslim University, Aligarh, India

Tasleem Arif Department of Information Technology, BGSB University, Rajouri, India

Bhavna Arora Central University of Jammu, Bagla (Rahya-Suchani), Samba, Jammu and Kashmir, India

Oroos Arshi School of Computer Science, University of Petroleum and Energy Studies, Dehradun, Uttarakhand, India

M. Azhagiri Department of Computer Science and Engineering, SRM Institute of Science and Technology, Ramapuram, Chennai, Tamil Nadu, India

Mohamed El Bakkali LPHE-MS, Faculty of Sciences, University Mohammed V in Rabat, Rabat, Morocco

Amul Batra Department of Electrical and Electronics Engineering, Birla Institute of Technology and Science, Pilani, Rajasthan, India

Boutaina Benhmimou LPHE-MS, Faculty of Sciences, University Mohammed V in Rabat, Rabat, Morocco

Neha Bhardwaj Sharda School of Basic Sciences and Research, Sharda University, Greater Noida, India

Chandradeep Bhatt Graphic Era Hill University, Dehradun, India

Ankur Singh Bist Department of CSE, Graphic Era Hill University, Bhimtal, India

Rohan Appasaheb Borgalli Department of Electronics and Telecommunication Engineering, Shah and Anchor Kutchhi Engineering College Chembur, Mumbai, India

Korhan Cengiz Department of Electrical-Electronics Engineering, Istinye University, Istanbul, Turkey

Amit Chanchal Department of Computer Science (AIML and DS), KIT's College of Engineering, Kolhapur, India

Saumya Chaturvedi Department of Computer Science and Engineering, School of Engineering and Technology, Sharda University, Greater Noida, India

Swapnil Chaudhari Department of Computer Science (AIML and DS), KIT's College of Engineering, Kolhapur, India

Aryan Chaudhary BioTech Sphere Research, Kolkata, India

Rahul Chauhan Computer Science and Engineering, Graphic Era Hill University Dehradun, Dehradun, Uttarakhand, India

Gunjan Chhabra Department of CSE, Graphic Era Hill University Dehradun, Dehradun, India;
Graphic Era Deemed to Be University, Dehradun, Uttarakhand, India

Namrata Dhanda Amity University, Lucknow, India

Neel H. Dholakia Marwadi University, Rajkot, Gujarat, India

Deepali Dixit School of Engineering and Technology, Sharda University, Greater Noida, India

A. E. M. Eljialy Department of Information Systems, College of Computer Engineering and Sciences, Prince Sattam Bin Abdulaziz University, Alkharj, Saudi Arabia

Ashish Garg Department of CSE, Graphic Era Deemed to be University, Dehradun, Uttarakhand, India

R. K. Garg Department of Mechanical Engineering, Deenbandhu Chhotu Ram University of Science and Technology, Murthal, India

Sangeeta Garg Mewar Institute of Management, Ghaziabad, Uttar Pradesh, India

Adamya Gaur SWINGER: Security, Wireless, IoT Network Group of Engineering and Research, University School of Automation and Robotics (USAR), Guru Gobind Singh Indraprastha University, Delhi, India

Rahul Gautam School of Computing, Graphic Era Hill University, Haldwani, Uttarakhand, India;
Department of Computer Application, Shri Ram College, Muzaffarnagar, UP, India

Richa Golash SCSET, Bennett University, Greater Noida, India

Drashti Gorasiya Department of Computer Science and Engineering, Devang Patel Institute of Advance Technology and Research, Charotar University of Science and Technology, Changa, India

Josep M. Guerrero CROM, Aalborg University, Aalborg, Denmark

Hardik Gulati Department of Computer Science and Engineering, Amity University, Noida, Uttar Pradesh, India

Gauranshi Gupta SWINGER: Security, Wireless, IoT Network Group of Engineering and Research, University School of Automation and Robotics (USAR), Guru Gobind Singh Indraprastha University, Delhi, India

Janhvi Gupta Department of Information Technology, KIET Group of Institutions, Delhi-NCR, UttarPradesh, Muradnagar, Ghaziabad, India

Kapil Kumar Gupta Amity University, Lucknow, India

Nancy Gupta ECE Department, Lyallpur Khalsa College Technical Campus, Jalandhar, Punjab, India

Prinima Gupta Department of Computer Science and Technology, Manav Rachna University, Faridabad, Haryana, India

Rahul Gupta Department of Information Technology, Delhi Technological University, New Delhi, India

Urshita Gupta Department of Computer Science and Engineering, Amity University, Noida, Uttar Pradesh, India

Vedant V. Gupta Department of CSE, National Institute of Technology, Delhi, India

Uma Gurav Department of Computer Science (AIML and DS), KIT's College of Engineering, Kolhapur, India

Alveera Hafeez Department of ECE, Dronacharya Group of Institutions, Greater Noida, India

Abdullah Hasmani Department of Computer Science and Engineering, Devang Patel Institute of Advance Technology and Research, Charotar University of Science and Technology, Changa, India

Himanshu Hora Department of Computer Application, Shri Ram College, Muzaffarnagar, UP, India

Niamat Hussain Department of Smart Device Engineering, Sejong University, Seoul, South Korea

Zamrooda Jabeen Department of Mathematics, National Institute of Technology, Srinagar, India

Anant Jain SWINGER: Security, Wireless, IoT Network Group of Engineering and Research, University School of Automation and Robotics (USAR), Guru Gobind Singh Indraprastha University, Delhi, India

Archi Jain Department of Computer Science and Engineering, Amity University, Noida, Uttar Pradesh, India

Shilpa Jain Department of CSE, Graphic Era Hill University, Bhimtal, India

Vishal Jain Department of Computer Science and Engineering, School of Engineering and Technology, Sharda University, Greater Noida, India

Aditi Jeena Computer Science and Engineering, Graphic Era Hill University Dehradun, Dehradun, Uttarakhand, India

Rahul Johari SWINGER: Security, Wireless, IoT Network Group of Engineering and Research, University School of Automation and Robotics (USAR), Guru Gobind Singh Indraprastha University, Delhi, India

Mukesh Joshi Department of CSE, Graphic Era Hill University, Bhimtal, India

Rohit Chandra Joshi Department of Computer Science and Engineering, School of Engineering and Technology, Sharda University, Greater Noida, India

Aayush Juyal Department of Computer Science and Engineering, School of Engineering and Technology, Sharda University, Greater Noida, India

Kashyap Kachhadiya Department of Computer Science and Engineering, Devang Patel Institute of Advance Technology and Research, Charotar University of Science and Technology, Changa, India

Raunak Kandoi School of Engineering and Technology, Sharda University, Greater Noida, India

Mohamed Karim LIMAS, Faculty of Sciences Dhar El Mahraz, Sidi Mohammed Ben Abdellah University, Fez, Morocco

Maneet Kaur Lovely Professional University, Jalandhar, Punjab, India

Manpreet Kaur Department of Electrical and Instrumentation Engineering, Sant Longowal Institute of Engineering and Technology, Longowal, Sangrur, Punjab, India

Sandeep Kaur Lovely Professional University, Jalandhar, Punjab, India

Keshav Kaushik Amity School of Engineering and Technology, Amity University, Punjab, India

P. Kaviya Department of Information Technology, Kamaraj College of Engineering and Technology, Madurai, Tamilnadu, India

Khalid El Khadiri Department of Physics, Faculty of Sciences El Jadida, University Chouaib Doukkali of El Jadida, El Jadida, Morocco

Fiza Khan Department of Information Technology, KIET Group of Institutions, Delhi-NCR, UttarPradesh, Muradnagar, Ghaziabad, India

Sumesh Kharnotia Central University of Jammu, Bagla (Rahya-Suchani), Samba, Jammu and Kashmir, India

M. Kishore Department of Information Technology, Kamaraj College of Engineering and Technology, Madurai, Tamilnadu, India

Bhawana Kothari Department of Commerce, Graphic Era (Deemed to Be University) Dehradun, Dehradun, India

Nishant Kothari Marwadi University, Rajkot, Gujarat, India

Adhithi Satish Kumar Department of Computer Science and Engineering, SRM Institute of Science and Technology, Ramapuram, Chennai, India

Akhil Kumar SCSET, Bennett University, Greater Noida, India

Anuj Kumar School of Computing, Graphic Era Hill University, Haldwani, Uttarakhand, India;
Department of Computer Application, Shri Ram College, Muzaffarnagar, UP, India

Pramod Kumar School of Computing, Graphic Era Hill University, Haldwani, Uttarakhand, India;
Department of Computer Application, Shri Ram College, Muzaffarnagar, UP, India

Satish Kumar Department of Information Technology, BGSB University, Rajouri, India

Sumit Kumar Department of Computer Science and Engineering, Amity University, Noida, Uttar Pradesh, India

Rachid Ahl Laamara LPHE-MS, Faculty of Sciences, University Mohammed V in Rabat, Rabat, Morocco

Vipul Ladva Marwadi University, Rajkot, Gujarat, India

Ambica Prakash Mani Department of Commerce, Graphic Era (Deemed to Be University) Dehradun, Dehradun, India

Suhail Manzoor Department of Computer Science and IT, Central University of Jammu, Jammu, India

Apoorva Mathur Amity University, Lucknow, India

Ajaz Hussain Mir Department of Electronics and Communication, National Institute of Technology, Srinagar, India

Gangeshwar Mishra Department of Computer Science and Technology, Manav Rachna University, Faridabad, Haryana, India

Shubhangi Mishra SWINGER: Security, Wireless, IoT Network Group of Engineering and Research, University School of Automation and Robotics (USAR), Guru Gobind Singh Indraprastha University, Delhi, India

Noor Mohd Graphic Era Deemed to be University, Dehradun, India

M. Muthuram Department of Information Technology, Kamaraj College of Engineering and Technology, Madurai, Tamilnadu, India

C. Nandhana Department of Computer Science and Engineering, SRM Institute of Science and Technology, Ramapuram, Chennai, Tamil Nadu, India

Abhishilpa Nandini Department of Computer Science and Engineering, IEC University, Solan, India

M. Niranjanee Department of Computer Science and Engineering, SRM Institute of Science and Technology, Ramapuram, Chennai, Tamil Nadu, India

Fouad Omari LPHE-MS, Faculty of Sciences, University Mohammed V in Rabat, Rabat, Morocco

Nissrine Oubahsis LPHE-MS, Faculty of Sciences, University Mohammed V in Rabat, Rabat, Morocco

Ali Asgar Padaria Department of Computer Science and Engineering, Institute of Technology, Nirma University, Ahmedabad, Gujarat, India

Aditya Pal Department of IT, Dronacharya Group of Institutions, Greater Noida, India

Vandana Pandey School of Computing, Graphic Era Hill University, Haldwani, Uttarakhand, India

Anchal Panwar Computer Science and Engineering, Graphic Era Hill University Dehradun, Dehradun, Uttarakhand, India

Bansari Patel Department of Computer Science and Engineering, Devang Patel Institute of Advance Technology and Research, Charotar University of Science and Technology, Changa, India

Dhyan Patel Department of Computer Science and Engineering, Institute of Technology, Nirma University, Ahmedabad, Gujarat, India

Jitali Patel Department of Computer Science and Engineering, Institute of Technology, Nirma University, Ahmedabad, Gujarat, India

Krishna Patel Department of Computer Science and Engineering, Devang Patel Institute of Advance Technology and Research, Charotar University of Science and Technology, Changa, India

Venkata Pavani Department of Computer Science and Engineering, SRM Institute of Science and Technology, Ramapuram, Chennai, India

Shreya Pungaonkar Department of Computer Science (AIML and DS), KIT's College of Engineering, Kolhapur, India

Hari Mohan Rai Department of Cyber Security, International IT University, Almaty, Kazakhstan;
Department of ECE, Dronacharya Group of Institutions, Greater Noida, India

Ishwari Singh Rajput National Institute of Technology Patna, Patna, Bihar, India;
School of Computing, Graphic Era Hill University, Haldwani, Uttarakhand, India

Suvita Rani Department of Computer Science and Engineering, Sant Longowal Institute of Engineering and Technology, Longowal, Sangrur, India

Amit Rathee Department of Computer Science, Sonipat, Haryana, India

Ruchira Rawat Computer Science and Engineering, Graphic Era Deemed to Be University, Dehradun, Uttarakhand, India

Abdul Razaque Department of Cyber Security, International IT University, Almaty, Kazakhstan

Nemi Rishi SCSET, Bennett University, Greater Noida, India

Ankur Rohilla School of Computing, Graphic Era Hill University, Haldwani, Uttarakhand, India;
Department of Computer Application, Shri Ram College, Muzaffarnagar, UP, India

Ambreen Sabha Department of Computer Science and IT, Central University of Jammu, Jammu, India

Bikash Chandra Sahana Department of Electronics & Communication Engineering, National Institute of Technology Patna, Patna, Bihar, India

Atir Sakhrelia Department of Computer Science and Engineering, Institute of Technology, Nirma University, Ahmedabad, Gujarat, India

Ashok Sapkota Lovely Professional University, Jalandhar, Punjab, India

G. Saranya Department of Computer Science and Engineering, Amrita School of Computing, Amrita Vishwa Vidyapeetham, Chennai, India

Arvind Selwal Department of Computer Science and IT, Central University of Jammu, Jammu, India

Kapil Sharma Department of Information Technology, Delhi Technological University, New Delhi, India

Suvita Rani Sharma Department of Computer Science and Engineering, Sant Longowal Institute of Engineering and Technology, Longowal, Sangrur, Punjab, India

Meetha V. Shenoy Department of Electrical and Electronics Engineering, Birla Institute of Technology and Science, Pilani, Rajasthan, India

Atharvaraj Shivudkar Department of Computer Science (AIML and DS), KIT's College of Engineering, Kolhapur, India

Kaustubh Kumar Shukla Department of ECE, Dronacharya Group of Institutions, Greater Noida, India

Madhu Shukla Marwadi University, Rajkot, Gujarat, India

S. Sibi Siddharthan Department of Information Technology, Kamaraj College of Engineering and Technology, Madurai, Tamilnadu, India

Simran Graphic Era Hill University, Dehradun, India

Abhinav Singh Lovely Professional University, Jalandhar, Punjab, India

Anupam Singh Department of Computer Science and Engineering, Graphic Era Hill University, Dehradun, Uttarakhand, India;
Department of Computer Science and Engineering, Graphic Era Deemed to be University, Dehradun, Uttarakhand, India

Arushi Singh Department of Information Technology, KIET Group of Institutions, Delhi-NCR, UttarPradesh, Muradnagar, Ghaziabad, India

Birmohan Singh Department of Computer Science and Engineering, Sant Longowal Institute of Engineering and Technology, Longowal, Sangrur, Punjab, India

Harpreet Singh Department of Computer Science and Engineering, Sant Longowal Institute of Engineering and Technology, Longowal, Sangrur, India

Raghuraj Singh School of Engineering and Technology, Sharda University, Greater Noida, India

Raj Pal Singh Department of Computer Science and Engineering, Pranveer Singh Institute of Technology, Kanpur, India

Randeep Singh Department of Computer Science and Engineering, IEC University, Solan, India;
IEC University Baddi, Baddi, India

Rashmi Singh Amity Institute of Applied Sciences, Amity University, Noida, Uttar Pradesh, India

Ravendra Singh Department of Computer Science and Engineering, Pranveer Singh Institute of Technology, Kanpur, India

Arieb Ashraf Sofi Department of Electronics and Communication, National Institute of Technology, Srinagar, India

Sompal IEC University Baddi, Baddi, India

A. G. Sreedevi Department of Computer Science and Engineering, Amrita School of Computing, Amrita Vishwa Vidyapeetham, Chennai, India

Jayanth Srinivasan Department of Computer Science and Engineering, SRM Institute of Science and Technology, Ramapuram, Chennai, Tamil Nadu, India

Subodh Srivastava Department of Electronics & Communication Engineering, National Institute of Technology Patna, Patna, Bihar, India

P. D. Surendar Department of Computer Science and Engineering, SRM Institute of Science and Technology, Ramapuram, Chennai, India

Sunil Surve Department of Computer Engineering, Fr. Conceicao Rodrigues College of Engineering, University of Mumbai, Bandra, Mumbai, India

Simrin Fathima Syed Marwadi University, Rajkot, Gujarat, India

Rohit Tanwar School of Computer Science, University of Petroleum and Energy Studies, Dehradun, Uttarakhand, India

Anand Shanker Tewari National Institute of Technology Patna, Patna, Bihar, India

Arvind Kumar Tiwari Kamla Nehru Institute of Technology, Sultanpur, Uttar Pradesh, India

Neeraj Tripathi Shri Mata Vaishnodevi University, Katra, India

Veena Tripathi Model Institute of Engineering and Technology, Jammu, India

Mihul Tyagi School of Engineering and Technology, Sharda University, Greater Noida, India

Sonam Tyagi Department of Electronics & Communication Engineering, National Institute of Technology Patna, Patna, Bihar, India;
School of Computing, Graphic Era Hill University, Haldwani, Uttarakhand, India

Abha Upreti School of Computing, Graphic Era Hill University, Haldwani, Uttarakhand, India

Vaishali Vadhavana Department of Computer Science and Engineering, Devang Patel Institute of Advance Technology and Research, Charotar University of Science and Technology, Changa, India

Deo Prakash Vidyarthi School of Computer and Systems Sciences, Parallel and Distributed System Lab JNU, Delhi, India

Surbhi Vijh Department of Computer Science and Engineering, Amity University, Noida, Uttar Pradesh, India

Om Vyas Department of Computer Science and Engineering, Institute of Technology, Nirma University, Ahmedabad, Gujarat, India

Rutuja Warke Department of Computer Science (AIML and DS), KIT's College of Engineering, Kolhapur, India

Communication and Networks Technologies

Incorporation of the Self-attention Mechanism into Convolutional Neural Networks for the Target Recognition

Gangeshwar Mishra, Prinima Gupta, and Rohit Tanwar

Abstract The study focuses on the increasing complexity of automated target recognition (ATR) in surveillance and reconnaissance. It highlights the need for quick and accurate identification in crucial situations and emphasizes the role of technical improvements in addressing this problem. The study incorporates a self-attention mechanism into Convolutional Neural Networks (CNNs) architectures such as YOLOv8, DenseNet-121, InceptionV4, ResNet-152, and VGG16, with a specific focus on categorizing ground targets in Synthetic Aperture Radar (SAR) pictures. This systematic improvement incorporates self-attention modules carefully positioned at critical spots in the network architectures. The research utilizes the renowned Moving and Stationary Target Acquisition and Recognition (MSTAR) Mixed Targets dataset. YOLOv8 is the best classifier, with an outstanding total accuracy of 97.65% and it has high specificity of 98.86% and precision of 96.62% with the testing. This study emphasizes the need to thoroughly assess CNNs architectures to make educated decisions when choosing models designed explicitly for complicated target identification tasks. It offers valuable insights that may contribute to developments in ATR technology.

Keywords CNN · Deep learning · Imaging processing · ML · VGG-16

G. Mishra · P. Gupta
Department of Computer Science and Technology, Manav Rachna University, Faridabad, Haryana, India
e-mail: prinima@mru.edu.in

R. Tanwar (✉)
School of Computer Science, University of Petroleum and Energy Studies, Dehradun, Uttarakhand, India
e-mail: rohit.tanwar.cse@gmail.com

P. J. S. Gonçalves et al. (eds.), *Proceedings of Fifth International Conference on Computing, Communications, and Cyber-Security*, Lecture Notes in Networks and Systems 1128, https://doi.org/10.1007/978-981-97-7371-8_1

1 Introduction

ATR which refers to the identification of targets without human involvement, is a substantial obstacle in modern surveillance and reconnaissance endeavors. This difficulty grows more prominent with technology improvements, requiring rapid and precise identification of objects, especially in military and disaster response scenarios. In order to tackle this difficulty, our study aims to develop a complete model that specifically focuses on classifying ground targets, with a particular emphasis on SAR images [1]. The analysis uses the esteemed MSTAR Mixed Targets dataset, which was made available by the Air Force Research Laboratory. The main goal is to create a model that can differentiate and classify ground objects using complex characteristics seen in SAR data. The MSTAR dataset consists of a wide range of SAR pictures that display several target categories, such as Infantry Fighting Vehicles (IFVs) and tanks. These photos were collected by X-band sensors operating in beam mode. The high-resolution pictures, characterized by their unique patterns, provide a great dataset for the training and validation of our proposed model [2, 3].

The use of CNNs for SAR analysis has become an attractive approach, motivated by the complex difficulties presented by SAR picture processing. The use of CNNs in SAR image processing demonstrates innovation by harnessing their outstanding capacities to capture intricate characteristics and correlations across diverse data categories. The implementation of transfer learning significantly enhances the transformational impact of CNNs on SAR image analysis. This approach uncovers hidden patterns and extracts important insights from raw SAR image data, eventually allowing precise and successful categorization [4].

The integration of the Self-attention Mechanism into CNNs entails a methodical approach targeted at improving the model's capacity to concentrate on relevant features throughout the process of feature extraction. The procedure starts with the modification of the CNN architecture to include self-attention layers. These layers provide varying weights to input characteristics depending on their relative importance. The attention weights are acquired during training, enabling the network to prioritize crucial spatial correlations within the data. During the forward pass, the attention-enhanced features are calculated by taking into account the weighted combination of input characteristics, highlighting important information that is essential for the current task. This method enhances the flexibility of the CNN by allowing it to effectively capture distant relationships and enhance its performance in tasks like image recognition and object identification. The integration of the Self-attention Mechanism represents a significant progress in deep learning, augmenting the comprehensibility and distinguishing capability of CNNs for diverse applications [5, 6].

The incorporation of the self-attention mechanism into CNNs like YOLOv8 [7], DenseNet-121 [8], InceptionV4 [9], ResNet-152 [10], and VGG16 [11] entails a methodical improvement of their structures to capture more informative characteristics. This technique effortlessly integrates self-attention modules into the current network topologies. Attention methods are intentionally included at crucial locations throughout the YOLOv8, DenseNet-121, InceptionV4, and VGG16 networks [5–7]. This enables the networks to concentrate on pertinent spatial characteristics when extracting data. The research emphasizes how the difficulty of ATR in modern surveillance and reconnaissance is increasing as technology progresses.

- A thorough approach for categorizing ground objects using SAR image is proposed in the work. This customized technique improves target identification.
- SAR analysis using CNNs is novel because CNNs can capture complex characteristics and correlations across data categories.
- The study integrates the Self-attention Mechanism into CNNs to enhance their architectures. This helps models concentrate on essential characteristics during feature extraction, improving picture recognition and object identification. The suggested technique may be applied to several CNN architectures, including YOLOv8, DenseNet-121, InceptionV4, ResNet-152, and VGG16.

2 Literature Review

The state-of-the-art CNN and Self-attention Mechanism SAR image identification research is summarized here.

Deep learning-based SAR automated target recognition algorithms perform better in real-world scenarios, according to the research, when trained with high-quality synthetic SAR images. Prominent MSTAR SAR data was utilized by the researchers. For posture synthesis and training data rife with debris, novel techniques for augmenting data were implemented utilizing target scattering center sparsity. Effectively managing out-of-distribution samples and enhancing the precision of target recognition, these techniques suggest a military or broader application [1].

The radar high-resolution range profile (HRRP) target detection job uses a CNN with a multiscale window self-attention technique. The presented self-attention module intelligently separates CNN features into equal-width bands utilizing sliding windows to extract multi-level features from various areas via continuous window expansion. The experimental findings suggest that the proposed technique is better in HRRP recognition, indicating its potential for radar target identification [5].

CNN-Bidirectional Encoder Representations from Transformers (CNN-BERT), uses High-Resolution Range Profiles' spatio-temporal structure. The model uses convolutional token embedding to create sequence features to characterize the target's local spatial structure. A multi-head self-attention mechanism allows the BERT module to acquire long-term temporal correlations across range cells in HRRP. Notably, the research introduced a cost function that covered recognition and rejection capabilities simultaneously [6].

CC-YOLO is the suggested highway driving scene-specific vehicle identification. It improved YOLOv5s architecture with numerous important changes. The C3 module in the Backbone and PAN-FPN is replaced with the C2F module from YOLOv8, improving vehicle detection while remaining lightweight. The PAN-FPN also has a lightweight upsampling operator called CARAFE to improved small target vehicle detection. An integrated attention module improved detecting head vehicle identification. CC-YOLO achieved a 6.8% better Precision than YOLOv5s [7].

RadarFormer is a suggested technique that performs human perception tasks by analyzing radar echoes without relying on picture technology. RadarFormer showed that the self-attention mechanism's radar echoes processing is as expressive as convolutional layers for image processing. The RadarFormer model, inspired on the Transformer idea, utilizes fast-/slow-time self-attention [12].

Bidimensional empirical mode decomposition (BEMD) and deep residual networks developed a new SAR target detection method. BEM extracts multi-mode representations, called bidimensional intrinsic mode functions (BIMFs), that described target features in SAR pictures. Each BIMF layer has a ResNet trained for it to use these BIMFs. Multiple BIMF choices are combined using a random weight matrix. With the MSTAR dataset, the suggested technique was extensively tested [13].

UAV surveillance data was used to detect real-world targets in the Russia-Ukraine conflict using YOLO V5 and V7 object identification algorithms. The object detection performance of YOLO V5 was enhanced by rigorous training. An extensive analysis of performance metrics provided valuable insights [14].

A novel CNN method combined pruning and knowledge distillation. A lightweight student network was developed via layer-by-layer systematic pruning and distilled using an 'unpruned teacher' network. Productivity improves with weight sharing. Experimental results showed that the suggested strategy compressed MSTAR recognition tasks and lowered computing expenses [15].

3 Material and Methods

MSTAR's SAR images were used as the primary source data. The study systematically integrated the self-attention mechanism into several CNNs such YOLOv8, DenseNet-121, InceptionV4, ResNet-152, and VGG16 for a comprehensive evaluation and comparison. The study delves into an in-depth analysis of several CNN structures integrated with self-attention to evaluate their specific effectiveness in target recognition tasks.

3.1 Dataset and Preprocessing

The analysis used the publicly available MSTAR dataset for analysis [3]. The MSTAR target dataset comprises a wide assortment of 8,688 SAR pictures, collected from seven land vehicles and a calibration target. The labels were automatically given according to the classification of the picture files (Table 1).

Table 1 Dataset distribution [3]

Assigned label	Number of images
2S1	1,164
BRDM-2	1,415
BTR-60	451
D7	573
SLICY	2,539
T62	572
ZIL131	573
ZSU-23/4	1,401

The dataset served as an indispensable foundational resource for target recognition, particularly for individuals engaged in CNN and image processing. Advancements in the automation of target detection and the analysis of radar signals were expedited by the availability of this technology [3, 5]. The research encompassed the creation of a comprehensive training set comprising 75% of the input images from the dataset, with the remaining 25% left aside for evaluation objectives. During the preprocessing stage, the dataset was artificially enhanced through the application of diverse modifications, including contrast adjustments, random rotations, and flips. The approach employed increased the size of the dataset, thereby bolstering the model's capacity to extrapolate findings from a wide range of conditions and viewing angles. Consequently, a substantial enhancement in the performance of target recognition was observed [16].

3.2 CNNs

CNNs are a pivotal element of deep learning and have significantly transformed the domain of computer vision and image processing. CNNs are very proficient in extracting hierarchical representations from visual data. This makes them exceptionally effective for tasks like picture identification, object detection, and segmentation [17]. CNNs are distinguished by their intricate, hierarchical structures, including convolutional layers that autonomously acquire spatial hierarchies of features via filters or kernels. These filters allow the network to detect and analyze patterns, edges, and textures, gradually creating intricate visual representations [18]. CNNs have played a crucial role in producing cutting-edge outcomes across many visual tasks, showcasing their adaptability and efficiency in addressing complex visual challenges.

YOLOv8. The acronym "YOLO" stands for "You Only Look Once." YOLOv8 known as version 8 of YOLO, is an advanced object identification system that has achieved notable recognition. YOLOv8 is highly regarded for its ability to analyze photos in real-time and accurately recognize and locate objects, building on the achievements of its previous versions. The architecture adheres to the YOLO principle, which entails partitioning a picture into a grid and making simultaneous predictions of bounding boxes and class probabilities for each grid cell. YOLOv8 is the result of progress in deep learning methods, integrating breakthroughs such a more scalable structure, upgraded backbone network, and improved training procedures [7].

DenseNet-121. DenseNet-121 used a densely linked layer topology. The notion was presented in an original way by establishing connections between each layer and all the subsequent layers in a feedforward manner. This methodology facilitated the reuse of features and enhanced the flow of gradients throughout the training process, hence addressing the problem of disappearing gradients. Consequently, DenseNet-121 demonstrated impressive performance while using fewer parameters, hence improving computing efficiency [8].

InceptionV4. InceptionV4 is a cutting-edge CNN architecture developed specifically to improve the performance of image processing jobs. InceptionV4 improved upon previous models by introducing a new inception module that includes filters of different sizes. This allows the model to efficiently capture elements of different scales. The integration of residual connections and factorized convolutions was implemented to tackle the issue of vanishing gradients and accelerate the convergence process during training [9].

ResNet-152. ResNet-152 is a well-known CNN architecture that is highly regarded for its ability to accurately identify images. It is distinguished by its impressive depth, consisting of 152 layers. By using residual connections, this approach effectively tackled the problem of vanishing gradients, hence enabling fast training even with complex models. The presence of skip connections in the design allowed for smooth propagation of gradients during training, hence enabling the creation of deeper networks [10].

VGG16. VGG16, or Visual Geometry Group 16, is a well-known convolutional neural network architecture for image recognition. The University of Oxford Visual Geometry Group created VGG16, which contains 13 convolutional layers, five max-pooling layers, and three completely connected layers. Because of its simplicity and interpretability, this model has become a deep learning standard and image categorization benchmark [11].

3.3 *Incorporation of the Self-attention Mechanism into CNNs*

3.3.1 Self-attention Mechanism

The Self-Attention Mechanism is a crucial element in deep learning systems, namely in the field of natural language processing and computer vision. The Self-Attention Mechanism is a technique that improves the capability of models to understand long-range connections and contextual information [5]. It enables a neural network to assign various levels of significance to distinct parts in a sequence. The Self-Attention Mechanism is used in computer vision applications, specifically in image analysis, to effectively capture the spatial connections among pixels. This enables the model to choose concentrate on important areas. The broad acceptance of this method is due to its adaptability and efficacy in capturing intricate patterns and dependencies [6, 12].

3.3.2 Incorporation

Integrating attention mechanisms into the CNNs entails including modules that enable the network to selectively concentrate on certain segments of the input data. Below is a Steps for integrating attention mechanism that explains the process of incorporating attention processes into the specified CNN architectures:

a. **Module for self-attention**

1. Input Transformation: Calculate three sets of linear transformations for the input features: Query (Q), Key (K), and Value (V).
2. Attention Scores: Compute attention scores by performing the dot product between the Query and Key matrices.
3. Rescaling Attention Weights: Adjust the attention scores and use the Softmax algorithm to calculate the attention weights.
4. Calculation of Weighted add: Perform a multiplication operation between the attention weights and the Value matrix, and then add up the resulting values.
5. Output: The output is the calculated sum of the characteristics, taking into account their respective weights [5, 7].

3.3.3 Application and Illustrations of CNN Architectures

- Incorporate the self-attention module into the CNN design, often after specific convolutional layers.
- Modify the characteristics of the input to match the requirements of the attention module, and then merge the attended features with the original ones.
- The ResNet-152 architecture employs residual blocks. Attention modules included into individual residual blocks and placed after certain blocks.

- DenseNet-121 exhibits dense connection. Attention modules included either inside tightly linked blocks or inserted between blocks.
- InceptionV4 consists of many parallel branches. Attention might be included either inside each branch or after certain blocks. It incorporates attention blocks to enhance spatial and channel-wise attention.
- The architecture of VGG16 is characterized by a simple but profound design. Attention modules may be included subsequent to designated convolutional layers. It presents the concept of SE blocks, which are used to implement channel-wise attention.
- YOLOv8 is specifically designed for the purpose of detecting objects. Attention methods, like the CSPNet (Cross-Stage Partial Network) used in YOLOv4, may be included into the backbone CNN. It incorporates cross-stage feature fusion and attention processes [7, 14].

3.4 *Performance Metrics*

Performance metrics measure ML model performance. Accuracy and F1 score are popular performance indicators. Accuracy is the key classification performance metric. It measures model prediction accuracy. So, it evaluates how well the model classifies data. The ratio of true positives and negatives to data points defines accuracy. Predicting the majority class may give the model great accuracy with imbalanced datasets, which may be deceiving [19]. F1 scores combine precision and recall [20]. It is computed with precision and recall. Precision is the percentage of true positive predictions over all positive predictions, whereas recall is the proportion over all positive occurrences. Since it counts false positives and negatives, the F1 score is a great performance metric for imbalanced datasets. High F1 scores indicate model correctness and recall [21, 22].

4 Results and Discussion

This investigation aimed to compare the performance of CNN architectures in target recognition. An assessment of different architectures using the specified metrics offers valuable information about their performance on the given job. The comparative examination of these architectures offers useful information for choosing a reliable model (Fig. 1).

The YOLOv8 CNN architecture exhibits the highest overall accuracy, reaching a remarkable 97.65%. The classifier exhibits outstanding performance in terms of specificity (98.86%) and precision (96.62%), underscoring its robustness across all evaluation metrics. The accuracy of the VGG-16 model is 94.21%, ranking it as the second highest among all models. With a remarkable specificity of 98.93% and precision of 95.45%, it successfully identifies authentic negatives while preventing

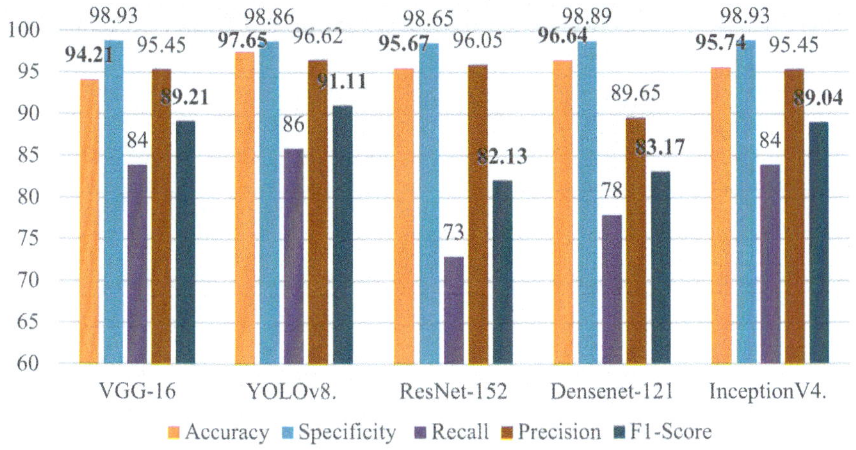

Fig. 1 Outcomes of the employed CNN architectures with Self-Attention

the occurrence of false positives. In comparison to alternative classifiers, ResNet-152 exhibits the lowest recall rate of 73%, indicating a marginally diminished capacity for precise true positive detection. While the achieved accuracy of 95.67% is deemed satisfactory, the reduced recall suggests an inherent challenge in accurately identifying every pertinent instance of the target class.

Figure 2 displays the superior performance of YOLOv8 in terms of accuracy and loss across 20 epochs. For the task, hyperparameters are critical to model performance. Learning rate is 0.001, batch size is 64, epochs are 20, optimizer is Adam, are employed hyperparameters. The learning rate affects optimization step size, impacting training speed and stability. Memory utilization and convergence speed depend on batch size and the amount of samples analyzed before updating model parameters. The model is trained by passing the full dataset through the model epochs. Optimizer choice affect optimization and model generalization. Grid search or random search should be used to determine the best hyperparameters for the dataset and model architecture.

This research investigation conducted a systematic evaluation of several CNN structures for the purpose of target identification. The objective was to determine their individual performances by applying specified metrics in a thorough assessment. The comprehensive evaluation of these architectures offers detailed insights essential for carefully selecting a robust model in the context of target recognition. The notable discoveries indicate that the YOLOv8 CNN architecture is the most skilled, showcasing outstanding accuracy. YOLOv8's exceptional performance establishes it as a leading classifier, showcasing its remarkable ability to detect and categorize targets reliably. The VGG-16 model effectively reduces false positives by achieving the second-highest accuracy. The scientifically rigorous results are essential for making educated decisions when choosing a suitable CNN architecture specifically designed for the complexities of target identification tasks.

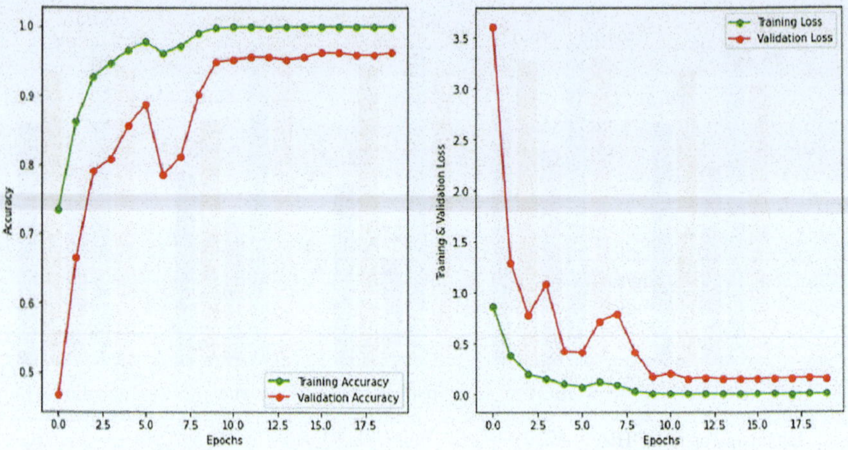

Fig. 2 Accuracy and loss of the YOLOv8

5 Conclusion

The study has focused on the daunting problem of Target Recognition in the contemporary period, highlighting the growing intricacy of this undertaking due to technological progress. The primary objective is to create a detailed model specifically designed for classifying ground targets, with a particular focus on SAR images. This highlights the dedication to improving accuracy in identifying objects. The self-attention technique has been systematically included into eminent CNN designs like as YOLOv8, DenseNet-121, InceptionV4, ResNet-152, and VGG16 to improve their ability to capture more informative features. The cautious incorporation of self-attention modules at strategic points throughout these networks has produced valuable outcomes. The YOLOv8 CNN architecture has been recognized as the leading performer, suggesting possible constraints in reliably identifying all relevant occurrences of the target category. While this study has achieved noteworthy accomplishments, it is crucial to recognize its limitations. The analysis largely concentrated on a certain set of CNN architectures, and future examination with supplementary designs may lead to more extensive findings. Moreover, the research primarily focused on SAR imaging, and expanding the assessment to additional modalities might enhance the comprehensiveness of target detection. To further enhance the work, there should be more focus on improving the integration of the self-attention mechanism and investigating its effects on other advanced CNN designs. Conducting more research on the generalization of the suggested model across other datasets and real-world situations will improve its practicality.

References

1. Geng Z, Xu Y, Wang B-N, Yu X, Zhu D-Y, Zhang G (2023) Target recognition in SAR images by deep learning with training data augmentation. Sensors 23:941. https://doi.org/10.3390/s23020941
2. Shi J (2022) SAR target recognition method of MSTAR data set based on multi-feature fusion. 2022 International conference on big data, information and computer network (BDICN). https://doi.org/10.1109/bdicn55575.2022.00120
3. SAR Target Classification using Deep Learning- MATLAB & Simulink, https://www.mathworks.com/help/radar/ug/sar-target-classification-using-deep-learn-ing.html#mw_rtc_SARTargetClassificationUsingDeepLearningExample_M_65EF9A8A. Accessed 26 December 2023
4. Vint D, Anderson M, Yang Y, Ilioudis C, Di Caterina G, Clemente C (2021) Automatic target recognition for low resolution foliage penetrating SAR images using CNNs and GANs. Remote Sensing 13:596. https://doi.org/10.3390/rs13040596
5. Diao Y, Liu S, Gao X, Liu A, Zhang Z (2022) CNN based on multiscale window self-attention mechanism for radar HRRP target recognition. 2022 7th International conference on signal and image processing (ICSIP). https://doi.org/10.1109/icsip55141.2022.9887058
6. Wang P, Chen T, Ding J, Pan M, Tang S (2022) Intelligent radar HRRP target recognition based on CNN-BERT model. EURASIP J Adv Signal Process, https://doi.org/10.1186/s13634-022-00909-9
7. Zhang S, Ge J (2023) Vehicle detection algorithm for highway driving scenarios. Fifth International conference on artificial intelligence and computer science (AICS 2023). https://doi.org/10.1117/12.3009232
8. Liuzhuang: GitHub - liuzhuang13/DenseNet: densely connected convolutional networks, In CVPR 2017 (Best Paper Award)., https://github.com/liuzhuang13/DenseNet
9. Kapoor A, Shah R, Bhuva R, Pandit T (2020) Understanding inception network architecture for image classification, https://doi.org/10.13140/RG.2.2.16212.35204
10. He K, Deep residual learning for image recognition, https://arxiv.org/abs/1512.03385v1.
11. Simonyan K, Zisserman A (2014) Very deep convolutional networks for large-scale image recognition, arXiv:1409.1556
12. Zheng Z, Zhang D, Liang X, Liu X, Fang G (2023) RadarFormer: end-to-end human perception with through-wall radar and transformers. IEEE Trans Neural Netw Learn Syst, 1–15, https://doi.org/10.1109/tnnls.2023.3314031
13. Miao X, Liu Y (2021) Target recognition of SAR images based on complex bidimensional empirical mode decomposition. Sci Program 2021:1–10. https://doi.org/10.1155/2021/6642316
14. Kumar S, Kumar C (2023) Deep learning based target detection and recognition using YOLO V5 algorithms from UAVs surveillance feeds. 2023 International conference for advancement in technology (ICONAT), https://doi.org/10.1109/iconat57137.2023.10080677
15. Zhang F, Liu Y, Zhou Y, Yin Q, Li H-C (2020) A lossless lightweight CNN design for SAR target recognition. Remote Sensing Lett 11:485–494. https://doi.org/10.1080/2150704x.2020.1730472
16. Mao C, Huang L, Xiao Y, He F, Liu Y (2021) Target recognition of SAR image based on CN-GAN and CNN in complex environment. IEEE Access. 9:39608–39617. https://doi.org/10.1109/access.2021.3064362
17. Kumar S, Kumar H (2023) Classification of COVID-19 X-ray images using transfer learning with visual geometrical groups and novel sequential convolutional neural networks. MethodsX 11:102295. https://doi.org/10.1016/j.mex.2023.102295
18. Kumar S, Kumar H (2023) PneuML: a novel sequential convolutional neural network-based X-ray diagnostic system for pneumonia in contrast to machine learning and pre-trained networks. The Scientific Bull Series C Electr Eng Comput Sci 85(4):119–136, https://www.scientificbulletin.upb.ro/SeriaC_-_Inginerie_Electrica_si_Stinta_Calculatoarelor.php?page=revistaonline&a=1&cat=C

19. Kumar S, Kumar H (2023) Lung cancer diagnosis using X-Ray and CT scan images based on machine learning approaches. Proceedings of fourth international conference on computing, communications, and cyber-security, 399–412, https://doi.org/10.1007/978-981-99-1479-1_30
20. Kumar S, Kumar H (2022) Lungcov: a diagnostic framework using machine learning and Imaging modality. Int J Techn Phys Problems Eng (IJTPE) 14(51), Number 2. http://mail.iotpe.com/IJTPE/IJTPE-2022/IJTPE-Issue51-Vol14-No2-Jun2022/23-IJTPE-Issue51-Vol14-No2-Jun2022-pp190-199.pdf
21. Kumar S, Kumar H, Agarwal R, Pathak VK (2022) Human disease prognosis and diagnosis using machine learning. Lecture Notes Electr Eng, 41–53, https://doi.org/10.1007/978-981-19-0284-0_4
22. Kumar S, Kumar H, Kumar G et al (2024) A methodical exploration of imaging modalities from dataset to detection through machine learning paradigms in prominent lung disease diagnosis: a review. BMC Med Imaging 24:30. https://doi.org/10.1186/s12880-024-01192-w

An Efficient Approach to Lip-Reading with 3D CNN and Bi-LSTM Fusion Model

Rohit Chandra Joshi, Aayush Juyal, Vishal Jain, and Saumya Chaturvedi

Abstract Communication occurs commonly between two or more individuals and often it involves the use of speech. However, under certain circumstances, the use of speech may be restricted causing a hindrance in communication. These circumstances may include noisy environments, people with hearing disability, and corrupt audio in forensic analysis. The alternative to speech in this situation can be to visually analyze the content from the patterns of lip movements to deduce the message. This phenomenon is known as lip-reading. Lip-reading can be intricate using the naked eye for an individual. Artificial Intelligence (AI) has gained recent attention in many fields, with speech recognition being one of them. The use of Machine Learning (ML) and Deep Learning (DL) algorithms, falling under the umbrella of Artificial Intelligence, can help build Automatic Lip-Reading (ALR) systems that can ease the process of lip-reading. This work aims to build an ALR system while obtaining a great performance compared to some of the state-of-the-art methods on the Grid corpus dataset. The proposed model is segregated into two parts: frontend and backend. The front end employs a 3D Convolutional Neural Network (CNN) and the back end uses a Bi-directional Long Short-Term Memory (Bi-LSTM) architecture. The model attained a Word Error Rate (WER) of 0.093, outperforming several baseline models.

Keywords Deep learning · Visual speech recognition · Lip-reading · Convolutional neural network · Long short-term memory

---

R. C. Joshi · A. Juyal (✉) · V. Jain · S. Chaturvedi
Department of Computer Science and Engineering, School of Engineering and Technology, Sharda University, Greater Noida 201310, India
e-mail: aayushjuyal12@gmail.com

R. C. Joshi
e-mail: rohit.chandra.joshi.rcj@gmail.com

V. Jain
e-mail: vishal.jain@sharda.ac.in

S. Chaturvedi
e-mail: saumyanmishra5@gmail.com

P. J. S. Gonçalves et al. (eds.), *Proceedings of Fifth International Conference on Computing, Communications, and Cyber-Security*, Lecture Notes in Networks and Systems 1128, https://doi.org/10.1007/978-981-97-7371-8_2

1 Introduction

6.1% of the world population suffers from a hearing impairment restricting them from understanding the speech spoken by other people [1]. Thus, communication between these individuals and the remaining population is hindered since it can be difficult to understand each other without speech. The common medium, for communication, in such situations, is to use sign language. However, sign language is tedious to learn and it is not commonly known around the world by everyone. Lip-reading is thereby a tool that can be utilized for communication by analyzing and perceiving lip movements to retrieve meaningful content. Often, lip-reading can be the only method for communication due to external circumstances i.e., noisy environments. Lip-reading is found to be applicable in numerous domains. For example, it can be used to resolve judicial events where the lip movements of individuals can be decoded from corrupted camera video recordings lacking audio [2]. Though lip-reading has the potential to be an extremely useful tool, it can be difficult to perform just using the naked eye as there are many challenges associated with it.

Humans perceive speech by integrating the waveforms corresponding to the speech information along with the lip positions and movement, referred to as visemes, that hold significant details about the content being spoken [3]. Lip-reading can be challenging as it is constant to geometric transformations and the rate of words being spoken (movements of lips) can be too high. The other challenges include environmental factors like background noise and the different poses of lips for the same word making it difficult to recognize the phoneme. These are just a few of the research gaps witnessed in previous work. Thus, the aim is to develop a method or an algorithm that is robust to such changes, and the system is indifferent to varying lighting conditions and different skin-colored people, as well as be person-independent.

AI can be further divided into two categories: ML and DL. Algorithms of these respective categories can be utilized to extract and classify the lip movements to get a result. It uses algorithms that fall under the umbrella of supervised, unsupervised, and reinforcement learning [4]. The common examples of some ML algorithms include Support Vector Machines (SVM) [5], Hidden Makarov Model (HMM) [6], Decision Tree (DT) [7], etc. DL uses computational models to represent abstract characteristics from a complex depiction of data [8]. While ML uses hand-crafted features to represent the data, DL automatically extracts the most relevant features, usually, resulting in greater results. The different algorithms often used are CNN [9], LSTM [10], Deep Belief Networks [11], etc.

Our work aims to propose a robust model that can perform lip-reading effectively and, in the future, have real-time applications to help the deaf and mute communicate with the rest of the population with ease. The main contribution of this work is summarized as follows:

- This work proposes a robust model that performs lip-reading using the grid corpus dataset.
- This paper reviews the work of existing authors using different DL algorithms to operate on the same dataset used.

- This paper analyses the results of the proposed model and compares it with the existing work proposed by other authors.

The rest of the paper is structured as follows: Sect. 2 reviews the existing work of other authors. Section 3 discusses the methodology behind lip-reading. Section 4 analyses the results and discusses them. Section 5 concludes the work and discusses the future scope.

2 Literature Review

Lip-reading is a tedious task to carry out using AI and its algorithms since there are many factors to consider to make this task accurate. In this article, several different AI models have been reviewed solving different issues while performing this task. Table 1 summarizes the findings of different researchers.

ML algorithms were the first AI algorithms to be applied to perform lip-reading. Puviarasan et al. [12] used Discrete Wavelet Transform (DWT) and Discrete Cosine Transform (DCT) to extract the features. The features then act as inputs to the HMM model to classify the visual speech.

The attention has now turned to DL models since ML algorithms are not suitable for dealing with continuously changing frames in video datasets. Vayadande et al. [13] used a combination of 3D CNN and LSTM to automate the process of lip-reading. This DL-based approach was highly effective in unseen datasets in comparison to ML-based approaches. Adeel et al. [14] proposed a novel approach using both the advantages of DL and acoustic modeling. It consists of two levels, where the first level employs a DL-based regression algorithm and the second level harnesses an enhanced visually derived Weiner filter (EVWF) to approximate a clean audio power spectrum using a stacked LSTM. The system failed to operate in noisy environments and also had poor generalizability. Qu et al. [15] had the objective of investigating the effect of cross-modal self-supervised pre-training for reconstructing speech by using audio-visual streams. An architecture, referred to as LipSound2 is proposed made up of an encoder-decoder framework and an attention mechanism to correspond sequences of face images to mel-spectrograms. Wang et al. [16] made use of transformers that have gained attention from researchers in recent years. To improve the performance of lip-reading systems, the authors developed a framework consisting of a 3D convolution and a vision transformer. This architecture takes maximum benefit of both the convolution and transformer in extracting local and global features via continuous images. Similarly, Koumparoulis et al. [17] introduced the EfficientNet framework to perform classification and improved the robustness of the backend module by using a transformer encoder. Using a transformer resulted in the system being five times less computationally expensive than some standard architectures. Similarly, Lu et al. [18] developed their system, 'LipPass' that can use the mouth of a user's speaking to extract distinctive behavior features using acoustic sensing. They evaluated the Doppler profiles of acoustic signals produced by the users and

Table 1 A summary of the previous work conducted by authors in recent years

Reference	Dataset	Major contribution	Algorithm
Puviarasan et al. [12]	Custom-built	Extracted features using DCT and DWT and compared their performance	HMM
Vayadande et al. [13]	Custom-built	Able to recognize spoken content in the case of corrupted audio and noisy environments	3D CNN + LSTM
Adeel et al. [14]	Grid and ChiME3	Uses the benefits of both DL and acoustic modeling for speech enhancement	LSTM
Qu et al. [15]	Grid and TCD-TIMIT	Investigated the effect of self-supervised pretraining for the conversion of video to audio by using the co-occurrence of audio and video data	LipSound (encoder-decoder + attention mechanism)
Wang et al. [16]	LRW and LRW-1000	Focused on extracting the subtle characteristics of small-scale images to improve the performance	3DCvt (3D Convolution + Vision Transformer)
Koumparoulis et al. [17]	LRW	Use the power of EfficientNet to classify images and propose the removal of the max-pooling layer in the 3D convolution to reach maximum performance	EfficientNet + Transformer encode
Lu et al. [18]	Custom-built	Discovered unique characteristics from the Doppler profile of users speaking	LipPass (Autoencoder)
Courtney et al. [19]	LRS2 and LRS3-TED	The model was able to selectively pick which spatiotemporal scales are most significant for a specific dataset	Convolutional LSTMs
Mesbah et al. [20]	AVLetters, OuluVS2 and LRW	Reduce the computational cost of training an algorithm by focusing on Hahn moments	HCNN
Chen et al. [21]	Mandarin lip-reading dataset	Developed a framework capable of performing lip-reading using Mandarin language	DenseNet + resBi-LSTM

discovered that there are distinctive mouth movements which get characterized using a DL method.

Many models suffer from being computationally expensive since they apply spatiotemporal modeling characteristics. For instance, Courtney et al. [19] explore the idea of learning spatio-temporal features using convolutional LSTMs, with training time becoming high. Mesbah et al. [20] developed an architecture to overcome such issues using Hahn moments along with CNN. The HCNN uses Hahn moments to retrieve the most significant information most effectively and less redundantly, along with the classification power of CNN.

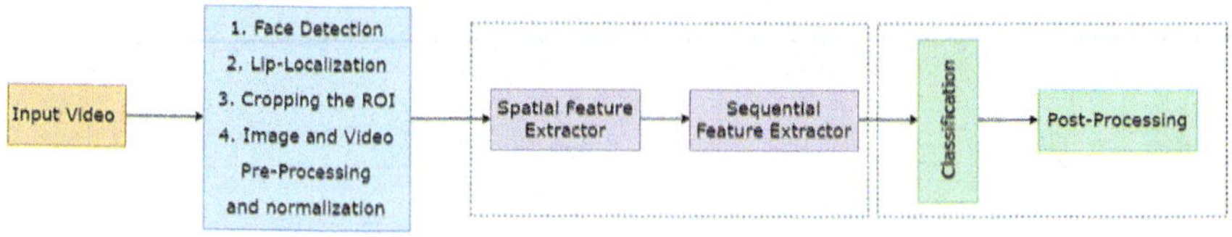

Fig. 1 Lip-reading pipeline describing the flow of information between the different modules. The input is a set of video frames, which then get pre-processed and sent to the feature extractors to capture temporal and spatial dynamics with classification and post-processing being the final steps

A lot of work has been done in the English language for lip-reading, and many authors have worked in different languages. For instance, Chen et al. [21] developed a dataset for the Mandarin language where a Chinese phonemic transcription referred to as Hanu Pinyin was used as a label. A 3D CNN integrated with DenseNet and a residual Bi-LSTM is used as the frontend and the multi-head attention stack gets applied to transform the Hayu Pinyin into Chinese characters.

Our proposed solution looks to overcome some of the basic problems that many authors have dealt with and propose a model surpassing the performance of some state-of-the-art architectures. Figure 1 illustrates the basic overview of the lip-reading process used. The system takes a sequence of video frames as input. Face detection and lip localization help retrieve the region of interest to perform the necessary operations. Preprocessing and normalization help the model reduce its variance in pixel intensities making it robust to variations in conditions, image quality, and environment. The spatial feature extractor involves using a 3D CNN and the sequential feature extractor aims to use a Bi-LSTM. A fully-connected layer is used to classify the input video sequence to textual content.

3 Methodology

The ALR system aims to predict the sentences being uttered by extracting the pattern of lip movements. This section discusses the architecture proposed to predict the sentences from a given video sample.

3.1 Proposed Architecture

The architecture developed consists of two modules: the front end and the back end. The frontend module consists of a 3D CNN model that acts as a spatial feature extractor. It takes a set of video frames of lip movements as input and applies the convolution process to extract spatial patterns to understand the content. The backend module involves performing temporal modeling with DL models that handle sequential data. These models can learn to capture the temporal dynamics

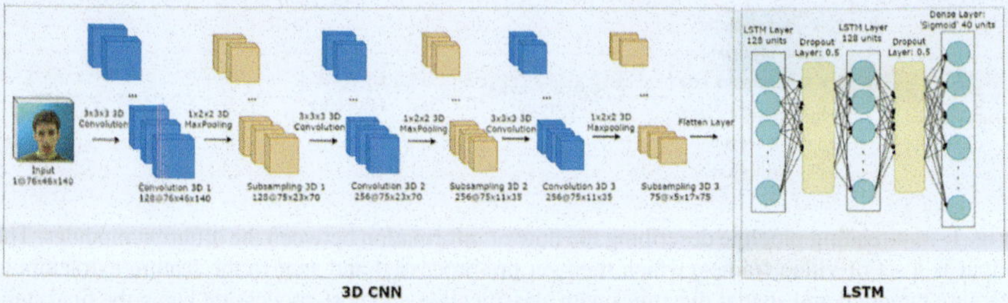

Fig. 2 Detailed design representing the front end, 3D CNN, to capture spatiotemporal features, and the backend, Bi-LSTM, to capture sequential information

of lip patterns, allowing consideration of long-term dependencies between distinctive frames. Finally, the fully connected layer is a classifier to predict the associated phonemes or sentences uttered in the input video. The choice for 3D CNN as compared to other models like 2D CNN is because the model is indifferent to temporal shifts, which enables them to deal with quick variations in speech tempo. 3D CNN allows for capturing spatial and temporal information and there is minimal information loss when processing the frames. Bi-LSTM has the power of capturing temporal dynamics and by processing video frames in both directions, it can learn both future and past context for the frames, which is crucial. Though the model is complex, it does take into account several different scenarios under which it has to undergo rigorous training e.g., environmental factors, poses, etc. Figure 2 illustrates all the modules in a detailed manner, showing how each frame gets processed along the front-end and back-end modules.

3.2 *Dataset*

The dataset used in this experiment is the Grid corpus containing video sequences of different speakers uttering different sentences [22]. There are over 500 videos for all the different 34 speakers (18 male and 16 female), where each video lasts approximately 6 s. Each video has a frame rate of 25 fps where there are variations in lighting conditions and poses to make the model generalize well to all situations. There were roughly 34,000 phrases uttered across all the videos. Figure 3 represents a sample view of the dataset being used throughout this process. Table 2 summarizes the statistics about the dataset.

3.3 *Front-End: 3D CNN*

CNN is a type of DL model that specializes in dealing with image datasets, where they have the convolution layer that can automatically extract features from the input

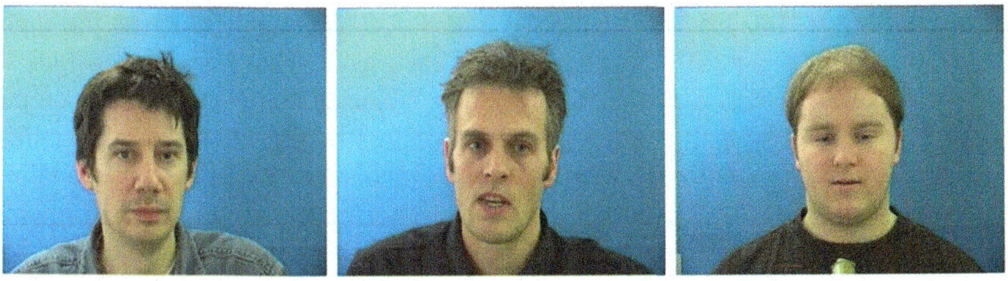

Fig. 3 Sample snapshot of Grid corpus dataset of three different speakers [22]

Table 2 Statistics of the Grid corpus dataset showing the details about the videos of the 34 speakers

Language	Speakers	Classes	Utterances	Resolution	Frame rate
English	34	34,000	34,000	720 × 576	25

data. Figure 4 represents all the layers used in a basic CNN model. The input layer will involve taking an image as input to forward it further to the convolution layers. The convolution layer has the core task of performing the convolution operation to produce a new feature map with the help of a filter or kernel. The filter gets moved over the input image matrix and the final result gets retrieved by multiplying each pixel value of the input image with the kernel values. This process gets repeated for all the values of the input image. Given the size of the image is $i \times j$, with filter size being f_i and f_j, the size of the feature map generated would be $(i - f_i + 1) \times (j - f_j + 1)$.

The second layer that gets used after the convolution layer is the pooling layer. The pooling layers help downsample the spatial dimensions of the output matrix from the convolution layer. This results in reducing the number of parameters being used and thus minimizes the computation and size of the model. Finally, the fully connected layer acts as a classifier to help output the corresponding label to the given input. The input video frame as seen in Fig. 2 has the dimensions of $1@76 \times 46 \times 140$, where 1 refers to a single channel, usually meaning the images are in grayscale, 76 is the height of the frame, 46 is the width and 140 is the temporal dimension

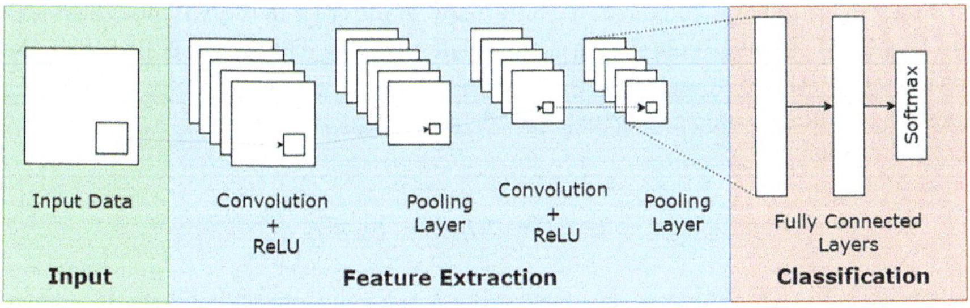

Fig. 4 CNN architecture consists of 2 convolution layers, using ReLU activation function, 2 pooling layers, and a fully-connected layer, using a SoftMax activation function for classification

corresponding to 140 frames. A filter of 3 × 3 × 3 gets applied across the input feature vector to generate 128 feature maps, each having the same dimensions as the input feature vector. A pool size of 1 × 2 × 2 is used to downsample the output feature vectors. The result is a set of feature vectors having been reduced in size. This process further continues where 2 more convolution layers and max pooling layers get applied, with the flatten layer being the final layer that transforms the multi-dimensional feature maps into a one-dimensional array so that it can be fed into a fully-connected layer. The choice for choosing 3 convolutions and 3 pooling layers was because it allowed for our model to capture the relevant information while ensuring the model does not get over complicated. The proposed model was experimented with different numbers of layers and it was discovered that by using too many layers the model was overfitting. Thus, the models best work came with 3 layers.

3.4 Back-End: Bi-LSTM

LSTM is a part of the RNN models that were introduced by two scientists in 1997 named Schmidhuber and Hochreiter. RNNs can deal with sequential data as they have a memory where the output produced in a preceding step gets used as input for the following step. The issue that comes with RNNs is that they are unsuccessful in capturing long-term dependencies from the sequential data because of the 'vanishing gradient' problem. The neural networks can remember both the recent and past information. As the time series data gets forwarded as a sequence to the input layer, it is sent to the output layer via the previous state. The classification process begins after the last piece of data, sent as a sequence, to the input layer finishes processing. Finally, a fully connected layer is used to transform the output obtained from the output state into probabilities. Bi-LSTM is an extension of the LSTM model where the input sequences get processed in both directions. The core components of Bi-LSTM are the gates namely, input gate, forget gate, update gate, and output gate. These gates hold the responsibility of controlling the flow of data between the forward and backward passes.

Figure 5 is a generalized figure to show the workings of a Bi-LSTM model. As seen the Bi-LSTM involves calculating a candidate value c_t, which refers to values that have the potential to be introduced to the cell state at a particular time. Equation (1) shows a candidate value can be calculated.

$$c_t = \tanh(W_c \cdot [h_{t-1}, x_t] + b_c) \tag{1}$$

where W_c is the weight matrix corresponding to the cell state update gate, h_{t-1} refers to the output data at time sequence $t - 1$, x_t is the input data at time t, and b_c is the bias value. The values calculated from the update, forget and output gates is shown as following respectively.

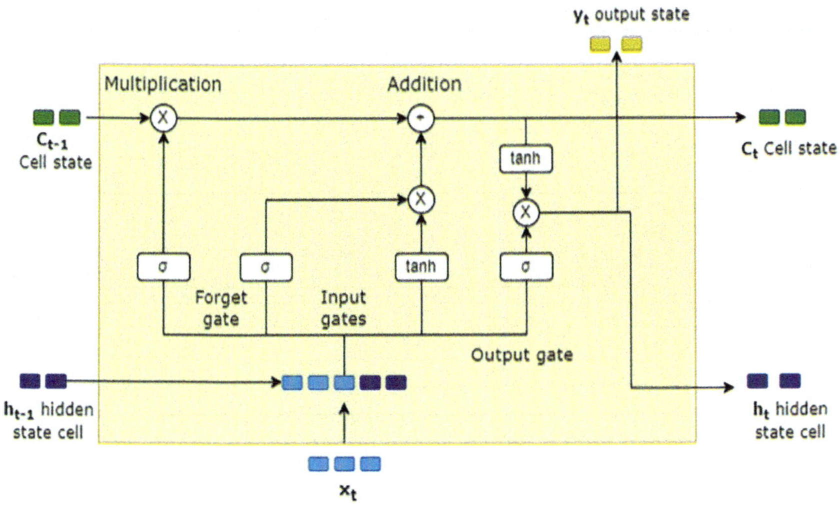

Fig. 5 Bi-LSTM architecture showing how information gets processed using the gating mechanisms including input, forget, and output gates and the hidden, cell and output state storing information

$$\Gamma_u = \sigma(W_u \cdot [h_{t-1}, x_t] + b_u) \tag{2}$$

$$\Gamma_f = \sigma(W_f \cdot [h_{t-1}, x_t] + b_f) \tag{3}$$

$$\Gamma_o = \sigma(W_o \cdot [h_{t-1}, x_t] + b_o) \tag{4}$$

where Γ_u, Γ_f, Γ_o refers to the update gate value, forget gate value and output gate value. The final cell state value is calculated as shown.

$$c_t = \Gamma_u * c_{\sim t} + \Gamma_f * c_{t-1} \tag{5}$$

$$h_t = \Gamma_o * c_t \tag{6}$$

As shown in Fig. 2, the output from the 3D CNN is a flattened one-dimensional array that gets into a Bi-LSTM network to perform temporal modelling. 2 LSTM layers get used, each containing 128 hidden units. 128 hidden units were used, since using a large number of hidden units results in an overfitting problem. A dropout layer gets used with a probability of 0.5, helping in tuning some of the neurons making the model generalize well, and reduce the chances of overfitting. The output from the LSTM layers gets fed into a dense layer to perform the classification and output the phrase said in the corresponding input.

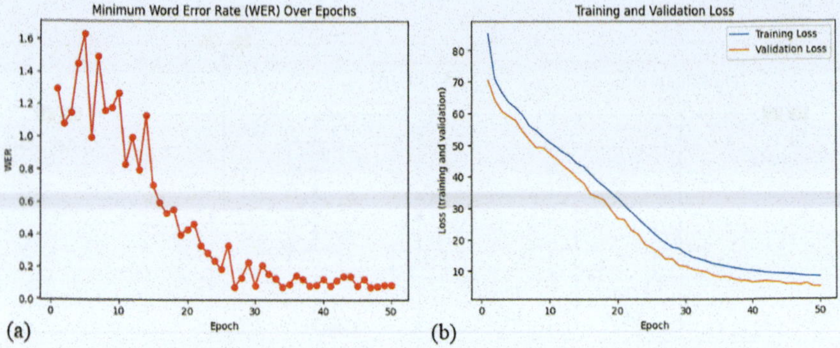

Fig. 6 **a** Illustration of WER of the proposed model, for speaker 1, over 50 epochs **b** Illustration of training and validation loss of the proposed model, for speaker 1, over 50 epochs

4 Results and Discussion

The efficiency of our model was tested on the Grid corpus dataset, consisting of 34 speakers. In this section, the results of our model tested on 3 speakers will be discussed and compared to some of the state-of-the-art models. The reason for choosing limited speakers is because of the time constraint. The time taken to train each speaker was long, thus the speakers trained were limited. The metric used to evaluate the performance of the model was WER, which is a metric that measures the percentage of incorrectly recognized words.

4.1 Results: Speaker 1

This section visualizes and discusses the results for speaker 1. Figure 6a represents the variation of WER over 50 epochs. It initiates with a value of 1.3 and has a huge spike in first 5 iterations, but then begins to decrease gradually over the next 45 iterations. The final WER value is around 0.15. Figure 6b depicts the validation loss and training loss over 50 epochs. It has a smooth and gradual decrease over the 50 epochs.

4.2 Results: Speaker 2

Speaker 1 had a massive increase in WER in the first few iterations. Speaker 2 seems to have its WER decrease gradually in the first few iterations and then decrease rapidly after the 10th epoch as shown in Fig. 7a. It begins with a value of 1.4 at the

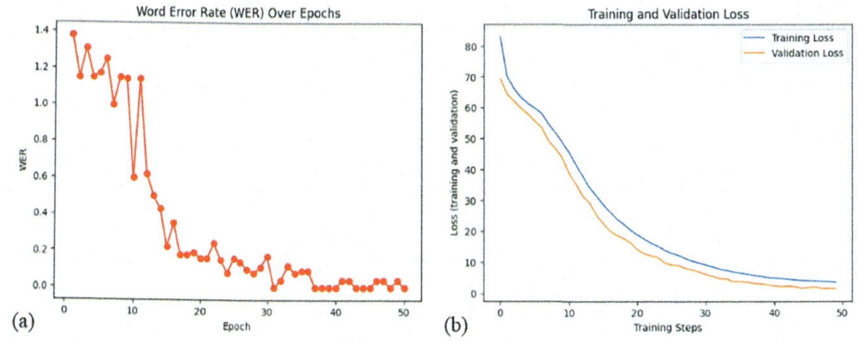

Fig. 7 **a** Illustration of WER of the proposed model, for speaker 2, over 50 epochs **b** Illustration of training and validation loss of the proposed model, for speaker 2, over 50 epochs

first epoch and then ends with a value of 0.04. The training and validation loss graphs of speaker 2, as illustrated in Fig. 7b, are quite similar to that of speaker 1.

4.3 Results: Speaker 3

Figure 8a gives the visual representation of WER for the third speaker. It starts off with the larger value, in comparison to the other two speakers, of 2.2. It then has a gradual decrease over the next 50 epochs. The final WER is 0.09. Figure 8b) visually depicts the validation and training loss, as it seems to be quite similar to that of speaker 1 and speaker 2.

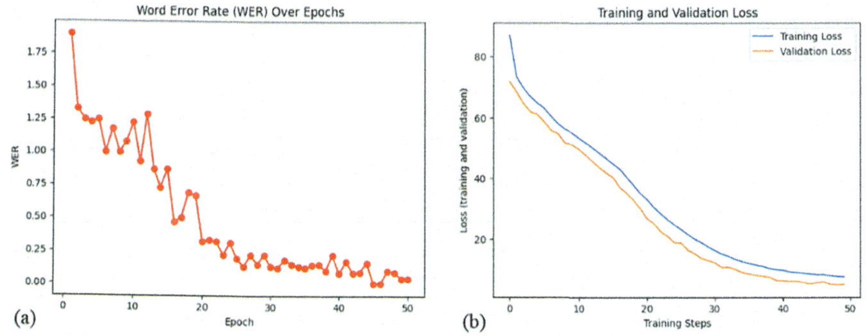

Fig. 8 **a** Illustration of WER of the proposed model, for speaker 3, over 50 epochs **b** Illustration of training and validation loss of the proposed model, for speaker 3, over 50 epochs

Table 3 WER of all three speakers

Speaker	WER
S1	0.15
S2	0.04
S3	0.09

Table 4 Performance of other researchers' proposed model on the Grid corpus using metrics such as WA and WER

Reference	Model	Evaluation Metric	Performance
Huang et al. [23]	Transformer	WA	45.81%
Sarhan et al. [24]	HLR-net	WER	0.097
Chen et al. [25]	DualLip	WER	0.0271
Li et al. [26]	CNN + GRU	WA	86%

4.4 Discussion

Our model aimed to perform lip-reading while tuning some of the parameters to make it as accurate as possible. Table 3 summarizes the results for all the three speakers. The average WER of all three speakers comes out to be 0.093. Table 4 highlights the work of other researchers performing the same task on the Grid corpus dataset. As shown, our proposed model outperformed some of them, whereas Huang et al. [23] only achieved an accuracy of 45.81%. Similarly, the WER of Sarhan et al. [24] had a WER of 0.097, which was 0.4 worse than ours. Li et al. [26] had a word accuracy (WA) of 86%, so in terms of WER, that is still greater than ours since our word accuracy would be approximately 90.7%. Overall, our model was able to outperform three of the four models proposed by other researchers. One of the challenges faced was the time required to train the system. To run 50 epochs, it was taking roughly 5 h. Also, the variation in WER seems to be too much, we aim to reduce that variation and have a gradual decrease to a low value. This outcome of this research was a fusion model that was trained on the Grid corpus dataset. It has a possible application to be intregated within an android application to be used by deaf and mute people to communicate with the rest of the population. It can find it use in forensic analysis to capture the context from a video of corrupt audio.

5 Conclusion and Future Scope

Visual speech recognition is a crucial task that can be performed via lip-reading by developing ALR systems using different ML and DL algorithms. Our proposed model uses a fusion of 3D CNN and Bi-LSTM RNN architecture to capture the spatial and temporal dependencies respectively. The model operated on 3 of the 34 speakers in the Grid corpus dataset. The performance of the model showed that our model had an error of 0.093 when trying to predict a word being uttered from a video, outperforming several existing models that operated on the same dataset. One of the limitations faced during the experiment was the time taken to train the model, thus the experiment could be performed only on 3 speakers. Another drawback was that the WER was varying too much over the epochs as a steadier decrease was expected. Therefore, as a future scope of work, several pre-trained models can be explored to possibly reduce training time so that it can be operated on many speakers. Some of the previous work used pre-trained models like ResNet and GoogleNet to perform lip-reading, so that can be some of the models that can be investigated in the future. Transformers have a self-attention mechanism that can help gain more accurate results, which can also be explored. Integrating these attention mechanisms can allow the model to focus on certain regions of the lip movement which aids in prioritizing relevant areas within the frames, allowing to improve the performance.

References

1. Bhaskar S, Thasleema TM (2023) LSTM model for visual speech recognition through facial expressions. Multimed Tools Appl 82:5455–5472. https://doi.org/10.1007/s11042-022-12796-1
2. Berkol A, Çolak M, Erdem H (2022) Visual lip reading dataset in Turkish. Data 8(1):15. https://doi.org/10.3390/data8010015
3. Miled M, Messaoud MAB, Bouzid A (2023) Lip reading of words with lip segmentation and deep learning. Multimed Tools Appl 82:551–571. https://doi.org/10.1007/s11042-022-13321-0
4. Singh R, Singh A, Bhattacharya P (2022) A machine learning approach for anomaly detection to secure smart grid systems. In: Research anthology on smart grid and microgrid development, pp 911–923. IGI global
5. Gupta A, Tadanki N, Berry N, Bardae R, Harikrishnan R, Wagle SA (2023) A comparative study on different machine learning algorithms for predictive analysis of stock prices. In: Tanwar S, Wierzchon ST, Singh PK, Ganzha M, Epiphaniou G (eds) Proceedings of fourth international conference on computing, communications, and cyber-security. CCCS 2022. Lecture Notes in Networks and Systems, vol 664. Springer, Singapore. https://doi.org/10.1007/978-981-99-1479-1_44
6. Yang H, Zhu D (2020) Improved detection algorithm for copy number variations based on hidden Markov model. Multimed Tools Appl 79:9237–9253. https://doi.org/10.1007/s11042-019-7368-z
7. Ghiasi MM, Zendehboudi S (2020) Application of decision tree-based ensemble learning in the classification of breast cancer. Comput Biol Med 128:104089. https://doi.org/10.1016/j.compbiomed.2020.104089
8. Baba A, Bonny T (2023) FPGA-based parallel implementation to classify Hyperspectral images by using a convolutional neural network. Integration 92:15–23. https://doi.org/10.1016/j.vlsi.2023.04.003
9. Anusuya V, Vignesh Saravanan K, Vishnu Praba V (2023) Real-time face mask detection using convolution neural network and computer vision. In: Tanwar S, Wierzchon ST, Singh PK, Ganzha M, Epiphaniou G (eds) Proceedings of fourth international conference on computing, communications, and cyber-security. CCCS 2022. Lecture Notes in Networks and Systems, vol 664. Springer, Singapore. https://doi.org/10.1007/978-981-99-1479-1_52

10. Nagabushanam P, Thomas George S, Radha S (2020) EEG signal classification using LSTM and improved neural network algorithms. Soft Comput 24:9981–10003, https://doi.org/10.1007/s00500-019-04515-0
11. Zhang H, Huang T, Liu S et al (2020) A learning style classification approach based on deep belief network for large-scale online education. J Cloud Comp 9:26. https://doi.org/10.1186/s13677-020-00165-y
12. Puviarasan N, Palanivel S (2011) Lip reading of hearing impaired persons using HMM. Expert Syst Appl 38(4):4477–4481
13. Vayadande K, Adsare T, Agrawal N, Dharmik T, Patil A, Zod S (2023) LipReadNet: a deep learning approach to lip reading. In: 2023 International conference on applied intelligence and sustainable computing (ICAISC), pp 1–6. IEEE
14. Adeel A, Gogate M, Hussain A, Whitmer WM (2019) Lip-reading driven deep learning approach for speech enhancement. IEEE Trans Emerg Topics Comput Intell 5(3):481–490
15. Qu L, Weber C, Wermter S (2022) LipSound2: self-supervised pre-training for lip-to-speech reconstruction and lip reading. IEEE Trans Neural Netw Learn Syst
16. Wang H, Pu G, Chen T (2022) A lip reading method based on a 3D convolutional vision transformer. IEEE Access 10:77205–77212
17. Koumparoulis A, Potamianos G (2022) Accurate and resource-efficient lipreading with efficientnetv2 and transformers. ICASSP 2022–2022 IEEE international conference on acoustics, speech and signal processing (ICASSP), Singapore, Singapore, pp 8467–8471, https://doi.org/10.1109/ICASSP43922.2022.9747729
18. Lu L, Yu J, Chen Y, Liu H, Zhu Y, Kong L, Li M (2019) Lip reading-based user authentication through acoustic sensing on smartphones. IEEE/ACM Trans Networking 27(1):447–460
19. Courtney L, Sreenivas R (2020) Using deep convolutional LSTM networks for learning spatiotemporal features. In: Palaiahnakote S, Sanniti di Baja G, Wang L, Yan W (eds) pattern recognition. ACPR 2019. Lecture Notes in Computer Science(), vol 12047. Springer, Cham. https://doi.org/10.1007/978-3-030-41299-9_24
20. Mesbah A, Berrahou A, Hammouchi H, Berbia H, Qjidaa H, Daoudi M (2019) Lip reading with Hahn convolutional neural networks. Image Vis Comput 88:76–83
21. Chen X, Du J, Zhang H (2020) Lipreading with DenseNet and resBi-LSTM. SIViP 14:981–989
22. Zhang T, He L, Li X, Feng G (2021) Efficient end-to-end sentence-level lipreading with temporal convolutional networks. Appl Sci 11(15):6975
23. Huang H, Song C, Ting J, Tian T, Hong C, Di Z, Gao D (2021) A novel machine lip reading model. Procedia Comput Sci 199:1432–1437. https://doi.org/10.1016/j.procs.2022.01.181
24. Sarhan AM, Elshennawy NM, Ibrahim DM (2021) HLR-net: a hybrid lip-reading model based on deep convolutional neural networks. Comput Mater Continua 68(2):1531–1549
25. Chen W, Tan X, Xia Y, Qin T, Wang Y, Liu TY (2020) DualLip: a system for joint lip reading and generation. In Proceedings of the 28th ACM international conference on multimedia, pp 1985–1993
26. Li X, Neil D, Delbruck T, Liu S-C (2019) Lip reading deep network exploiting multi-modal spiking visual and auditory sensors. 2019 IEEE international symposium on circuits and systems (ISCAS), Sapporo, Japan, pp 1–5,https://doi.org/10.1109/ISCAS.2019.8702565

Deep Residual Networks Including Transfer Learning for Facial Emotion Identification

Abhinav Singh and Maneet Kaur

Abstract Facial expression recognition by computer vision is an important task with many possible uses. To improve the classification effect, we developed and created a lightweight convolutional neural network (CNN) in this research that can detect face expressions both in real-time and in bulk. We build a real-time vision system to test the effectiveness of our approach. Automated facial impression recognition with real-time accurate interpretation is one of the assistive technologies that can be used to address the problem mentioned previously. With this method, seven facial expressions can be clearly identified and communicated to the relevant parties. This model employs a partial transfer of learning methodology, employing a CNN that has been specially trained to recognise face emotions. A new model is developed that can transfer characteristics between datasets. Making use of the freshly trained CNN at the core of the suggested method is a lightweight, portable facial expression recognition system that has excellent detection accuracy and wireless connection. The model also employs global average pooling instead of traditional fully connected layers. Finally, the FER-2013 dataset is used to test our model. Merely 2.9% of the 16 GB of RAM, or 0.396 GB memory, are required to do the face expression classification assignment. The weight of our model can be stored in 872.9 kilobytes files and has achieved 85% accuracy on the FER-2013 dataset. Additionally, it has strong recognition and detection capabilities for figures that are not included in the collection.

Keywords ResNet50 · Facial emotion recognition · Transfer learning · Light weight CNN model · Emotion classification

---

A. Singh (✉) · M. Kaur
Lovely Professional University, Jalandhar, Punjab, India
e-mail: tabhinavsingh222@gmail.com

M. Kaur
e-mail: maneet.15709@lpu.co.in

P. J. S. Gonçalves et al. (eds.), *Proceedings of Fifth International Conference on Computing, Communications, and Cyber-Security*, Lecture Notes in Networks and Systems 1128, https://doi.org/10.1007/978-981-97-7371-8_3

1 Introduction

One of the most important components of human emotion communication is facial expression, which is utilised in a variety of contexts including advertising, healthcare, human–computer interaction, and education. Furthermore, excellent results are obtained when facial expression recognition (FER) is performed with RGB images that are gathered in a lab setting. Large-scale and real-world photographs are among the RGB datasets that have been employed in a number of recent research. However, differences in unrelated facial expressions need to be addressed in order to increase FER effectiveness. In computer vision, there are two main ways to use FER methods. The initial method carries out identification. Whereas the other is an end-to-end method that carries out recognition, employing manually created feature vectors [1–3] using the use of automatically derived deep neural network-based features [4, 5]. In Shan et al.'s study [6], two manually designed feature-based methods—"A local binary pattern (LBP) and a Gabor filter" —were evaluated for feature extraction process time and memory costs. It was demonstrated that the "LBP"-based approach may reduce the space and time requirements by around ten times when it's compared to the "Gabor filter". A fusion feature was employed by Liu et al. [7] that comprised an "LBP and a histogram of the oriented gradient (HOG)" that was taken from prominent regions of a face. Making use of the "SVM" classifier and "dense scale-invariant feature" transform descriptor, the maximum accuracy of 75.43% was achieved for the four facial emotions. Additionally, they categorised eight different facial expressions with the maximum accuracy of 46.7% using the HOG and SVM algorithms. One of the most significant ways that people communicate their personal feelings is through their facial expressions. Sometimes a person's facial expressions may reveal a lot about their inner thoughts. Capturing the subject's emotional changes through facial emotions is the primary purpose of facial expressions. There is greater diversity in facial expressions as compared to other communication modalities. It is simpler to unintentionally convey someone's actual sentiments. Ekman [8] initially classified expressions into six categories in 1971. These categories included happiness, sorrow, fear, disgust, surprise, and rage. The FER-2013 dataset has been updated with a normal expression [7]. Furthermore, as we can see, manually sorting them is challenging. Additionally, humans have an accuracy of 63%–65% when classifying photographs of faces according to the seven emotions. The most sophisticated techniques for working with images, like object recognition and image classification, rely on pooling layers rather than fully connected layers and combine depth-wise separable convolution with residual modules to subtly reduce many parameters and simplify our network architecture. The most sophisticated image processing techniques, such object recognition and picture classification, rely on convolution neural networks. To address computer-aided detection problems, Shin et al. [2] use deep convolutional neural networks; their model involves 5 thousand to 160 million parameters, requiring a high level of computer hardware; Chang et al. [3] built a convolution neural network for feature extraction from the input images. Laerence and Giles [6] propose a hybrid neural network that combines a self-organizing map (SOM) neural network, a convolutional neural network, and local image sampling in combination. Furthermore, using the FER-2013 dataset, the recognition rate accuracy was 67%. Tang [8] presented a framework integrating a convolutional neural network (CNN) and SVM in deep neural network-based techniques. In the FER2013 competition, the suggested model had the greatest identification rate (71.2%) [9]. The CNN architecture of such tasks requires millions of parameters [10], making deployment on embedded devices challenging. When applying large convolution kernels to high-dimensional feature graphs in order to directly reduce dimensionality, GoogleNet [11] and AlexNet [12] do not overcome. Additionally, the number of parameters can be further compressed and the model's complexity reduced by using continuous big convolution kernels rather than tiny convolution kernels. They demonstrated that their model could increase accuracy by 2.98% in comparison to a prior work [4], and on the FER2013 dataset, with an accuracy of 85%, the group of seven models generated the highest results. However, a number of CNN learning techniques have been researched in order to lower computing costs [11], enhance the generalisation performance of deep neural network models, and solve class-imbalance issues. Dropout was used by Wu et al. [5] to illustrate how it affects max-pooling and convolutional layers. In this study, we propose and develop a real-time, bulk face emotion detection system using convolution neural networks. In order to gradually minimize a huge number of parameters and simplify our network topology, our model combines the residual module and depth-wise separable convolution and uses a GlobalAverage Pooling layer in place of a fully connected layer.

2 Related Works

Facial expression recognition (FER) has emerged as a significant area of research, with applications in various fields, including human–computer interaction, emotion analysis, and psychological assessment. The two primary types of facial expression identification techniques available now are network models that use deep learning and the conventional manual method. The practical uses of the old technique are quite restricted, despite its widespread use [10, 11]. In order to reflect the emotional characteristics of excellent sample data, deep learning techniques for facial expression classification typically need understanding how to use robust supervision techniques [12, 13]. Mostly, these datasets are derived from four studies [14]. Convolutional neural networks (CNNs) have demonstrated remarkable success in FER due to their ability to extract and process complex spatial patterns from facial images. However, traditional CNNs often suffer from high computational complexity and memory requirements, limiting their deployment in real-time applications. In response to these challenges, lightweight CNNs have been developed specifically for FER, aiming to achieve efficient and accurate expression recognition while maintaining low computational overhead. These lightweight models typically employ various techniques to reduce model complexity, such as depth wise separable convolutions, group convolutions, and efficient activation functions. The two primary categories of the existing facial expression recognition techniques are the network model employing deep learning and the conventional manual technique. Despite being extensively utilised, the old technique has very few real-world applications [15, 16]. In order to reflect the emotional characteristics of excellent sample data, deep learning techniques for facial expression classification typically need knowing how to use robust supervision techniques [13, 16]. These datasets primarily originate from the four publications [17, 18]. Barros et al. formalised the facial channel neural network for facial expression recognition (FER) using a network model based on the topological structure of VGG-16. [17]. A CNN that could identify human emotions from a single facial picture was proposed by Koujan et al. [19]. The pretrained model can be transferred to a new model for task training as a method of transfer learning. For instance, AlexNet, a model pre-trained as ImageNet, was employed for FER transfer learning by Sekaran et al. [20]. The work at hand involved object identification in the source domain, ImageNet, and emotion classification in the destination domains, FER2013 and CKC. Model training was conducted using the Adam optimizer [2] and picture data augmentation.

3 Methodology

Figure 1 depicts the general structural outline of the FER process. The four stages of the proposed FER scheme are as follows: (i) preprocessing for face detection and normalisation; (ii) CNN model optimisation; (iii) using confusion matrix analysis to group similar facial expressions; and (iv) classifying the final facial expressions from the categorised similar facial expression groups. The next section discusses specifics of the methods employed in the FER process.

Dataset

The open-source dataset FER-2013 is used in this work [2]. As the We must use pan-to parse and extract the photos because the source dataset is in CSV format. 35,887 facial expressions make up the dataset after parsing. The train set, the public validation set, and the private validation set are the three of them. Both sets include 3589 entries. Every figure is made up of a 48 by 48 grayscale picture. There are seven phrases, denoting digital labels 0–6, namely: 0, anger; 1, disgust; 2, fear; 3, happiness; 4, sadness; 5, surprise; and 6, normal. The seven types of expressions are represented by the following figures in the train set: 3995, 436, 4097, 7215, 4830, 3171, and 4965 (Fig. 2).

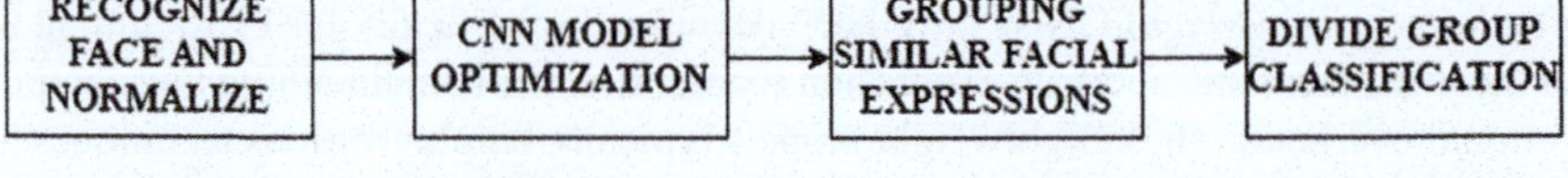

Fig. 1 Structural outline of FER process

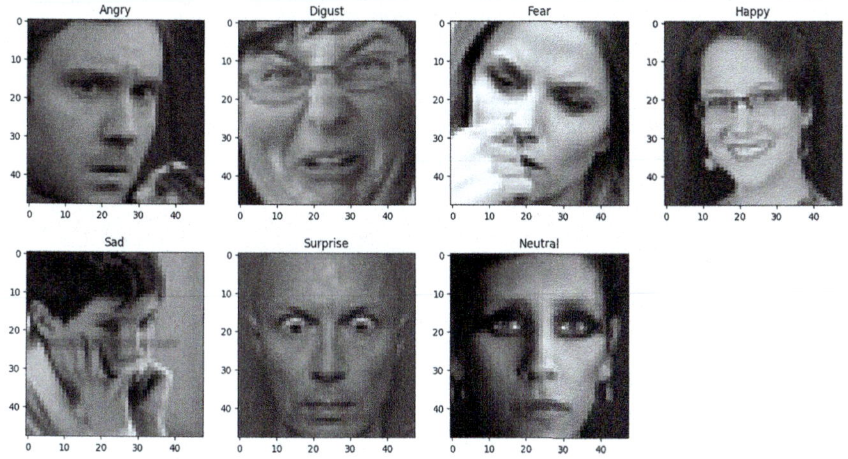

Fig. 2 Sample of FER-2013 dataset

3.1 Normalization and Face Detection

There are facial and non-facial sections in the majority of facial expression datasets, and the non-facial parts are most likely not necessary area. Consequently, in order to eliminate the unnecessary portions of the input image, face detection is required. Retina Face, which shows remarkable accuracy and fast performance on "RGB datasets" with varying head positions, blurring, and lighting fluctuations, was used as the face detector in this work. Excellent performance on grayscale thermal photos is also anticipated with the Retina Face and it is also used because of its unique feature which is retina face can work perfectly with low quality images. The facial photos were manually cropped in situations when the facial region could not be recognised automatically. Upon identifying the face region, every picture was normalised to a same resolution as the identified facial regions vary throughout individuals. Additionally, it was determined that data augmentation was necessary to keep untrained data's generalisation performance from declining. There are a few techniques that may be used to create augmented data such as rotation which includes rotating of the image around some specific point, flipping which includes mirror image, and colour-space modification include changing of RGB colour palette or contrast or brightness.

3.2 CNN Model Optimization

It is difficult to train the CNN model from start with limited data. The initial "ResNet-50" model weights, which was trained on "ImageNet", were applied to RGB pictures of 224 $\times$ 224 pixels. The pre-trained model's layers were changed to enhance FER

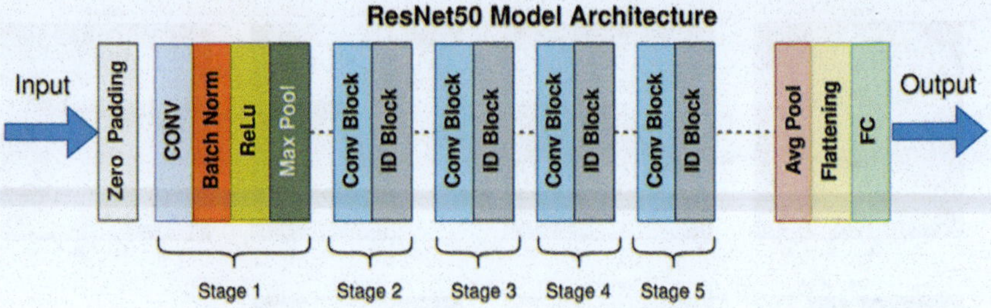

Fig. 3 Architecture of ResNet-50

performance. First, the filter size was reduced while maintaining the input convolution layer's output size. This is a result of the features being retrieved from the face photos at a lower spatial resolution than ImageNet. Secondly, in order to extract additional features during training, "Convolution" was used in place of "max-pooling" in the second layer. Ultimately, the fully connected layer's output node determines the quantity of expressions on the face that require agreement on classification (Fig. 3 and Table 1).

Using Global Average Pooling, our first model eliminates the fully linked layer entirely. This is accomplished by employing the SoftMax activation function to solve the classification issue and inserting a feature map in the final convolution layer that is consistent with the number of classes. The ADAM optimizer [2] is used to train our model. Figure 5 depicts the model structure of our expression classification (Figs. 4, 6, 7 and Table 2).

3.3 *Utilizing Confusion Matrix Analysis, Grouping Faces with Similar Expressions*

Training and testing neural networks using tagged data is a popular CNN learning technique. Handling the FER issue the commonalities make it challenging to use all of the marked data and differences in the photos of the faces' expressions. It is expected that training CNN models with subgroups of lower complexity will increase classification accuracy. The suggested algorithm is divided into two primary sections. Phase 1: Combining identical emotions on the face. First, the training's inference outcome produced a confusion matrix. The CNN model that has been optimised using every training dataset. Within the "confusion matrix", the percentage of erroneously anticipated test pictures is represented by "false positives (FPs) and false negatives (FNs)". Grouping facial expressions in the matrix that have a high mistake rate should lead to performance improvement. Phase 2: Categorising comparable emotions on the face. In this stage, the subproblems are the main emphasis. A further optimised CNN model was trained using the generated Set S. between the initial set E and set S, there are exactly the same number of pictures in total; however, the objective

Table 1 Optimized CNN architecture

Type	Filter	Channel	Stride	Output Size
Input image	–	1	–	130 × 160
Conv1	3 × 3	64	1	130 × 160
Conv2	3 × 3	64	2	65 × 80
Conv3_x	3 × 3	64	1	65 × 80
	3 × 3	64	1	65 × 80
	3 × 3	64	1	65 × 80
	3 × 3	64	1	65 × 80
Conv4_x	3 × 3	128	2	33 × 40
	3 × 3	128	1	33 × 40
	3 × 3	128	1	33 × 40
	3 × 3	128	1	33 × 40
Conv5_x	3 × 3	256	2	17 × 20
	3 × 3	256	1	17 × 20
	3 × 3	256	1	17 × 20
	3 × 3	256	1	17 × 20
Conv6_x	3 × 3	512	2	9 × 10
	3 × 3	512	1	9 × 10
	3 × 3	512	1	9 × 10
	3 × 3	512	1	9 × 10
Average Pool	–	512	–	1 × 1
FC layer	–	512	–	Expressions

labels that the CNN must categorise have fallen to m. Moreover, any set S includes classifying two or more final facial expressions (Fm) is required. For example, when Fm > 2 and even, Smaller sections of half the size of the set can be created. Every face expression in the set was categorised, with the exception of this one. Ultimately, the Phase 2 classification findings.

4 Experimental Results

The RGB dataset used in the studies to assess the suggested approach was given in full. The tests employed images from the FER2013, which had all of the face expressions identified beforehand. The purpose of the tests was to demonstrate that the suggested approach may enhance accuracy performance. Each dataset's categorised facial expressions are described, along with comparative data with earlier techniques. In addition, we used TensorFlow and ResNetV50 model to unfreeze last 25 layers to use data augmentation and the CNN model that has been optimised.

Fig. 4 Structure of emotion classification used in the study

Lastly, the FER2013 dataset was used to evaluate the suggested methodology. All facial expressions were classified with a low accuracy of 69.04% by the optimised CNN mode. We grouped the four most comparable facial expressions: anger, fear, neutrality, and sorrow. We then separated the groupings into two subgroups: anger and fear, and sadness and neutrality. As a result, the suggested approach had the best accuracy, coming in at 85%. Because of this, the experimental findings on the four "FER datasets" indicated that it is challenging to enhance performance using the prior CNN learning approach due to comparable facial features and variances that are unrelated to facial features. Our study team used an Intel (R) Core (TM) i5-8400 CPU @ 2.80 GHz processor, 16 GB of RAM, an NVIDIA GeForce GTX 1060 GPU, and the Ubuntu 16.04 operating system to execute their experiment.

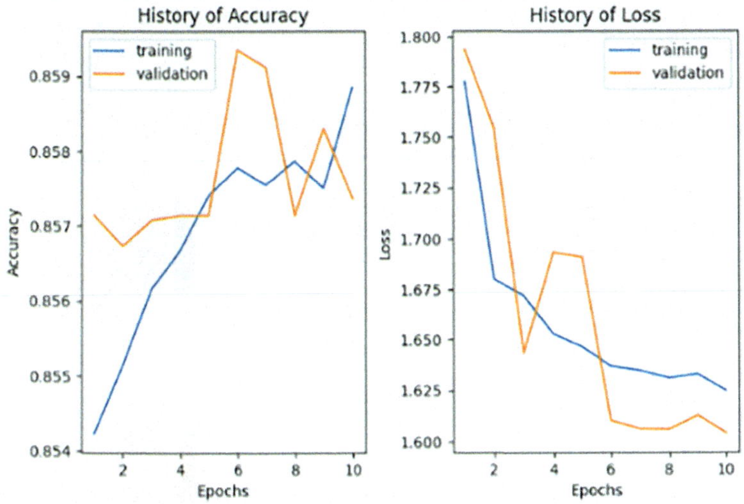

Fig. 5 Accuracy and loss graph

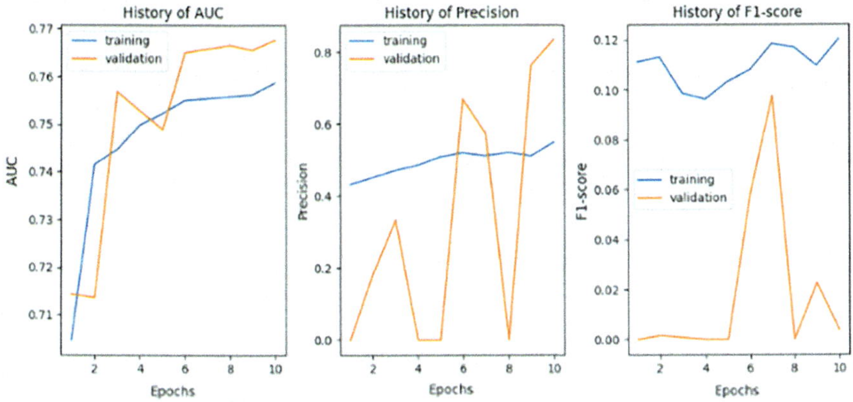

Fig. 6 AUC, Precision and F1-score graph

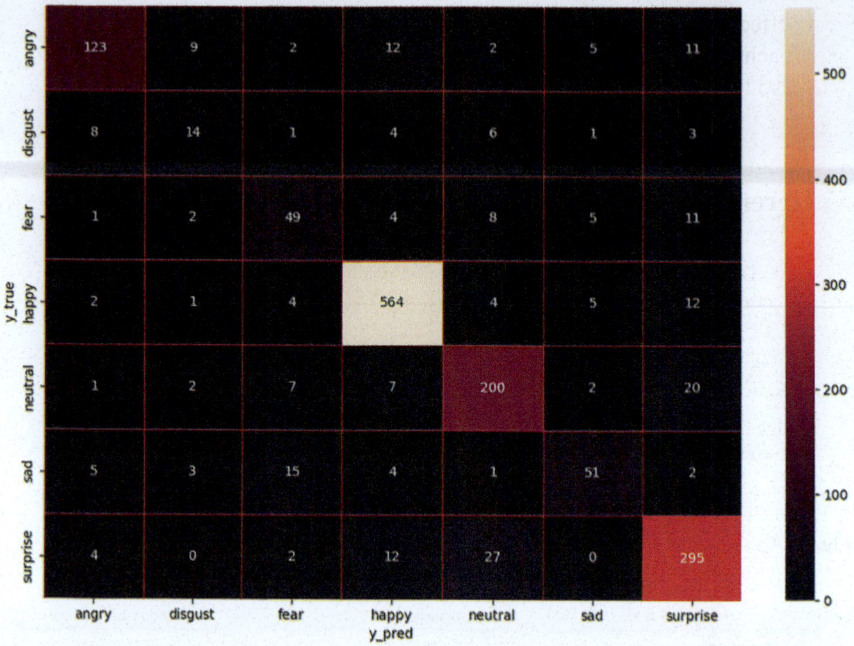

Fig. 7 Correlation of features

Table 2 Model's evaluation metrics

Parameters	Percentage
Training Accuracy	0.85
Training Loss	0.77
Training AUC	0.8
Training Precision	0.7
Training F1-Score	0.5

5 Conclusion

We have created a lightweight CNN for face expression recognition in this study. Furthermore, there is no overt negative impact of our approach on classification and detection. Our model has decent detection performance successfully recognising pictures that are not included in the dataset, demonstrating the applicability of the model created in this research for multiple facial emotion categorization. Generally speaking, we have real-a visual system that can be incorporated on low-end computer devices in order to perform parameter reduction and face expression categorization recent comparisons of the models show that ours is more accurate than theirs, that the loss has gone down, and that it produces good detection results in pictures that are not part of the experimental dataset. With the help of pre-trained weights, the model was effectively initialised and updated for FER. We anticipate that the suggested approach will be applied in the future not just to Vision Transformer-based models but also to speech, activity and text recognition tasks.

References

1. Lee D-H, Yoo J-H (2023) CNN learning strategy for recognizing facial expressions. IEEE Access 11
2. Shehada D, Turky A, Khan W, Khan B, Hussain A (2023) Lightweight facial emotion recognition system using partial transfer learning for visually impaired people. IEEE Access 11
3. Zhang H, Jolfaei A, Alazab M (2019) Face emotion recognition method using CNN and image edge computing. IEEE Access 7
4. Pinto LV, Alves AV, Medeiros AM, da Silva Costa SW, Pires YP, Costa FA, da Rocha Seruffo MC (2023) Systematic review of facial expression detection methods. IEEE Access 11
5. Humpe L, Murillo SF, Muires J, Gloor P (2020) The Rhythm of flow :detecting Facial expressions of flow experiences using CNNs. IEEE Trans Affect Comput
6. Yen C-T, Li K-H (2022) Discussions of different deep transfer learning models for emotion recognitions. IEEE Access 10
7. Zhou N, Liang R, Shi W (2021) Lightweight CNN for real-time facial expression detection. IEEE Access 9
8. Ab Wahab MN, Nazir A, Ren ATZ, Noor MHM (2019) Efficientnet-lite and hybrid CNN-KNN implementation for FER on raspberry Pi. IEEE Access 7
9. Muhammad G, *Senior Member, IEEE*, Shamim Hossain M (2021) Emotion recognition for cognitive edge computing using deep learning. IEEE Internet Things J 8(23)
10. Putro MD, Nguyen D-L, Jo K-H (2022) Fast CPU real-time facial expression detector using sequential attention network. IEEE Trans Indus Inf 18(11)
11. Samadiani N, Huang G, Hu Y, Li X (2021) Happy emotion recognition from unconstrained videos using 3D hybrid deep features. IEEE Access 9
12. Mohan K, Seal A, Krejcar O, Yazidi A (2021) Facial expression recognition using local gravitational force descriptor-based deep convolution neural networks. IEEE Trans Instrum Measur 70
13. Hassan T, Seuß D, Wollenberg J, Weitz K, Kunz M, Lautenbacher S, Garbas J-U, Schmid U (2021) Automatic detection of pain fromfacial expressions: a survey. IEEE Trans Pattern Anal Mach Intell 43(6)
14. Kim B-S, Seo S (2022) Intelligent digital human agent service with deep learning based-face recognition. IEEE Access 10
15. Chang Y, Mukai N (2022) Color feature based dominant color extraction. IEEE Access 10
16. Aly M, Ghallab A, Fathi IS (2023) Enhancing facial expression recognition system in online learning context using efficient deep learning model. IEEE Access 11
17. Lan J, Jiang X, Lin G, Zhou X, You S, Liao Z, Fan Y (2023) Expression recognition based on multi regional coordinate attention residuals. IEEE Access 11
18. Aleisa HN, Alrowais F, Negm N (2023) Henry gas solubility optimization with deep learning based facial emotion recognition for human computer interface. IEEE Access 11
19. Karnati M, Seal A, Yazidi A (2023) Understanding deep learning techniques for recognition of human emotions using facial expressions: a comprehensive survey. IEEE Trans Instrum Measur 72
20. Ye J, Yu Y, Fu G, Zhu Y (2023) Analysis and recognition of voluntary facial expression mimicry based on depressed patients. IEEE J Biomed Health Inf 27(8)

Plant Leaves Disease Detection Using Integrated Convolutional Neural Networks: A Case Study on Potato

Anuj Kumar, Abha Upreti, and Vandana Pandey

Abstract Agriculture is very important for India, making up a big part (17–18%) of the country's economy. Many people rely on farming for their livelihood. India has a lot of farmland, even more than the U.S. and China. But keeping crops safe from pests and diseases is crucial for getting good amount of produce. The research focuses on using technology like image processing and machine learning to help with this. Specifically, we looked at potato plants and their leaves. The model uses convolutional neural network (CNN) and got really good results. Our model could tell with 99.78% accuracy if a potato leaf was healthy or had diseases like early blight or late blight. The cool thing about our research is that it's not just for the lab—it can be super helpful for farmers with big fields. The technology we used can quickly check large areas of farmland for any issues. This means we can find and deal with problems early, keeping crops healthy. Using technology like image processing, machine learning, and deep learning makes it easier and more accurate to identify diseases. This is a big deal for farmers with large farms, as it helps them keep an eye on everything without spending too much time. Our research is like giving a boost to India's farming strength. The high accuracy of the model and its ability to measure the affected leaf area are big steps forward in making farming more precise. As India stays on top in global farming, adopting these advanced technologies is crucial for making sure the future of agriculture is strong and sustainable.

Keywords Machine learning · CNN · Augmentation · Potato blight

Abha Upreti and Vandana Pandey—The authors made equal contributions to this project.

A. Kumar (✉) · A. Upreti · V. Pandey
School of Computing, Graphic Era Hill University, Bareilly Road, Haldwani, Uttarakhand 263139, India
e-mail: anujkumardixit@gehu.ac.in

P. J. S. Gonçalves et al. (eds.), *Proceedings of Fifth International Conference on Computing, Communications, and Cyber-Security*, Lecture Notes in Networks and Systems 1128, https://doi.org/10.1007/978-981-97-7371-8_4

1 Introduction

Potatoes are important worldwide and are considered the basic vegetable. Potatoes rank as the third most significant food crop globally for human usage, following rice and wheat [1]. In India, which relies heavily on agriculture, potatoes are a big deal. Since there's a growing demand for potatoes everywhere, it's crucial for us to increase our production and export capabilities in our region. Plant diseases are believed to decrease global crop production by 16%, presenting a significant risk to global food security [2]. Unfortunately, potato blight diseases have caused a decline in both potato export and production, impacting farmers. Late blight, caused by the oomycete Phytophthora infestans, has resulted in the greatest harm to potato crops ever documented [3]. In this project, we explain a way to identify potato blight by looking at pictures of their leaves. Image processing helps us get useful information from these images. Machine learning, a part of artificial intelligence, helps automate tasks and gives instructions. The goal is to comprehend the training data thoroughly and develop models capable of making smart decisions and predictions using a lot of data. We use leaf color, damage extent, leaf area, and texture parameters to classify leaves. We've analyzed various images to identify different types of plant leaves, focusing on potato leaves affected by diseases like early and late blight. Early blight shows small black spots, while late blight causes blistered and rotted leaves. Farming success depends a lot on the monsoon, which affects productivity. Even though India has the most people globally, farm lands are often small and not very economical. A vast amount of land is utilized for food crops, and the rest for commercial crops. India has two main cropping seasons, Kharif and Rabi, relying on monsoon. Indian agriculture has faced many problems such as using old methods, too many fertilizers and pesticides, not enough water, not much growth in crop production, tired soil, expensive farming tools, and problems from global climate change. Plant diseases also cause huge crop losses. Solving these issues needs a thorough approach, looking at the biochemical and physical aspects of plants. Traditional ways of checking, where farmers look with their eyes, take too much time and are costly for big areas. Modern technologies offer better and more efficient alternatives. This research looks at potato plant leaves, which are prone to diseases like late blight and early blight. Potatoes are important for our diets, giving us Vitamin C and B1, along with starch. Healthy potato leaves are crucial for later-stage food production, helping potatoes grow underground. To tell these dis- eases apart, we're using a CNN algorithm, which will be a big help for farmers. CNNs differ from other methodologies in several ways. Unlike traditional methods, CNNs do not rely on manually engineered features or human segmentation, and they typically require a substantial amount of data due to their large number of trainable parameters.

However, strategies like"data augmentation" or employing"pretrained models" can help mitigate this data requirement issue. CNNs are commonly employed in tasks such as image identification, object recognition, classification, and face recognition, where they excel at recognizing and categorizing items with minimal preprocessing, leading to superior performance in object analysis. Moreover, their multilayered architecture makes it easier to discern features in complex scenarios. We've split the total photos among training and testing. About 75% of the pictures are used as development, and remaining are for testing. Our model aims to accurately classify normal and diseased potato leaves. One potential research gap in this project is the exploration of transfer learning techniques specifically tailored for potato leaf disease detection. While convolutional neural networks (CNNs) have shown promise in various image recognition tasks, including plant disease detection, there is a lack of research specifically focusing on transfer learning approaches optimized for the characteristics of potato leaf dis- eases. Transfer learning involves leveraging pre-trained models on large datasets and fine-tuning them for specific tasks with smaller datasets, which could be particularly useful in agricultural contexts where labeled data might be limited. Research could delve into optimizing transfer learning methods, such as selecting appropriate pre- trained models, designing effective transfer learning strategies, and investigating how to mitigate issues like domain shift between source and target datasets in the context of potato leaf disease detection. Additionally, there is a need for research on inter-pretability and explainability techniques tailored to agricultural stakeholders. While CNNs can achieve high accuracy in disease detection, their black-box nature often makes it challenging for farmers or agronomists to trust and understand the model's decisions. Exploring techniques for providing interpretable explanations of the model's predictions, such as saliency maps or attention mechanisms, could bridge this gap and facilitate the adoption of AI-based solutions in agriculture.

1.1 Objectives of This Research

The crucial function of potato crops in assessing the plant's health prompts the research goal of developing a system capable of identifying and categorizing potato crop infection with the help of CNN.

2 Related Work

Mohanty et al. [4] trained their deep learning model using 54,306 images gathered from an openly available dataset and achieved an accuracy of 99.35%. Manya Afonso et al. [5] proposes a method for detecting blackleg disease in potato plants utilizing CNNs. The authors utilize CNNs, a technique, for automated and error free identification of blackleg symptoms in images of potato plants. In a controlled laboratory environment, Ghosh et al. [6] used RGB photographs of potato crop to conduct an experiment. They observed differences in performance under various environmental circumstances while training CNN and SVM models. Islam et al. [7] suggested an image processing based approach combined with machine learning for detecting potato leaf diseases. They trained their model on an openly accessible dataset with

300 images using support vector machine and achieved an accuracy of 95%. In order to detect plant illnesses, Sumit Kuar et al. [8] suggested deep learning project and a reign-based fully connected RCNN network, achieving a validation accuracy of 94.6% Ushadevi et al. [9]. "A survey on utilizing machine and deep learning techniques for predicting plant infection." This survey investigates the prediction of plant diseases using deep learning and machine learning. It talks about several methods and strategies in relation to plant disease forecasting. Mohameth et al. [10] presents a method for plant disease detection using deep learning and feature extraction techniques, utilizing the Plant Village dataset. It aims to improve accuracy in identifying plant diseases through advanced computational methods. Acharjee et al. [11] focuses on potato leaf disease detection using deep learning techniques. It likely provides a review of existing literature on similar topics, exploring previous methods and approaches for detecting plant diseases, particularly in the context of potato plants. The inclusion of articles from different years suggests a comprehensive review of the evolution of computational intelligence in plant disease detection, highlighting advancements over time. Bonik et al. [12] "A deep learning related potato infection detection with a Sequential Model." In this paper, a particular use of CNNs for identifying infected potato crops is presented. The study examines the efficacy of the suggested method for illness diagnosis using a sequential model. Sinshaw et al. [13] utilized methods reliant on computer vision and examined various kinds of machine learning methods for detecting potato leaf diseases. They reviewed 39 previous researches and found out that deep learning methods were more often used than the traditional machine learning methods. Lastly, the review identified open research issues in the field and classified the state-of-the-art methods. Singh et al. [14] suggested an approach based on k means and multi class support vector machine for image partitioning and grouping of potato crops respectively. They used an openly available, standardized, and dependable dataset for the same and achieved a precision of about 95.99%. Tarik et al. [15] processed and segmented 2,034 potato leaf images which they took from an open source platform and made a 75:25 train-test split. They achieved a precision of 99.23%. Mahum et al. [16] suggested an enhanced deep learning method that groups the dataset into five groups. They used an openly available dataset and collected some data manually too. They utilized a denseNet model that has been trained beforehand and achieved a precision of 97.2%. Alberta et al. [17] suggested a ResNet-9 model for the detection of diseases in potato and tomato leaves. They utilized an openly available dataset consisting of 5,321 images and trained the model based on some hyperparameters such as total epochs, weight decay, momentum, gradient clipping, optimizer, and learning rate. They achieved an accuracy of 99.25%. Gardie et al. [18] suggested a transfer learning based approach using a pre trained inception v3 model for detecting potato leaf diseases. They utilized rgb picture dataset with color channels for training their model and achieved an accuracy of 98.7% (Table 1).

Table 1 Comparison of several Techniques for identifying plant Damage

Ref. no	Dataset	Technology used	Accuracy (%)
4	Plant Village dataset for 26 different disease with 38 classes	AlexNet	85.83
5	Potato leaf data collected in an experiment consisting of 532 images	ResNet 18	95
6	Potato Leaf dataset in-field images 3387, Open source 2152	Pre-trained SNet	93.78
7	PlantVillage Dataset	Segment and Multi	95.0
8	PlantVillage Dataset	SVM	94.6
10	Open source dataset with 2152 potato images	CNN	94.6
		Transfer learning	97.82
11	Plantvillage dataset	CNN	98
12	Potato fields in a village	CNN	94.2
14	Plantvillage dataset	k-means and multiclass SVM	95.99
15	Potato fields in a village	CNN	99
16	Plantvillage dataset	modified DenseNet-201	97.2
17	Plantvillage dataset	ResNet-9	99.25
18	Plantvillage dataset	inception-V3	98.7

3 Methodology

This research collect the right data set and train deep learning model to determine the infected crops in early stage. One type of Deep learning model used for identifying and classifying particular features in images is the Convolutional Neural Network (CNN). It is essential to the study of visual images and consists of two main parts: Feature extraction and Fully Connected Layer. **Feature extraction** is used to separates and recognizes various features in an image. **Fully Connected (FC) Layer** uses the features that were extracted in the previous step to forecast the image's category. **FC layers, pooling layers, and convolutional layers** make up CNN's three primary layer types. **Convolutional Layer** is the first layer, pulls various information from the image between the input image and a NxN filter, creating a feature map for the layer above. Furthermore, **Pooling Layer** is used to reducing the convolved feature map's dimensions to improve execution. The two types used are max pooling and average pooling. Two further layers are included as well; the activation **layer**, which is essential for making decisions when forwarding input, and the **Dropout layer**, which removes specific neurons to combat over-fitting. And activation Layer is essential to decision-making because it decides which information is appropriate to move forward and which is not. The CNN model depict in in Fig. 1. We studied previously existing works and found that CNN is the most suitable architecture for

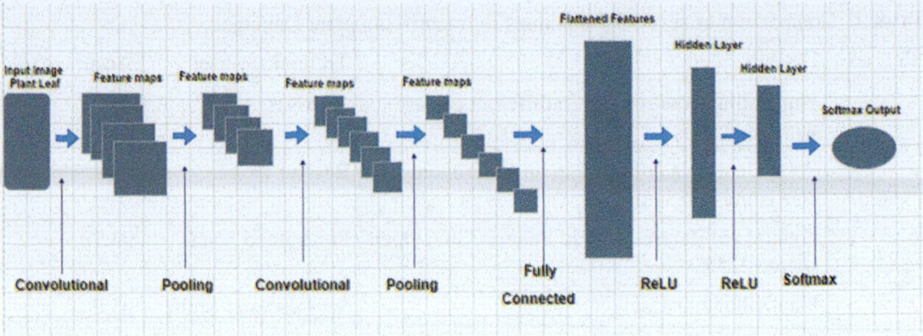

Fig. 1 CNN architecture

detecting plant diseases. In this paper, Motive of the author is to proposed a more accurate CNN model for potato leaf disease detection and try to proposed model will higher than the other deep learning models. Firstly, Model perform data acquisition process on the plant leaf dataset which is large and diverse than previously used datasets. Innovative Feature Engineering techniques tailored to the characteristics of potato leaf disease. Next, pre-processing and data augmentation is performed to reduce overfitting problems and improving the model's robustness. We have used categorical cross entropy to deal with multiclass classification problem and perform classification to classify leaves into three categories such as Late Blight(LB), Early Blight(EB) and Healthy. The proposed model depict in Fig. 2.The proposed model is scalable and adaptable to various crops and illnesses with few alterations, making it useful for a broader range of agricultural applications (Table 2).

4 Dataset

This research utilized images sourced from Kaggle, an open repository offering the Plant Village dataset. This dataset comprises more than 4000 potato's leaf images, showcasing various species, both healthy and diseased. Specifically focusing on potato leaves, we downloaded images from the Plant Village dataset and resized them to 256 × 256 pixels. The dataset includes images of healthy, EB, and LB leaves. Sample images from each class are depicted in Fig. 3. The table describes the total no of leaves used in training and testing section in the proposed system.

5 Experimental Setup

Our work centers on a deep learning algorithm trained to identify potato leaf illnesses, and it has demonstrated a remarkable capacity to discriminate between diseased and healthy leaves. TensorFlow had to be installed using Python libraries, among other

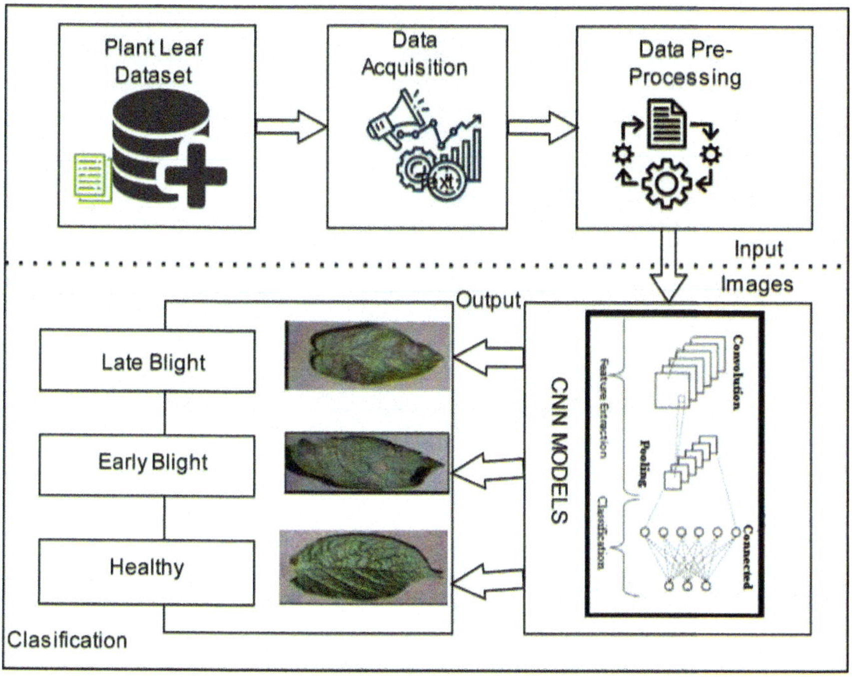

Fig. 2 Proposed Model for Potato Leaf Disease

Table 2 Data set for training process

Sno	Category	Total sample	Training sample	Testing sample
1	Early Blight	1628	1303	325
2	Healthy	1020	816	204
3	Late Blight	1424	1139	285
	Total	4072	3528	814

techniques and system dependencies that were leveraged in the implementation. We focused on three main types of leaf diseases on potatoes and grouped our annotated photos accordingly. The network became non-linear with the addition of rectified linear units (ReLU), and the final layer's softmax function made disease classification easier (Table 3).

We used adaptive moment estimation to improve the model's efficiency. Our research is unique in the farming field since it significantly improves upon existing machine learning techniques. Notably, we ensured effective training times by streamlining the deep learning method by lowering the total parameters. Our investigation used a single 8 GB RAM and the Google Colab (Jupyter Notebook) deep learning framework on a Windows 10 computer with a Core i5 processor.

Fig. 3 Potato leaf disease dataset

Table 3 Performance metrices across Epochs

Execution time in minutes	82 min 5 s
Total Epochs	50
Iteration per Epoch	102
Accuracy	99.78
Loss	0.22

6 Results

The analysis of proposed model depict in Fig. 4 which show an amazing precision of 99.78%.

This work is an innovative attempt to use cutting edge techniques to quickly detect plant leaf disease using neural networks. The efficiency of the suggested paradigm is noteworthy, as Figs. 4 and 5 demonstrate its potential to completely transform the agriculture sector. Historically, image classification tasks have leaned on manually crafted features which were then coupled with learning algorithms. The efficacy of these methods heavily depended on predefined features. Feature engineering, being laborious and intricate, necessitates reevaluation with substantial shifts in circumstances or datasets. Traditional computer vision techniques for identifying plant diseases predominantly hinge on handcrafted features, image augmentation, and other time-consuming techniques, exacerbating this challenge (Fig. 6).

The great degree of accuracy gained highlights the model's capacity to handle the difficulties of the current issue. The training procedure is shown in Fig. 5, which also includes two graphs for accuracy and loss. When the epoch reaches its maximum level, the accuracy graph peaks at 100% and the loss graph reaches 0%. It illustrates

Fig. 4 Final prediction of the proposed model

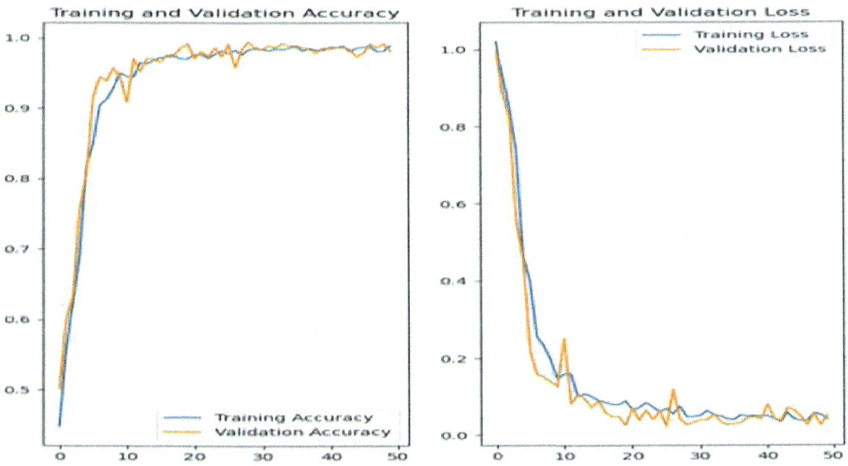

Fig. 5 Graphs showing the Training And Validation accuracy and loss

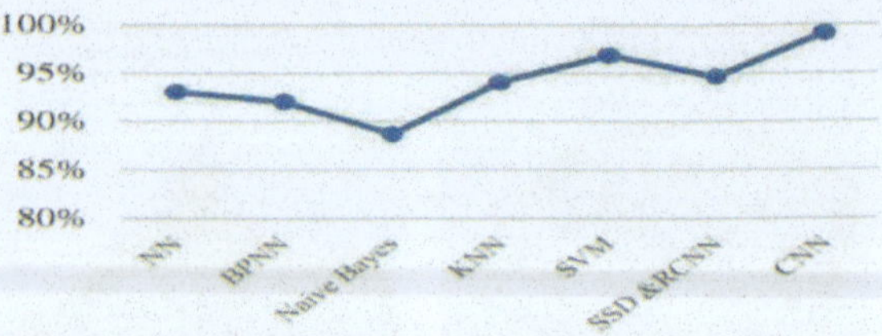

Fig. 6 Model accuracy comparison

how to train a Convolutional neural network to recognize leaf photos of potato plants. Table 4 offers more information about this training procedure. The table 6 provides details on the performance metrics across epochs during the training process. It includes the execution time, total number of epochs, and the number of iterations per epoch. Additionally, it presents accuracy and loss values for each epoch, indicating how the model's performance evolves over time. The algorithm effectively identifies potato diseases without the need for feature engineering. While training demands considerable time on a high-performance GPU cluster, the classification task itself is rapid, taking less than a second on a CPU, enabling seamless implementation on smartphones. This paves the way for widespread adoption of smartphone-based crop disease diagnosis on a global scale. A comparison between our suggested method and the classification of potato plant diseases is presented in Table 1 which demonstrates that the proposed CNN model gave the highest accuracy in testing samples.

7 Conclusion

Diseases in plants is one of the major factors for more than 20% of crop output losses on an annual basis. Deep learning techniques are essential for preserving product quality, increasing productivity, and detecting diseases early. With an astounding 99.78% accuracy rate, this study's Convolutional Neural Network (CNN Model), that is quick and customized, successfully recognizes and categorize illnesses in potato plant leaves. Our all-inclusive neural network system can accurately classify leaf photos and differentiate between leaves that are diseased and healthy. Convolutional neural network (CNN) encoder-decoder is incorporated into the model to improve its leaf image categorization performance. This paper summarizes findings from multiple investigations that applied deep learning methods to the diagnosis of illnesses in potato leaves. When com- pared to alternative deep neural networks, CNN demonstrated the highest accuracy in recognizing and categorizing illnesses in potato leaves, encompassing early blight, late blight and non-infected leaves. As per research, Artificial Neural Networks (ANN) achieved an 85% accuracy, Support Vector Machines (SVM) achieved an 88.89% accuracy, while CNN, with a remarkable 99.07% accuracy [3], outperformed them. This study underscores that Convolutional Neural Networks offer superior accuracy levels. In the domain of Deep Learning, CNN represents a relatively recent technology adept at recognizing images and identifying distinctive features within them. Our approach can help farmers spot infections early on and increase crop production. Figure 5 compares the categorization accuracy of the proposed technique with different implementations. This research demonstrates that the utilization of deep learning methodologies enhances the precision and effectiveness of identifying crop diseases.

This advancement has the potential to enhance disease control measures and increase agricultural productivity. The study establishes a foundation for further exploration and implementation of these models within precision agriculture frameworks.

8 Applications and Contributions

The project's principal application is the early detection of potato leaf diseases, which allows farmers to intervene on time and avoid disease spread, resulting in increased crop yields and lower economic losses. Implementing the created CNN model in precision agricultural systems can help farmers better monitor the health of their potato crops, allowing for focused interventions like pesticide application or irrigation. Integrating disease detection systems into crop monitoring platforms can offer farmers with real-time information about the health of their potato plants, allowing for more informed decision-making and resource allocation. The initiative can serve as a foundation for future research and development in computer vision and agricultural technology, resulting in new breakthroughs and advances in plant disease diagnosis and control. The initiative helps to automate the process of detecting potato leaf dis- eases, which saves time and effort compared to human inspection, resulting in greater agricultural efficiency. Early detection of infections allows farmers to take proactive actions to avoid disease transmission, resulting in increased agricultural yields and lower economic losses. The proposed model is scalable and adaptable to various crops and illnesses with few alterations, making it useful for a broader range of agricultural applications. The initiative advances research in computer vision and agricultural technology by investigating the use of CNNs to identify plant diseases. This could spur future research and innovation in the field.

9 Limitations and Challenges

The availability and quality of labeled datasets for training the model may be a restriction. Limited or biased datasets can produce erroneous results. The model may not be able to generalize well to new data or variations in leaf diseases that were not included in the training set. Convolutional Neural Networks (CNNs) can be computationally demanding, necessitating sophisticated gear for training and inference that may not be available to everyone. CNNs are frequently regarded black-box models, making it difficult to understand how the model generates decisions, particularly for stakeholders unfamiliar with deep learning techniques. Obtaining a large and diverse dataset of potato leaf photos with precise disease diagnoses can be difficult, necessitating significant time, effort, and money. Ensuring that the CNN model generalizes well to previously unknown differences in potato leaf diseases and environmental variables is critical for its practical application in various geographical regions and growing seasons. Training and deploying CNN models for disease detection may necessitate large computational resources, providing difficulties for farmers or organizations with limited access to high-performance computing equipment. Addressing the difficulty of interpreting and communicating the model's decisions to non-expert stakeholders, such as farmers or agricultural advisors, is critical to fostering trust and encouraging the adoption of AI-based solutions in agriculture. Integrating the disease detection system into existing agricultural workflows and technologies while maintaining usability and scalability necessitates meticulous planning and consideration of stakeholders' demands and limits. Addressing ethical concerns such as data privacy, bias in AI algorithms, and the possible influence on conventional farming techniques is critical to ensuring the responsible development and deployment of AI-based agriculture technology.

10 Future Work

Looking ahead, the challenging task of early-stage plant disease detection prompts future work to implement the system on smartphones, facilitating timely results for farmers. Also, we can provide remedies to treat the plant diseases. Further evolution of a voice-activated mobile application designed exclusively for people or agriculturist with less knowledge of technology presents a groundbreaking approach to enhance farming.

References

1. Akash HS, Rahim MA, Miah ASM, Okuyama Y, Tomioka Y, Shin J (2023) Generalized technique for potato leaves disease classification using convo- lutional neural network. In: International conference on ICT for sustainable development, pp 589–601. Springer
2. Zhou C, Bucklew VG, Edwards PS, Zhang C, Yang J, Ryan PJ, Hughes DP, Qu X, Liu Z (2023) Portable diffuse reflectance spectroscopy of potato leaves for pre-symptomatic detection of late blight disease. Appl Spectrosc 77(5):491–499
3. Fern´andez CI, Leblon B, Haddadi A, Wang K, Wang J (2020) Potato late blight detection at the leaf and canopy levels based in the red and red-edge spectral regions. Remote Sensing 12(8):1292
4. Mohanty SP, Hughes DP, Salath´e M (2016) Using deep learning for image-based plant disease detection. Frontiers Plant Sci 7:215232
5. Afonso M, Blok PM, Polder G, Wolf JM, Kamp J (2019) Blackleg detection in potato plants using convolutional neural networks. IFAC-PapersOnLine 52(30):6–11

6. Ghosh S, Rameshan R, Dinesh DA (2021) An empirical study on machine learning models for potato leaf disease classification using rgb images. In: ICPRAM, pp 515–522
7. Islam M, Dinh A, Wahid K, Bhowmik P (2017) Detection of potato diseases using image segmentation and multiclass support vector machine. In: 2017 IEEE 30th Canadian conference on electrical and computer engineering (CCECE), pp 1–4. IEEE
8. Kumar S et al (2021) Plant disease detection using cnn. Turkish J Comput Math Educ (TURCOMAT) 12(12):2106–2112
9. Ushadevi G et al (2020) A survey on plant disease prediction using machine learning and deep learning techniques. Intel Artif 23(65):136–154
10. Mohameth F, Bingcai C, Sada KA (2020) Plant disease detection with deep learning and feature extraction using plant village. J Comput Commun 8(6):10–22
11. Acharjee T, Das S, Majumder S (2023) Potato leaf diseases detection using deep learning. Int J Digital Technol 2(1)
12. Bonik CC, Akter F, Rashid MH, Sattar A (2023) A convolutional neural network based potato leaf diseases detection using sequential model. In: 2023 international conference for advancement in technology (ICONAT), pp 1–6. IEEE
13. Sinshaw NT, Assefa BG, Mohapatra SK, Beyene AM et al (2022) Applications of computer vision on automatic potato plant disease detection: a systematic literature review. Comput Intell Neurosci 2022
14. Singh A, Kaur H (2021) Potato plant leaves disease detection and classification using machine learning methodologies. In: IOP conference series: materials science and engineering, vol 1022, p 012121. IOP Publishing
15. Tarik MI, Akter S, Al Mamun A, Sattar A (2021) Potato disease detection using machine learning. In: 2021 Third international conference on intelligent communication technologies and virtual mobile networks (ICICV), pp 800–803. IEEE
16. Mahum R, Munir H, Mughal Z-U-N, Awais M, Sher Khan F, Saqlain M, Mahamad S, Tlili I (2023) A novel framework for potato leaf disease detection using an efficient deep learning model. Human Ecol Risk Assess Int J 29(2):303–326
17. Anim-Ayeko AO, Schillaci C, Lipani A (2023) Automatic blight disease detection in potato (solanum tuberosum l.) and tomato (solanum lycopersicum, l. 1753) plants using deep learning. Smart Agric Technol 4:100178
18. Gardie B, Asemie S, Azezew K, Solomon Z (2022) Potato plant leaf diseases identi- fication using transfer learning. Indian J Sci Technol 15(4):158–165

Electrocardiogram Classification Using Convolutional Neural Network

Kashyap Kachhadiya, Abdullah Hasmani, Drashti Gorasiya, Bansari Patel, Krishna Patel, and Vaishali Vadhavana

Abstract Circulation of Blood relies significantly precise electrical initiation. Arrhythmias can arise from any disturbance in the excitatory wave's regular pattern in the heart. Electrocardiograms (ECG) are a popular diagnostic tool due to their speed, affordability, and non-invasive nature. This article presents a method for classifying ECGs that makes use of Convolutional Neural Network (CNNs) and the Continuous Wavelet Transform. Using Continuous Wavelet Transform (CWT), ECG signals are broken down into different time–frequency parts, while CNN extracts feature from the resulting 2D-scalogram. To further enhance the model considering the RR interval for arrhythmia diagnosis, 4 RR interval parameters are taken out and connected to CNN parameters to provide an ECG that is completely integrated classifier layer. This proposed model examines the impact of CNN and CWT on ECG classification alongside RR intervals. After training Accuracy of 97.27% was achieved. Based on the proposed method further models can designed for efficient clinical applications.

Keywords Electrocardiogram · Convolutional neural network · Arrhythmias · Continuous wavelet transform

1 Introduction

An essential component of the circulatory system is the heart that can lead to Cardiovascular Diseases (CVDs), which are increasing globally, particularly in low- and middle-income countries. Symptoms indicative of heart disease such as light headedness, fainting, shortness of breath, and chest discomfort can be brought on by cardiac arrhythmias, which cause the heart to beat abnormally. Certain arrhythmias, including ventricular fibrillation, ventricular escape, and atrial fibrillation, can cause strokes and cardiac arrest and therefore require immediate medical intervention.

K. Kachhadiya · A. Hasmani · D. Gorasiya · B. Patel (✉) · K. Patel · V. Vadhavana
Department of Computer Science and Engineering, Devang Patel Institute of Advance Technology and Research, Charotar University of Science and Technology, Changa, India
e-mail: bansaripatel797@gmail.com

P. J. S. Gonçalves et al. (eds.), *Proceedings of Fifth International Conference on Computing, Communications, and Cyber-Security*, Lecture Notes in Networks and Systems 1128, https://doi.org/10.1007/978-981-97-7371-8_5

In order to provide the medical attention timely diagnosis of cardiac arrhythmia becomes vital. Electrocardiography is a non-invasive method to detect cardiac arrhythmia using electric activity with the help of two electrodes placed on the skin. With the help of Electrocardiography, the ECG is generated which is representation cardiac wave generated by the heart.

Long-term monitoring is not feasible using the 12-lead format commonly used in ECG-based biometrics research. For Holter monitoring, a practical technique for long-term continuous cardiac monitoring, a 2-lead arrangement is typically employed. In this study, the MIT-BIH arrhythmia dataset and a 2-lead ECG configuration are used to evaluate the performance of arrhythmia detectors.

A CNN [1] is an advanced artificial neural network designed to handle grid-like data like images and videos. Its unique feature is its ability to autonomously and flexibly acquire spatial hierarchies of features from input images through interconnected layers, each serving a specific purpose [2]. The core components of a CNN include convolutional layers, where tiny filters capture local patterns and elements, ReLU activation introduces non-linear elements, enabling complex network understanding. Pooling layers reduce spatial dimensions, preserve information, and reduce computational load. Completely linked layers gather acquired features, facilitating advanced reasoning processes. CNNs perform remarkable feats like image recognition, object detection, and image segmentation, transforming fields like computer vision and pattern recognition, and more. Here is the CNN architecture, as seen in Fig. 1.

Complex and Non-stationary Signal Analysis (ECG) uses the Convolutional Wavelet Transform (CWT) [3] to analyze the temporal localization and frequency of ECG data. Unlike the conventional Fourier Transform, CWT splits ECG data into distinct frequency components over time scales, allowing for a thorough examination of features like arrhythmias or irregularities in the cardiac cycle. Researchers and clinicians use CWT in combination with advanced classification algorithms like neural networks or support vector machines to enhance the accuracy of automated ECG diagnosis systems. A DL-based classification model is introduced to solve the drawbacks of conventional ECGs for identifying arrhythmias, such as noise and

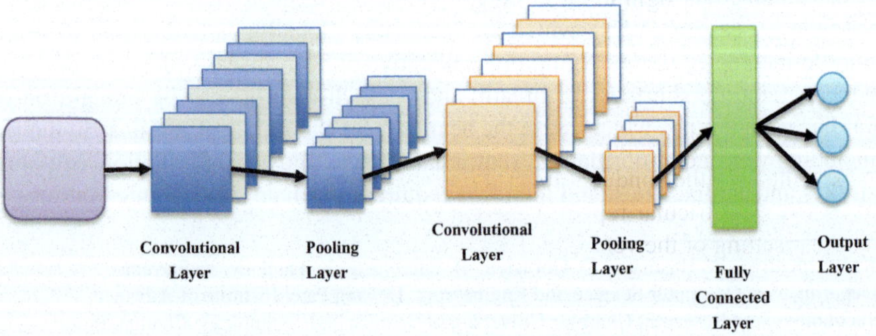

Fig. 1 Architecture of CNN

unexpected events that may result in misinterpretations and mistakes. The model has two dense layers and three convolution layers, adjusting the hyperparameters of a 2D-CNN. The main objective is to provide [16] a rapid and automated substitute for state-of-the-art arrhythmia diagnostic techniques. From the ECG MLII lead data, the model effectively separates Fusion (F), Normal (N), Suprathecal (S), and Ventricular-Electroventricular (V) [4] beats are the four types of arrhythmias.

2 Background Theory and Literature Review

The Electrocardiogram (ECG) is a crucial tool in cardiology and healthcare, demystifying Electrical Activity of Human Heart [5]. This non-invasive diagnostic technique captures the electrical impulses produced by the heart during contractions. It provides valuable information about its functionality as it pumps blood throughout the body. The ECG is essential for assessing heart health, serving various clinical purposes from routine check-ups to emergency care and long-term monitoring. Previous models for arrhythmia classification include the suggested seven-layer CNN method, which, based on the MIT-BIH arrhythmia database, achieves an overall accuracy of 99.43%, sensitivity of 99.2%, with positive predictivity of 99.4% in VEB identification and a sensitivity of 97.5% with positive predictivity of 99.1% in SVEB detection[6]. By adjusting hyperparameters, the optimized K-Nearest Neighbour (O-KNN) Classifier attains an accuracy of 99.03%. Processing vast volumes of data, the suggested approach increases the KNN classifier's accuracy for classification. The recommended approach raises accuracy of the classification of the KNN classifier [7]. The RNN models produced higher diagnosis results, reaching 83.1% testing classification accuracy with specified attributes. The ECG signals' behaviour is analyzed, and the SVM classifier's average accuracy for the gathered data is calculated. The recorded signals have an average accuracy of 98.61 [9].

The depolarization and repolarization of the heart's chambers are essentially what the ECG depicts. These electrical events are represented on a graph, usually with time on the right-hand side and voltage on left-hand side. The resultant ECG waveform includes numerous significant elements, including:

1. P-wave: the P-wave signifies atrial depolarization, represents electrical activity of atriums as they expand and contract to pump blood to chambers.
2. QRS complex: QRS complex is related to Ventricular Depolarization (VDP) [10], indicating the initiation of ventricular contraction. Its shape and duration offer insights into the conduction system's integrity.
3. T-wave: ventricular repolarization is reflected in the T-wave [11], representing the resetting of the ventricles for the next cardiac cycle.

The gaps and segments of these waveforms, such as the PR interval and the ST segment, provide further diagnostic information about the condition of the heart. A longer QT interval may indicate a greater risk of arrhythmias, but ST-segment changes may indicate myocardial ischemia or injury.

Accurate diagnostics for a broad spectrum of heart diseases, including arrhythmias (abnormal heart rhythms) [4], myocardial infarctions (heart attacks), conduction abnormalities, and structural heart defects, is possible with ECGs. ECGs are essential for risk assessment, medication selection guidance, and patient monitoring during critical care and surgery.

The Table 1 of literature reviews offers a brief overview of research studies together with information on the datasets characteristics and accuracy of each paper. This collection is an invaluable resource for comprehending the many approaches and performance indicators used in the industry. This extensive Table 1 raises credibility by providing a solid foundation for the research paper's contributions.

3 Dataset Description

This study makes use of the MIT-BIH Dataset. The algorithms for arrhythmia identification using ECG are evaluated and tested using the publicly accessible standard dataset known as MIT-BIH. 48-half hour dual channel EKG recordings from 47 patients are included in the MIT -BIH dataset. A total of 25 participants were added to accommodate unusual but clinically significant readings, with 23 subjects chosen at random to be part of the dataset. There were 25 males and 22 women among the subjects, making up around 60% of the population. Most recordings use a modified limb lead II (ML II) [15] on one channel and V1, V2, V4, or V5 on the [16] other, depending on the patient. Association for Progress of the Medical Technology [4] (AAMI's) proposal in ANSI/AAMI-EC57 served as the basis for this. 4 samples containing paced beats have been removed. Based on the EC57 recommendation, arrhythmia can be divided into 5 categories as shown in the Table 2.

4 State of the Art Algorithms for ECG Classification

4.1 Convolutional Neural Network

A CNN consists of multiple layers, each enhancing the quality and detail of features extracted from an input image. These layers use filters or kernels to detect basic and rudimentary features, gradually focusing on identifying distinctive characteristics that define the input object. Consequently, the output of each convolutional layer represents a partially recognized rendition of input image and serves as the input for the subsequent layer. Ultimately, in the final fully connected (FC) layer, the CNN [1] amalgamates these acquired features to achieve recognition of the image or object it represents. Classification of ECG can be performed by using CNN alongside Continues wavelet transform. CNN allows classification without need for manual feature extraction.

Table 1 Literature review table

Paper Title	Published by	Algorithm used	Dataset used	Accuracy (%)
Generalization of convolutional neural networks for ECG classification using generative adversarial networks [12]	IEEE, 2020 [12]	GANs and CNN	MIT-BIH arrhythmia dataset	>98.0
ECG heartbeat classification using convolutional neural networks	IEEE, 2020	CNNs [13]	MIT-BIH arrhythmia dataset	99.43
Electrocardiogram heartbeat classification using convolutional neural networks for the detection of cardiac arrhythmia	IEEE, 2020	Convolutional Neural Network [14] (CNN)	MIT-BIH Arrhythmia Dataset	95.2
Automatic ECG diagnosis using convolutional neural network [4]	MDPI, 2020	Convolutional neural networks (CNN)	MIT-BIH arrhythmia database	98.33
ECG classification using a hybrid deeplearning approach	IEEE, 2021	Hybrid structure of Convolution network with LSTM-Network (CNN-LSTM)	MIT-BIH Arrhythmia Database	98
A hybrid deep learning approach for ECG-based arrhythmia classification [12]	MDPI, 2022	2D CNN and LSTM [4]	MIT–BIH arrhythmia database	≈98.7
A novel method for ECG signal classification via onedimensional [12] convolutional neural network	Springer, 2020	1D CNN [13]	MIT-BIH arrhythmia dataset	97
Deep convolutional neural network application to classify the ECG arrhythmia	Springer, 2020	Convolutional neural network (CNN)	MIT-BIH arrhythmia database	99.84
Classifying cardiac arrhythmia from ECG signal using 1D CNN deep learning model	MDPI, 2023	1D-CNN [5]	MIT-BIH dataset	99
ECG recurrence plot-based arrhythmia classification using two-dimensional deep residual CNN Features [12]	MDPI, 2022 [12]	Deep residual convolutional neural network [14]	MIT-BIH Arrhythmia Database	98.36
ECG based heart disease classification and validation using 2D CNN	IEEE, 2023	Convolution neural network (CNN)	MIT BIH arrhythmia database	98.10

(continued)

Table 1 (continued)

Paper Title	Published by	Algorithm used	Dataset used	Accuracy (%)
ECG heartbeat classification using CNN	IEEE, 2021	1D convolution neural network (CNN)	MIT-BIH dataset	97.36
ECG heartbeat classification based on multi-scale wavelet convolutional neural networks	IEEE, 2020	CNN and stationary wavelet transform (SWT)	MITBIH Arrhythmia database	99.11

Table 2 Five categories of the arrhythmia based on EC57

Symbol	Class
N	Normal
S	Supraventricular ectopic
V	Ventricular ectopic
F	Fusion
Q	Unknown

4.2 Support Vector Machine

SVMs are widely used in machine learning for classification tasks due to their ability to determine optimal hyperplanes and maximize data margin. They are adaptable to both linear and non-linear data, making them useful in fields like bioinformatics, computer vision, and natural language processing. Renowned for their robustness and versatility, Several classification problems are better suited for SVMs. In the ECG heart rhythms, the Discrete Wavelet Transform is used to [12] preprocess signals, removing noise and extracting coefficients. Subsequently, SVMs with Gaussian kernels are employed to classify different ECG rhythms.

4.3 Artificial Neural Network

Artificial Neural Networks (ANNs) [17] are neural networks that mimic the human brain's structure and function. They excel in interpreting complex data relationships and are used in machine learning tasks like classification, regression, and pattern recognition. Deep Neural Networks (DNNs) with multiple hidden layers drive advancements in Deep Learning in natural language processing, computer vision, and speech recognition. In arrhythmia classification, a Multilayer Perceptron (MLP) with static backpropagation distinguishes between normal and abnormal cases.

5 Methodology

It can be more challenging to extract usable [18] information from ECG signals is challenging due to noise like starting point wandering, electromyography (EMG) interference, and power line artifacts. Filtering is essential before processing, focusing on removing baseline wandering, which affects ECG classification [3], while avoiding excessive filtering. Patient movement or respiration are the primary causes of baseline wandering. 2 median filters, one with a width of 200 ms and the other 600 ms, are used to [16] generate baseline-corrected ECG signals by subtracting them from the raw signals. After preprocessing the data is split into 80% training data and 20% testing data.

This paper presents a time–frequency analytics technique for translating ECG signals into time frequency domain, enabling easier extraction of feature. The Continuous Wavelet Transform (CWT) is the primary tool for time frequency analysis [19], offering high frequencies with limited frequency resolution and high time resolution. However, the Short-Time Fourier Transform [20] can only resolve frequencies at low frequencies and with high temporal resolution.

A Neural Network (NN) is a complex system consisting of several layers, including the batch normalization layer, convolution layer, and pooling layer. The convolution kernels use backpropagation (BP) mistake for training [21], while other layers like the Rectified Linear Unit (RLU) and batch normalization layer reduces training time and offers nonlinear capabilities, while the ReLU layer [16] increases the network initialization sensitivity. Pooling layer, which was also known as subsampling layer [16], reduces feature dimension and accelerates training by calculating average or maximum convolution characteristics among nearby neurons. The final layer of the CNN [13] have one or more neurons that are fully linked [22] calculate the class scores. The CNN utilized in this work, depicted in Fig. 4, integrates convolution network layer, the batch normalization, and the ReLU layer into the convolution unit to provide a clear representation. The convolution kernels are 7×7, 5×5, and 3×3, with three sequential convolution and pooling processes applied to the scalogram to extract features.

Arrhythmia alters RR intervals and heartbeat structure, so our CNN-based ECG classification includes RR interval information [3]. Key RR interval features used are the previous-RR, post-RR, ratio-RR, and local RR [16], representing the time between the present and preceding heartbeat, present and subsequent heartbeat, the ratio of previous-RR to post-RR, and an average of preceding 10 RR intervals of present heartbeat [22]. To adjust for inter-patient variance, we deduct the average of RR interval from previous-RR, post-RR, and the local-RR. Two completely connected layers receive these fusion features as input for classification [17].

The proposed model is displayed in the accompanying Figs. 2 and 3:

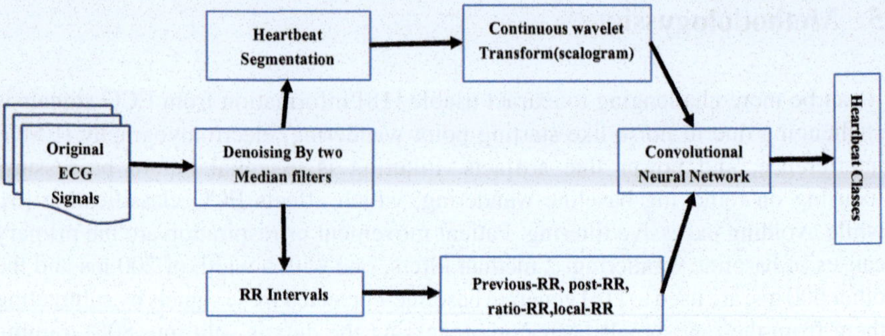

Fig. 2 Proposed model

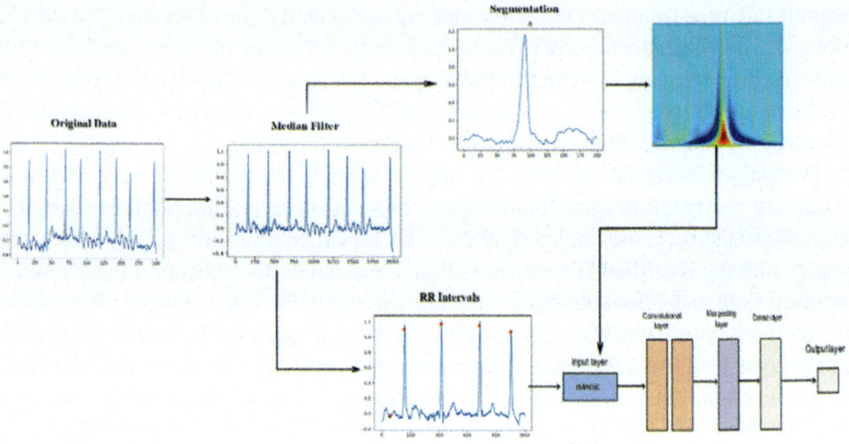

Fig. 3 Output of various steps of proposed model

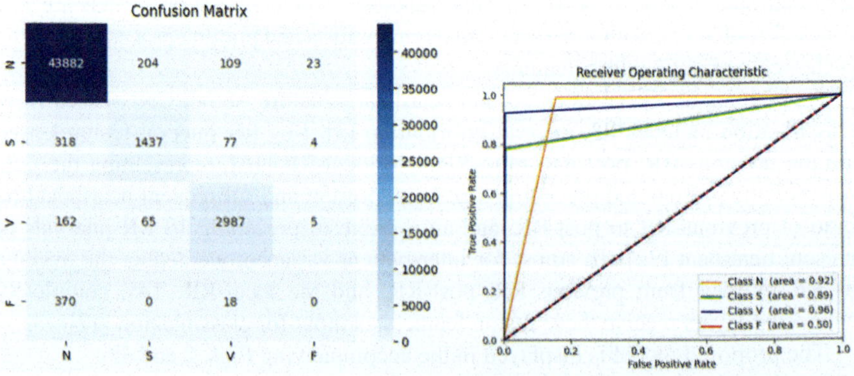

Fig. 4 Confusion matrix and ROC Curve after the final testing

6 Result Discussions

In this paper the evaluation of model is done using metrics Precision, Recall, Accuracy and F1-score to account for the imbalance in the data. The equations of the given metrics is as follows:

$$\text{Precision} = \frac{\text{True Positive (TP)}}{\text{True Positive (TP)} + \text{False Positive (FP)}} \tag{1}$$

$$\text{Recall} = \frac{\text{True Positive (TP)}}{\text{True Positive (TP)} + \text{False Negative (FN)}} \tag{2}$$

$$\text{Accuracy} = \frac{\text{True Positive (TP)} + \text{True Negative (TN)}}{\text{True Positive (TP)} + \text{False Positive (FP)} + \text{True Negative (TN)} + \text{False Negative (FN)}} \tag{3}$$

$$F1\text{Score} = 2 * \frac{\text{Precision} * \text{Recall}}{\text{Precision} + \text{Recall}} \tag{4}$$

We meticulously evaluated various metrics, including validation performance, F1 score, and duration, to guarantee a thorough evaluation of our model. The confusion matrix findings from the final training and testing procedures are also included in the figure. Furthermore, the classification outcomes obtained from the model's final training and testing phases are also presented in this figure. Notably, our model achieved an impressive accuracy of 97.27%. These findings underscore the robustness and efficacy of our approach, making a significant contribution to the field as documented in this research paper. The classes S and V represent the Supraventricular ectopic and Ventricular ectopic respectively which are also classified very well by the model. But model faces issue when classifying the Fusion which is encoded as class 4 which is due to the nature of the Fusion rhythm which is a combination of two rhythms generally consisting of the Normal rhythm and any other rhythms (Table 3).

The Confusion Matrix as shown in Fig. 4 can be interpreted as class 0 which us the Normal rhythm class has a huge amount of samples in the dataset and they are classified successfully by the model. The Receiver Operating Characteristic (ROC) Curve is as shown in Fig. 4.

In the table given below we try to analyze and compare the result of the proposed model with other model. The scope of the comparison is based on the model performance on the MIT-BIH dataset. The model proposed by Farag uses combination of

Table 3 Final Classification report after the final testing of model

Class	Precision	Recall	F1-score	Support
N	0.9810	0.9924	0.9867	44,218
S	0.8423	0.7827	0.8114	1836
V	0.9361	0.9279	0.9320	3219
F	0.0000	0.0000	0.0000	388

Table 4 Comparison of the results of the proposed model with other model

Work	Features	Approach	ACC%	Precision	Recall	F1-score
Proposed model	ECG Segment + RR	Conv2D + CWT	97.27	96.53	97.27	96.89
Farag et al. [23], 2022	ECG Segment + RR	Conv1D + Conv2D + STFT	99.08	99.09	99.08	99.09
Rahul et al. [22] 2021	ECG Segment + RR	SVM	99.51	99.28	99.28	–
Ahamed et al. [8] 2020	ECG Segment	KNN	94.95	92.62	94.79	–
Ahamed et al. [8] 2020	ECG Segment	ANN with CLASS weights	97.286	96.158	96.890	–

Conv1D and Conv2D alongside short-time Fourier Transform [23] (STFT) while the proposed model uses Conv2D with CWT. Three sorts of classes are classified using SVM in Rahul's model: normal, Premature Ventricular Contraction (PVC), and premature atrial contraction [22]. Ahamed's work is a comparative study of various approach in which the class imbalance is tackled by assigning class weights (Table 4).

7 Conclusion and Future Work

The study used a dataset from MIT-BIH for testing and training deep learning models for cardiac arrhythmias. The dataset had an uneven distribution of categories, affecting the model's generalization. The model achieved 98.10% precision, 98.67% F1 score, and 97.27% accuracy. The project plans to integrate with Electronic Health Records (EHR) for quick access to ECG classification results, integrating personalized treatment recommendation systems for precision medicine. Collaboration with wearable technology manufacturers is also planned to embed ECG classification models into biosensor devices for continuous monitoring and real-time feedback for individuals with chronic cardiac conditions. However, the study acknowledges limitations in using CNNs for ECG classification, such as the lack of labelled ECG data, unpredictability, interpretability, and computational demands. Imbalances in class distribution within ECG datasets may bias model predictions towards dominant classes, impacting performance for minority groups. These limitations emphasize the need for comprehensive strategies addressing data quality, interpretability, generalization, class balance, and noise robustness in using CNNs in ECG classification jobs.

References

1. Jayswal D et al (2022) Study and develop a convolutional neural network for MNIST handwritten digit classification. In: Proceedings of third international conference on computing, communications, and cyber-security: IC4S 2021. Springer Nature Singapore, Singapore
2. Bohara M et al (2021) An AI based web portal for cotton price analysis and prediction. In: 3rd international conference on integrated intelligent computing communication & security (ICIIC 2021). Atlantis Press
3. Ramoliya D, Goel P, Vaidya J, Vadaliya P (2022) University indoor scene classification using transfer learning. In: 2022 4th international conference on inventive research in computing applications (ICIRCA), Coimbatore, India, pp 793–797. https://doi.org/10.1109/ICIRCA54612.2022.9985516
4. https://www.mdpi.com/
5. Ahmed AA, Ali W, Abdullah TA, Malebary SJ (2023) Classifying cardiac arrhythmia from ECG signal using 1D CNN deep learning model. Mathematics 11(3):562
6. Xu X, Liu H (2020) ECG heartbeat classification using convolutional neural networks. IEEE Access 8:8614–8619. https://doi.org/10.1109/ACCESS.2020.2964749
7. Mohebbanaaz RK, L V, Padma Sai Y (2021) Classification of arrhythmia beats using optimized K-nearest neighbor classifier. In: Udgata SK, Sethi S, Srirama SN (eds) Intelligent systems. Lecture notes in networks and systems, vol 185. Springer, Singapore. https://doi.org/10.1007/978-981-33-6081-5_31
8. Ahamed MA, Hasan KA, Monowar KF, Mashnoor N, Hossain M (2020) ECG heartbeat classification using ensemble of efficient machine learning approaches on imbalanced datasets. In: 2020 2nd international conference on advanced information and communication technology (ICAICT), Dhaka, Bangladesh, pp 140–145. https://doi.org/10.1109/ICAICT51780.2020.9333534
9. Thilagavathy R, Srivatsan R, Sreekarun S, Sudeshna D, Priya PL, Venkataramani B (2020) Real-time ECG signal feature extraction and classification using support vector machine. In: 2020 international conference on contemporary computing and applications (IC3A), Lucknow, India, pp 44–48. https://doi.org/10.1109/IC3A48958.2020.233266
10. Lopez AD, Joseph LA (2013) Classification of arrhythmias using statistical features in the wavelet transform domain. In: 2013 international conference on advanced computing and communication systems. IEEE, pp 1–6
11. Moody GB, Mark RG (2001) The impact of the MIT-BIH arrhythmia database. IEEE Eng Med Biol Mag 20(3):45–50
12. https://www.scilit.net/
13. Çakal K, Efe MÖ (2023) Cardiovascular anomaly detection with heterogeneous wave segment harmonization for lightweight systems. Cybern Syst 1–20
14. Nannavecchia A, Girardi F, Fina PR, Scalera M, Dimauro G (2021) Personal heart health monitoring based on 1D convolutional neural network. J Imaging 7(2):26
15. Yang P et al (2021) Ensemble of kernel extreme learning machine based random forest classifiers for automatic heartbeat classification. Biomed Signal Process Control 63:102138
16. https://www.ncbi.nlm.nih.gov/
17. Sahetai M et al (2023) Abnormal human behavior detection from a video sequence using deep learning. In: International conference on information and communication technology for intelligent systems. Springer Nature Singapore, Singapore
18. Sumathi S, Balaji Ganesh A. Secure healthcare monitoring of arrythmias in internet of things with deep learning and elgamal encryption. J Intell Fuzzy Syst 1–16
19. Xie H et al (2023) A lightweight 2-D CNN model with dual attention mechanism for heartbeat classification. Appl Intell 53(13):17178–17193
20. Toledo E et al (1998) The use of a wavelet transform for the analysis of nonstationary heart rate variability signal during thrombolytic therapy as a marker of reperfusion. Comput Cardiol 25

21. Wang L, Zhang X (2023) Anomaly Detection for automated vehicles integrating continuous wavelet transform and convolutional neural network. Appl Sci 13(9):5525
22. Farag MM (2022) A self-contained STFT CNN for ECG classification and arrhythmia detection at the edge. IEEE Access 10:94469–94486. https://doi.org/10.1109/ACCESS.2022.3204703
23. Rahul J, Sora M, Sharma LD, Bohat VK (2021) An improved cardiac arrhythmia classification using an RR interval-based approach. Biocybern Biomed Eng 41(2):656–666

From Pixels to Truth: Unveiling Deepfakes with a Meso-5 and XceptionNet Fusion Network

S. Abhinandhan, A. G. Sreedevi, and G. Saranya

Abstract The growth of deepfakes, or hyper-realistic synthetic media, demands effective detection algorithms to ensure the validity of visual information. This study presents a unique deepfake detection model based on a hybrid architecture that takes advantage of the characteristics of both Meso-5 and XceptionNet. Meso-5, which is well-known for its capacity to efficiently capture mesoscopic picture characteristics, is carefully paired with Xception-Net's outstanding feature extraction capabilities. This novel strategy tries to achieve higher deepfake detection accuracy than previous approaches. We assess the proposed model's performance using the publicly available Deepfake Detection Challenge (DFDC) dataset, which was published by Facebook in 2018. Our results show that this hybrid technique works well, with an impressive accuracy of 81.24% in detecting deepfakes. These findings demonstrate our model's ability to mitigate the deepfake danger and build a more dependable future for visual information.

Keywords Deepfake detection · Meso-5 · XceptionNet · Hybrid architecture · DFDC dataset · Hyper-realistic synthetic visuals · Visual threat · Image detection

1 Introduction

Digital media has brought forth a new era of data consumption in which visual content stands out. From the latest news footage to famous Instagram reels, everything we see helps us learn about our surroundings. However, the dependence on visual evidence raises a significant concern: the proliferation of fakes. Deepfakes are synthetic creations that use artificial intelligence (AI) to simulate the look or voice of real humans. These hyper-realistic products can be extremely convincing,

S. Abhinandhan (✉) · A. G. Sreedevi · G. Saranya
Department of Computer Science and Engineering, Amrita School of Computing, Amrita Vishwa Vidyapeetham, Chennai, India
e-mail: ch.en.u4cse20001@ch.students.amrita.edu

P. J. S. Gonçalves et al. (eds.), *Proceedings of Fifth International Conference on Computing, Communications, and Cyber-Security*, Lecture Notes in Networks and Systems 1128, https://doi.org/10.1007/978-981-97-7371-8_6

merging the boundaries between what is real and what is not while offering a serious challenge to the basic foundation of visual information [1].

Deepfakes are easily made and circulated online, creating a fertile environment for inaccurate information and disinformation. Intentional criminals can use this technology to create fake events, create strife among groups, and influence the general public to fulfil private benefits. A fabricated media portrayal of a political leader using incendiary words, for example, might have disastrous consequences during an election. The decline of faith in visual evidence goes beyond politics. Deepfakes may be used to harm reputations, with disastrous implications for those in the public spotlight. A celebrity deepfake showing scandalous misbehaviour might permanently damage their professional and personal lives. Deepfakes might be used to mimic executives or influence financial markets, posing a significant risk of fraud [2].

The repercussions of deepfakes are far-reaching and constitute an existential threat to the future of imagery. Traditional techniques of media verification, such as relying on established news channels or fact-checking corporations, may be ineffective in the face of error-free deepfakes. This loss of trust can have a domino effect, causing a breakdown of confidence in establishments and the media altogether. Furthermore, history itself becomes vulnerable to alteration. Deepfakes might be used to alter ages, concealing the truth and generating a false account of history [3].

This troubling scenario highlights the critical need for efficient strategies to tackle the deep-fake danger. A holistic approach is required, which includes not just technological breakthroughs in deepfake identification but also a strong general educational plan. On the technology front, researchers continue to innovate on novel and enhanced deepfake detection algorithms. These methods use visual and audio signals to identify altered information [4].

However, technology alone can't fix the "deep fake" problem. Offering individuals, a knowledge of media and analytical skills is vital for understanding the increasingly complicated information ecosystem. Public education campaigns must concentrate on instructing individuals on how to see possible suspicious elements in audio and video recordings, how to verify material from numerous sources, and how to be more cautious viewers of internet content. By cultivating a verification culture and equipping individuals with the ability to recognise deepfakes, we may begin to limit the existential danger they represent to the integrity of visual knowledge and its role in society as a whole [5].

This research goes into the realm of deepfakes, investigating the technology that powers them, the potential harm they might cause, and the numerous remedies we can devise to protect the future of visual information. Our suggested deepfake detection methodology, which draws on the benefits of both the Meso-5 and XceptionNet architectures, intends to contribute to the technical breakthroughs required to resist this emerging danger. Finally, by encouraging a mix of innovative technology and traditional schooling, we may guarantee that image-based data remains a trustworthy medium of reality in modern times.

2 Related Works

Researcher Hu et al. address this gap by proposing a two-stream convolutional neural network (CNN) for detecting compressed deepfakes in social networks. Their method tackles the challenges introduced by video compression. The frame-level stream analyses individual frames using a pruned MesoNet architecture to extract mesoscopic properties while mitigating the influence of compression artifacts. The temporality-level stream focuses on the time-dependent features present in P-frames, exploiting the inherent inconsistencies in temporal information between real and deepfake videos. This two-pronged approach demonstrates superior performance compared to existing methods in compressed deepfake detection [6].

Liao et al. address this challenge in "FAMM: Facial Muscle Motions for Detecting Compressed Deepfake Videos Over Social Networks". They propose a Facial-Muscle-Motion-based (FAMM) framework that analyses facial features across consecutive frames. FAMM first extracts facial landmarks from video frames. These landmarks are then used to construct facial muscle motion features by modelling five facial regions. Finally, the system employs Dempster-Shafer theory to combine the evidence from different features and arrive at a final detection result. The authors demonstrate that compression has minimal impact on FAMM's ability to analyze facial muscle motion. They also show that the features extracted from real and deepfake videos exhibit distinct characteristics. Extensive experiments validate that FAMM outperforms existing methods in detecting deepfakes within compressed videos, particularly those encountered on social networking platforms [7].

MRE-Net tackles deepfake detection by capturing spatial and temporal inconsistencies within the video. It achieves this through a three-pronged approach: Bipartite Group Sampling Strategy (BGS), Momentary Inconsistency Excitation (MIE) module, and Longstanding Inconsistency Excitation (LIE) module. BGS samples video sequences at various rates to capture diverse facial motion dynamics. The MIE module focuses on short term inconsistencies within a group of frames, examining both spatial artifacts and motion patterns. The LIE module, on the other hand, analyzes inconsistencies between frames across different sampling rates, capturing long-term temporal discrepancies. By combining these modules, MRE-Net aims to achieve robust deepfake detection by identifying various types of inconsistencies introduced during the manipulation process [8].

Yu et al. propose the Multiple Spatiotemporal Views Transformer (MSVT) to address this challenge. MSVT tackles the limitations of prior methods by incorporating both local and global spatiotemporal information. It achieves this through two key components: Local Spatiotemporal View (LSV) and Global Spatiotemporal View (GSV). LSV captures fine-grained details by analysing local groups of consecutive frames within a video. GSV, on the other hand, focuses on the broader temporal context by examining all frames collectively. Finally, MSVT integrates these multi-level features using a novel global–local transformer (GLT) for robust deepfake detection. The experimental results demonstrate that MSVT outperforms existing state-of-the-art methods, highlighting the effectiveness of its multi-view approach [9].

Mahmud et al. propose a cost-sensitive deep learning approach for effective deepfake detection in videos. Their method utilizes four pre-trained convolutional neural networks (XceptionNet, InceptionResNetV2, EfficientNetV2S, and EfficientNetV2M) to extract features from keyframes within videos. This approach addresses the class imbalance issue commonly encountered in deepfake datasets by employing a cost-sensitive neural network. The authors achieve high accuracy rates (XceptionNet: 98% on CelebDf-V2, InceptionResNetV2: 94% on FaceForensics++) on benchmark datasets, demonstrating the effectiveness of their method for real-world deepfake detection [10]. Saha et al. address a critical gap in deepfake detection: identifying manipulated segments within real videos. Their work tackles this challenge by proposing a new method for deepfake prediction at both the frame and video level. To evaluate their method, they introduce a benchmark dataset featuring videos with subtly altered segments. Their approach leverages Vision Transformers for spatial features and Timeseries Transformers for temporal features, facilitating the detection of potential deepfakes. This research paves the way for improved deepfake moderation tools by enabling targeted human oversight on suspicious video segments [11].

The growing sophistication of deepfake technology necessitates the development of robust detection methods, particularly for manipulated facial images. Traditional approaches often rely on handcrafted features or rule-based systems, which are susceptible to evasion tactics. To address this challenge, Sudharson et al. propose a novel deep learning architecture that leverages the strengths of both CapsNet and EfficientNet for deepfake detection. CapsNet excels at capturing spatial relationships between features, making it well-suited to identify geometric transformations and variations often present in deepfakes. EfficientNet, a powerful and efficient deep learning model, serves as a feature extractor, learning high-level image characteristics for accurate analysis. Their experiments demonstrate an impressive 99.64% accuracy in detecting various deepfakes, including face swaps and re-enactments. This method holds promise for real-world applications in media forensics, law enforcement, and online media verification [12].

Deep learning has emerged as a promising approach for image forgery detection. Suresh et al. propose a system utilizing a Convolutional Neural Network (CNN) with ResNet50v2 architecture and pre-trained YOLO weights. Their model, implemented in Python, achieved an accuracy of 85% on the CASIA image forgery detection benchmark datasets. This research exemplifies the potential of deep learning for image forensics, offering a powerful tool to combat the spread of manipulated images. However, further exploration is necessary to improve accuracy and robustness against evolving forgery techniques [13].

3 Proposed System

The research presented here proposes a unique deepfake detection method based on a hybrid architecture that combines the characteristics of two strong models: Meso-5 and XceptionNet. This section looks into each model's separate features, investigates the reason for their integration, and describes the suggested hybrid model's design.

3.1 A Deep Dive into Meso-5: Extracting Mesoscopic Features

The Meso-5 model is a convolutional neural network (CNN) that is meant to extract mesoscopic features from pictures. Mesoscopic qualities fall in between low-level pixel details and high-level semantic information. They include important details like textures, edges, and local object sections, which provide essential clues regarding the content's legitimacy.

Meso-5 is an important deepfake detection technique because it identifies tiny discrepancies that are frequently present in altered videos. Deepfakes may include abnormalities in skin texture, hair patterns, or lighting effects that differ from their natural looks. Meso-5's convolutional layers function as feature detectors, gradually extracting mesoscopic information from video frames (input pictures). Each convolutional layer utilizes filters to scan the image, recognising and activating certain patterns. The feature maps are then down sampled using the max-pooling layers, which capture the most noticeable properties while minimizing computing complexity.

Meso-5 is used in the proposed system above its predecessor, Meso-4, for numerous reasons. While Meso-4 provides a solid basis for mesoscopic feature extraction, Meso-5 improves upon this critical skill. Meso-5's incorporation of one additional convolutional block has the ability to extract a richer and more nuanced collection of mesoscopic properties. This new layer allows the model to go a little further into the feature hierarchy, perhaps catching even more minor anomalies in deepfakes [14].

In simple terms, Meso-5 is an improved version of Meso-4. It inherits its predecessor's fundamental assets, such as its capacity to recognise textures, edges, and local object components. Meso-5, on the other hand, improves this procedure by adding another convolutional block (see Fig. 1), which might result in a more comprehensive and informative feature representation. This increased feature set may then be coupled with features derived by another model, resulting in a more effective deepfake detection system [15].

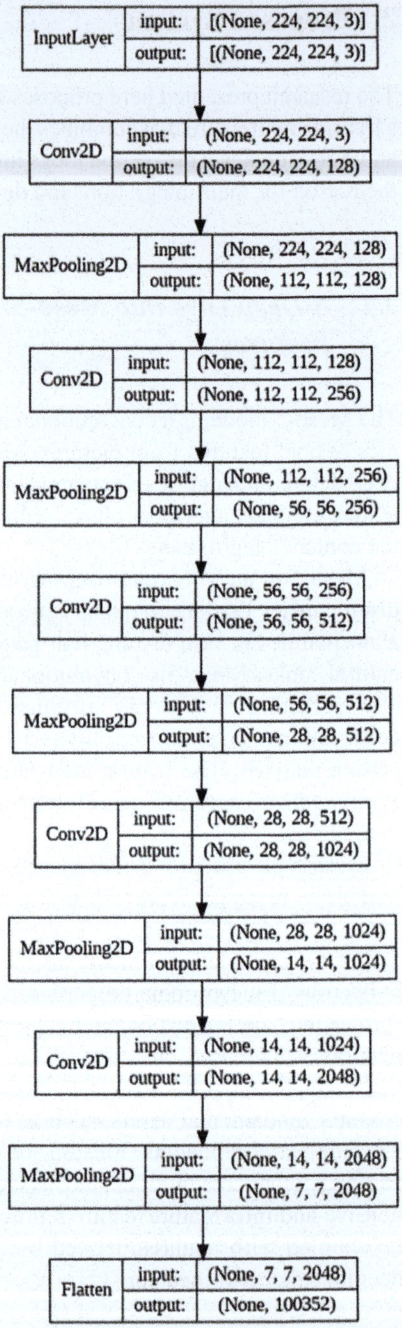

Fig. 1 The Meso-5 architecture—an extension to the existing Meso-4

3.2 XceptionNet: A Pre-Trained Powerhouse for Feature Extraction

The next core of the proposed framework is XceptionNet, a pre-trained deep convolutional neural network known for its superior feature extraction capabilities. Unlike Meso-4, which is primarily developed for mesoscopic characteristics, XceptionNet has a more general-purpose architecture and was trained on a large dataset, such as ImageNet [16].

This training gives XceptionNet the capacity to extract a variety of characteristics from photos, from low-level minutiae to higher-level semantic notions.

In the field of deepfake detection, XceptionNet can collect useful information such as face landmarks, emotions, and lighting conditions. Inconsistencies typical of deepfakes can be identified by analysing these properties across successive video frames. For example, slight alterations in face expressions or lighting anomalies across frames may indicate tampering.

The prepared nature of XceptionNet provides a huge benefit. Using the massive information, it has previously gathered from the ImageNet dataset, XceptionNet can easily extract important features from video frames without requiring lengthy retraining on the specialized job of deepfake detection. This not only decreases training time, but also enhances generalization capabilities, allowing the model to perform well on previously unknown deepfakes. Xception-Net's output, a tensor of size (batch_size, 2048), provides this entire feature collection, which is ready for integration with Meso-4's extracted features.

3.3 Merging Strengths: The Hybrid Model Architecture

The suggested hybrid Meso-XceptionNet model (see Fig. 2) combines Meso-4 and XceptionNet to improve the feature extraction process. By using the same input for both models, the hybrid model may analyze the same visual data, resulting in a more thorough feature extraction process. Meso-4 and XceptionNet operate separately, with Meso-4 focused on mesoscopic details and XceptionNet collecting a larger variety of visual data. This strategy enables the model to extract a more diversified set of features since each model specializes on a distinct component of the input data.

The outputs of both models are integrated, yielding a more comprehensive and usable feature representation that contains both mesoscopic and broader visual qualities. This merged feature vector is then sent into a fully connected layer with 1024 units that serves as a classifier.

Using the combined feature representation, the classifier learns to distinguish between deepfakes and genuine video frames. The hybrid model's last layer is dense, with a sigmoid activation function. The sigmoid function provides a value between 0 and 1 indicating the likelihood of the input video frame being classified as 'FAKE'.

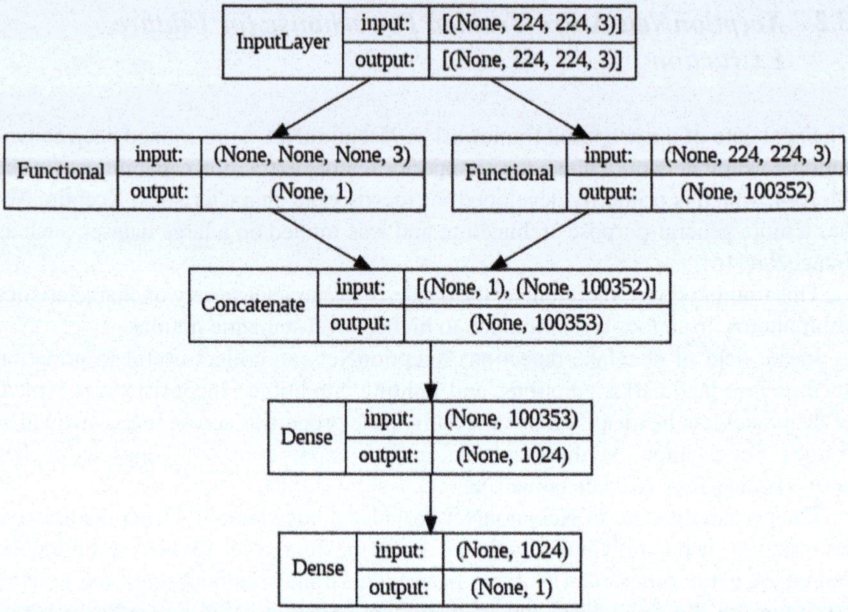

Fig. 2 The Meso-5 architecture combined with the predefined Xception-Net's Architecture making it a hybrid model with extraction properties of both Meso5 & Xception-Net

A number closer to one suggests a greater likelihood of being a deepfake, whereas a value closer to zero indicates a real video frame.

The inclusion of Meso-4 and XceptionNet in the proposed hybrid model enables a more extensive feature extraction procedure, resulting in a more accurate and robust deepfake detection system. The combination of mesoscopic characteristics with bigger visual attributes results in a fuller and more diversified feature representation, allowing the model to reliably distinguish between deepfakes and real video frames.

3.4 The DFDC (Deepfake Detection Challenge) Dataset

The DFDC (Deepfake Detection Challenge) dataset is a large dataset intended for training and testing deepfake detection algorithms. The dataset was created through a crowdsourcing initiative in which actors of various genders, ages, and skin tones were requested to submit films of themselves answering predetermined questions. The dataset has a wide variety of lighting situations, head positions, and backdrops, making it a valuable resource for deepfake detection studies.

The DFDC collection comprises 5000 videos, both altered and original. The dataset's films were converted into picture frames using Python's face_recognition module and maintained in another dataset named deepfake_faces (because of hardware restrictions, we had to use another dataset). This dataset represents approximately 80% of the training set. Each face was scaled to 224 × 224. This arbitrary possibility additionally represents the average, indicating that fifty percent of the faces are adjusted up and the remaining half are scaled down. Because this impacts blurriness and may influence categorization, the original size was included as information [17].

4 Implementation and Results

4.1 Data Cleaning and Preparation

Due to computational restrictions, a representative subset of "Deepfake faces" was retrieved from the DFDC dataset. Stratified sampling guarantees that the selected subset appropriately represents the distribution of actual and deepfake images in the original dataset. Python's face_recognition library simplifies video frame extraction, resulting in the "Deepfake Faces" dataset for training and assessment.

4.2 Data Preprocessing

To ensure smooth integration with the neural network architecture, all retrieved frames are scaled to the same size (224 × 224 pixels). Additionally, pixel values are normalized to a range of 0 to 1, which streamlines the training process.

4.3 Data Augmentation

Various data augmentation strategies are used to artificially increase the training data volume and improve the model's capacity to generalize to previously unknown data.

Random Cropping: During training, picture parts are randomly cropped to expose the model to a variety of image compositions.

Flipping: Randomly flips photos horizontally to simulate real-world settings in which faces may be taken from various perspectives.

Color Jittering: Creates minor random fluctuations in picture brightness, contrast, and saturation to resemble natural lighting discrepancies.

Figure 3 depicts the visual impact of data augmentation on example photos from the validation dataset. By incorporating these modifications, the model is exposed

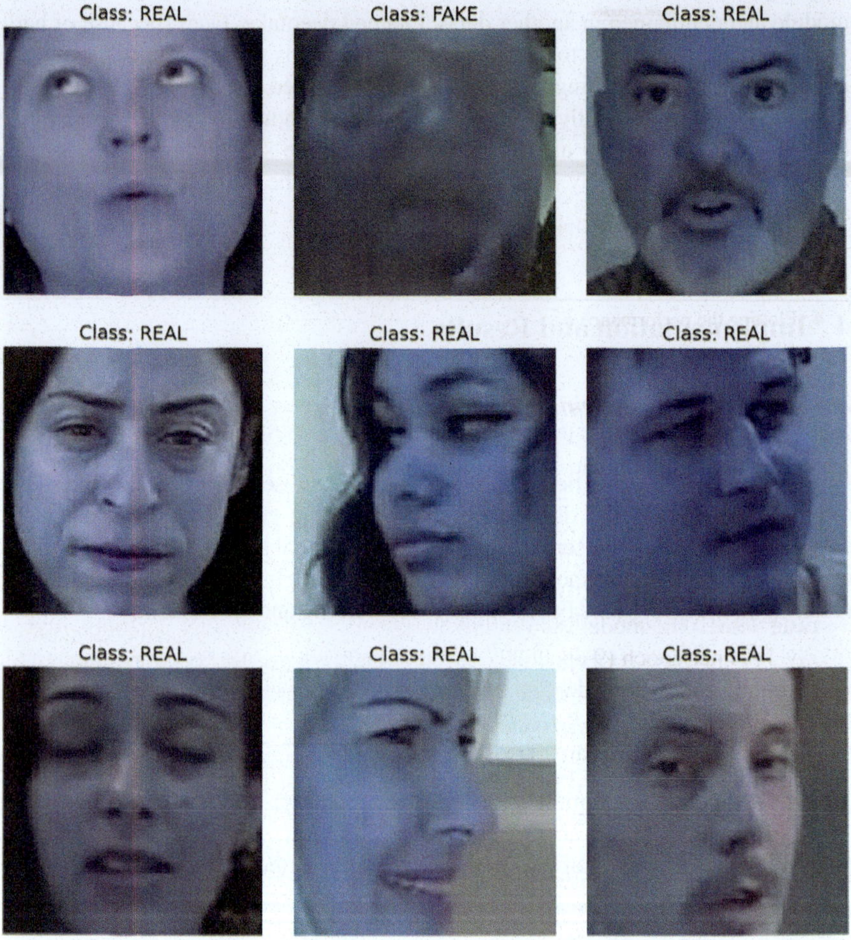

Fig. 3 Displaying the first 9 images from the validation set, after augmentation of data

to a greater variety of probable picture attributes, which improves its capacity to generalize and identify deepfakes.

4.4 Computational Resources

Training the proposed deepfake detection model requires substantial computing resources. This section describes the hardware and software configurations utilized for this study.

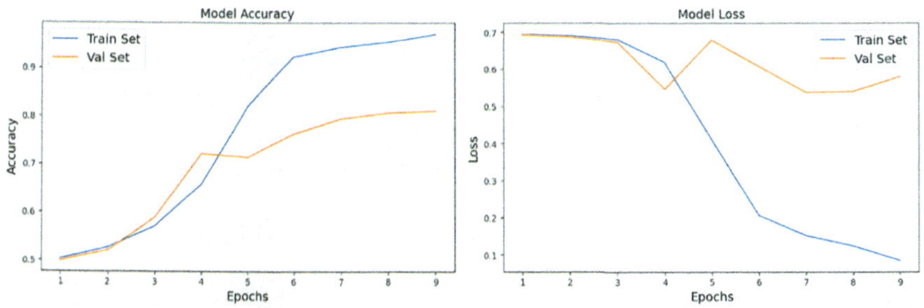

Fig. 4 Plotting the performance on model's accuracy, loss and epochs

Hardware: Two NVIDIA Tesla T4 GPUs were used to speed up the training procedure. GPUs are especially well-suited for deep learning applications because of their parallel processing capabilities, which drastically reduce training time when compared to CPUs.

Software: The TensorFlow deep learning framework was used to build the model architecture and training process.

Training Time: The model training procedure, as shown in the code snippet, takes roughly 199s per epoch (9 epochs total) on the aforementioned hardware combination. Training time varies depending on the hyperparameters (learning rate, batch size) and dataset size.

The supplied code explains how to construct data augmentation using TensorFlow's tf.keras.Sequential API. The code sample also shows how augmented photos are visualized in comparison to their originals for better comprehension.

4.5 Evaluation and Inference

The obtained accuracy of 81.24% (see Fig. 4) on the test set demonstrates the efficiency of the proposed hybrid model in deepfakes identification. However, it is critical to recognise that deepfake production processes are continually changing. Continuous progress through additional data augmentation tactics, hyperparameter tweaking, and the investigation of more complex deep learning architectures is critical for maintaining effective deepfake detection capabilities.

5 Conclusion

This study introduced a unique deepfake detection system based on a hybrid Meso5-XceptionNet architecture. This model outperforms traditional techniques, such as VGG19, by including a custom-designed Meso5 block. The Meso5 block focuses on mesoscopic properties including textures, edges, and local object sections, which are typically affected in deepfakes. The Meso5-XceptionNet model is potentially more suited for deepfake detection because to the combination of focused feature extraction and Xception-Net's wider feature collection capabilities. This theoretical advantage is supported by the DFDC dataset's 81.24% accuracy rate.

However, understanding limits is critical. Overfitting is still an area for development. Future research might look into improved data augmentation methodologies and hyperparameter tweaking to improve generalization and resilience. Furthermore, more advanced deep learning architectures may enhance performance.

The practical ramifications of this study are enormous. The proposed open-source software application based on this model has enormous potential for widespread adoption, community contributions, and ongoing model evolution. Furthermore, educational modules based on the Meso5-XceptionNet architecture might provide students with critical thinking abilities for navigating the information landscape and distinguishing between authentic and manipulated content. Looking ahead, study will focus on:

Scalability and Generalizability: Investigating how the model scales to bigger datasets and applies to various types of deepfakes.

Computational Efficiency: Investigating methods to optimize the model for deployment on resource-constrained devices.

Audio Deepfake Detection: Addressing the rising problem of altered audio by creating detection algorithms particularly for this domain.

Real-Time Detection: Adding real-time processing capabilities to the model architecture to enable quick deepfake detection.

Transfer Learning: Using transfer learning techniques to extend the model to a variety of deepfake detection tasks beyond its present scope.

This study aims to protect the integrity of information in the digital age by integrating technology breakthroughs with public education campaigns.

References

1. Ba Z et al (2024) Exposing the deception: uncovering more forgery clues for deepfake detection. arXiv preprint arXiv:2403.01786
2. Lin L et al (2024) Preserving fairness generalization in deepfake detection. arXiv preprint arXiv:2402.17229
3. Timm Doloriel C, Cheung N-M (2024) Frequency masking for universal Deepfake detection. arXiv e-prints arXiv-2401
4. Wang H, Liu Z, Wang S (2023) Exploiting complementary dynamic incoherence for deepfake video detection. IEEE Trans Circ Syst Video Technol
5. Alnaim NM et al (2023) DFFMD: a deepfake face mask dataset for infectious disease era with deepfake detection algorithms. IEEE Access 11:16711–16722

6. Hu J et al (2021) Detecting compressed deepfake videos in social networks using frame-temporality two-stream convolutional network. IEEE Trans Circ Syst Video Technol 32(3):1089–1102
7. Liao X et al (2023) FAMM: facial muscle motions for detecting compressed deepfake videos over social networks. IEEE Trans Circ Syst Video Technol
8. Pang G et al (2023) Mre-net: multi-rate excitation network for deepfake video detection. IEEE Trans Circ Syst Video Technol
9. Yu Y et al (2023) Msvt: multiple spatiotemporal views transformer for deepfake video detection. IEEE Trans Circ Syst Video Technol
10. Mahmud F et al (2023) Unmasking deepfake faces from videos using an explainable cost-sensitive deep learning approach. In: 2023 26th international conference on computer and information technology (ICCIT). IEEE
11. Saha S et al (2023) Undercover deepfakes: detecting fake segments in videos. In: Proceedings of the IEEE/CVF international conference on computer vision
12. Sudharson S, Kokil P, Sasikala D, Swamy PA (2023) Deepfake detection system using a hybrid model. In: 2023 14th international conference on computing communication and networking technologies (ICCCNT), Delhi, India, pp 1–6. https://doi.org/10.1109/ICCCNT56998.2023.10308279
13. Suresh HR et al (2024) Deep learning-based image forgery detection system. Int J Electron Secur Digit Forensics 16(2):160–172
14. Afchar D et al (2018) Mesonet: a compact facial video forgery detection network. In: 2018 IEEE international workshop on information forensics and security (WIFS). IEEE
15. Xia Z et al (2022) Deepfake video detection based on MesoNet with preprocessing module. Symmetry 14(5):939
16. Saxena A et al (2023) Detecting deepfakes: a novel framework employing XceptionNet-based convolutional neural networks. Traitement du Signal 40(3)
17. Dolhansky B et al (2019) The deepfake detection challenge (dfdc) preview dataset. arXiv preprint arXiv:1910.08854

Wireless Networks and the Internet of Things: Empowering Connectivity and Smart Applications

Hari Mohan Rai, Aditya Pal, Kaustubh Kumar Shukla, Alveera Hafeez, Abdul Razaque, and Saule Amanzholova

Abstract This paper explores the crucial role of wireless networks in facilitating connectivity for smart applications within the Internet of Things (IoT). Objectives include examining wireless network protocols for IoT, addressing security challenges, exploring design considerations, investigating energy efficiency, and analyzing standardization efforts. The study involves an extensive review of literature, research articles, and industry reports. It comprehensively analyzes wireless network protocols in IoT applications, detailing their characteristics. Security challenges are emphasized, advocating robust authentication, encryption, and access control. Design considerations and scalability challenges are explored, with potential solutions discussed. Energy efficiency is investigated, covering power consumption, energy-efficient protocols, and harvesting methods. The study also scrutinizes standardization efforts' impact on interoperability and IoT adoption. Through these insights, the paper aims to advance the seamless integration of wireless networks and IoT, unlocking the full potential of smart applications in diverse domains.

Keywords IoT · Security system · Smart applications · Wireless system

1 Introduction

In the symphony of technological progress, wireless networks and the Internet of Things (WN-IoT) harmonize as avant-garde instruments, shaping the interconnected landscape that defines our modern era. The IoT, a ballet of interconnected devices,

A. Pal (✉)
Department of IT, Dronacharya Group of Institutions, Greater Noida, India
e-mail: aditya.16202@gnindia.dronacharya.info

H. M. Rai · A. Razaque · S. Amanzholova
Department of Cyber Security, International IT University, Almaty, Kazakhstan

H. M. Rai · K. K. Shukla · A. Hafeez
Department of ECE, Dronacharya Group of Institutions, Greater Noida, India

P. J. S. Gonçalves et al. (eds.), *Proceedings of Fifth International Conference on Computing, Communications, and Cyber-Security*, Lecture Notes in Networks and Systems 1128, https://doi.org/10.1007/978-981-97-7371-8_7

pirouettes on the stage of the internet, gracefully exchanging data and orchestrating a mesmerizing performance of automation and communication. The ethereal realm of wireless networks, a sonnet of seamless connectivity, forms the very stage upon which the IoT's enchanting ballet unfolds [1]. This synergy holds a spellbinding promise, poised to metamorphose industries and infuse magic into our daily routines. As mundane objects, from household appliances to vehicles, waltz into the embrace of the internet, the IoT orchestrates a grand spectacle of automation, insights gleaned from data, and the enchantment of remote control [2]. With domains as diverse as smart homes, industrial automation, and healthcare systems taking center stage, the transformative allure of this technological ballet is undeniable. Within the pages of this research paper, we embark on an exploration of the intricate pas de deux between WN-IoT, unraveling their interdependencies while unmasking the challenges and opportunities that unfold in this mesmerizing dance. The paper aims to shed light on the following key areas:

- Wireless Network Protocols for IoT: This paper will provide an in-depth analysis of commonly used wireless protocols in IoT applications, focusing on their unique characteristics, advantages, and limitations. By understanding the strengths and weaknesses of these protocols, their suitability for different IoT use cases can be evaluated.
- Security Challenges in Wireless IoT Networks: The research will delve into the distinct security challenges encountered in wireless IoT networks. It will examine authentication mechanisms, encryption methods, and access control measures that portray a crucial responsibility in safeguarding IoT devices and data. Addressing these challenges is vital for fostering the widespread adoption of IoT technologies.
- Network Infrastructure for IoT: This section will explore the design and architecture considerations involved in establishing robust network infrastructure for IoT deployments. Scalability challenges, edge computing, and fog computing will be investigated as potential solutions to optimize IoT connectivity and enhance overall system performance.
- Energy Efficiency in Wireless IoT: The research will investigate the energy consumption patterns of IoT devices and wireless network communication. It will explore energy-efficient protocols, low-power communication techniques, and energy harvesting methods to minimize power consumption and prolong the battery life of IoT devices.

This paper will analyze the standardization efforts aimed at establishing unified wireless network protocols for IoT connectivity. It will delve into the role of standardization bodies in defining and promoting these protocols, while also examining their impact on interoperability, compatibility, and the wider adoption of IoT technologies.

In pursuit of these goals, this study paper endeavors to illuminate novel perspectives on the intricate interplay between WN-IOT. It will contribute to a comprehensive understanding of the challenges, advancements, and future directions within this dynamic field. Figure 1 presents a detailed block diagram illustrating the symbiotic

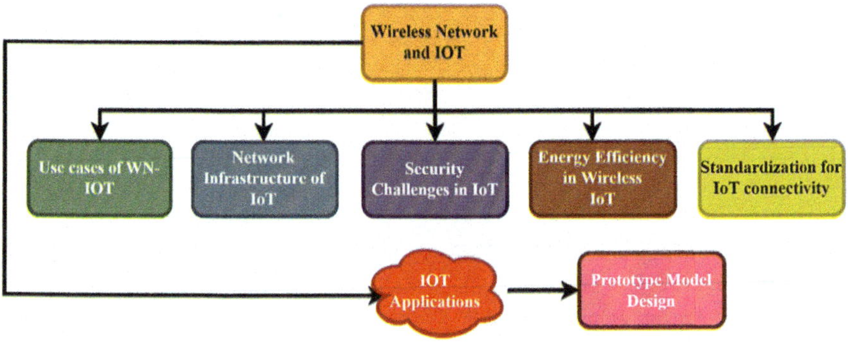

Fig. 1 Applications of wireless network and IOT

relationship between WN-IoT. This visual representation encapsulates the interconnected components and key elements that constitute the collaborative framework between WN-IoT.

2 Use Cases

2.1 Wireless Network Protocols for IoT

In the domain of the IoT, various wireless network protocols have emerged to facilitate seamless connectivity among IoT devices. This section discusses several frequently and mostly utilized protocols in the applications of IoT, including Wi-Fi, Z-Wave, Bluetooth, Zigbee, and LoRaWAN. Each protocol possesses unique characteristics, advantages, and limitations, making them suitable for specific IoT use cases [3]. Figure 2 illustrates a comprehensive array of use cases for various wireless network protocols. The figure serves as a valuable reference, showcasing the diverse applications of these protocols in real-world contexts.

- Zigbee: Within the ethereal realms of the 2.4 GHz frequency band, Zigbee, a wireless protocol artfully tailored for the nuanced landscape of low-complexity IoT applications, extends its graceful hand to bestow upon us the enchanting gift of mesh networking capabilities. Its key strengths include low power consumption, robustness against interference, and support for large-scale deployments. However, it has a limited data rate and may involve higher implementation costs compared to other protocols.
- Z-Wave: Designed with precision for the intricate choreography of home automation, Z-Wave stands as a wireless protocol, gracefully pirouetting within the sub-GHz frequency band. Its performance takes center stage, orchestrating a harmonious ballet of reliable and secure communication, elegantly spanning short to medium distances. Z-Wave offers excellent interoperability among devices, low

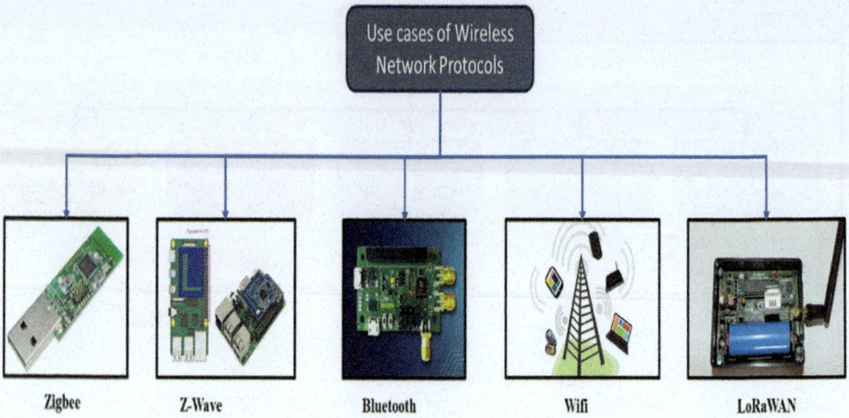

Fig. 2 Use cases of wireless network protocols

power consumption, and robust mesh networking capabilities. However, its range is comparatively limited, and the selection of Z-Wave devices may be relatively smaller.

- Bluetooth Low Energy (BLE): Renowned for its prowess in energy efficiency, BLE emerges as a wireless protocol tailor-made for the delicate ecosystem of low-power IoT devices, including but not limited to wearables and sensors. Within the melodic resonance of the 2.4 GHz frequency band, BLE orchestrates succinct yet impactful short-range communication, aligning seamlessly with the nuanced needs of modern devices. Its advantages include easy integration with smartphones and tablets, low implementation costs, and support for beacon technology.
- Wi-Fi: It is a broadly implemented known as wireless fidelity offering high-speed data transfer and broader coverage. Incorporating itself into the established fabric of internet infrastructure, Wi-Fi seamlessly operates in the range of 2.4–5 GHz frequency bands. It enables high-bandwidth applications and supports a large number of concurrent connections. However, its higher power consumption compared to other protocols may limit its usage in battery powered IoT devices.
- LoRaWAN: Enveloped in the sub-GHz frequency band, a dance of long-range communication with low power consumption is orchestrated by LoRaWAN, a wireless protocol exquisitely crafted for IoT applications seeking expansive coverage. It provides connectivity for IoT devices in remote areas and offers excellent penetration capabilities through buildings and urban environments. However, it has lower data rates compared to other protocols, making it more suitable for applications transmitting small amounts of data over longer intervals [4].

When comparing these wireless protocols, aspects including range, power consumption, data rate, and suitability for different IoT applications come into play. Zigbee and Z-Wave excel in low-power home automation applications, with

Zigbee supporting larger-scale deployments and Z-Wave emphasizing interoperability. BLE is ideal for energy-efficient wearables and proximity-based applications. Wi-Fi provides high-speed connectivity for applications requiring large data transfers but consumes more power. LoRaWAN is suitable for wide-area IoT deployments with low-power requirements. Essential in the selection of an apt protocol for specific IoT use cases is the comprehension of the characteristics, advantages, and limitations inherent in these wireless protocols.

3 Network Infrastructure for IoT

Designing and implementing a vigorous network infrastructure is vigorous for supporting IoT implementations. This section discusses the key considerations in designing the network architecture for WN-IoT applications. It explores scalability challenges, potential solutions like edge computing and fog computing, network topologies, network coverage optimization, and Quality of Service (QoS) requirements [5].

3.1 Design Considerations

Within the intricate dance of IoT deployments, where a multitude of devices takes the stage, a symphony arises, demanding a network infrastructure of grand scalability to harmonize with the crescendo of data and device connections. Certain IoT applications, akin to the choreography of real-time monitoring and control systems, crave the grace of low-latency communication, ensuring their movements unfold with a timely precision that resonates through the digital ballet. In this performance, where the spotlight is on operational fluidity and data's graceful pirouette, the network infrastructure must wear the cloak of robust security, warding off the lurking shadows of unauthorized access and cyber threats, thus safeguarding the delicate choreography of IoT devices and their data.

3.2 Scalability Challenges and Solutions

In the labyrinth of IoT networks, scalability emerges as a formidable challenge. As the ensemble of IoT devices takes center stage, traditional centralized network architectures find themselves waltzing on the precipice of struggle, grappling with the swell of increased data traffic. In a visionary pas de deux, edge computing pirouettes into the scene, elegantly shifting the ballet of data processing and analysis closer to the edge of the network. Yet, the symphony of innovation doesn't cease there.

Edge computing is an extended version of Fog computing, weaving into the narrative intermediate computing nodes—ethereal fog nodes—strategically placed within the network's tapestry. The result is a ballet of localized data processing, a dance that curtails latency, ensuring the rhythmic flow of information in the grand IoT performance [6].

3.3 Network Topologies

In the kaleidoscope of IoT deployments, diverse network topologies unfurl like petals, each embracing a unique role tailored to the whims of the application. Within the constellation of possibilities, the star topology emerges as a luminary. Here, IoT devices, akin to celestial satellites, gravitate toward a central hub or gateway, creating a cosmic ballet of interconnected nodes. This architecture simplifies management and allows for centralized control but may create a single point of failure. A mesh topology enables direct communication between IoT devices, creating a self-forming and self-healing network. A hybrid topology combines multiple network topologies to leverage their respective strengths and address specific requirements of the IoT deployment.

- Network Coverage Optimization: Network coverage optimization is essential to ensure reliable connectivity across the IoT deployment area. Considerations include:
- Signal Strength: Analyzing signal strength to identify areas with weak connectivity and implementing solutions such as signal boosters or additional access points.
- Channel Allocation: Optimizing the allocation of channels to minimize interference and maximize network capacity.
- Range Extension: Employing range-extending technologies like repeaters or extenders to overcome limitations of wireless connectivity.
- Quality of Service (QoS) Requirements: Different IoT applications may have varying QoS requirements. In the diverse tapestry of applications, some, like real-time monitoring or video streaming, seek the symphony of low latency and high bandwidth, while others prioritize the quiet efficiency of energy or the steadfast reliability of operation.

3.4 IoT Applications and Use Cases

IoT has found widespread adoption among several industries and fields, revolutionizing business processes and enabling innovative solutions. This section highlights specific industries where IoT is widely adopted, discusses the wireless networking requirements and challenges unique to each application, and provides case studies or examples to illustrate successful IoT implementations [7]. Figure 3 succinctly

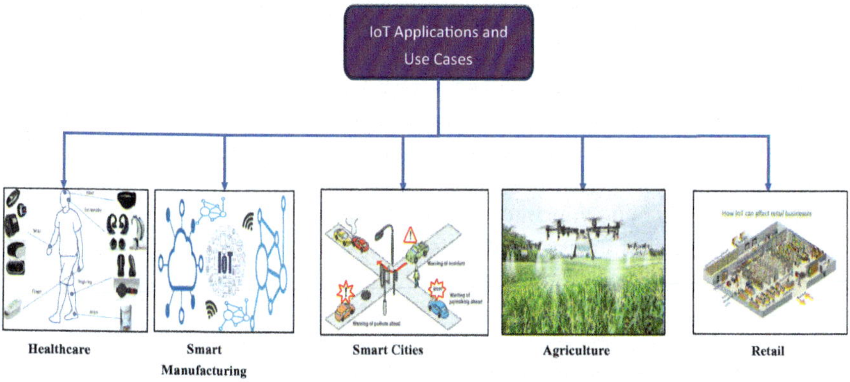

Fig. 3 Application of IoTs and their use cases

captures the array of practical applications and use cases of Internet of Things (IoT) technology. It provides a visual overview of IoT's versatile impact across diverse sectors, including healthcare, manufacturing, smart cities, agriculture, and retail.

- Healthcare: In the healthcare industry, IoT is transforming patient care, remote monitoring, and operational efficiency. Wireless networking requirements include Reliable and secure data transmission for real-time patient monitoring, Low-latency communication for telemedicine and remote consultations, Interoperability among medical devices and systems, Scalable and robust network infrastructure for managing a great quantity of linked devices. The integration of IoT-enabled wearable devices and remote monitoring systems in healthcare empowers providers to observe patients' vital signs in real-time, facilitating early detection of health issues and ultimately enhancing patient outcomes.
- Smart Manufacturing: IoT is revolutionizing manufacturing processes, optimizing production, and enhancing supply chain management. Wireless networking requirements includes high-speed data transfer for simultaneous supervision and governance of machines, Seamless connectivity and interoperability among devices and systems, Reliable and low-latency communication for industrial automation and robotics, secure data transmission to protect intellectual property and sensitive information [8]. IoT empowers smart factories to interconnect and monitor production lines, ushering in an era of predictive maintenance, diminished downtime, and heightened operational efficiency.
- Smart Cities: The integral role of the IoT in the establishment of sustainable and efficient urban environments is paramount. The requisites for wireless networking in this context encompass expansive coverage to interconnect myriad sensors and devices throughout the city, low-power connectivity to ensure prolonged battery life for sensor nodes, secure communication protocols to safeguard citizen privacy and sensitive data, and a scalable network infrastructure capable of accommodating the burgeoning number of connected devices [9]. Leveraging

IoT sensors, these solutions facilitate real-time monitoring of parking space availability [10, 11].
- Agriculture: IoT is transforming agriculture practices, enabling precision farming, and optimizing resource utilization. Wireless networking requirements include Long-range connectivity for large agricultural areas, Low-power communication to maximize the lifespan of battery-powered devices, Environmental monitoring for real-time data on temperature, humidity, soil moisture, etc. Data analytics for intelligent decision-making and resource optimization. Leveraging IoT, irrigation systems meticulously scrutinize soil moisture levels and weather conditions, orchestrating the precise delivery of water to crops.
- Retail: IoT is reshaping the retail industry, enhancing customer experiences, and improving supply chain management. IoT-enabled smart shelves in retail stores automatically track inventory levels and send notifications for restocking, reducing manual effort and improving inventory management [12]. These case studies exemplify the successful implementation of IoT in various industries. By meeting specific wireless networking requirements and addressing challenges, organizations can leverage IoT to enhance operations, improve efficiencies, and unlock new opportunities in their respective domains [13].

3.5 *Energy Efficiency in Wireless IoT*

Addressing power consumption issues is imperative in safeguarding the continuing sustainability and efficient operation of IoT devices and wireless networks. This section focuses on the power consumption challenges in IoT devices and wireless networks and discusses energy-efficient protocols, low-power communication techniques, energy harvesting methods, strategies for prolonging device battery life, and optimizing network communication to minimize energy consumption [14].

- Power Consumption Challenges: Managing power efficiently stands as a critical challenge for IoT devices, characterized by constraints in battery life, particularly due to limited energy sources. Transmitting data wirelessly, especially in scenarios requiring constant connectivity or handling large data volumes, represents a significant power-consuming aspect. Additionally, the operation of sensors for environmental data collection imposes substantial power demands during data acquisition and processing [15]. The persistence of idle power consumption, even during periods of inactivity, necessitates the implementation of power-saving techniques to optimize energy usage.
- Energy-Efficient Protocols: Implementing energy-efficient protocols stands as a pivotal strategy to substantially mitigate power consumption in both IoT devices and wireless networks. Several noteworthy examples include LoWPAN, which achieves low-power communication by compressing IPv6 packets, thereby minimizing overhead. MQTT-SN (MQTT for Sensor Networks) is a lightweight

protocol specifically crafted for resource-constrained devices, effectively minimizing energy consumption during communication. Another noteworthy standard is Zigbee Green Power, facilitating ultra-low-power communication for batteryless IoT devices, consequently extending battery life or eliminating the need for batteries altogether.

- Low-Power Communication Techniques: Minimizing energy consumption in IoT devices necessitates the implementation of low-power communication techniques. Duty cycling, a pivotal strategy, involves devices periodically alternating between active and sleep modes, conserving power during idle periods. Wake-up radio is another effective technique that employs a separate low-power radio to awaken a device only, when necessary, thereby reducing overall power consumption. Additionally, adaptive transmission power proves advantageous by dynamically adjusting the transmission power established on the proximity of devices and the strength of the wireless signal.
- Energy Harvesting Methods: Energy harvesting techniques offer viable alternatives or supplements to traditional battery power in IoT devices. Solar energy harvesting involves utilizing solar panels to convert sunlight into electrical energy, providing a sustainable and renewable power source for devices. Kinetic energy harvesting taps into mechanical motion or vibrations, employing sensors or piezoelectric materials to convert this energy into electricity. Meanwhile, thermal energy harvesting leverages temperature differences and thermoelectric materials to generate electrical energy.
- Optimizing battery life and network communication: To enhance the efficiency of wireless IoT devices and extend their battery life, various strategies can be employed. Power management techniques, such as implementing sleep or hibernate modes during periods of inactivity, help minimize power consumption. Additionally, optimizing network communication through data compression, aggregation, and traffic optimization techniques reduces the amount of transmitted data and minimizes energy usage. Offloading computation tasks to edge devices, known as edge computing, further conserves energy by reducing the need for data transmission to centralized servers. These strategies collectively aim to minimize environmental impact while prolonging device operation [16].

4 Standardization Efforts for IoT Connectivity

4.1 Role of Standardization Bodies

Crucially, standardization bodies play a pivotal role in delineating wireless network standards for the IoT. This section discusses the significance of standardization, focusing on interoperability, compatibility, and widespread adoption. It also analyzes the impact of standards on the development and deployment of IoT technologies [17]. Table 1 presents an outline of the roles, uses, features, and challenges associated with standardization bodies in the development of wireless network standards for IoT.

Table 1 Standardization bodies in IoT—roles, uses, features, and challenges

Bodies	Uses in IoT	Features	Challenges
Institute of electrical and electronics engineers (IEEE)	– Defining protocols: IEEE contributes by developing protocols for seamless communication and data exchange in IoT	– Ensures standards for interoperability and compatibility, allowing diverse devices to work together	– Ensuring inclusivity and representation from diverse regions and industries in standard development
International telecommunication union (ITU)	– Establishing Guidelines: ITU sets guidelines for the implementation of interoperable and compatible IoT solutions	– Promotes global collaboration and harmonization among industry stakeholders	– Navigating the complexities of aligning diverse interests and perspectives within the standardization process
Internet engineering task force (IETF)	– Addressing Security and Privacy: IETF focuses on developing security and privacy standards for safeguarding IoT devices, data, and networks	– Contributes to the creation of unified global standards for IoT connectivity	– Balancing the need for detailed standards with the rapidly evolving IoT landscape
Bluetooth special interest group (SIG)	– Defining bluetooth standards: SIG establishes standards for Bluetooth connectivity in IoT devices	– Enables seamless wireless communication between a wide range of devices	– Addressing potential security vulnerabilities and ensuring backward compatibility
Thread group	– Standardizing mesh networking: thread group focuses on standards for IoT mesh networking	– Facilitates reliable and scalable communication in IoT networks	– Overcoming challenges related to interoperability with other networking technologies
Zigbee alliance	– Defining Zigbee standards: Zigbee Alliance sets standards for low-power, low-data-rate IoT applications	– Supports interoperability for smart home and industrial IoT devices	– Adapting to the evolving needs of diverse IoT use cases

4.2 Essential IoT Aspects

Interoperability and compatibility stand as pivotal elements in IoT connectivity, with standards playing a crucial role in facilitating effective communication and collaboration among devices from various manufacturers. The significance of these aspects lies in their ability to ensure seamless integration, allowing diverse IoT devices to work together in a unified ecosystem. This not only simplifies deployment but also provides flexibility and scalability. Additionally, standards promote vendor neutrality, enabling organizations to choose devices based on their specific needs without vendor lock-in. The widespread adoption of standards fuels a vibrant IoT ecosystem, fostering innovation, reducing costs, and driving market growth. Moreover, standards help mitigate fragmentation risks in the IoT landscape, ensuring a cohesive environment that enhances user experiences, improves efficiency, and boosts productivity. Security is also heightened through standardization efforts, establishing best practices and mechanisms to ensure consistent security levels across IoT deployments.

4.3 IoT Standards Impact

Standards wield a profound influence on the trajectory of IoT technologies, ushering in a realm of accelerated innovation, reduced development costs, and an interoperable ecosystem. The widespread embrace of standards not only fosters market confidence but becomes a catalyst for Increased Investment and Adoption in IoT technologies. Stakeholders, buoyed by proven and tested solutions, find assurance in standardized approaches, thereby propelling the market towards greater heights [18]. In essence, the concerted standardization efforts led by organizations like IEEE, ITU, and IETF serve as the bedrock for defining wireless network standards in the IoT domain. The triumvirate of interoperability, compatibility, and widespread adoption, orchestrated through these standards, not only fuels growth and innovation but also ensures the robust security of the IoT ecosystem. Adhering to these standards becomes the linchpin for organizations, paving the way for Seamless Integration, Reduced Fragmentation, and the swift Acceleration of Development and Deployment in the dynamic landscape of IoT technologies.

5 Conclusion

This research delves into the intricacies of WN-IoT, scrutinizing prevalent protocols, security challenges, and practical applications. It provides a comprehensive assessment of wireless protocols like Zigbee, Z-Wave, BLE, Wi-Fi, and LoRaWAN, offering insights into their characteristics and guiding factors for selection. The paper emphasizes the critical security challenges intrinsic to wireless IoT networks and advocates for proactive measures such as authentication, encryption, and access control. It further explores the design considerations, scalability challenges, and power consumption issues in the context of IoT deployments, suggesting strategies for energy efficiency and network optimization. Looking forward, the study underscores the transformative potential of the convergence between wireless networks and IoT, heralding new possibilities for connected devices and data-driven decision-making. However, the paper recognizes persistent challenges, including the need for standardization, interoperability, and robust security. It calls for continued research efforts to explore novel wireless protocols, enhance security frameworks, and foster collaborative initiatives among industry stakeholders, standardization bodies, and academia to ensure a cohesive and sustainable IoT ecosystem. In essence, the research provides a nuanced understanding of wireless networks and IoT, paving the way for a connected and intelligent future while highlighting the imperative of addressing ongoing challenges.

References

1. Truong TV, Nayyar A, Bilal M, Kwak KS (2023) System design and optimization of mobile edge computing in the NOMA wireless tactile internet of things network. Alex Eng J 73:737–749. https://doi.org/10.1016/j.aej.2023.04.056
2. Gulec O (2022) Extending lifetime of Wireless nano-sensor networks: an energy efficient distributed routing algorithm for internet of nano-things. Futur Gener Comput Syst 135:382–393. https://doi.org/10.1016/j.future.2022.05.009
3. Bozorgi SM, Golsorkhtabaramiri M, Yazdani S, Adabi S (2023) A smart optimizer approach for clustering protocol in UAV-assisted IoT wireless networks. Internet of Things 21:100683. https://doi.org/10.1016/j.iot.2023.100683
4. Abid K, Lakhlef H, Bouabdallah A (2021) A survey on recent contention-free MAC protocols for static and mobile wireless decentralized networks in IoT. Comput Netw 201:108583. https://doi.org/10.1016/j.comnet.2021.108583
5. Mathalai Raj J, Siva Ranjani S (2023) A secured blockchain method for multivariate industrial IoT-oriented infrastructure based on deep residual squeeze and excitation network with single candidate optimizer. Internet of Things 22:100823. https://doi.org/10.1016/j.iot.2023.100823
6. Chin WS, Kim H, Heo YJ, Jang JW (2015) A context-based future network infrastructure for IoT services. Procedia Comput Sci 56:266–270. https://doi.org/10.1016/j.procs.2015.07.207
7. Veiga T, Asad HA, Kraemer FA, Bach K (2023) Towards containerized, reuse-oriented AI deployment platforms for cognitive IoT applications. Futur Gener Comput Syst 142:4–13. https://doi.org/10.1016/j.future.2022.12.029
8. Wang X, Zhang X, Gong H, Jiang J, Rai HM (2021) A flight control method for unmanned aerial vehicles based on vibration suppression. IET Collab Intell Manuf 3:252–261. https://doi.org/10.1049/cim2.12027
9. Ma S, Chen J, Zhang Y, Shrivastava A, Mohan H (2021) Cloud based resource scheduling methodology for data-intensive smart cities and industrial applications. Scalable Comput Pract Exp 22. https://doi.org/10.12694/scpe.v22i2.1899
10. Rai HM, Goswami B, Majumdar S, Gupta K (2023) Intelligent post-lockdown management system for public transportation. In: Advanced computer science applications. Apple Academic Press, pp 175–189

11. Rai HM, Atik-Ur-Rehman PA, Mishra S, Shukla KK (2023) Use of internet of things in the context of execution of smart city applications: a review. Discover Internet Things 3:8. https://doi.org/10.1007/s43926-023-00037-2
12. Patel K, Mistry C, Gupta R, Tanwar S, Kumar N (2023) A systematic review on performance evaluation metric selection method for IoT-based applications. Microprocess Microsyst 101:104894. https://doi.org/10.1016/j.micpro.2023.104894
13. Rai HM, Goswami B, Majumdar S, Gupta K (2022) COVID-19 TravelCover: post-lockdown smart transportation management system. In: Assessing COVID-19 and other pandemics and epidemics using computational modelling and data analysis. Springer International Publishing, Cham, pp 19–43. https://doi.org/10.1007/978-3-030-79753-9_2
14. Bani Irshaid M, Bany Salameh H, Jararweh Y (2023) Intelligent multichannel cross-layer framework for enhanced energy-efficiency in 6G-IoT wireless networks. Sustain Energy Technol Assess 57:103211. https://doi.org/10.1016/j.seta.2023.103211
15. Rai HM, Gupta A, Kulshrestha PM (2017) A novel technique (trick) to determine the output and draw its waveform of diode clamper circuits. Int J Adv Res Comput Sci 8:1017–1022
16. Olatinwo SO, Joubert T-H (2020) Energy efficiency maximization in a wireless powered IoT sensor network for water quality monitoring. Comput Netw 176:107237. https://doi.org/10.1016/j.comnet.2020.107237
17. Chavhan S (2022) Shift to 6G: exploration on trends, vision, requirements, technologies, research, and standardization efforts. Sustain Energy Technol Assess 54:102666. https://doi.org/10.1016/j.seta.2022.102666
18. Feijoo-Añazco A, Garcia-Carrillo D, Sanchez-Gomez J, Marin-Perez R (2023) Innovative security and compression for constrained s networks. Internet of Things 24:100899. https://doi.org/10.1016/j.iot.2023.100899

Intelligent IoT System for Irrigation and Water Management

Swapnil Chaudhari, Rutuja Warke, Shreya Pungaonkar, Atharvaraj Shivudkar, Uma Gurav, and Amit Chanchal

Abstract Water is one of the most crucial aspects of one's life and as the population is increasing drastically, the need for water is also increasing. However freshwater resources are scarce, and freshwater is the main component for irrigation. As the technology is growing rapidly the agriculture sector needs to use advanced irrigation systems. This study addresses the water control project by exploring the combination of Artificial Intelligence, Machine Learning, and the Internet of Things. Drip irrigation, a key generation in water conservation, ensures managed water release to plant roots. However, traditional methods fall short in optimizing water use, necessitating the utility of device studying for sensible irrigation. By the use of the Kaggle dataset Irrigation Scheduling.csv comprising soil moisture, temperature, altitude, etc. The observations of this paper employ four ML models—LR, Naive Bayes, SVM, and XGBoost. This model is evaluated based on accuracy scores; results exhibit the effectiveness of XGBoost which emerges as the top-performing model with a perfect accuracy of 0.9979. This comparative evaluation presents treasured insights, setting up a basis for the mixing of devices and gaining knowledge of enhancing irrigation structures. The research contributes to sustainable water useful resource control in agriculture, important for addressing the worldwide project of water shortage.

Keywords Artificial intelligence · Machine learning · Internet of Things

S. Chaudhari (✉) · R. Warke · S. Pungaonkar · A. Shivudkar · U. Gurav · A. Chanchal
Department of Computer Science (AIML and DS), KIT's College of Engineering, Kolhapur, India
e-mail: srchaudhari324@gmail.com

U. Gurav
e-mail: gurav.uma@kitcoek.in

A. Chanchal
e-mail: chanchal.amit@kitcoek.in

P. J. S. Gonçalves et al. (eds.), *Proceedings of Fifth International Conference on Computing, Communications, and Cyber-Security*, Lecture Notes in Networks and Systems 1128, https://doi.org/10.1007/978-981-97-7371-8_8

1 Introduction

Water, often regarded as the lifeblood of planet Earth, is assuming an increasingly critical role as the global population approaches 8 billion. Despite Earth's abundance, with 75% covered in water, only a mere 2.5% constitutes freshwater, and a staggering 70% of the world's drinking water is allocated to agriculture. The impending demographic surge, expected to reach 8 billion by mid-November 2023, emphasizes the urgency of addressing water scarcity, particularly in regions where the average use of freshwater is devoted to agricultural pursuits. Agriculture has transitioned from its historical focus on crop cultivation to a modernized system. The integration of AI into agriculture has become feasible due to technological breakthroughs such as big data analytics, automation, IoT, robotics, affordable sensors, Drones, hydroponics, artificial lighting, and AI-powered cameras. AI, characterized by systems that operate and think like humans, is demonstrating its effectiveness across various technological sectors in real-time applications. Its role in agriculture is seen as a crucial factor in ensuring the enduring success of high-quality food production [1].

Traditionally the farmers used to provide irrigation by manually turning on and off the water pumps but with the help of this project the farmers can automatically provide the water based on the detected soil moisture content in soil using machine learning models which provide accurate results. This can help farmers save their time and efforts which they can utilize for other work. Also, the water will be provided directly to the root zones of the plants this will help save excessive water usage during irrigation. Farmers can use Drip irrigation, a method recognized for its water-saving attributes by delivering water directly to the roots by row cropping. However, the research seeks to surpass the limitations of traditional practices by harnessing the capabilities of AI and ML. The focal point of this study lies in leveraging these advanced technologies to augment the efficiency of drip irrigation, transforming it into a smart, data-driven mechanism capable of optimizing water usage [2]. It has been observed that machine learning models exhibit notable accuracy in predicting soil moisture requirements for the upcoming years, relying significantly on atmospheric temperature and humidity content. The accuracy levels, however, display variability across different machine-learning algorithms. Looking ahead, there is potential for expanding this research by incorporating additional raw data related to various crops, considering the soil quality variations across the country. The future direction could involve refining calculations with a more technical emphasis on features, thereby contributing to a more comprehensive and nuanced understanding of soil moisture prediction in diverse agricultural contexts.

2 Related Work

In recent years, there has been a growing interest in the application of artificial intelligence (AI), machine learning (ML), and the Internet of Things (IoT) in water control projects. These technologies have shown great potential in optimizing water use in agriculture, which is crucial for addressing the global issue of water shortage. In a systematic literature review published in the journal Water, Ghobadi and Kang (2023) [3] examined the application of machine learning in water resources management. The authors found that machine learning algorithms have been widely used in various areas of water resources management, including water quality monitoring, water demand forecasting, and water distribution systems. They also highlighted the potential of machine learning in improving the efficiency of irrigation systems. In a similar study, Krishnan et al. (2022) [4] conducted a review of smart water resource management using artificial intelligence. The authors discussed the role of AI in various aspects of water resource management, including water quality monitoring, water demand forecasting, and water distribution systems. They also emphasized the potential of AI in improving the efficiency of irrigation systems and reducing water waste. Both studies emphasized the importance of integrating AI, ML, and IoT in water control projects, particularly in drip irrigation systems. These technologies can help optimize water use, reduce water waste, and improve the efficiency of irrigation systems. In conclusion, the application of AI, ML, and IoT in water control projects has shown great potential in optimizing water use in agriculture.

Table 1 provides a comprehensive examination of various methodologies within the domain of intelligent irrigation systems using IoT. By exploring different methodologies, this review aims to provide a solid foundation for researchers and practitioners seeking to navigate the complex landscape of intelligent irrigation systems.

3 Proposed Methodology

The motivation of the paper is to reduce the time and efforts of the farmers and they can utilize that time in some other work. Water should reach the root zones of the plants in a specific amount resulting in less wastage of the water. The envisioned system seeks to revolutionize modern agriculture by an approach which integrates IoT devices, machine learning algorithms, and automation to elevate the efficiency and sustainability of crop irrigation practices. In proposed methodology, the machine learning model takes the input features such as temperature, altitude, pressure, and soil moisture which predict the status of the soil moisture and then compares it with the actual soil moisture value, if the actual value is less than the predicted value then the pump gets on and if the actual value is greater than the predicted value then then the pump gets off.

Figure 1 illustrates the comprehensive tasks and processes in this proposed model.

Table 1 Related work

Sr. No	Year	Authors/ Applications	Title	Methodology
1	2024	Reddy, et al. smart irrigation system [5]	IoT and cloud based sustainable smart irrigation system	The paper employs a sensors and cloud-based platform to monitor and control an irrigation system. Moisture sensors are placed in the field to monitor in real-time. The system uses online weather prediction data to provide precise watering schedule optimization
2	2024	Patil, et al. precision agriculture model [6]	Machine learning based precision agriculture model for farm irrigation to optimize water usage	The paper developed a precision agriculture model for farm irrigation using ML algorithms, including SVM, RF, and Linear regression. Three ML techniques were used: traditional linear regression, boosted trees classifiers, and gradient boosted regression trees
3	2024	Kranthi, et al. classification model [7]	Internet of Things sensors and support vector machine integrated intelligent irrigation system for agriculture industry	IoT-based humidity and soil sensors are used to collect soil-related data, which is then processed using feature selection and clustering algorithms. A K-means SVM hybrid classifier is used to build a classification model
4	2023	Archana, et al. smart cultivation system [8]	Advanced irrigation and cultivation system based on machine learning in IOT Environment	The paper discusses the implementation of an advanced irrigation and cultivation system based on machine learning in an IoT environment
5	2023	Anoop, et al. IoT based smart irrigation [9]	IoT and Ml-based automatic irrigation system for smart agriculture system	Analysis of hardware modules in smart irrigation systems. Evaluation of machine learning methods for prediction
6	2023	Deepak Kumar, et al. [10] IoT based Smart Irrigation	Long Short-Term Memory and KNN based Approach for Smart Irrigation System using IoT	LSTM-RNN-trained recommendation system for effective water use. Preprocessing with M3 filter, feature selection with RFE, LSTM-RNN model training
7	2023	Arthi, et al. [11] IoT based Smart Irrigation	Smart Agriculture System Using IoT and ML	A low-cost system that uses Internet of Things (IoT) and Machine Learning (ML) to maximize crop yield and productivity is proposed, which consists of three key components: an IoT device, a mobile application, and servers
8	2023	Palash, et al. [12] IoT based Smart Irrigation	Application of IOT and Machine Learning in Smart Agriculture System	Smart irrigation system using machine learning and IoT uses decision tree algorithm for forecasting water supply

(continued)

Table 1 (continued)

Sr. No	Year	Authors/ Applications	Title	Methodology
9	2023	Anitha, et al. [13] Smart Solar Irrigation System	Design and Implementation of a Smart Solar Irrigation System Using IoT and Machine Learning	Smart solar irrigation system using IoT and ANN algorithms makes utilization of machine learning algorithms and solar-powered IoT device
10	2023	Rohan, et al.[14] Smart Irrigation System	Precise Prediction of Water Level Using the Naïve Bayes Algorithm for Smart Irrigation in Pomegranate Farms	Utilizes drip irrigation technology for precise water delivery directly to plant roots. Facilitates remote monitoring and has the potential for GSM module integration
11	2022	Khongdet, et al. [15] IoT based Smart Irrigation	Application of IoT and Cloud Computing in Automation of Agriculture Irrigation	Utilizes IoT and cloud computing for a smart irrigation framework. Employs machine learning algorithms (Support Vector Machine, Random Forest, and Na¨ıve Bayes) for data analysis
12	2020	Ali, et al. [16] Transfer Learning, Smart Irrigation	Smart Irrigation IoT Solution using Transfer Learning for Neural Networks	Applies artificial neural networks for predicting soil moisture and controlling irrigation. Utilizes transfer learning, enhancing the model's adaptability across different soil types
13	2020	Wakhare, et al. [17] IoT based Smart Irrigation	Automatic Irrigation System Based on Internet of Things for Crop Yield Prediction	Utilizes soil pH values to select the most suitable crops from a comprehensive crop database. Displays values of various monitoring parameters on mobile apps for user reference and analysis
14	2020	Kashinath, et al. [18] IoT based Smart Irrigation	IoT based Smart Irrigation System using Raspberry Pi3	Incorporates sensors measuring humidity, temperature, and light, integrated with Raspberry Pi3 and Arduino for comprehensive data collection. Utilizes ZigBee for wireless control, enabling remote monitoring of the entire system
15	2019	Aggarwal, et al. [19] IoT based Smart Irrigation	A Smart Irrigation System to Automate Irrigation Process Using IOT and Artificial Neural Network	Utilizes IoT technology, featuring an Arduino-based device equipped with various sensors for real-time data. The collected data undergoes analysis through ANN, enabling the system to recognize patterns and make decisions

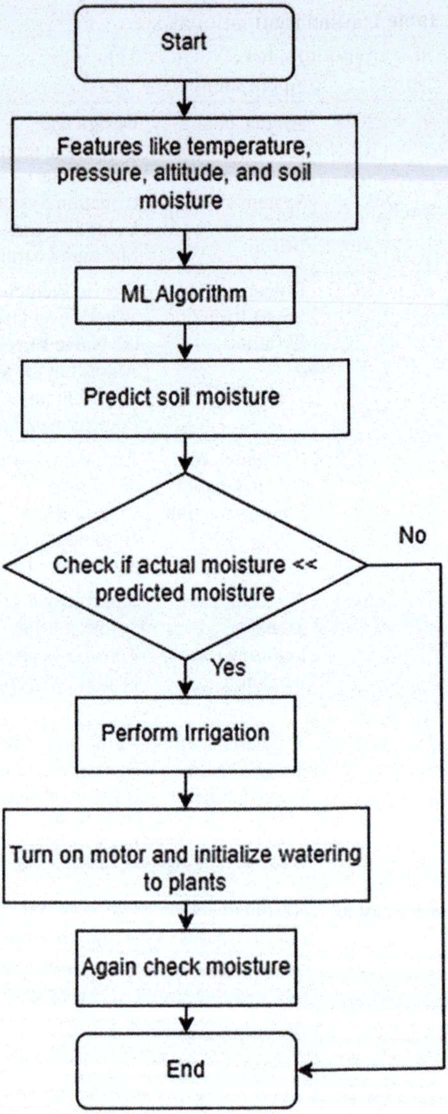

Fig. 1 Working flowchart of model

3.1 Data Collection

Data collection for the research involved meticulous sourcing and curation of the Kaggle dataset named Irrigation Scheduling.csv [20]. This dataset, comprising 10 columns, provided essential information, including humidity, temperature, altitude, and pressure. These parameters were chosen due to their relevance in understanding the dynamics of soil moisture in agricultural settings. The selection of this dataset served as the foundational step for training and evaluating four distinct machine learning models: LR, Naive Bayes, SVM, and XGBoost regressor. The datasets richness was crucial for the effectiveness of the machine-learning models in predicting soil moisture [21]. By using the time series supervised learning format, the research aimed to enhance the accuracy of predictions over time. This approach allowed the models to capture temporal patterns and trends, thereby refining their ability to forecast soil moisture requirements. In future extensions of this work, the potential for augmenting the dataset with additional raw data on diverse crops across the country was highlighted. This envisioned expansion could offer a more nuanced understanding of soil moisture dynamics, considering variations in soil quality and agricultural practices (Table 2).

3.2 Data Storage and Monitoring

This paper consists, of the web application platform to store all data, including soil moisture levels (SMC), temperature (T), humidity (H), pH, and the status of the irrigation process. Web app is integrated using Flask API, Flask API is a web framework for Python that is highly used for the building of web application due to its capabilities, developers can seamlessly integrate various functionalities, making it well-suited for constructing web applications that serve as frontends for the system. This infrastructure ensures smooth management and accessibility of crucial agricultural data.

3.3 Working of the Proposed System

The system operates in a continuous cycle, commencing with the continuous monitoring of soil moisture levels through dedicated sensors. Dataset taken serves as the foundation for all subsequent decision-making. The machine learning model M processes the collected data and generates predictions (P) for upcoming soil moisture conditions. These predictions are instrumental in the system's automatic irrigation control. When the model predicts that SMC is higher than the actual plant conditions, a relay is activated (R on). This relay transmits a signal to the irrigation motor, initiating the supply of water to the crops. Conversely, if the model predicts lower SMC than the actual plant conditions, another relay (R off) signals the motor to cease the water supply. This control mechanism ensures the judicious use of water resources, promoting efficiency and sustainability in agricultural irrigation practices. The mathematical representation of the system can be summarized as follows:

$$P = M(H,\ T,\ pH,\ SMC)$$

Table 2 Dataset

Id	Temp	Pressure	Altitude	Moisture	Note	Status	Class	Date	Time
1	28.82	9983.34	–12.31	381	0	0	Very Dry	08-10-2022	22:06:24
2	28.81	9982.83	–12.35	384	0	0	Very Dry	08-10-2022	22:06:24
3	28.8	9982.52	–12.38	383	0	0	Very Dry	08-10-2022	22:06:24
4	28.79	9982.52	–12.38	385	0	0	Very Dry	08-10-2022	22:06:24
5	28.79	9982.17	–12.4	383	0	0	Very Dry	08-10-2022	22:06:24
6	28.78	9981.97	–12.42	382	0	0	Very Dry	08-10-2022	22:06:24
7	28.77	9981.83	–12.43	381	0	0	Very Dry	08-10-2022	22:06:24
8	28.77	9982	–12.42	382	0	0	Very Dry	08-10-2022	22:06:24
9	28.76	9981.83	–12.43	381	0	0	Very Dry	08-10-2022	22:06:24
10	28.75	9981.48	–12.46	382	0	0	Very Dry	08-10-2022	22:06:24
11	28.75	9981.66	–12.45	380	0	0	Very Dry	08-10-2022	22:06:24
12	28.74	9981.98	–12.45	383	0	0	Very Dry	08-10-2022	22:06:24
13	28.74	9982.02	–12.42	389	0	0	Very Dry	08-10-2022	22:06:24
14	28.73	9982.02	–12.42	320	1	0	Very Dry	08-10-2022	22:06:24
15	28.73	9982.02	–12.42	284	1	0	Very Dry	08-10-2022	22:06:24
16	28.82	9981.97	–12.42	250	2	1	Wet	08-10-2022	22:06:24
17	29.08	9982.14	–12.41	205	2	1	Wet	08-10-2022	22:06:24
18	29.4	9983.11	–12.33	188	3	1	Very Wet	08-10-2022	22:06:24
19	29.7	9983.69	–12.28	183	3	1	Very Wet	08-10-2022	22:06:24
20	29.92	9983.96	–12.25	181	3	1	Very Wet	08-10-2022	22:06:24

$$R_{on} = \begin{cases} 1 \text{ if } P > \text{Threshold} \\ 0 \quad \text{otherwise} \end{cases}$$

$$R_{off} = \begin{cases} 1 \text{ if } P > \text{Threshold} \\ 0 \quad \text{otherwise} \end{cases}$$

These equations encapsulate the decision-making process of the system, ensuring optimal irrigation control based on predicted soil moisture conditions.

4 Implementation and Training

To implement the intelligent irrigation system, a carefully planned hardware configuration is essential. This setup primarily revolves around the utilization of IoT devices, with a focus on Arduino microcontrollers and an assortment of sensors for actual soil moisture prediction. In terms of data collection, the system is designed to utilize an existing dataset. This data is seamlessly integrated into the machine learning model, which serves as the core intelligence of the system. The efficacy of this model in predicting soil moisture levels relies on the availability of a substantial dataset. The machine learning model undergoes a thorough training phase utilizing appropriate algorithms [22]. During the training process, iterative adjustments are made to minimize prediction errors, ensuring the model's accuracy in forecasting soil moisture levels based on the input variables.

Once effectively trained, the machine learning model takes on the role of the decision-making component within the system. It autonomously evaluates predicted soil moisture levels and makes informed determinations regarding the necessity of irrigation. In cases where the model forecasts surplus soil moisture compared to actual plant conditions, the irrigation motor is activated. Conversely, if the model predicts insufficient soil moisture, the motor is instructed to halt water supply. This control logic ensures precise water delivery, thereby enhancing irrigation efficiency.

To complete the process, all relevant data generated by the sensors, machine learning model predictions, and the operation status of the irrigation system are seamlessly integrated and stored within a dedicated web application developed using Flask API. This Flask-based web application serves as a centralized data repository, offering an intuitive interface for monitoring. The inclusion of Flask API enhances the flexibility and responsiveness of the web application, allowing for efficient data flow and user interaction. The focus remains on the system's functionality and the seamless integration of data without delving into explicit hardware details. This holistic approach ensures that stakeholders have access to a comprehensive and user-friendly tool for monitoring and analyzing the dynamic aspects of the proposed irrigation control system.

5 Machine Learning Models

This research explored the application of various machine learning models for accurate prediction of soil moisture value [23]. The performance of the following models was specifically evaluated.

5.1 *Logistic Regression*

In this system, the logistic regression model emerges as a pivotal tool for classifying soil types and determining whether irrigation is needed. Logistic regression, a widely used algorithm for binary classification tasks, predicts the probability of an event occurring. In this system, the event of interest is whether the soil is wet (Y = 1) or dry (Y = 0). The logistic regression equation is defined as:

$$P(Y = 1) = \frac{1}{1 + e^{-}(\beta_0 + \beta_1 X_1 + \beta_2 X_{2+} \ldots + \beta_n X_n)} \tag{1}$$

Here, P(Y = 1) represents the probability of the soil being wet, and $1 - (Y = 1)$ is the probability of it being dry.

This equation ensures that the predicted probabilities lie between 0 and 1. The coefficients $\beta 0$, $\beta 1$, …, βn are estimated through methods like maximum likelihood to optimize the model fit to the observed data. Training the logistic regression model involves adjusting these coefficients to maximize the likelihood of observing the actual soil types in the dataset. Once trained, the model can predict the probability of wet or dry soil based on input features such as humidity, temperature, elevation, and pressure. By selecting a threshold probability (e.g.0.5), the model classifies whether irrigation is needed.

5.2 *Gaussian Naive Bayes Algorithm*

Gaussian Naive Bayes (GNB) is a probabilistic classifier based on Bayes theorem, making it particularly suited for scenarios involving multiple features. The GNB algorithm assumes that the features are conditionally independent given the class label, and it models the distribution of each feature for each class using a Gaussian(normal) distribution. The probability density function for the GNB classifier is expressed as:

$$P(X_i|y) = \frac{1}{2\pi\sigma_y^2} \exp\left(\frac{(X_i - \mu_y)^2}{2\sigma_y^2}\right) \tag{2}$$

Here, Xi represents a specific feature, y is the class label (wet or dry soil), μ_y is the mean of the feature for class y, and $2\sigma_y^2$ is the variance of the feature for class y. The classifier calculates the probability of observing a particular feature value given the class and multiplies these probabilities across all features, assuming independence. The training process involves estimating the mean and variance of each feature for each class based on the observed data. During prediction, the GNB algorithm calculates the probability of each class given the input features using the Bayes theorem and assigns the class with the highest probability as the predicted class. The interpretability of GNB lies in its simplicity and efficiency, making it a valuable tool for predicting soil conditions.

5.3 *Support Vector Machine*

Support Vector Machines operate by identifying the optimal hyperplane that maximally separates distinct classes within the feature space. In the context of irrigation, SVM proves useful for categorizing soil conditions, such as determining whether irrigation is required, based on features like soil moisture levels.

Mathematically, the decision function for an SVM is represented as:

$$f(x) = \text{sign}(w \cdot x + b) \tag{3}$$

In this equation, w represents the weight vector, x is the input feature vector, and b is the bias term. The sign function dictates the class assignment of a data point based on its position relative to the hyperplane. During the training phase, SVM identifies support vectors, and pivotal data points crucial for defining the optimal hyperplane. The objective is to maximize the margin, which signifies the distance between the hyperplane and the nearest data point from either class. Mathematically, this entails solving an optimization problem to determine the optimal w and b. Once trained, the SVM model can predict outcomes for new data points by ascertaining their position relative to the hyperplane. In the context of the system, these predictions would correlate with whether the soil moisture levels indicate a need for irrigation.

5.4 *XGBoost Regressor*

The XGBoost Regressor serves as a robust machine-learning model employed for predicting water requirements in agriculture. Representing extreme Gradient Boosting,

XGBoost is an ensemble learning algorithm esteemed for its efficiency and superior predictive capabilities. Specifically tailored for regression tasks, where the aim is to forecast continuous outcomes like the volume of water needed for irrigation, the XGBoost Regressor operates by combining predictions from multiple weak learners, often in the form of decision trees. This combination results in a resilient and precise predictive model.

Mathematically, the prediction of the XGBoost Regressor for a given data point can be represented as:

$$y_i = \sum_{k=1}^{k} f_k(x_i) \tag{4}$$

where y_i is the predicted value for the ith data point, K is the number of weak learners, and $f_k(x_i)$ represents the prediction of the kth weak learner.

Throughout the training phase, XGBoost fits a sequence of weak learners to the data, with each subsequent learner rectifying the errors of the combined ensemble. The objective function of XGBoost incorporates a regularization term, contributing to controlling the model's complexity and preventing overfitting. The hyperparameters of the XGBoost Regressor, encompassing variables like the learning rate, maximum tree depth, and the number of boosting rounds, can be fine-tuned to enhance the model's performance in predicting irrigation needs based on input features To sum up, the XGBoost Regressor excels in managing intricate data relationships, providing precise predictions for regression tasks in the realm of agriculture.

6 Results and Discussions

6.1 Result

The Kaggle dataset "Irrigation Scheduling.csv" was employed for training the machine learning models, comprising 10 columns containing soil moisture, temperature, altitude, pressure data, class, note and status of irrigation. The dataset was synthesized and preprocessed to remove irrelevant features from the dataset and predictions were generated using four distinct machine learning models which were, Logistic Regression, Gaussian Naive Bayes, Support Vector Machine, and XGBoost Regressor. Then performance of these models was calculated by using accuracy scores as an important metric to compare their predictive efficacy within the irrigation system dataset. The outcomes are as follows (Table 3).

XGBoost Regressor achieved the highest accuracy, F1-score, and ROC-AUC scores, SVM achieved the highest precision, and Gaussian Naive Bayes achieved the highest recall, indicating XGBoost has overall superior performance. Logistic Regression and SVM also performed well, with comparable scores in most metrics. Gaussian Naive Bayes showed the lowest performance across all metrics compared to other models. This comparative analysis furnishes valuable insights into the strengths of each algorithm, laying a solid foundation for the application of machine learning

Table 3 Model performance comparison

Metric	Logistic Regression	Gaussian NB	SVM	XGBoost Classifier
Accuracy	0.994781	0.949896	0.996868	0.997912
Precision	1.000	0.859	1.000	0.997
Recall	0.983	1.000	0.990	0.997
F1-score	0.991	0.924	0.995	0.997
ROC-AUC	0.991468	0.963910	0.994881	0.997542

to enhance irrigation systems and foster sustainable water resource management in agriculture.

Figure 2 compared the accuracy of four different machine learning algorithms: Logistic Regression, GaussianNB, SVM, and XGBoost Regressor. Remarkably, XGBoost achieved highest accuracy level. This near-perfect accuracy demonstrates the potential of these models to make highly precise predictions, which could be invaluable in various applications such as agriculture sectors. Even the models with lower accuracy.

GaussianNB showed commendable performance. These results highlight the effectiveness of these machine learning models and their potential to contribute significantly to future research and real-world applications. The study also underscores the importance of choosing the right model based on the specific dataset and the computational efficiency and interpretability of the model.

Figure 3 includes a depiction of the learning curve for the logistic regression model, demonstrating the training and cross-validation scores across varying numbers of training examples. It reveals a convergence of scores as the number of

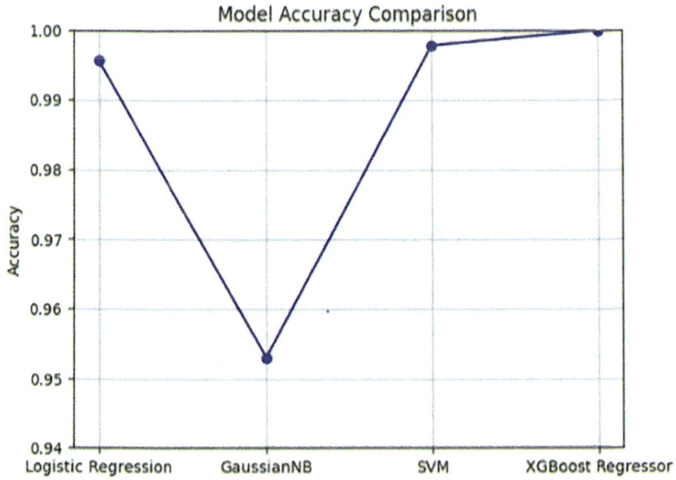

Fig. 2 Graphical representation of accuracy results of machine learning algorithms

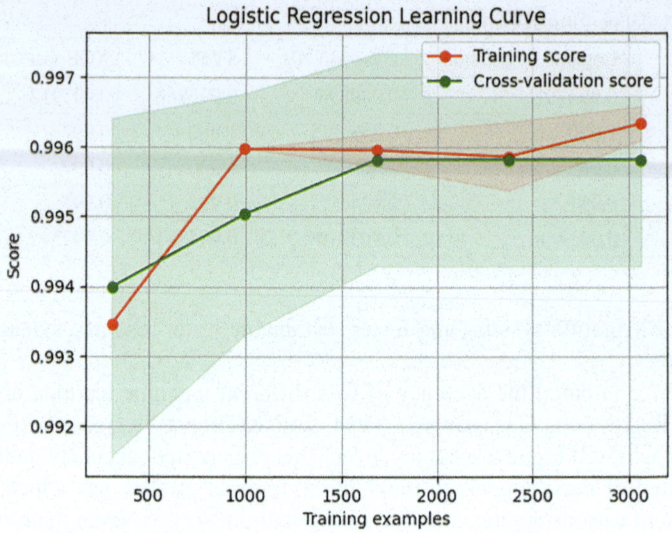

Fig. 3 Logistic regression learning curve

training examples increases, suggesting that the model does not suffer from overfitting. This observation underscores the model's capacity to generalize well to unseen data, enhancing its reliability and applicability in real-world scenarios.

Figure 4 illustrates the Root Mean Squared Error (RMSE) for the XGBoost regressor model across various epochs, encompassing both training and validation sets. It was observed that the RMSE diminishes with an increasing number of epochs, suggesting a continual improvement in the model's performance over time. This trend underscores the effectiveness of the model in minimizing prediction errors and enhancing its overall predictive capability.

6.2 Limitations

Despite its numerous strengths, the intelligent irrigation system is not without limitations. One notable constraint is its reliance on the quality and availability of environmental data. Inaccurate or incomplete sensor data can potentially impact the precision of the machine-learning model, leading to sub-optimal irrigation decisions. Ensuring data accuracy and sensor reliability remains a critical challenge. Another limitation pertains to the energy requirements of the system, especially in remote or off-grid agricultural settings. The continuous operation of IoT devices and the machine learning model demands a consistent power supply. Implementing sustainable and energy-efficient solutions, such as solar power or low-power components, is an avenue for future exploration.

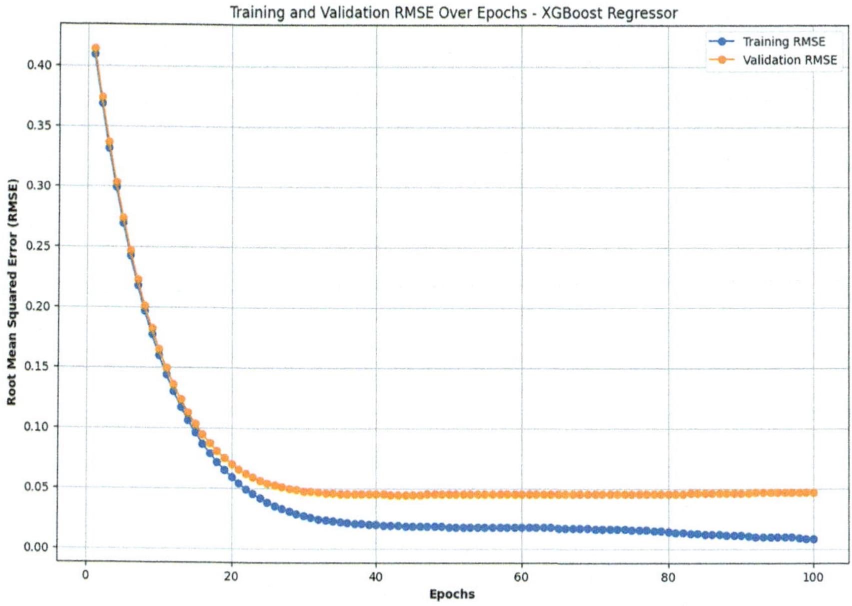

Fig. 4 XGBoost training and validation curve

7 Conclusion

In conclusion, the integration of AI into irrigation systems offers a promising solution for efficient and sustainable water management in modern agriculture. By utilizing AI, machine learning, and IoT devices, this system automatically controls irrigation based on real-time soil moisture data, reducing water wastage and conserving resources. The system's machine learning model, trained on historical sensor data, accurately predicts soil moisture levels and makes informed irrigation decisions. This approach enhances irrigation efficiency, promotes environmental sustainability, and contributes to addressing the global issue of water shortage. Overall, the integration of AI into irrigation systems fosters a more resilient, efficient, and environmentally conscious approach to water management in agriculture, benefiting farmers, conserving resources, and cultivating a greener and more sustainable planet.

References

1. Neha G, et al. (2023) Artificial intelligence in agriculture (February 2, 2023)
2. Fawibe Oluwasegun O, et al. (2022) Fundamentals of irrigation methods and their impact on crop production. Irrig Drain-Recent Adv. IntechOpen
3. Ghobadi F, Kang D (2023) Application of machine learning in water resources management: a systematic literature review. Water 15(4):620

4. Krishnan SR, Nallakaruppan MK, Chengoden R, Koppu S, Iyapparaja M, Sadhasivam J, Sethuraman S (2022) Smart water resource management using artificial intelligence—A review. Sustainability 14(20):13384
5. Reddy, Sreelatha V, Harivardhagini S, Sreelakshmi G (2024) IoT and cloud based sustainable smart irrigation system. E3S Web of Conferences. Vol. 472. EDP Sciences
6. Patil SB, et al. (2024) Machine learning based precision agriculture model for farm irrigation to optimize water usage. IOP conference series: earth and environmental science, vol. 1285. No. 1. IOP Publishing
7. Kranthi KG, et al. (2024) Internet of Things sensors and support vector machine integrated intelligent irrigation system for agriculture industry. Discov Sustain 5(1): 6
8. Archana C, Gupta P (2023) Advanced irrigation and cultivation system based on machine learning in IOT environment. In: 2023 international conference on communication system, computing and IT applications (CSCITA). IEEE
9. Anoop EG, JoseminBala G. IoT and Ml-based automatic irrigation system for smart agriculture system. Agron J
10. Deepak Kumar S, et al. (2023) Long short-term memory and KNN based approach for smart irrigation system using IoT. In: 2023 international conference on sustainable computing and smart systems (ICSCSS). IEEE
11. Arthi R, Nishuthan S, Deepak Vignesh L (223) Smart agriculture system using IoT and ML. In: 2023 international conference on signal processing, computation, electronics, power and telecommunication (IConSCEPT). IEEE
12. Palash V, Shreesh S, Yatharth L (2023) Application of IOT and machine learning in smart agriculture system. Interantional J Sci Res Eng Manag 07. https://doi.org/10.55041/IJSREM17750
13. Anitha B, et al. (2023) Design and implementation of a smart solar irrigation system using IoT and machine learning. E3S web of conferences, vol. 387. EDP Sciences
14. Rohan C, et al. (2018) Smart drip irrigation system using moisture sensors. In: 2018 international conference on smart city and emerging technology (ICSCET). IEEE
15. Khongdet P, et al. (2022) Application of IoT and cloud computing in automation of agriculture irrigation. J Food Qual 2022:1–8
16. Ali R, Jalili A, Nazerfard E. (2020) Smart Irrigation IoT solution using transfer learning for neural networks. In: 2020 10th international conference on computer and knowledge engineering (ICCKE). IEEE
17. Wakhare, Prashant B, Neduncheliyan S, Sonawane GS (2020) Automatic irrigation system based on Internet of Things for crop yield prediction. In: 2020 international conference on emerging smart computing and informatics (ESCI). IEEE
18. Kashinath B, et al. (2020) IoT based smart irrigation system using raspberry Pi3
19. Sachin A, Kumar A (2019) A smart irrigation system to automate irrigation process using IOT and artificial neural network. In: 2019 2nd international conference on signal processing and communication (ICSPC). IEEE
20. https://www.kaggle.com/code/triptmann/irrigation-scheduling-for-smart-agriculture
21. Prakash S, Sharma A, Sahu SS (2018) Soil moisture prediction using machine learning, In: 2018second international conference on inventive communication and computational technologies (ICICCT), Coimbatore, India, pp 1–6, https://doi.org/10.1109/ICICCT.2018.8473260
22. Meghna R, et al. (2021) A survey on the role of Internet of Things for adopting and promoting Agriculture 4.0. J Netw Comput Appl 187:103107
23. Gurav U, Sidnal N (2018) Predict stock market behavior: Role of machine learning algorithms. In: Bhalla S, Bhateja V, Chandavale A, Hiwale A, Satapathy S (eds) Intelligent computing and information and communication. Advances in Intelligent Systems and Computing, vol 673. Springer, Singapore. https://doi.org/10.1007/978-981-10-7245-138

5G Technology: From Evolution to Current Time

Simran, Chandradeep Bhatt, and Noor Mohd

Abstract More and more people are using smartphones and laptops, which need a lot more high-speed data to be sent over mobile networks. This has led researchers and designers to find ways to improve your overview of what 5G technology is! in which principle it works, what was the main concept in driving its evolution, the good impacts of 5G on Artificial Intelligence (AI), Internet of Things (IoT), and in the concept of virtual reality. Also, some bad impacts on health concerns, privacy, and security risks. But before getting started we must note that the word "5G" is not yet officially mentioned in any of the organizations yet, as currently, 5G is in the testing phase in most of the counties, with researchers facing the challenge of security issues. Once all the issues with 5G networks are successfully resolved, they can serve as the cornerstone of the Smart City concept, offering solutions to various ongoing problems related to infrastructure, public transport, the environment, and more.

Keywords 5G · Security risks · IoT (Internet of things) · Virtual reality · Artificial Intelligence (AI)

1 Introduction

5G is a wireless technology whose main objective is to connect everything and everyone, i.e., devices and machines. It's an upgrade from old network generations that are 1G, 2G, 3G, and 4G. The primary goals of 5G wireless technology include everything ultrafast data speeds in the range of multi-gigabits per second, ultra-low latency, enhanced reliability, massive network capacity, increased availability, and a consistent user experience for a larger user base. The need for 5G arises from

Simran · C. Bhatt (✉)
Graphic Era Hill University, Dehradun, India
e-mail: bhattchandradeep@gmail.com

N. Mohd
Graphic Era Deemed to be University, Dehradun, India

P. J. S. Gonçalves et al. (eds.), *Proceedings of Fifth International Conference on Computing, Communications, and Cyber-Security*, Lecture Notes in Networks and Systems 1128, https://doi.org/10.1007/978-981-97-7371-8_9

the explosive growth of wireless data, largely driven by devices like smartphones, tablets, and video streaming. By 2020, it is projected that the amount of IP data handled by wireless networks by surpass 500 exabytes, with potentially billions of devices connected to the network.

To meet these demands, engineering and researchers are actively working on innovative technologies and collaborative projects such as METIS and 5GNOW. Standardized activities for 5G are already underway in the industry. Several intriguing aspects of 5G include the utilization of millimeter wave (mmWave) spectrum, novel methods for bandwidth allocation, the virtualization of network infrastructure, the integration of cellular and WIFI standards, and the concept of an "Internet of Things" with billions of interconnected devices. In addition to technical considerations, addressing regulatory and standardization issues is crucial for the successful implementation of 5G, particularly in terms of spectrum regulation and fostering innovation. Overall, 5G is set to revolutionize communication networks by delivering superior performance, improved efficiency, and enabling novel user experiences across various industries (see Fig. 1).

5G networks are cellular networks that divide the service area into small cells. Within each cell, 5G devices communicate with a cellular base station using radio waves and fixed antennas. These base stations, also known as nodes, are connected to switching centers and internet routers through high-bandwidth optical fiber or wireless connections. When a mobile device transitions from one cell to another, a seamless handoff takes place. The Third Generation Partnership Project (3GPP), an industry consortium, defines "5G" as any system using 5G New Radio (5G NR) software, which became widely adopted in late 2018.

To enhance capacity, millimeter waves are used by some network operators which are referred to as FR2 in 5G, which have shorter ranges and smaller cell sizes. However, millimeter waves face challenges, in penetrating building walls and obstacles. Compared to previous cellular networks, 5G achieves higher data rates by utilizing additional higher-frequency radio waves alongside low-and medium-band frequencies.

There are majorly three types of 5Gs: low-band, mid-band, and high-band millimeter-wave. Low-band 5G is somewhat similar to 4G's features, in terms of frequency but can offer faster download speeds ranging from 5 to 250 megabits per second (Mbps). Low-band coverage and range resemble that of 4G towers, they are usually designed to deal with wide coverage areas but have the limitation of having limited carrier capacity. Whereas the Mid-band 5G uses microwaves within the range of 1–6 GHz, is highly suitable for 5G networks due to its ability to transmit large amounts of data over long distances. According to the GSMA, the frequency range between 3.3 and 3.8 GHz is particularly ideal for 5G deployment, and in many countries, this spectrum has already been allocated specifically for 5G use, also provides speeds of 100–900 Mbit/s, with each cell tower covering several kilometers (in radius). This is the most widely deployed level of service, especially in metropolitan areas. High-band 5G networks are operated in frequencies of 24–47 GHz, achieving gigabit-per-second (Gbit/as) download speeds. However, high-band millimeter requires numerous small cells due to which they are highly used in a

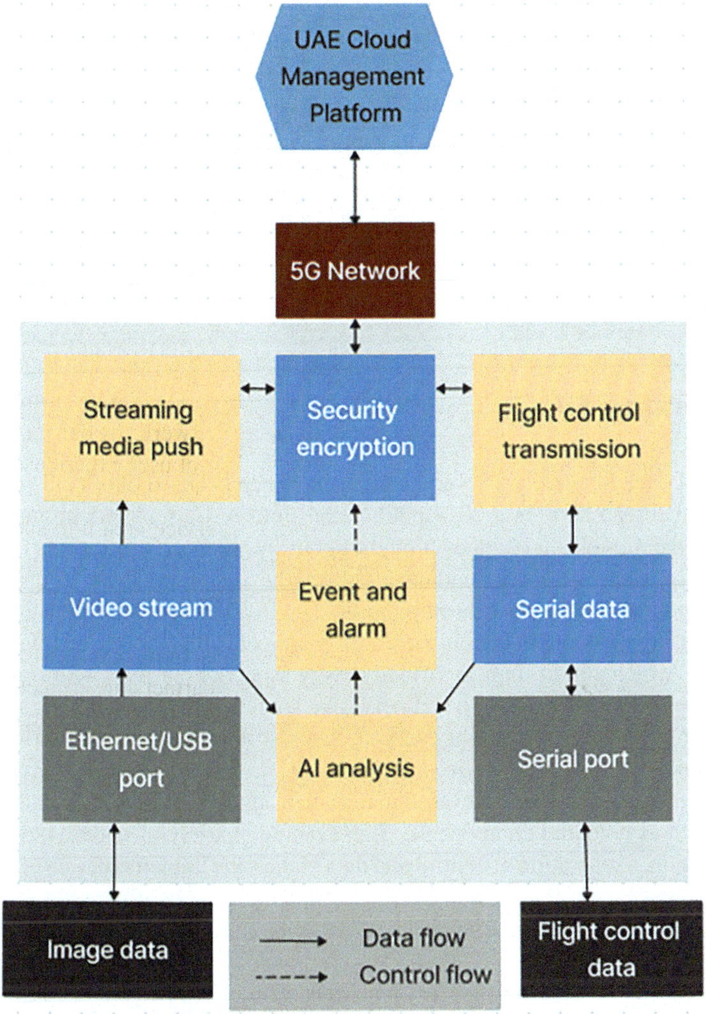

Fig. 1 Data processing flowchart of 5G

limited range of areas. They can be obstructed by materials, such as walls, windows, or by pedestrians. Because of their higher cost, high-band cells are primarily deployed in dense urban areas and places with large crowds, like stadiums and convention centers. The most widely licensed and deployed 5G mm. Wave spectrum range worldwide spans from 24.25 to 29.5 GHz (see Fig. 2).

The rollout of 5G technology has sparked debates concerning security and the involvement of Chinese vendors. It has also generated health concerns and misinformation, including discredited conspiracy theories associating it with the COVID-19 pandemic.

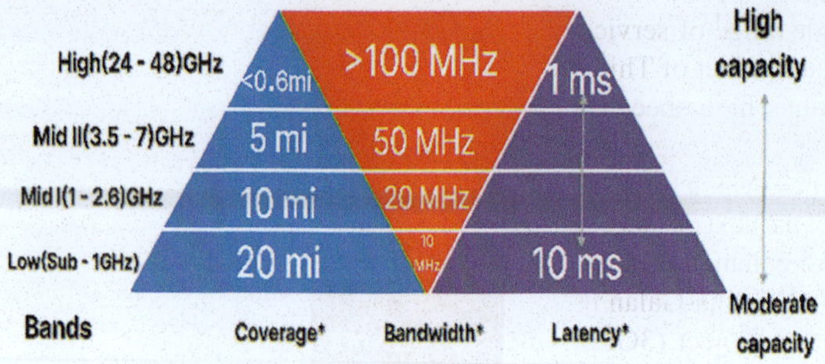

Fig. 2 Low latency requirements

2 Literature Survey

In the last decade, there are several studies have been carried out in 5G technology by several researchers. In a 2014 publication, Jeffrey G. Andrews prompted us to consider the significant changes brought about by the four generations of cellular technology, each of which marked a substantial departure from its predecessor without maintaining backward compatibility [1]. In 2014, Jude Okwuibe highlighted the increasing momentum behind the development of Fifth Generation (5G) wireless networks, which aim to connect virtually every aspect of our lives with significantly higher speeds, minimal latency, and pervasive connectivity by giving the significant impact of these networks on our everyday lives, it becomes crucial to prioritize the protection of components, services, and users [2]. In 2014, Ali, Imran suggested that a definitive 5G framework had yet to materialize, and several key elements could be anticipated as potential components. These include variety of node types, network densification, a separation of control and data planes, the implementation of extensive and localized caching, network virtualization, simultaneous utilization of different medium access control and physical layers, infrastructure sharing, and adaptable spectrum allocations [3].

In 2017, Naser AI-Falahy explored the transformation of mobile communication, which has evolved from a limited voice-only service to a sophisticated and interconnected environment. This ecosystem now encompasses numerous services that rely on a robust system capable of supporting a wide range of applications. Additionally, this system facilitates high-speed connectivity for an extensive user base comprising both subscribers and machines [4]. According to Sriganesh K. Rao in 2018, Industry 4.0 sets the standard that companies should strive to meet. In this era, manufacturing systems surpass mere connectivity and instead involve communication, analysis, and utilization of gathered data to facilitate intelligent decision-making. Industry 4.0 encompasses the combination of various technologies that are the Internet of Things (IoT), additive manufacturing, analytics, artificial intelligence, robotics, augmented reality, and advanced materials [5]. In 2020, MM Silva and J Guerreiro asserted that while 5G technology plays a crucial role in the field of development, 6G (the subsequent generation) will become ready to enter this country with its emerging services.

This wide range of services includes augmented reality (AR), virtual reality (VR), advanced Internet of Things (IoT) applications, holographic services, AI-integrated applications, high-speed mobility, and wireless brain-computer interaction (BCI) [6].

In 2020, Yu Tang emphasized the significance of agriculture, highlighting the transformative potential of the superfast 5G network in the farming industry. Over the course of the next decade, this advanced technology is poised to revolutionize farming practices, enhancing crop yields and quality with reduced labor requirements [7]. Alejandro Molina-Galan in 2021 provides detailed insights into the 3rd Generation Partnership Project (3GPP) and its recent publication, Release 16. This release is particularly noteworthy as it incorporates the initial Vehicle-to-Everything (V2X) standard based on the 5G New Radio (NR) air interface. The introduction of 5G NR V2X brings forth enhanced features that cater to the specific demands of connected and automated driving scenarios [8].

In 2021, Jasneet Kaur focused on integrating Machine Learning (ML) with 5G networks. Providing an up-to-date analysis, she presents a comprehensive overview on the concepts of future wireless systems, including 6G, and explores the role on techniques of ML within these emerging wireless networks. The paper introduces a conceptual 6G model, highlighting the significance and utilization of ML techniques across each layer of the architecture [9]. Markus Rup directed his attention toward the advancement of Smart Railways, with a focus on addressing design challenges related to high reliability and high throughput in High-Speed Rail (HSR) communications. Recognizing the significance of safety, eco-friendliness, comfort, transparency, predictability, and reliability in this context, Markus suggests that fifth-generation (5G) technologies hold promising potential as a solution [10].

In 2021, Justus Rischke developed a comprehensive experimental setup to accurately measure the time it takes for packets to travel from a 5G end-device, via a radio access network (RAN), to a packet core. This setup allows for precise measurement of one-way packet delays with sub-microsecond precision for the RAN, and with nanosecond precision for the packet core [11]. In 2020, Dinh C. Nguyen conducted a comprehensive survey that examines the advancements and current state of integrating blockchain technology with 5G and higher networks. The survey primarily focuses on discussing the potential of blockchain in enabling various 5G technologies such as edge computing, cloud computing, Network Function Virtualization, Device-to-Device (D2D) communications, and Network Slicing [12].

Yushan Siriwardhana emphasized the significance of 5G-enabled e-health solutions and 5G systems. This paper focuses on methodologies that productively leverage 5G technology for the use of e-health and its role in facilitating appropriate digital services. Additionally, this paper provides a thorough discussion on implementation challenges, potential solutions to overcome those challenges, and future developmental research for 5G to address health provocations associated with COVID-19 [13]. In 2021, Huan X. Nguyen highlights the relatively new concept of digital twins in the context of 5G/6G networks. Despite the apparent potential in aiding the development and deployment of complex 5G environments, digital twins are not yet widely explored. This article delves into the subject matter and explores how digital twins can serve as a robust tool to unlock the full potential of 5G networks and other higher networks [14]. In 2022, Hanliang Fu conducted a study on the acceptance of 5G base stations by the public. The study employed eye-tracking techniques, questionnaires, and the Stimulus-Organism-Response Model (SOR) and Technology Acceptance Model (TAM) framework to examine the factors that influence people's acceptance. By utilizing these methodologies, the study aimed to gain insights into the various factors that impact the perception and acceptance of public on 5G base stations [15].

2.1 Evolution of 5G

This part will give you the long roadmap of evolution from 1 to 5G mobile networks. Everything in this networking world starts with 1G networks which are now upgraded to 5G and this all is a continuous 30 years of hard work of researchers and developers. In this world of evolving web technology, it's way hard to compare the web with anything else. Let's go straight back in space–time, from 1 to 5G networks (see Fig. 3).

Marconi, whose era was 1895, was the first to use radio waves to send signals. The first Generation (1G) was introduced in 1970 in Tokyo, however, since then, significant changes have been introduced in wireless communications. During the 1980s, 1G gained popularity worldwide, providing analog switching, a bandwidth

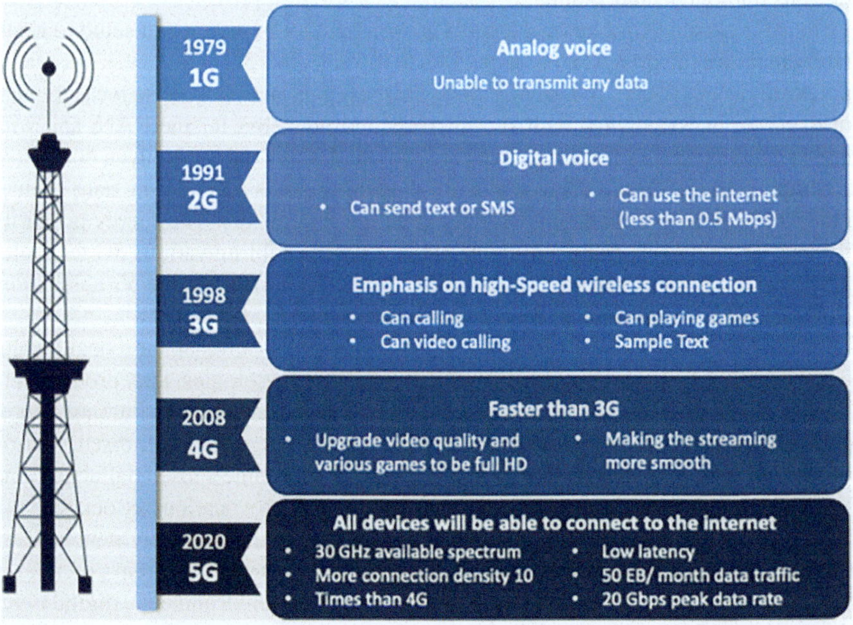

Fig. 3 Guide to mobile IOT

of 10 MHz, and voice-only services. In the 1990s, 2G brought GSM technology, offering SMS services, roaming, encrypted voice transmission, digital switching, and enhanced security.

The third generation (3G) introduced UMTS, which provided video calling, higher data rates, high-quality games, multimedia messaging, and location tracking. The fourth generation (4G), which is widely used today, was introduced in 2012 with TD-LTE technology, offering high-definition features, data rates of up to 1Gbps, HD gaming enhanced security, HD gaming, and more. Now, the fifth generation (5G) is expected to be the quickest and most connected in developing countries like India. It is an advanced wireless technology that offers increased capacity and enables new experiences for users. It provides high speeds, reliable connections, and minimal delay, expanding the possibilities in various industries. With 5G, we can expect advancements in healthcare, improved security, HD streaming, ultra-fast internet, cloud-based infrastructure, and even autonomous driving. Like this, the evolution of 5G took place and revolutionization in the world of communication came.

3 Discussion

I believe that once the security issues associated with 5G networks are effectively resolved, they can play an important role in realizing the vision of Smart Cities. 5G can offer innovative solutions to ongoing challenges related to infrastructure, public transportation, environmental sustainability, healthcare, smart manufacturing, IoT, internet and media, agriculture, and many more.

Imagine a seamless shopping experience where your AR glasses or smartphone guide you to the exact location of the item you're looking for in a store. Visualize trying on clothes virtually with virtual reality technology, displaying how they would look on you. Digital signage and smart tags make shopping smoother and more enjoyable. In the future, clothing printers could even produce clothes instantly based on your measurements. In the restaurant industry, 5G and IoT enable your phone to connect to the network, transmitting important information like party size and food allergies to the hostess before you arrive. Integrated supply chains powered by 5G and IoT allow for real-time tracking, reducing costs, improving customer service, and minimizing product loss during transit. This technology also aids in identifying production bottlenecks and enhancing industrial processes.

The high-speed and low-latency connectivity provided by 5G can facilitate efficient management of city resources, enabling effective monitoring and control systems. 5G networks and AI technology offer numerous advantages and solutions to wireless challenges. The combination of AI and 5G can enhance device efficiency, extend battery life, and provide superior user experiences. AI improves network security, network efficiency, service quality, and deployment simplicity for 5G networks. Similarly, AI machine learning enhances system performance, security, and device experience for 5G devices. 5G transforms AI by democratizing its functionality and enabling distributed AI processing, thanks to its high capacity and low latency. This

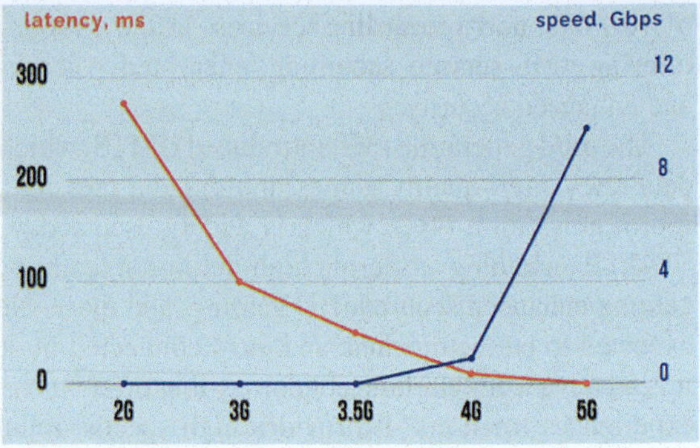

Fig. 4 Latency versus speed graph of different networks

convergence allows for personalized and engaging experiences in various fields, such as healthcare, customer service, autonomous vehicles, augmented reality, and industrial automation. These technologies are revolutionizing everyday life by enabling remote surgeries, personalized recommendations, immersive entertainment, reliable autonomous vehicles, and efficient manufacturing processes. This, in turn, can lead to improved public transportation operations, optimized energy usage, and enhanced environmental monitoring. The introduction of 5G networks enables the integration of IoT devices, sensors, and AI systems, promoting smarter and interconnected city infrastructure. This advancement can lead to the creation of sustainable, efficient, and connected cities. One of the best examples of advanced technology is China's area of development, as they have set up artificial SUN, moon, and many more amazing masterpieces by 7G and 8G (see Fig. 4).

On the other hand, Extensive research conducted thus far has found no conclusive evidence linking wireless technologies, including 5G, to adverse health effects. Studies encompassing the entire radio spectrum have not identified any health concerns, although only a limited number of studies have specifically examined the frequencies used by 5G. The primary interaction between radiofrequency fields and the human body is tissue heating, but current exposure levels from wireless technologies do not cause significant temperature increases. Higher frequencies in 5G result in less penetration into body tissues, with energy absorption primarily confined to the skin and eyes. As long as exposure remains within international guidelines, no detrimental effects on public health are anticipated.

Also, 5G networks are being deployed in industries like agriculture, manufacturing, transportation, and healthcare. Security concerns need to be addressed as organizations consider using existing providers or building their own 5G capabilities. It is crucial to prioritize security measures to effectively mitigate the potential risks associated with the implementation of 5G networks in these sectors. It is essential to prioritize security measures to mitigate the risks associated with the implementation of 5G networks in these industries. Despite varying levels of market penetration and adoption, most enterprise security leaders are already expressing concerns about the potential risks associated with 5G. Surveys have shown high levels of concern regarding the emergence of new security threats and the impact of 5G on organizations' security postures.

4 Conclusion

The impact of AI on education is undeniable, benefiting individuals from various professions in unimaginable ways. AI has significantly influenced teaching, management, and individual educational institutions. Rather than fearing AI, our focus should be on learning skills to better utilize this technology. Ethical considerations, such as accountability for grading mistakes, must be addressed to build trust in AI. While certain skills like interpersonal skills cannot be learned through AI, a collaboration between educators and engineers can narrow the gap between teachers and technology. The progress in AI and 5G technology, coupled with the emergence of smart cities, brings significant advantages to our daily lives. These advancements result in improved AI capabilities, self-driving vehicles, IoT devices, and enhanced security measures. However, it is crucial to address any potential health concerns related to 5G networks to ensure their safe integration. The widespread development of smart cities worldwide demonstrates the positive impact of technology on urban environments. Continued efforts to promote understanding and support for smart city initiatives will contribute to their success in various regions, such as Serbia.

References

1. Andrews JG, Buzzi S, Choi W, Hanly SV, Lozano A, Soong AC, Zhang JC (2014) What will 5G be. IEEE J Sel Areas Commun 32(6):1065–1082
2. Vimal V, Maheshwari S (2014) Energy management routing in wireless sensor networks. In: Proceedings of the third international conference on soft computing for problem solving: SocProS 2013, vol 1. Springer India, pp 161–168
3. Ali I, Zoha A, Abu-Dayya A (2014) Challenges in 5G: how to empower SON with big data for enabling 5G. IEEE Netw 28(6):27–33
4. Al-Falahy N, Alani OY (2017) Technologies for 5G networks: challenges and opportunities. IT Prof 19(1):12–20
5. Vimal V, Nigam MJ, Verma H (Dec 2017) Route assortment procedure for mobile Ad Hoc networks using a novel modified mobility factor. In: 2017 14th IEEE India council international conference (INDICON). IEEE, pp 1–5
6. da Silva MM, Guerreiro J (2020) On the 5G and beyond. Appl Sci 10(20):7091
7. Tang Y et al (2021) A survey on the 5G network and its impact on agriculture: challenges and opportunities. Comput Electron Agric 180:105895
8. Garcia MHC et al (2021) A tutorial on 5G NR V2X communications. IEEE Commun Surv Tutor 23(3):1972–2026
9. Balkus SV, Wang H, Cornet BD, Mahabal C, Ngo H, Fang H (2022) A survey of collaborative machine learning using 5G vehicular communications. IEEE Commun Surv Tutor 24(2):1280–1303
10. Ai B, Molisch AF, Rupp M, Zhong ZD (2020) 5G key technologies for smart railways. Proc IEEE 108(6):856–893

11. Rischke J, Sossalla P, Itting S, Fitzek FHP, Reisslein M (2021) 5G campus networks: a first measurement study. IEEE Access 9:121786–121803. https://doi.org/10.1109/ACCESS.2021.3108423
12. Nguyen DC, Pathirana PN, Ding M, Seneviratne A (2020) Blockchain for 5G and beyond networks: a state of the art survey. J Netw Comput Appl 166:102693
13. Siriwardhana Y, Gür G, Ylianttila M, Liyanage M (2021) The role of 5G for digital healthcare against COVID-19 pandemic: opportunities and challenges. Ict Express 7(2):244–252
14. Nguyen HX, Trestian R, To D, Tatipamula M (2021) Digital twin for 5G and beyond. IEEE Commun Mag 59(2):10–15
15. Fu H, Zhu H, Xue P, Hu X, Guo X, Liu B (2022) Eye-tracking study of public acceptance of 5G base stations in the context of the COVID-19 pandemic. Eng Constr Arch Manag

An Ai Powered 6 g Wireless Network

Anchal Panwar, Rahul Chauhan, Aditi Jeena, and Ruchira Rawat

Abstract The widespread rollout of 5G networks represents a critical turning point in the development of wireless technology. With 5G networks now beginning to appear everywhere, we are enticed to go farther into the future of wireless communication. This study explores the conception of 6 g, a new generation of wireless networks that will be very different from their predecessors and will experience extraordinary change. 6 g is expected to be a significant divergence from its predecessors, offering an unparalleled revolution that will distinguish it in a number of ways. AI emerges as the cornerstone in this imaginative voyage, essential to the creation and optimization of the 6 g architecture. The study examines prospective 6 g technologies, emphasizing how they can strengthen mobile AI applications. Moreover, it unravels AI-enabled techniques that influence the architecture of the 6 g network from the ground up, demonstrating the cooperation between AI and the upcoming wireless technology. As we move from 5 to 6 g, it's clear that AI and ML solutions will be crucial for a variety of applications that are carried out at this cutting-edge network's edge nodes. This research illustrates the broad reach of artificial intelligence in transforming 6 g communication technologies. Through an examination of the specifications, uses, and technical features of 6 g wireless networks within the framework of mobile edge computing, it illuminates the enormous possibilities and 6 g's revolutionary potential. In essence, 6 g communication networks are poised to become the sixth sense of the next-generation communication landscape, heralding a new era wireless connectivity.

Keywords Wireless · Architecture · Artificial intelligence · Machine learning · Methodologies

---

A. Panwar · R. Chauhan (✉) · A. Jeena
Computer Science and Engineering, Graphic Era Hill University Dehradun, Dehradun, Uttarakhand, India
e-mail: chauhan14853@gmail.com

R. Rawat
Computer Science and Engineering, Graphic Era Deemed to Be University, Dehradun, Uttarakhand, India
e-mail: ruchira.rawat.cse@geu.ac.in

P. J. S. Gonçalves et al. (eds.), *Proceedings of Fifth International Conference on Computing, Communications, and Cyber-Security*, Lecture Notes in Networks and Systems 1128, https://doi.org/10.1007/978-981-97-7371-8_10

1 Introduction

The transition from 1 to 5G wireless connectivity has revolutionized our ability to connect and communicate.

With severe specifications than previous generations, 6 g will be revolutionary and revolutionize the wireless development by linking objects to generate intelligent connections.

2 Overview

The wireless communications industry have kept a fast growing trend with creative features for a number of decades. In this section we discuss about 1G-5G transition [1].

First Generation (1G): The first generation of wireless (cellular) network technology is referred to as 1G. These are standards for mobile telecommunications that were first used in the 1980s. 1G networks used analog audio transmissions.

Second Generation (2G): With digital circuit-switched solutions, the second generation of wireless networks aims to outperform analog ones. In the early 1990s, 2G was developed. GSM (Global System for Communication) is the technology that powers the majority of 2G networks, including those in the UK. Mobile data services, starting with SMS text messaging, were also made available via 2G.

Third Generation (3G): The third generation of mobile wireless technology is known as 3G. Better speech quality and faster data transfer are two features that set it apart from 2G. The first two generations of the standard, which were implemented on mobile networks and in mobile phones, are followed by the third generation. In order to link mobile phones and handsets to the internet or other IP networks for the purposes of making voice and video conversations, downloading and uploading files, and Web browsing, 3G was primarily utilized with these devices.

Fourth Generation (4G/LTE): The 4G network, which replaces the 3G network, is referred to as the fourth generation of mobile communication or wireless communication technology. All IP-based cellular communications are offered by it. They provided far quicker data rates, lower latency, and more efficient networks. The utility of internet gaming, streaming videos, and mobile video conferencing was enhanced by 4G networks. Long-Term Evolution is currently the most widely used 4G technology (LTE) [2].

All things considered, 4G networks, with their high-speed data access, low latency, and compatibility with a wide range of services and apps, have fundamentally revolutionized mobile communication.

As the shift to 5G and beyond occurs, continual advancements in network technology are intended to tackle the limitations and challenges associated with them.

5G (Fifth Generation): The most recent development is 5G, or fifth generation, which began to appear in the late 2010s and is still growing.

By providing better coverage, capacity, energy efficiency, security, and support for a variety of use cases, it overcomes many of the drawbacks of 4G networks. With its ability to transmit data at ultra-high speeds, low latency, enormous device connection, and network slicing, it makes technologies like augmented reality, driverless cars, and the Internet of Things (IoT) possible.

Every generation improved on the strengths of the one before it, removing obstacles and raising the potential applications of wireless communication.

Even with its improvements, 5G still has drawbacks, such as poor coverage in remote regions, latency that makes it unsuitable for real-time applications, and difficulties handling large numbers of IoT devices. 6 g is required because it offers terabit data rates, almost low latency, and unmatched connectivity. Initial standards for 6 g are anticipated by the late 2020s, according to the development roadmap [2, 3].

Every generation improved on the strengths of the one before it, removing obstacles and raising the potential applications of wireless communication.

Even with its improvements, 5G still has drawbacks, such as poor coverage in remote regions, latency that makes it unsuitable for real-time applications, and difficulties handling large numbers of IoT devices. 6 g is required because it offers terabit data rates, almost low latency, and unmatched connectivity.

The next generation of cellular technology is called 6 g, or sixth-generation wireless. Supporting communications with a latency of one microsecond is one of the objectives of the 6 g internet. Expected to offer data speeds of up to 1 terabyte per second, the 6 g network will be 100 times faster than the existing 5G network. Additionally, millions of devices will be able to be supported simultaneously by the 6 g network. Figure 1 discuss the key features and Fig. 2 shows the key objectives of 6 g network.

With the aid of emerging technologies like distributed computing, blockchain, edge computing, and artificial intelligence, 6 g will provide 3D forms of support [4]. The 6 g communication network will be integrated and widely distributed. Calculation, routing, and detection are all planned to be combined in 6 g for the communication organization. Regarding security, privacy, and safeguarding of the massive amount of data generated by billions of smart devices, 6 g will address these issues. Gadgets that are currently clever will soon become intelligent.

Motivated by these developments, we try to conceptualize 6 g in this article as an intelligent information system that both drives and is driven by contemporary AI technologies. A 6 g roadmap that is projected based on the 5G State and plotted according to the strategic plans of different standard bodies. There are also illustrations of service categories and key performance indicators (KPIs).

We believe that AI will enable close-loop optimization to support the previously mentioned new service types and significantly improve network operators' situational awareness [5]. Considering this, 6 g will fully utilize mobile computing, communications, and control in a wide range of fascinating applications, such as wearable computing, autonomous vehicles, smart cities, seamless virtual and augmented reality, Internet of Things, space-air ground integrated networks, and much more.

Fig. 1 Key Features of 6 g architecture

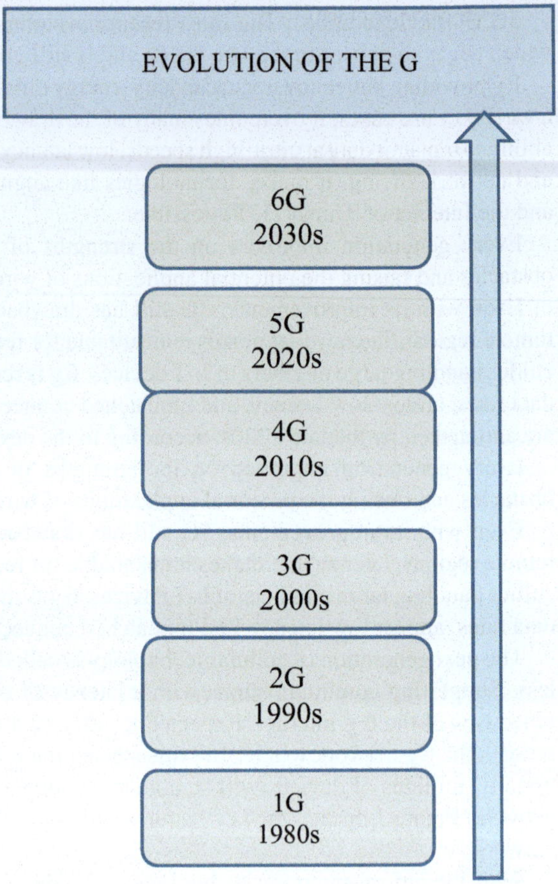

This paper represents a modest effort to offer a future-focused research roadmap for 6 g. This is how the rest of the article is structured. There will be a presentation on the architecture of 6 g. We'll demonstrate how 6 g uses the development of AI to power its main functions. To meet the expected stringent requirements of such applications, hardware-aware communications will be embraced in 6 g.

Fig. 3 Shows the overview of 6 g Wireless Communication Network and Table 1 represent the 6 g Technology and Description.

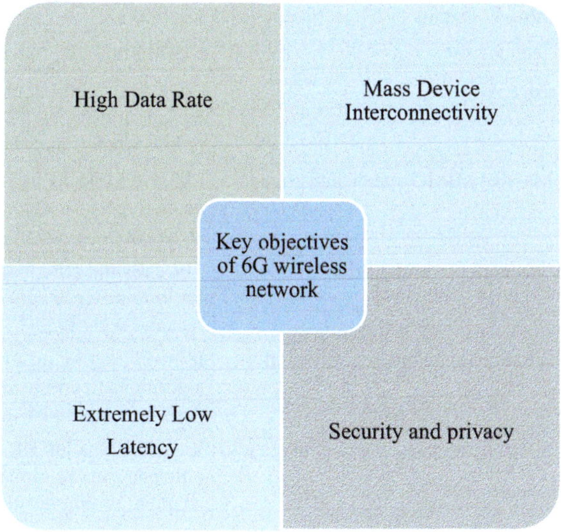

Fig. 2 Key objective of the 6 g network

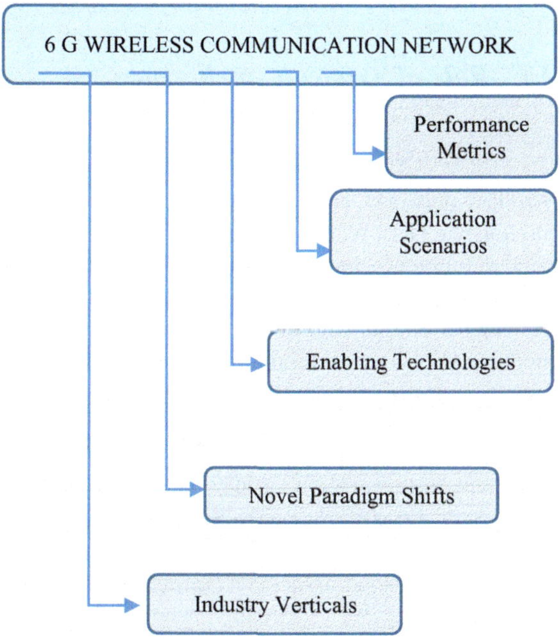

Fig. 3 Overview of 6 g wireless communication network

Table 1 Details of 6 g technology and description

Technology	Description
Edge AI	Edge AI involves deploying AI algorithms on local devices at the network's edge to enhance data processing, reduce latency, and support real-time communication
Massive MIMO and beamforming	Massive MIMO employs numerous antennas to enhance network performance, while beamforming optimizes signal delivery and customized wireless experiences
Quantum communication	The essential criteria of 6 g use quantum communication for long-distance communication. Also, for high data rate and security
Intelligent network management	It involves utilizing advanced artificial intelligence. algorithms to optimize network resources, enhance performance and ensure seamless connectivity
Spectrum sharing and cognitive radio	It enable dynamic allocation of frequencies, enhancing efficiency and minimizing interference for seamless, adaptive communication

3 Discussion

3.1 Role of Artificial Intelligence

A) AI in wireless network

Wireless networks are undergoing a fundamental transformation due to artificial intelligence (AI), which makes them more intelligent, flexible, and efficient. Artificial Intelligence has been applied to almost every facet of wireless communication, including network administration, security, optimization, and user experience improvement. Algorithms driven by artificial intelligence and machine learning models are utilized to monitor network conditions on their own, forecast network outages or congestion, and proactively optimize resource allocation for better quality of service [6].

Moreover, AI is utilized to identify threats and detect anomalies in real time, enhancing the security of wireless networks.

6 g will differ significantly from other generations because to its unparalleled revolution in wireless networks, as it will have a high degree of heterogeneity in several areas, including network infrastructures, radio access technologies, and RF gadgets, storage and processing capacity, kinds of applications, etc. Additionally, there is a noticeable increase in the amount and type of data generated by wireless networks. In order to achieve real-time additivity to dynamic network environments, this creates excellent potential for data-driven network planning and operation.

B) Machine learning in 6 g networks

Machine learning (ML) is gaining popularity because of its many uses and powerful features. Computer programs that are used to learn the traits or patterns of a system that is an explicit mathematical model are called machine learning models. These models are applied to tasks like regression, categorization, and intelligent agent-environment interactions. With ultra-low latency, pervasive connection, and previously unheard-of data rates predicted for 6 g networks, effective and flexible network management is critical. Network optimization, resource allocation, and intelligent decision-making could all be revolutionized by machine learning algorithms, especially deep learning models. Machine learning that is compatible with 6 g connection technology has enormous influence in every field and will do a lot of tasks in the near future. 6 g networks are also expected to include massive MIMO (Multiple Input, Multiple Output) and beamforming technology; machine learning (ML) can enhance these technologies by refining beamforming patterns and adapting to the dynamic environment. Furthermore, ultra-low latency and ultra-reliable communication—which are necessary for mission-critical applications like remote surgery and autonomous vehicles—can only be achieved with machine learning (ML).

3.2 Ai Powered 6 g Technologies

A) Edge AI:
This AI-driven 6 g technology has the capacity to fundamentally alter the way humans utilize and engage with wireless networks. It turns becomes an essential component, enabling decentralized, on-the-spot decision-making at the edge of the network [7].

B) Massive MIMO and Beamforming:
Multiple-input multiple-output, or MIMO, is a technology that increases the data throughput and dependability of wireless communication by using several antennas at the transmitter and the receiver. With this technology, a radio link's capacity can be increased by utilizing multiple send and multiple receive antennas. Multiple data streams can be delivered over the same radio channel because of multipath propagation and decorrelated routes between the transmitter and receiver. This increases the peak data rate per user and the cellular network's capacity.

C) Quantum Communication:
It will contribute to two key 6 g requirements: security and data throughput.

The Fig. 4 shows the components of the Quantum Enabling in 6 g Network.

D) Spectrum Sharing and Cognitive Radio:
Cognitive radio and spectrum sharing are two essential components that have the power to alter wireless communication networks. A fascinating area of study that looks toward the creation of intelligent and adaptable communication systems is the evolution of wireless networks using cognitive radio technology. A technology known as cognitive radio (CR) allows wireless devices to efficiently

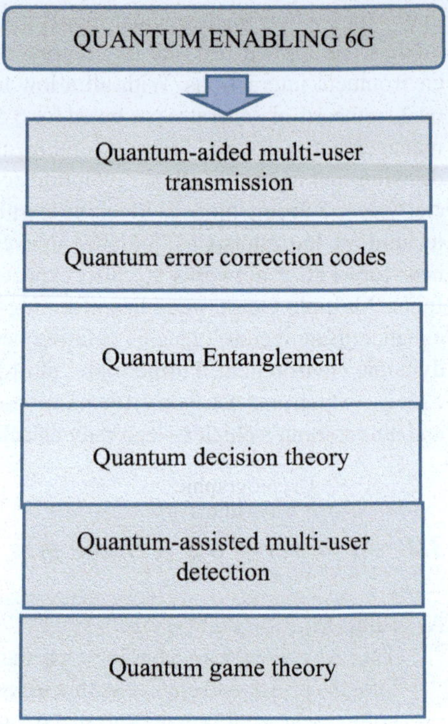

Fig. 4 Overview of quantum enabling in 6 g network

access accessible spectrum bands in real-time. Spectrum sharing refers to the flexible and efficient use of radio frequency spectrum resources among multiple users, services, and apps.

3.3 Challenges and Considerations

The above Fig. 5 shows the components of the 6 g Security Considerations.

A) Security and privacy:

Security and privacy are areas which will require top priority in advancement and implementation of 6 g networks. As 6 g technology will develop, it will have several advantages in different field as well as risk factors because of its low latency and ever-present access. This also raises concerns about the security and privacy of the massive amounts of data that will be generated and handled.

People and smart items should relate to 6 g, which makes network security considerably more challenging [8]. Unfortunately, in order to obtain services, mobile customers may have to assume risks, and they might not be aware of these risks until something bad happens. Another common mistake is believing that gathering

Fig. 5 Overview of 6 g security considerations

personal data is pointless since most users believe data is unimportant or presents little danger. Robust authentication procedures, intrusion detection systems, and encryption will be necessary to defend against cyberthreats including AI-driven assaults and data breaches. Furthermore, privacy will be a crucial factor to take into account since the AI algorithms that run on 6 g networks will have to carefully balance the needs of individual users with the requirement to provide individualized services.

Generally speaking, 6 g has the option to transmit at the highest security level. The aforementioned security issues will be exacerbated by integrating the Internet of Everything with 6 g and offering additional management features like smart homes, emergency clinics, transit, and smart grids, among others. Quantum communications will also draw attention to the crucial security requirements of 6 g. In 6 g, the unified AI will fend off hostile attacks. Notably, physical layer security will be offered by 6 g [10, 12]. In general, 6 g is able to determine the most important transmission security level due to the terahertz (THz). The 6 g security offer is the only factor that determines the mystery and protection. Either way, some kind of mystery is necessary when dealing with sensitive information, like the secret term for financial balances. The enigma can be solved via quantum cryptography.

Human services in particular demand the greatest level of security assurance. With 6 g, new concepts for coordinating comfort and safety will become possible. These concepts include central management, use cases, KPIs, enabling innovation, models, factory operations, challenges that now exist, possible solutions, transparency, and avenues for research.

B) Scalability and infrastructure requirements:

The need for sophisticated network management and high-speed, low-latency connection will increase as we get closer to the 6 g future. In this context, scalability refers to the network's capacity to support an ever-growing number of devices and applications. With artificial intelligence (AI) at its foundation, 6 g can instantly improve security, optimize resource allocation, and adjust to changing traffic patterns. As a result, the infrastructure needs must be both adaptable and strong. To achieve the claimed performance, 6 g networks will need to install a large number of tiny cells, millimeter-wave spectrum, and edge computing nodes. To allow the ultra-fast data transmission and reception that 6 g promises, the infrastructure must also handle enormous MIMO (multiple-input, multiple-output) antenna arrays and sophisticated beamforming algorithms.

C) Energy efficiency and sustainability:

Energy efficiency and sustainability are major factors in advancement of an AI driven 6 g wireless network. The increment in using internet of things devices and leveling demands for data intensive applications have caused a momentous problem in energy consumption of wireless network. As in 6 g technology, we have ultra-high latency transmission so this could intensify the issue. Sustainable use of energy efficiency is very crucial [9]. Here AI plays an important role by providing dynamic power control, reduce energy consumption and improve resource allocation and network performance. By using machine learning algorithms to tract traffic patterns and alter network designs, 6 g network will be able to manage resources more efficiently [11].

4 Applications of AI Powered 6 g

The application of the AI powered 6 g are listed below:

1) Smart Cities and IOT
2) Autonomous Vehicles and Transportation
3) Augmented and Virtual Reality
4) Environmental Monitoring and disaster Response
5) Telemedicine and Remote Surgery
6) Holographic Communication

5 Research and Development

A) Current Project and Initiatives: Key 6 g projects include tech company contributions, industry-academia alliances, and government-funded research. [13] In order to maximize network performance, these projects are concentrated on investigating new spectrum bands, cutting-edge antenna technologies, energy-efficient solutions, and the incorporation of artificial intelligence. Additionally, research is being done to create innovative networking architectures and materials that can support the lofty 6 g objectives.

B) Industry Collaborations: The creation of 6 g networks with AI capabilities requires close industry cooperation in many different areas. Leading tech and telecommunications firms, such as Qualcomm, Samsung, Huawei, Nokia, Ericsson, and Ericsson, have teamed together to conduct research, exchange knowledge, and influence the direction of 6 g technology. These collaborations include work with academic and research institutions, which frequently results in exciting and standard quality research. Nowadays, governments have started to support and fund projects and programs that encourage and promote industry cooperation in order to support the development of 6 g networks worldwide [14].

C) Predictive data analysis in 6 g: AI was seen as bringing the world into our hands in the form of a tiny device called a mobile phone. Proper use of artificial intelligence in established environments allows for more accurate data predictions. Model optimization and model refinement become much easier with 6 g communication network technology.

The use of statistics and models to evaluate future outcomes based on current and past information is presented in predictive analytics. Predictive analytics technology provides potential insights with a high level of accuracy. Advanced predictive analytics software and models enable any organization to use historical and current data to accurately predict patterns and behaviors milliseconds, days, or years into the future. Predictive analytics is a form of data analysis that focuses on historical data and analytical techniques such as machine learning to make predictions about future outcomes. The data-driven network architecture of 6 g communications technology focuses specifically on connectivity and data availability [15]. You can also speculate how the 6 g and its AI with intelligent performance will facilitate the marketing of his more economical model.

D) Deep Learning in 6 g: All areas of the deep learning system would be impacted by the 6 g intelligent communication system. In the field of artificial intelligence, deep learning is a subset of machine learning that mostly works with unsupervised data. Together, supervised and/or unsupervised learning are components of deep learning.

Instead of depending solely on features created by humans, a system that uses automatic learning features can learn intricate functions that convert inputs at multiple levels of abstraction into the output. To diagnose network anomalies, faults, identify and stop intrusions, configure networks, and optimize them, deep learning has been investigated. The system can gather and process data in real-time thanks to 6 g connectivity technologies.

Some foundational Concepts in the field of deep learning in 6 g are:

(i) Artificial Neural Network (ANN).
(ii) Deep Learning Network (DNN).
(iii) Federated Learning.

6 Conclusion

Based on the agreement, we may infer that 6 g communications will soon become a basic human necessity and that it will be combined with artificial intelligence to create an intelligent communication system. Artificial general intelligence, artificial superintelligence, and limited artificial intelligence are only a few of the aspects that will be impacted by the cognitive model of the network architecture, which will drive the intelligent network to its maximum potential. The AI-enabled architecture and AI-centric communication strategies for 6 g networks have been presented in this article. The 6 g evolution's new features were noted, and enabling technologies were covered. Furthermore, revolutionary applications like augmented reality, virtual reality, and real-time edge computing will be made possible by 6 g's AI-driven capabilities. The convergence of AI and 6 g will not only transform communication, but also drive innovation across several industries and pave the path for a more interconnected and sustainable society.

According to the agreement, 6 g communications is going to be the daily need of ordinary people, and it will be combined with AI to create an intelligent communication system within the next few years. The intelligent network is going to be 100% AI-driven. The cognitive model of the network architecture is going to influence every aspect, including artificial general intelligence (AGI), narrow artificial intelligence (NAI), and artificial super intelligence (ASI). More importantly, the different potential enabling technologies are going to promise high QoS and high QoE to take society to an AI-powered smart city. So, we can predict that in the next few years, AI with 6 g will be more efficient and reliable for the end user. The combination of artificial intelligence with 6 g communications networks will usher in a new era of digital communication technology.

References

1. Dang S, Amin O, Shihada B, Alouini M-S (2020) What should 6 g be? Nature Electronics 3(1):20–29
2. Qamar F, Siddiqui MUA, Hindia M, Hassan R, Nguyen QN (2020) Issues, challenges, and research trends in spectrum management: A comprehensive overview and new vision for designing 6 g networks. Electronics 9(9):1416

3. Ajay Bandi A Review towards AI empowered 6 g communication requirements, applications, and technologies in mobile edge Computing (ICCMC 2022)
4. Letaief KB, Chen W, Shi Y, Zhang J, Zhang YA (2019) The roadmap to 6 g: AI empowered wireless networks IEEE Commun Mag (ISSN: 1558–1896) 57(8):84–90
5. Ahammed TB, Patgiri R, Nayak S (2022) A vision on the artificial intelligence for 6 g communication 'KICS"
6. Hussain B, Student Member, IEEE, Du Q, Member, IEEE, Imran A, Senior Member, IEEE, Ali Imran M, Senior Member, IEEE. Artificial intelligence-powered mobile edge computing-based anomaly detection in cellular networks
7. Zhang Z, Xiao Y, Ma Z, Xiao M, Ding Z, Lei X, Karagiannidis GK, Fan P (2019) 6 g wireless networks: Vision, requirements, architecture, and key technologies. IEEE Veh Technol Mag (ISSN: 1556–6080) 14(3):28–41
8. Kumar I, Bhatt C, Vimal V, Qamar S (2022) Automated white corpuscles nucleus segmentation using deep neural network from microscopic blood smear. J Intell Fuzzy Syst 42(2):1075–1088. https://doi.org/10.3233/JIFS-189773
9. Vimal V, Maheshwar S (2014) Energy management routing in wireless sensor networks. Adv Intell Syst Comput 258:161–168. ISBN: 978–813221770–1. https://doi.org/10.1007/978-81-322-1771-8_14
10. Letaief KB, Chen W, Shi Y, Zhang J, Zhang YJA (2019) The roadmap to 6 g: AI empowered wireless networks. IEEE Commun Mag 57(8):84–90
11. Singh R, Sicker D (2019) Beyond 5G: THz spectrum futures and implications for wireless communication. Proc. 30th European Conf. Int. Telecommunication Society (ITS), June 16–19, [online] https://www.econstor.eu/bitstream/10419/205213/1/Singh-Sicker.pdf
12. Sharma MK, Zappone A, Assaad M, Debbah M, Vassilaras S (2019) Distributed power control for large energy harvesting networks: A multi-agent deep reinforcement learning approach. IEEE Trans Cogn Commun Netw 5(4):1140–1154
13. Singh PR, Singh VK, Yadav R, Chaurasia SN (2023) 6 g networks for artificial intelligence-enabled smart cities applications: a scoping review. Telemat Inform Rep 9:100044
14. Gupta R, Shukla A, Tanwar S (2020) BATS: A blockchain and AI-empowered drone-assisted telesurgery system towards 6 g. IEEE Trans Netw Sci Eng 8(4):2958–2967
15. Bhat JR, Alqahtani SA (2021) 6 g ecosystem: Current status and future perspective. IEEE Access 9:43134–43167

Advanced Computing Technologies

An Efficient Method to Deal with Missing Values and Class Imbalance in Healthcare Data

Harpreet Singh, Birmohan Singh, Manpreet Kaur, and Suvita Rani

Abstract The present era belongs to the age of digital devices, where everything is going to be digitized. This makes the massive production of data at faster rates and brings Big Data to light. The arrival of big data has influenced and reformed many domains like Healthcare, Bioinformatics, Agriculture, etc. Healthcare big data has gained the attention of various researchers as it aids in the early diagnosis and prediction of diseases. Generally, real-world healthcare data contains missingness and imbalance distribution of classes which makes classification a challenging problem. Therefore, an efficient method is introduced to deal with the problems of data incompleteness and class imbalance in healthcare data. A class mean and standard deviation-based imputation strategy is proposed to impute the missing values within the region of a particular class. Further, an Improved Elephant Herding Optimization (IEHO) algorithm is utilized to apply the optimal weights for improving the imbalanced learning in healthcare data. A new fitness function is proposed which considers the data missingness and class imbalance along with the feature weighting. The optimized k-Nearest Neighbor (k-NN) learning algorithm is used to perform the classification. The performance of the proposed method is evaluated on four healthcare datasets: Pima diabetes, Wisconsin Breast Cancer (Original), Hepatitis, and HCC Survival. A comparative analysis with state-of-the-art methods is performed based on four-performance metrics: accuracy, balanced accuracy, g-mean, and f-score. The proposed method obtained higher classification accuracy for Wisconsin Breast Cancer (Original), Hepatitis, and HCC Survival datasets than state-of-the-art methods. Experimental results validate the effectiveness of the proposed method in dealing with the missing values and class imbalance problems in healthcare data.

H. Singh (✉) · B. Singh · S. Rani
Department of Computer Science and Engineering, Sant Longowal Institute of Engineering and Technology, Longowal, Sangrur, India
e-mail: singh.harpreet.801@gmail.com

M. Kaur
Department of Electrical and Instrumentation Engineering, Sant Longowal Institute of Engineering and Technology, Longowal, Sangrur, India

P. J. S. Gonçalves et al. (eds.), *Proceedings of Fifth International Conference on Computing, Communications, and Cyber-Security*, Lecture Notes in Networks and Systems 1128, https://doi.org/10.1007/978-981-97-7371-8_11

Keywords Data Imputation · Imbalanced learning · Improved elephant herding optimization · Feature weighting · Healthcare big data

1 Introduction

In this modern era of electronic devices, applications, and technologies where everything is going to be digitized. It accelerates the data generation rates and produces a massive amount of raw data (known as Big Data) [1]. Generally, the data with the only characteristic of hugeness is considered big data. Several characteristics jointly define the term Big data. *Volume* represents the hugeness of data, *velocity* defines the speed of data, *variety* demonstrates different forms, designs, and structures of data, *veracity* depicts missing values or noisy features in data and the *value* shows the insightful information hidden inside the data [2]. Big data is not directly handled through manual procedures. It requires powerful storage, processing, and tools for data analysis. The real value of big data depends on the possibility of extracting useful information from it to make further decisions. The impact of the emerging Big Data field influenced many domains like Healthcare, Business Intelligence, Bioinformatics, Astrology, etc., and raises different challenges in these domains [3]. The healthcare domain has been reshaped due to the impact of big data since the high dimensional, large volume, and heterogeneous variety of data sources have been involved in healthcare organizations. Health is precious for every human life and necessary for the growth of nations. Therefore, the healthcare domain has more attracted researchers worldwide in recent years. Still, this domain has many challenges and needs to be explored. Healthcare data contains various data irregularities like missing values, the imbalanced distribution of samples, noisy samples, irrelevant and redundant features, etc. [4].

Imbalance distribution of samples and missing values are general problems for real-world healthcare data. The incomplete information on patient records may result in inappropriate prediction. A record is incomplete when values of one or more independent variables are unavailable. Data preprocessing is a crucial step of the knowledge discovery process in which data imputation plays an important role by estimating the missing values efficiently to maintain the overall distribution of the dataset [5]. An imbalanced dataset contains an uneven distribution of samples over the classes, while one class has more data samples than others. Generally, the positive (abnormal or patient) class samples are fewer than the negative (normal) class samples and treated as minority and majority classes [6]. Therefore, analyzing healthcare data with imbalanced class distribution is a challenging task for machine learning classifiers. Many machine learning algorithms are developed to improve general classification accuracy. Most of the machine learning algorithms expect a balanced class distribution with equal costs. Therefore, these algorithms tend to bias toward the majority class while dealing with imbalanced class distribution and result in classification errors regarding minority class. Misclassification of minority class samples is more probable and cost-sensitive than the majority class samples in healthcare data. The positive (minority) class samples are the region of interest, and a false negative is much costlier than a false positive while diagnosing diseases and detecting suffering patients [6, 7].

In recent years, various methods have been proposed to deal with missing values and class imbalance problems. The traditional imputation methods simply ignore or delete the samples with missing values whereas statistical learning-based imputation methods such as K-Nearest Neighbor Imputation (KNNI), Simple Imputer (mean, mode, median), and Regression-based methods impute the missing values based on the correlation among existing samples. The advent of machine learning-based imputation methods has shown a great impact on the efficacy of imputation methods [8]. The imbalanced learning methods are categorized into three categories: algorithm-level methods, data-level methods, and cost-sensitive methods [9]. In algorithm-level methods, learning algorithms have been designed in such a way that their learning reinforced the minority class to tackle the data imbalance problem [9]. Several oversampling and under-sampling techniques, such as Random Under Sampling (RUS), Random Oversampling (ROS) [10], Synthetic Minority Oversampling Technique (SMOTE) [11], Borderline-SMOTE (BSMOTE) [12], Adaptive Synthetic sampling technique (ADASYN) [13], and Safe-level-SMOTE [14], have been used in data-level methods to achieve a balanced dataset by resampling the original data distribution. The cost-sensitive approaches are dataset-dependent and require some prior knowledge of data to compute the misclassification costs for each class. It biases the classifier toward a class with high misclassification costs. [7, 15]. The data level methods degrade the classification by generating synthetic samples in overlapped regions of classes [16]. In literature [17–20], the metaheuristic algorithms have shown a great impact in dealing with imbalanced healthcare data by performing feature weighting. Feature weighting is a widely used method in the field of machine learning to improve imbalanced learning. It is a continuous search problem that assigns more weights to important information, hence boosting prediction accuracy. It is appropriate for data in which the relevance of attributes varies [21]. However, the metaheuristic algorithms suffer from the problem of local optima stagnation and premature convergence. Therefore, there is a need to design such an efficient method that can deal with missing values and class imbalance problems to improve the performance of machine learning algorithms for the classification of imbalanced healthcare data with missing values.

In the proposed method, a class mean and standard deviation-based imputation strategy is introduced to impute the missing values by utilizing the information of complete samples of each class. Further, an Improved Elephant Herding Optimization (IEHO) algorithm is utilized to perform feature weighting on imbalanced healthcare data with imputed values. The IEHO algorithm is an improved version of the standard Elephant Herding Optimization (EHO) algorithm, proposed by Singh et al., in 2021 [22]. It has provided promising results in solving global optimization problems, real-world engineering design problems, and feature selection problems [22, 23]. Therefore, this work presents the IEHO algorithm for feature weighting to improve the imbalanced learning for healthcare data. In this work, k-Nearest Neighbor (k-NN) is employed as a machine learning algorithm to classify healthcare data. The value of k parameter impacts the performance of the k-NN algorithm. Therefore, IEHO algorithm has been used to tune the value of k parameter along with feature weighting for improve the classification performance.

The rest of the research article is structured as follows: Sect. 2 presents the proposed methodology. Section 3 presents the Experimental setup of this work. Results analysis and discussion are presented in Sect. 4. Section 5 summarizes this work and suggests future research directions.

2 The Proposed Method

In the proposed work, the healthcare data is preprocessed by imputing the missing values with the help of existing complete samples. After that, feature weighting is performed on the complete healthcare data after imputation to improve the imbalanced learning by using Improved Elephant Herding Optimization (IEHO) algorithm. Also, a new fitness function is introduced to scrutinize the data incompleteness and class imbalance while performing feature weighting. Moreover, the k parameter of the k-NN algorithm has been optimized simultaneously by using the IEHO algorithm to improve the classification performance. The working of the proposed method is presented in Fig. 1 where *iter* represents the current iteration and *max_iter* represents the maximum number of iterations.

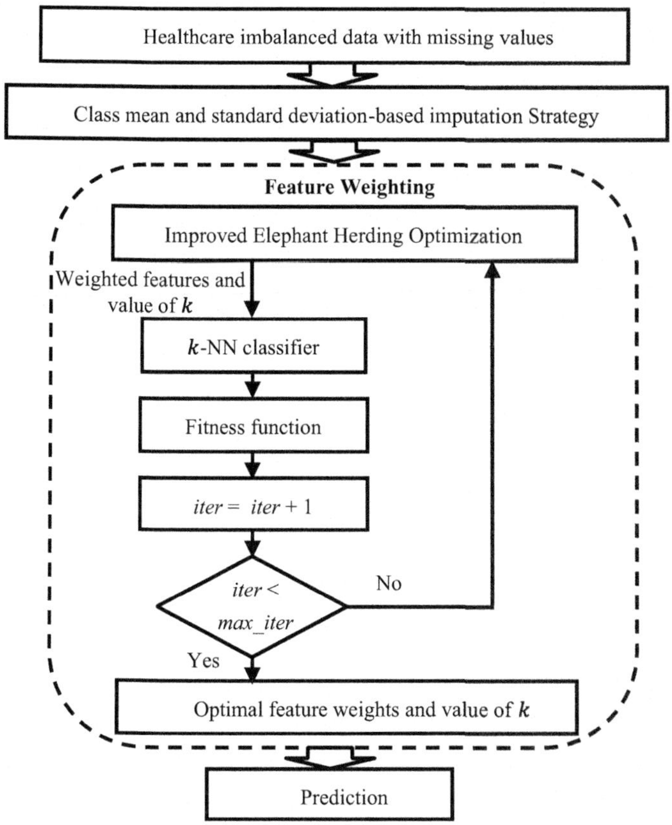

Fig. 1 Working of the proposed method for data imputation and imbalanced learning in healthcare data

2.1 Class Mean and Standard Deviation-Based Imputation Strategy

Data pre-processing is a key step in achieving excellent categorization metrics. It addresses many data abnormalities, including missing values, noisy samples, and irrelevant and duplicated features [24, 25]. Data imputation is a crucial pre-processing step that involves predicting missing values in the data in order to increase classification performance of the learning algorithm. In the proposed imputation technique, missing values are imputed based on the class mean and standard deviation of the complete samples so that the imputed values would be estimated in the region of the specific class. An algorithm of the proposed imputation technique is presented in Algorithm 1.

Algorithm 1: Clan mean and standard deviation-based imputation strategy

Input: Dataset with missing values (D), Labels (L)

Output: Complete Dataset with imputed values (D')

1. Identify the complete samples from each class of dataset (D) and create a set of complete samples $(CD) = \{CD_1, CD_2 \dots CD_n\}$, where n represents the number of classes.
2. Calculate the class mean $(CM) = \{CM_1, CM_2 \dots CM_n\}$ and standard deviation $(CSD) = \{CSD_1, CSD_2 \dots CSD_n\}$ of each class for complete samples.
3. Identify the incomplete samples from the dataset (D) and create a set of incomplete samples (ICD).
4. Initialize $i = 1$ and repeat steps 5 to 8 until all incomplete samples are not imputed
5. Identify the nearest class mean (CM_j) from i^{th} sample of an incomplete set (ICD_i) by calculating the Euclidean distance between ICD_i and CM.
6. Identify the attributes with missing values from ICD_i and store its indices in idx
7. Impute the missing values using

   $if(rand1 < 0.5)$

   $$ICD_{i,idx} = CM_{j,idx} + rand2 \times CSD_{j,idx}$$

   $else$

   $$ICD_{i,idx} = CM_{j,idx} - rand2 \times CSD_{j,idx}$$

   end

   where $rand1$ represents the uniform random number in a range of [0, 1] and $rand2$ represents the uniform random number in a range of [0, 0.5] to reduce the weight of standard deviation for imputing the missing values.
8. $i = i + 1$
9. $D' = ICD \cup CD$

2.2 *Improved Elephant Herding Optimization (IEHO) Algorithm*

The Improved Elephant Herding Optimization (IEHO) [22] algorithm utilizes the opposition-based learning mechanism in its initialization phase to commence from a better initial solution. It improves its solutions over the iterations by using a sine cosine-based clan updating operator and a modified separating operator. The sine–cosine-based clan updating operator updates the position of clan individuals either toward or outwards the clan leader by using sine–cosine mathematical functions. The Levy flight-based mutation strategy enhances the exploration and exploitation ability of the algorithm. The modified separating operator provides a guided diversification in the whole population by replacing the worst clan search individuals with newly initialized search individuals. Moreover, an elitism strategy has been introduced to retain the best search individuals in consecutive iterations. A detailed explanation of the IEHO algorithm is presented in the original research article [22]. The promising results of the IEHO algorithm in solving various optimization and feature selection problems motivate this work to investigate its performance in the domain of feature weighting. Therefore, the IEHO algorithm is utilized in this work to provide optimal weights to the preprocessed healthcare data for improving imbalanced learning. It simultaneously optimizes the k parameter of the k-NN algorithm in the range of [1, 10] to improve the classification performance.

2.3 Fitness Function for Feature Weighting

Feature weighting is a continuous search challenge in which continuous solutions are utilized to provide optimal weights to the features in the range of values [0, 1]. The performance of the feature weighting directly depends on the fitness function of the searching algorithm. Therefore, in this work, a novel minimization fitness function is introduced in which both class imbalance and data incompleteness are scrutinized to improve the classification performance of the proposed method. The fitness function is evaluated as shown in Eq. (1).

$$fit(X_i) = 1 - \left(\mu \times Bal\_acc + \vartheta \times \left(\frac{\sigma}{\rho}\right)\right) \tag{1}$$

where σ defines the number of correctly classified samples that contain imputed values, ρ represents the total number of samples that contains imputed values, *Bal_ acc* defined as an average of recall obtained on each class, μ, and ϑ are the weights to make a balance between the average of recall obtained on each class and the ratio of the number of correctly classified samples that contain imputed values with the total number of samples that contains imputed values, where $\mu = 0.8$ and $\vartheta = 1 - \mu$ is fixed in this work based on the experiment analysis.

3 Experimental Setup

This section presents detailed information about the experimental setup, datasets, and performance metrics used to execute the experiments. Python 3.8 programming platform have been utilized to implement the experiments and Intel Xeon E5-2650 CPU 2.30 GHz with 8 GB RAM system has been used to perform the experiments. In addition, 30 independent runs of all the experiments have been executed, and the mean of these results is considered for comparative analysis.

In this work, the performance of the proposed imputation method is validated first and then the effectiveness of the proposed method to deal with missing values and class imbalance problems in healthcare data is demonstrated. The performance of the proposed imputation method is evaluated on two healthcare datasets by generating missingness of four different percentages (5, 15, 25, and 35%) using the Missing Completely at Random (MCAR) mechanism. Also, a comparison between the proposed imputation strategy and existing statistical imputation techniques has been performed on the basis of Root Mean Square Error (RMSE) metric. The RMSE score will be calculated by using Eq. (2).

$$RMSE = \sqrt{\frac{\sum_{i=1}^{n} (x_i - \widehat{x_i})^2}{n}} \tag{2}$$

where x_i is the original value and $\widehat{x_i}$ is the imputed value for the ith observation in the dataset with n number of instances.

The details of the dataset used to validate the proposed data imputation method are presented in Table 1.

The performance of the proposed method to deal with missing values and class imbalance is evaluated on four healthcare datasets. All the datasets have been collected from the University of California, Irvine (UCI) machine learning dataset repository and contain missing values. The details of the dataset are presented in Table 2. The number of search agents and the maximum number of iterations are set equal to 30 and 500 for the IEHO algorithm, respectively. In this experiment, the stratified tenfold cross-validation scheme has been utilized to divide the imputed dataset into training and testing sets. The stratified cross-validation scheme maintains the original class distribution in each fold, and it helps in building a more generalized classification model. The performance of the proposed method is evaluated on four widely used performance metrics: accuracy, balanced accuracy, g-mean, and f-score. Accuracy represents the ratio of correctly classified samples to the total number of samples. However, it stays biased toward the majority class. Therefore, balanced accuracy, g-mean, and f-score metrics have been considered for the imbalanced data classification. These metrics are the balanced measures for both classes. Balanced accuracy provides information about that how precise the classification model is to recognize each class. G-mean represents the square root of the sensitivity multiplied by specificity. F-score is the weighted harmonic mean of precision and recall, representing the classifier's capability to predict positive class accurately.

Table 1 Description of healthcare datasets

Sr. no.	Dataset name	Samples	Features	Class	Dataset repository
1	Indian Liver Patient Dataset (ILPD)	583	10	2	UCI[a]
2	Brainstroke	4981	11	2	Kaggle[b]

[a] https://archive.ics.uci.edu/
[b] https://www.kaggle.com

Table 2 Description of healthcare datasets used in this work

Sr. no.	Dataset name	Samples	Features	Class	Missing percentage (%)
1	Pima diabetes	768	6	2	56.25
2	Wisconsin Breast Cancer (Original)	699	9	2	2.29
3	Hepatitis	155	19	2	48.39
4	HCC Survival	165	49	2	95.15

4 Results and Discussion

This section presents a result analysis of the proposed and existing methods on healthcare data.

4.1 Comparative Analysis of the Proposed and Existing Imputation Methods

In this section, the results of the proposed class mean and standard deviation-based imputation strategy are evaluated on two healthcare datasets (described in Table 1) and compared with four statistical imputation methods: KNNI, Mean, Median, and Frequent One. Figure 2 presents the mean results of RMSE scores for all the imputation methods.

It is clear from the results of the RMSE score of data imputation methods presented in Fig. 2a and b for ILPD and Brainstroke datasets, respectively that the proposed imputation method has provided promising and comparative results for different missing rates (10, 15, 25, and 35%) on both healthcare datasets with existing statistical imputation techniques. This demonstrates the validity of the proposed imputation method for estimating the missing values in healthcare datasets.

4.2 Comparative Analysis of the Proposed Method with State-of-the-Art Methods to Deal with Missing Values and Class Imbalance Problem

This section presents the results of the proposed method to deal with missing values and class imbalance problems in healthcare datasets. In this analysis, the performance of the proposed method is evaluated on four healthcare imbalanced datasets with missing values based on the four-performance metrics: accuracy, balanced accuracy, g-mean, and f-score. The performance of the proposed method is compared with Anter and Ali [26], Hegazy et al. [27], El-Shair et al. [28], and Christo et al. [29].

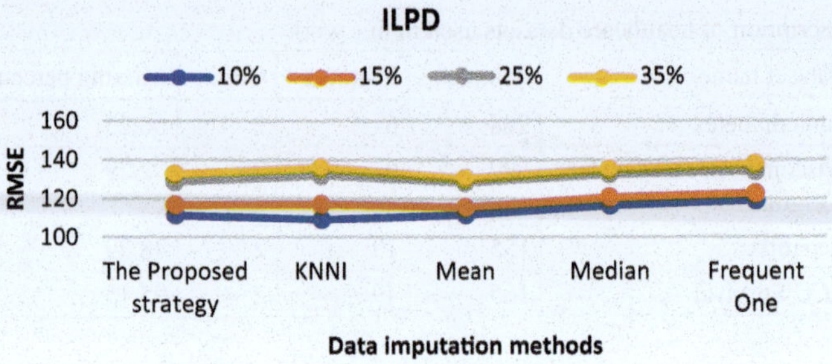

(a)

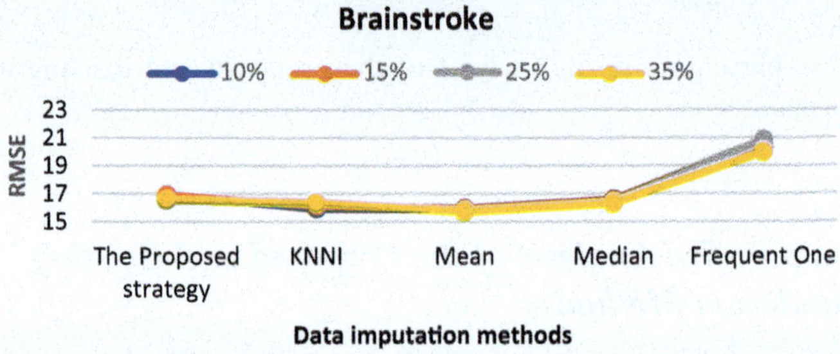

(b)

Fig. 2 Represents the mean RMSE scores of data imputation methods for 10, 15, 25, and 35 missing percentage

The results of all methods are presented in Table 3. The values in boldface represent the best results.

It is clear from Table 3 that the proposed method has obtained a higher classification accuracy than other state-of-the-art algorithms for Wisconsin Breast Cancer (Original), Hepatitis, and HCC Survival datasets. Also, the proposed method obtained the comparative classification results for the Pima diabetes dataset. The proposed method has shown high potential in providing promising results for the classification of imbalanced healthcare datasets with missing values than existing state-of-the-art algorithms.

Table 3 Results of the proposed and state-of-the-art methods for imbalanced healthcare datasets with missing values

Dataset name	Metrics	Anter and Ali [26]	Hegazy et al. [27]	El-Shair et al. [28]	Christo et al. [29]	The Proposed method
Pima diabetes	Accuracy	0.755	–	–	**0.8101**	0.7971
	Balanced accuracy	–	–	–	–	0.7748
	G-mean	–	–	–	–	0.7689
	F-score	–	–	–	–	0.7051
Wisconsin Breast Cancer (Original)	Accuracy	–	0.9570	0.9377	–	**0.9771**
	Balanced accuracy	–	–	–	–	0.9784
	G-mean	–	–	–	–	0.9782
	F-score	–	–	0.90	–	**0.9674**
Hepatitis	Accuracy	0.680	0.9123	–	0.8230	**0.9283**
	Balanced accuracy	–	–	–	–	0.9119
	G-mean	–	–	–	–	0.9042
	F-score	–	–	–	–	0.8345
HCC Survival	Accuracy	–	–	–	0.7220	**0.8088**
	Balanced accuracy	–	–	–	–	0.8034
	G-mean	–	–	–	–	0.7928
	F-score	–	–	–	–	0.8415

5 Conclusion

This work proposes an efficient method to deal with the problems of missing values and class imbalance in healthcare datasets. The proposed method utilizes a class mean and standard deviation-based strategy to preprocess the healthcare data by imputing the missing values in class-specific regions. Subsequently, an Improved Elephant Herding Optimization (IEHO) algorithm is applied to the preprocessed healthcare data with complete records to provide optimal weights for reducing the bias toward the majority class during the classification. A novel fitness function is introduced which scrutinizes the data incompleteness and class imbalance while searching for the optimal weights. The k parameter of the k-Nearest Neighbor (k-NN) machine learning algorithm is optimized along with feature weighting to improve the classification performance. The performance of the proposed imputation method is evaluated on two healthcare datasets for four different missing percentages (10, 15, 25, and 35%) and compared with existing data imputation techniques based on the RMSE score. Results have shown the efficiency of the proposed imputation method than existing techniques. Afterward, four imbalanced healthcare datasets with missing values are utilized to evaluate the performance of the proposed method to deal with data incompleteness and class imbalance problems. A comparative analysis between the proposed and state-of-the-art methods is made based on four-performance metrics: accuracy, balanced accuracy, g-mean, and f-score. The proposed method obtained higher classification results than state-of-the-art methods and has shown great potential in dealing with missing values and class imbalance problems in healthcare data. In the future, investigating the impact of data imputation, feature selection, and weighting methods to deal with the problems of class imbalance and missing values in healthcare data will be the region of interest.

References

1. Stylianou A, Talias MA (2017) Big data in healthcare: a discussion on the big challenges. Health Technol (Berl) 7:97–107. https://doi.org/10.1007/s12553-016-0152-4
2. Martin-Sanchez F, Verspoor K (2014) Big data in medicine is driving big changes. Yearb Med Inform 9:14–20. https://doi.org/10.15265/IY-2014-0020
3. Philip Chen CL, Zhang CY (2014) Data-intensive applications, challenges, techniques and technologies: a survey on big data. Inf Sci (Ny) 275:314–347. https://doi.org/10.1016/j.ins.2014.01.015
4. Lee C, Luo Z, Ngiam KY et al (2017) Big healthcare data analytics: challenges and applications. In: Khan SU, Zomaya AY, Abbas A (eds) Handbook of large-scale distributed computing in smart healthcare. Springer, Springer International Publishing, Cham, pp 11–42
5. Qin Y, Zhang S, Zhu X et al (2009) POP algorithm: Kernel-based imputation to treat missing values in knowledge discovery from databases. Expert Syst Appl 36:2794–2804. https://doi.org/10.1016/j.eswa.2008.01.059
6. Leevy JL, Khoshgoftaar TM, Bauder RA, Seliya N (2018) A survey on addressing high-class imbalance in big data. J Big Data 5:1–30. https://doi.org/10.1186/s40537-018-0151-6
7. Haixiang G, Yijing L, Shang J et al (2017) Learning from class-imbalanced data: review of methods and applications. Expert Syst Appl 73:220–239. https://doi.org/10.1016/j.eswa.2016.12.035
8. Behera MP, Sarangi A, Mishra D (2021) K-medoids crazy firefly algorithm for unsupervised data clustering. In: 2021 1st Odisha international conference on electrical power engineering, communication and computing technology (ODICON). IEEE, pp 1–6
9. Fotouhi S, Asadi S, Kattan MW (2019) A comprehensive data level analysis for cancer diagnosis on imbalanced data. J Biomed Inform 90:103089. https://doi.org/10.1016/j.jbi.2018.12.003
10. Batista GEAPA, Prati RC, Monard MC (2004) A study of the behavior of several methods for balancing machine learning training data. ACM SIGKDD Explor Newsl 6:20–29. https://doi.org/10.1145/1007730.1007735
11. Chawla NV, Bowyer KW, Hall LO, Kegelmeyer WP (2002) SMOTE: synthetic minority over-sampling technique. J Artif Intell Res 16:321–357. https://doi.org/10.1613/jair.953
12. Han H, Wang W-Y, Mao B-H (2005) Borderline-SMOTE: a new over-sampling method in imbalanced data sets learning. In: Advances in intelligent computing: international conference on intelligent computing, ICIC 2005, Hefei, China, pp 878–887
13. He H, Bai Y, Garcia EA, Li S (2008) ADASYN: adaptive synthetic sampling approach for imbalanced learning. In: 2008 IEEE international joint conference on neural networks (IEEE world congress on computational intelligence). IEEE, pp 1322–1328
14. Bunkhumpornpat C, Sinapiromsaran K, Lursinsap C (2009) Safe-level-SMOTE: safe-level-synthetic minority over-sampling technique for handling the class imbalanced problem. In:

Lecture notes in computer science (including subseries Lecture Notes in Artificial Intelligence and Lecture Notes in Bioinformatics). pp 475–482
15. Douzas G, Bacao F, Last F (2018) Improving imbalanced learning through a heuristic oversampling method based on k-means and SMOTE. Inf Sci (Ny) 465:1–20. https://doi.org/10.1016/j.ins.2018.06.056
16. Douzas G, Bacao F (2017) Self-Organizing Map Oversampling (SOMO) for imbalanced data set learning. Expert Syst Appl 82:40–52. https://doi.org/10.1016/j.eswa.2017.03.073
17. Li J, Fong S, Mohammed S, Fiaidhi J (2016) Improving the classification performance of biological imbalanced datasets by swarm optimization algorithms. J Supercomput 72:3708–3728. https://doi.org/10.1007/s11227-015-1541-6
18. Jiang F, Zhu Q, Tian T (2022) Breast cancer detection based on modified Harris hawks optimization and extreme learning machine embedded with feature weighting. Neural Process Lett 55:3631–3654. https://doi.org/10.1007/s11063-021-10700-w
19. Dalwinder S, Birmohan S, Manpreet K (2020) Simultaneous feature weighting and parameter determination of Neural Networks using Ant Lion Optimization for the classification of breast cancer. Biocybern Biomed Eng 40:337–351. https://doi.org/10.1016/j.bbe.2019.12.004
20. Al-Zoubi AM, Hassonah MA, Heidari AA et al (2021) Evolutionary competitive swarm exploring optimal support vector machines and feature weighting. Soft Comput 25:3335–3352. https://doi.org/10.1007/s00500-020-05439-w
21. Wettschereck D, Aha DW, Mohri T (1997) A review and empirical evaluation of feature weighting methods for a class of lazy learning algorithms. Artif Intell Rev 11:273–314. https://doi.org/10.1007/978-94-017-2053-3_11
22. Singh H, Singh B, Kaur M (2021) An improved elephant herding optimization for global optimization problems. Eng Comput 1–33. https://doi.org/10.1007/s00366-021-01471-y
23. Singh H, Singh B, Kaur M (2022) An efficient feature selection method based on improved elephant herding optimization to classify high-dimensional biomedical data. Expert Syst 39:e13038. https://doi.org/10.1111/exsy.13038
24. Singh D, Singh B (2020) Investigating the impact of data normalization on classification performance. Appl Soft Comput 97:105524. https://doi.org/10.1016/j.asoc.2019.105524
25. Tsai CF, Chen YC (2019) The optimal combination of feature selection and data discretization: an empirical study. Inf Sci (Ny) 505:282–293. https://doi.org/10.1016/j.ins.2019.07.091
26. Anter AM, Ali M (2020) Feature selection strategy based on hybrid crow search optimization algorithm integrated with chaos theory and fuzzy c-means algorithm for medical diagnosis problems. Soft Comput 24:1565–1584. https://doi.org/10.1007/s00500-019-03988-3
27. Hegazy AE, Makhlouf MA, El-Tawel GS (2020) Improved salp swarm algorithm for feature selection. J King Saud Univ - Comput Inf Sci 32:335–344. https://doi.org/10.1016/j.jksuci.2018.06.003
28. El-Shair ZA, Sanchez-Perez LA, Rawashdeh SA (2020) Comparative study of machine learning algorithms using a breast cancer dataset. In: 2020 IEEE international conference on electro information technology (EIT). IEEE, pp 500–508
29. Christo VRE, Nehemiah HK, Brighty J, Kannan A (2022) Feature selection and instance selection from clinical datasets using co-operative co-evolution and classification using random forest. IETE J Res 68:2508–2521. https://doi.org/10.1080/03772063.2020.1713917

Improving Machine Learning Algorithm's Accuracy for Detecting Code Smell Using Data Balancing and Parameter Optimization

Abhishilpa Nandini, Randeep Singh, and Amit Rathee

Abstract Detecting code smells is vital for improving software quality, improving its maintainability, and reducing software failures and faults. Code smell detection using machine learning-based methods has become more prevalent in recent years, as they easily overcome the limitations of heuristic-based methods that are based mostly on rules that must be manually created. Data imbalance and parameter optimization are found to be directly affecting performance of underlying machine learning models. However, to the best of our knowledge about literature of the field, little experimental evaluation is carried out in the direction of impact of data balancing and parameter tuning on machine learning models. Therefore, this paper considers it important to propose a new data balancing and parameter optimization technique. Further, the proposed approach is experimentally validated by considering standard code smell datasets. The empirical evaluation considers four code smells taken from two categories (two at method level and rest of the two are considered at class level) and studied four machine learning models, viz., Decision Tree, Random Forest, kNN, and SVM. The obtained experimental results justify the suitability of proposed data balancing and parameter optimization techniques in improving performance of machine learning models.

Keywords Code smell detection · Classifiers · Parameter tuning · Data imbalance · Machine learning models

A. Nandini (✉) · R. Singh
Department of Computer Science and Engineering, IEC University, Solan, India
e-mail: abhishilpa05@gmail.com

A. Rathee
Department of Computer Science, GC Barota, Sonipat, Haryana, India

P. J. S. Gonçalves et al. (eds.), *Proceedings of Fifth International Conference on Computing, Communications, and Cyber-Security*, Lecture Notes in Networks and Systems 1128, https://doi.org/10.1007/978-981-97-7371-8_12

1 Introduction

During software development, negligence for the non-functional requirements, viz., testability, evolution, reusability, maintainability, etc., causes serious quality issues in terms of enhanced complexity, increased maintenance cost, and weakness in the evolution of the underlying software [1]. These design flaws are termed in literature as code/bad smells and mitigating them can significantly improve software quality [2]. In literature, researchers developed heuristic-based design rules based on code smell definitions in order to identify the various smells present in code and mitigate them using various refactoring techniques [3–6]. They conclude that classes found infected by code smells undergo more frequent changes and require more maintenance effort. The difficulty of having validated and unique definitions for code smells is one reason and challenge that hinders the application of these heuristic-based techniques. In addition, different heuristic-based tools aren't able to find all code smells, as they are confined to detecting only a few specific code smells.

Machine learning-based techniques are one of the methods to overcome the shortcomings caused by heuristic-based methods [7–11]. We observe that Machine Learning (ML) approaches are inaccurate unless they are trained on high-quality datasets. Several other factors that are correlated with the efficiency of ML models are as follows:

1. The manner to which the code smell dataset contains balanced instances and how various metrics are distributed.
2. Modeling classifiers with hyper-parameter (ML algorithm's parameter tuning) optimization.
3. The manner in which different software features are selected and used for training and testing a ML model.

To the best of our knowledge, parameter optimization is the most ignored aspect that may improve efficiency of a ML model. The authors in [12] applied the Bayesian Optimization method of parameter tuning for only decision tree and random forest ML models. However, the Grid Search strategy for optimizing parameter tuning used by us in the paper is found to outperform Bayesian Optimization in terms of complexity and performance measure (AUC) [13]. Further, applying data balancing techniques during ML models helps in achieving more accuracy [8]. Therefore, this paper considers applying both data balancing and parameter tuning techniques together to obtain a more accurate ML model for code smell detection. To our best knowledge, no existing work targets carrying out empirical investigation of data balancing and parameter optimization on ML model accuracy. The rest of this paper gives details about related works, proposed solution approach, experimental setup, and results analysis, followed by conclusion and future work section.

2 Literature Survey

One of the most widely used methods in code smell detection is heuristics or rule-based analysis. The authors in [6] proposed a multi-stage approach to handle code smell by analyzing different quality metrics of software systems along with refactoring opportunities. The proposed approach detects all candidate solutions for code smells and subsequently removes all false positive solutions by analyzing different refactoring opportunities. The main limitation is the use of very trivial quality metrics. The authors in [5] provided a semi-automatic IntelliJ Idea tool support for code smell and refactoring activity. The main limitation is the limited support for code smell varieties (only four). Further, several ML-based models were also proposed for code smell mitigation. The authors in [8, 14] carried out extensive empirical study to identify the role of various data-balancing techniques in detecting code smells using ML-based models. They concluded that existing ML approaches based on data balancing are not effective at detecting code smells because existing techniques are not adequate. The authors in [7] studied whether optimizing hyper-parameters of six ML algorithms for detecting feature envy and data class code smells had a positive impact on performance of the considered ML model. According to their findings, hyper-parameter optimization can greatly enhance the detection of code smells. The authors in [15] studied the impact of chi-square feature selection technique on the performance of five considered ML-based classifier models. The authors in [16] carried out evaluation on 31 different types of imbalanced strategies using 7 ML classifiers. They concluded that deep forest generally excel other ML techniques and they observed mixed responses for various data imbalanced strategies.

Based on the study of existing literature, we identified the following research gaps that are relevant to justify our carried out research work in this paper:

1. No existing work targets applying data balancing and parameter optimization together and study their impact on code smell detection using various ML models/classifiers.
2. Non-standard or commercial software systems are mostly used for empirical validation in literature that hinders reproducibility and/or validation of the proposed approach.

3 Proposed Methodology

Figure 1 depicts the structure of the proposed methodology that aims at improving prediction accuracy of the ML algorithms for detecting code smells using data balancing and ML algorithm's parameter optimization.

The proposed approach works in two phases. In the first phase, we build code smell datasets from different software systems that are part of the well-known Qualitas Corpus. In the second phase, we first applied different data balancing techniques to the code smell datasets that were obtained in the first phase. Then, we build different

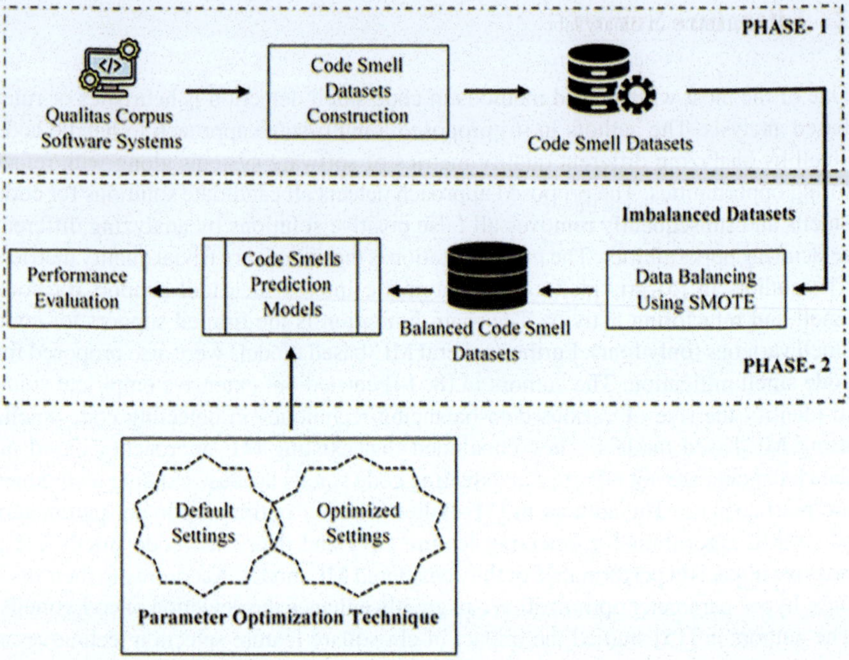

Fig. 1 Proposed data balancing and parameter optimization approach for improving machine learning algorithm's performance

ML prediction models for detecting code smells. At last, we applied various models' performance evaluation measures to study the effect of the proposed methodology. Each of these phases is further elaborated in subsequent subsections of this paper.

3.1 Code Smell Datasets Construction

The quality of the considered datasets during experimentation and evaluation plays an important role. Therefore, this paper considers already tested and verified Qualitas Corpus (QC) as proposed in [17, 18] to build our datasets. The QC datasets consist of 51,826 classes and 3420 modules. The 74 Java open-source software are considered from this QC and 61 class-level and 82 method-level object-oriented software metrics are collected by parsing source-code of these software systems using well-known structural analysis tools, viz., PMD, iPlasma, DFMC4J, and Structure101. The computed metrics measure various dimensions of the software, viz., size, encapsulation, cohesion, inheritance, coupling, and complexity. These metrics are used to detect two class-level code smells, viz., God Class (GC) and Data Class (DC), and two method-level code smells, viz., Feature Envy (FE) and Long Method (LM). The

Table 1 Summary of binary labeled obtained datasets

Dataset ID	Code smell type	# Smelly instances	# Non-smelly instances
REF_DS_1	God Class	4,065 (8%)	47,761 (92%)
REF_DS_2	Long Method	10,237 (20%)	41,589 (80%)
REF_DS_3	Data Class	1,897 (4%)	49,929 (96%)
REF_DS_4	Feature Envy	7,231 (14%)	44,595 (86%)

reason behind considering these code smells is that these are among the most critical code bad symptoms that highly affect quality and maintainability efforts [19]. Further, all the classes are labeled into "Smelly" and "Non-Smelly" ones depending upon whether any of a class is found affected with considered four code smells. This labeling is done by applying heuristic-based code smell tools, viz., Fluid Tool, iPlasma, and Anti-pattern Scanner. During this labeling a class that is found affected with a given type of code smell is counted under smelly instance and rest of the cases are counted under non-smelly instance. This task is performed by 3 M.Tech. scholars and the smelly and non-smelly number of classes in the obtained datasets are summarized in Table 1. From this dataset, it is clearly observed that the obtained datasets are significantly imbalanced in nature as the ratio of smelly and non-smelly number of classes in each smell category is very huge and such a dataset can negatively affect ML model during training and prediction phase.

3.2 Data Balancing, Parameter Tuning, and Model Evaluation

The imbalance in class types (smelly and non-smelly instances in a dataset) in classification models occurs when one category of a class has fewer examples than another category. Classification models cannot predict minority modules effectively due to a class imbalance problem. In literature, several data balancing techniques exist, viz., Oversampling (algorithm randomly adds minority class instances to the dataset), Undersampling (in this technique we randomly remove majority class instances), SMOTE (algorithm systematically adds minority class instances based upon the nearest neighbor concept), Cost-Sensitive Classifier (algorithm penalize every false negative detection), and One-Class Classifier (algorithm that considers only minority class instances for learning) [8]. Based on the results of large-scale empirical analysis conducted by Pecorelli et al. [8], this paper adopts SMOTE for dataset balancing. The SMOTE technique uses the concept of k nearest neighbors of the given minority class instance and based on these neighbors a new minority class instance is created to reduce class imbalance problem. For the original minority instance (x), and randomly selected one of its k nearest neighbors (x^k), a new instance of the minority class ($x\prime$) is computed using the following formula:

$$x\prime = x - \delta * \left|x - x^k\right|;\ \text{where}\ \delta \in (0, 1) \quad (1)$$

3.3 *Data Balancing, Parameter Tuning, and Model Evaluation*

ML algorithms are among the most suitable techniques with the help of which a computer is capable of making predictions, enhancing predictive accuracy, and forecasting behavior patterns using data sets [20]. Therefore, this paper considers applying and evaluating four different supervised ML algorithms namely Support Vector Machine (SVM), Decision Tree (DT), Random Forest (RF), and k-Nearest Neighbor (kNN). Table 2 provides a brief description about these algorithms and their corresponding parameters that are tuned in this paper. This table specifies a particular version of the ML algorithm implemented during experimentation along with a short description of the nature of the corresponding algorithm. Next tuned parameters are specified for each of the considered ML algorithms along with specifying starting value, end value, and stepping used during experimental evaluation. In this paper, all the considered ML algorithms are implemented using scikit-learn Python library.

3.4 *Parameter Optimization Technique*

The accuracy, classification performance, and learning process of each ML algorithm are influenced by a set of parameters. Moreover, training data determines which parameters are the best for each algorithm. Further, different ML-based classification models can accurately predict the testing dataset by testing each combination of parameters for the underlying ML algorithm. Therefore, this paper considers applying an exhaustive search strategy called Grid Search [21] for optimizing different parameter values from parameter sets as depicted in Table 2. Applying grid search is among the easiest and most intuitive methods. Mathematically, let $v = (v_1, v_2, \ldots, v_n)$ is the set of parameters that we want to optimize. The simplest way to do a grid search is to define lower bounds $x = (x_1, x_2, \ldots, x_n)$ and upper bounds $y = (y_1, y_2, \ldots, y_n)$ for each element of the set v. In grid search, we created m equally spaced points in the interval of $[x_i, y_i]$. Thus, in grid space, we generally have m^n possible grid points for analysis. At the end of the calculation process, we select the maximum value from each pair of m^n possible points. Figure 2 diagrammatically depicts the steps of the proposed optimization process using Grid Search strategy.

Table 2 Statistics about different ML algorithms and their tuning parameters

S. No	ML algorithm name	Brief description	Parameter name	Lower bound	Upper bound	Stepping
1	Random Forest (criterion = "Gini Impurity")	Creates decision tree forest and performs classification based on maximum voting	num.trees (number of generated trees)	1	2000	100
			mtry	0	1	0.10
			n_estimators (number of trees needed to perform classification)	1	100	10
2	kNN (distance = "Euclidean")	Uses feature similarity mechanism to classify unknown instance to known clusters	k	1	30	3
3	SVM (kernel = "rbf")	Uses kernel tricks to perform transformations such as to identify optimal hyperplane that separates possible outputs	cost (a penalty parameter)	-10	10	2
			gamma (kernel coefficient)	1	100	10
4	Decision Tree (criterion = "Gini Index")	A non-parametric hierarchical tree model that performs classification based on features and decision rules represented in the form of internal nodes and edges respectively	max_depth (maximum depth allowed in the generated decision tree)	1	36	3
			max_features	1	60	3
			min_split (least features needed for dividing internal node of tree)	1	60	5

3.5 Performance Evaluation

For determining the efficiency of different trained ML models, to test proposed data balancing and parameter optimization technique, we considered five state-of-the-art performance measures, namely, precision (percentage measure of how many code smell instances are correctly detected), recall (percentage of instances of code smells

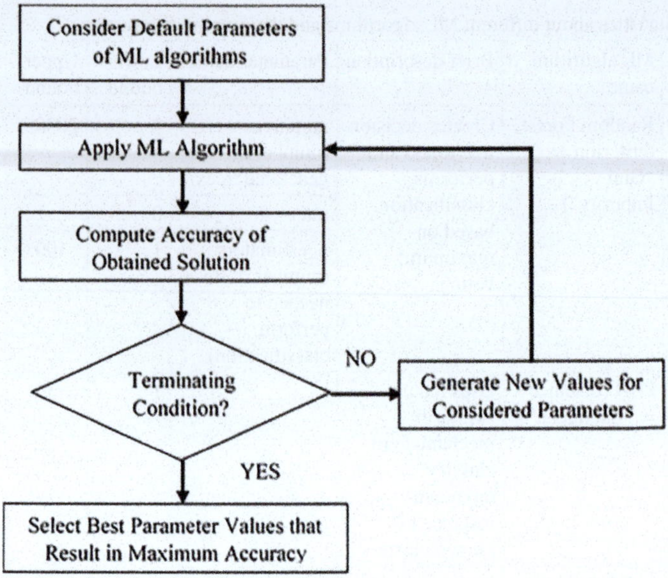

Fig. 2 Working process of parameter optimization using grid search strategy

that are actually detected), f-measures (denotes harmonic mean of precision and recall measures), accuracy (denotes percentage of correctly identified code smells), and Area Under the ROC Curve (AUC) (measures performance as a trade-offs among correct predictions and false predictions) [2, 22]. Computation of these measures depends upon four variables, viz., True Positive (TP), True Negative (TN), False Positive (FP), and False Negative (FN) that are measurable from the confusion matrix. Finally, we made a conclusion based on the obtained experimental results along with different values of used performance evaluation metrics.

4 Discussion and Analysis

The purpose of this study is to perform a series of experiments and assess each experiment's performance using a validation scenario. In the first step of our training process of a ML model, we applied tenfold cross-validation in which the given dataset is partitioned into 10 parts and repeated validation 10 times by considering different training data in each repetition. The trained models are then tested on a dataset that has not been seen before. Finally, the following Research Questions (RQs) are answered based on the obtained experimental results.

RQ1. Are there any implications of carrying out data balancing on the accuracy for code smell detection?

RQ2. Is there a ML algorithm that is most effective at detecting code smells?

RQ3. What is the impact of parameter optimization on the performance of various code smell ML models?

To answer RQ1, we applied different ML models on original and balanced versions of the datasets. Table 3 and Table 4 depict experimental results obtained using different considered performance measures. Table 4 also depicts percentage improvement in corresponding values with respect to original values. Here, the depicted values are represented as the average of various experimental values obtained for different considered datasets. From the depicted results it can be clearly observed that the balancing step significantly improves performance of every ML model.

To answer RQ2, we studied performance of different ML models for identifying individual code smells. Figure 3 displays box-plots of three performance measures, viz., accuracy, f-measure, and AUC against four considered ML models for detecting different code smell. From the plots, it is clear that the decision tree classifier outperforms in terms of accuracy and AUC parameters. The SVM classifier is at second position in terms of higher f-measure score (higher precision and recall score) and less scattered AUC values. The KNN classifier is kept at third position and DT classifier is found to be least performing during experimentation.

To answer RQ3, we applied Grid Search to perform parameter optimization and carried out experimental evaluation on the obtained balanced datasets. Table 5 depicts experimental average AUC and Average performance criteria score for optimized (as

Table 3 Obtained experimental results for original datasets

Original dataset						
S.No	ML classifier model name	Performance measures				
		Accuracy	F-measure	Precision	Recall	AUC
1	Random Forest	0.866	0.821	0.762	0.889	0.866
2	KNN	0.864	0.790	0.743	0.844	0.864
3	SVM	0.832	0.774	0.737	0.815	0.832
4	Decision Tree	0.820	0.766	0.715	0.824	0.704

Table 4 Obtained experimental results for balanced datasets

Balanced dataset						
S. No	ML classifier model name	Performance measures				
		Accuracy	F-measure	Precision	Recall	AUC
1	Random Forest	0.991 (14%)	0.945 (15%)	0.992 (30%)	0.903 (2%)	0.952 (10%)
2	KNN	0.993 (15%)	0.927 (17%)	0.981 (32%)	0.880 (4%)	0.982 (14%)
3	SVM	0.989 (19%)	0.928 (20%)	0.991 (34%)	0.873 (7%)	0.960 (15%)
4	Decision Tree	0.981 (20%)	0.920 (20%)	0.961 (34%)	0.883 (7%)	0.929 (30%)

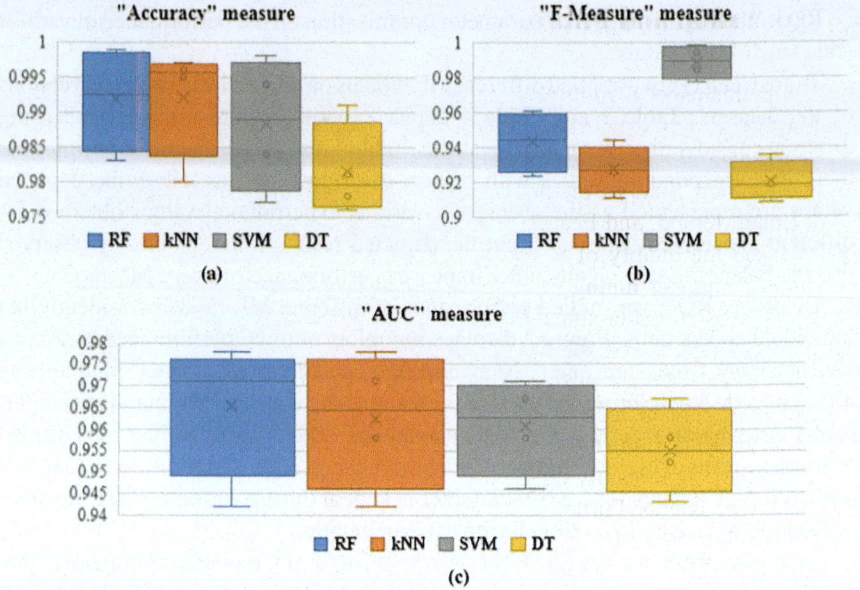

Fig. 3 Performance measures comparison for different ML models

depicted in Table 2) and default parameters (considered default values for PYTHON-based implementation of ML algorithms) of ML models. The table depicts average score for AUC and Accuracy measures for different ML models/classifiers. The presented values depict the performance of different classifiers with default parameters and Grid Search-based optimized set of parameter values of corresponding ML algorithm. From the presented results, it is clear that the Grid Search-based optimization method outperforms and hence is necessary to improve efficiency of the ML model. The percentage improvement is also shown in Table 5.

Table 5 Average AUC and average measure score for optimized and default parameters ML models

Classifier	Average "AUC" score for considered datasets		Average "Accuracy" score for considered datasets	
	Grid search	Default	Grid search	Default
Random forest	**0.952 (6%)**	0.896	**0.991 (12%)**	0.882
KNN	**0.982 (7%)**	0.920	**0.993 (7%)**	0.930
SVM	**0.960 (12%)**	0.859	**0.989 (10%)**	0.899
Decision tree	0.929 (1%)	0.921	**0.981 (14%)**	0.861

5 Conclusion and Future Work

Our aim in this paper is to investigate the performance of four ML classifiers, viz., Random Forest, KNN, SVM, and Decision Tree for code smell detection based on employing data imbalanced learning technique along with ML algorithm's parameter optimization. State-of-the-art datasets of four code smells, viz., Data Class, God Class, Long Method, and Feature Envy were used during empirical evaluation and we enhanced the quality of these datasets by employing SMOTE data balancing technique. Parameter tuning technique is also proposed in this paper and is used to enhance prediction/classification accuracy of different ML models. The effect of parameter tuning and considering default values of parameters is also investigated. Further, five performance measures are used to evaluate performance of different ML models, viz., precision, recall, f-measure, accuracy, and AUC. According to our findings, imbalanced learning techniques have different effects on different kinds of code smells and they significantly improve the performance of each and every ML model but to different extents. Further, the decision tree classifier is found to outperform in terms of accuracy and AUC evaluation measures. Finally, the Grid Search parameter tuning strategy is found to have a positive impact on the performance of considered ML models/classifiers. In the future, different data resampling techniques should be considered for preprocessing different code smell datasets. In order to verify the generalization of our findings, in future, we plan to conduct large-scale empirical analysis on commercial systems also. Moreover, we are interested in finding out if parameter tuning can be used for cross-project code smells. ML algorithms, such as neural networks and deep learning, will also be used to enhance the accuracy of code smell prediction models and new data balancing techniques and feature selection strategies will be tested in the future.

References

1. Alibrahim H, Ludwig SA (2021) Hyperparameter optimization: Comparing genetic algorithm against grid search and bayesian optimization. In: 2021 IEEE Congress on Evolutionary Computation (CEC), pp 1551–1559
2. Alkharabsheh K, Alawadi S, Kebande VR, Crespo Y, Fernández-Delgado M, Taboada JA (2022) A comparison of machine learning algorithms on design smell detection using balanced and imbalanced dataset: A study of God class. Inf Softw Technol 143:106736
3. Arcelli Fontana F, Mäntylä MV, Zanoni M, Marino A (2016) Comparing and experimenting machine learning techniques for code smell detection. Empir Softw Eng 21(3):1143–1191
4. Bafandeh Mayvan B, Rasoolzadegan A, Javan Jafari A (2020) Bad smell detection using quality metrics and refactoring opportunities. J Softw: Evol Process 32(8):e2255
5. Bergstra J, Bengio Y (2012) Random search for hyper-parameter optimization. J Mach Learn Res 13(2)
6. Chung L, Nixon BA, Yu E, Mylopoulos J (2012) Non-functional requirements in software engineering, vol. 5. Springer Science & Business Media
7. De Stefano M, Gambardella MS, Pecorelli F, Palomba F, De Lucia A (2020) cASpER: A plug-in for automated code smell detection and refactoring. In: Proceedings of the international conference on advanced visual interfaces

8. Dewangan S, Rao RS, Mishra A, Gupta M (2021) A novel approach for code smell detection: an empirical study. IEEE Access 9:162869–162883
9. Dewangan S, Rao RS, Mishra A, Gupta M (2022) Code smell detection using ensemble machine learning algorithms. Appl Sci 12(20):10321
10. Jain S, Saha A (2021) Improving performance with hybrid feature selection and ensemble machine learning techniques for code smell detection. Sci Comput Program 212:102713
11. Jain S, Saha A (2022) Rank-based univariate feature selection methods on machine learning classifiers for code smell detection. Evol Intell 15(1):609–638
12. Khleel NAA, Nehéz K (2023) Detection of code smells using machine learning techniques combined with data-balancing methods. Int J Adv Intell Informat-Ics 9(3):402–417
13. Li F, Zou K, Keung JW, Yu X, Feng S, Xiao Y (2023) On the relative value of imbalanced learning for code smell detection. Softw: Pract Exp 53(10):1902–1927
14. Mhawish MY, Gupta M (2020) Predicting code smells and analysis of predictions: Using machine learning techniques and software metrics. J Comput Sci Technol 35:1428–1445
15. Nandini A, Singh R, Rathee A (2024) Code smells and refactoring: a tertiary systematic literature review. Int J Syst Syst Eng 14(1):83–143
16. Pecorelli F, Di Nucci D, De Roover C, De Lucia A (2020) A large empirical assessment of the role of data balancing in machine-learning-based code smell detection. J Syst Softw 169:110693
17. Singh R, Bindal AK, Kumar A (2022) Improving software design by mitigating code smells. Int J Softw Innov (IJSI) 10(1):1–21
18. Singh R, Bindal A, Kumar A (2020) A framework to improve quality of a Java system by performing refactoring. Int J Syst Syst Eng 10(4), pp 324–336
19. Shen L, Liu W, Chen X, Gu Q, Liu X (2020) Improving machine learning-based code smell detection via hyper-parameter optimization. In: 2020 27th Asia-pacific software engineering conference (APSEC)
20. Sukkasem P, Soomlek C (2023) Enhanced machine learning-based code smell detection through hyper-parameter optimization. In: 2023 20th international joint conference on computer science and software engineering (JCSSE)
21. Singh LK, Garg H, Khanna M, Bhadoria RS (2021) An analytical study on machine learning techniques. In: Multidisciplinary functions of Blockchain technology in AI and IoT applications, pp 137–157
22. Tempero E, Anslow C, Dietrich J, Han T, Li J, Lumpe M, Noble J (2010) The qualitas corpus: A curated collection of Java code for empirical studies. In: 2010 Asia pacific software engineering conference

Enhanced Multimodal Conversational AI Using Speech and Image Integration

Deepali Dixit, Mihul Tyagi, Raunak Kandoi, and Raghuraj Singh

Abstract The development of conversational artificial intelligence (AI) is examined in this research paper, with a focus on how speech and image recognition technologies can be combined to transform and interact with systems. The emphasis is on cutting-edge developments in multimodal conversational AI, particularly the smooth fusion of sophisticated image and speech processing methods to improve user experiences. The study addresses the applications, difficulties and potential applications of multimodal conversational AI by conducting a thorough review of the existing literature and case studies. By emphasizing the significance of enhancing voice recognition algorithms and utilizing advanced image processing techniques, the study promotes contextual understanding and customization. We have developed an innovative multimodal conversational AI system that integrates speech, text and image processing capabilities for seamless human–computer interactions. This study presents an improved multimodal conversational AI system that integrates several techniques, such as Google Text-to-Speech (gtts), speech recognition, sound playback and Convolutional Neural Networks (CNN). Our model architecture combines RNNs for sequential data analysis, CNNs for image processing and attention mechanisms for context extraction. Trained on diverse datasets, our model is optimized for accuracy and robustness. Evaluation against existing models demonstrates superior performance in terms of accuracy, speed and resource efficiency. Our study highlights how improved multimodal conversational AI can advance the field of human–computer interaction.

Keywords Multimodal conversational AI · Speech and image integration · User experience enhancement

---

D. Dixit · M. Tyagi · R. Kandoi · R. Singh (✉)
School of Engineering and Technology, Sharda University, Greater Noida 201310, India
e-mail: raghuraj.yadav@sharda.ac.in

P. J. S. Gonçalves et al. (eds.), *Proceedings of Fifth International Conference on Computing, Communications, and Cyber-Security*, Lecture Notes in Networks and Systems 1128, https://doi.org/10.1007/978-981-97-7371-8_13

1 Introduction

The discipline of Multimodal Conversational AI is a fascinating area of study that integrates several sensory modalities to improve interaction and communication. A kind of artificial intelligence known as "conversational AI" is powered by text- and voice-based agents that mimic human discourse. The fast-developing field of conversational AI has the potential to completely change how humans interact with technology. The main application of conversational AI has been in sophisticated chatbots. Conventional chatbots are only equipped to handle a restricted range of functions. They answer a predetermined set of questions and are presented on business websites and mobile applications [1]. The introduction provides a thorough examination of conversational AI and its evolution, emphasizing the growing importance of multimodal approaches. The primary objective of this research is to clearly define the goals, with a specific focus on the importance of exploring the possible collaboration between voice and picture recognition [2]. This study aims to enhance speech and picture recognition in conversational AI systems through technology integration. We explore advancements in machine learning and computer vision to improve accuracy and efficiency. By investigating core elements of integrating these technologies, we address challenges in developing sophisticated multimodal conversational AI systems. Our goal is to advance AI-driven dialogue systems, enhancing user experience and interaction with AI-enabled interfaces. The progress in voice and picture recognition technologies have a substantial impact on a range of disciplines, such as computer science, artificial intelligence and human–computer interaction. The advancement of these technologies has significantly transformed the manner in which individuals engaged with computers, hence presenting novel opportunities for the development of applications, including virtual assistants. This section provides a comprehensive analysis of voice and image recognition systems, encompassing a detailed exploration of the underlying algorithms and methodologies that support their functionality [3].

The phenomenon of multimodal integration pertains to the cognitive mechanism that human brain amalgamates and synthesizes information derived from several sensory modalities, including but not limited to vision. This examines various approaches, such as neural networks and deep learning models [4], which are utilized to successfully merge audio and visual data. It elucidates the advantages of multimodal integration, including improved understanding of context and high-end user engagement. The application of Multimodal Conversational AI has garnered considerable attention and acclaim across several disciplines and industries. The utilization of this advanced technology has been observed in various domains, demonstrating its capacity to augment user experiences and optimize communication efficacy [5]. This study analyses various case studies encompassing healthcare, automotive, customer service and education sectors to demonstrate the efficacy of multimodal AI technology in enhancing user experiences across many domains [6]. Another study analysed by Kostas [7] for the diverse challenges that emerge within the domain of multimodal conversational AI. These challenges encompass issues pertaining to data privacy, computational complexity and ethical considerations. This study examines many approaches, including privacy-preserving algorithms and efficient hardware architectures, to tackle these challenges and facilitate the widespread use of multimodal technology. The future prospects and advancements within the discipline are of considerable interest and importance [8]. This study investigates the prospective trajectory of multimodal conversational AI, with a particular emphasis on the identification of new trends and prospective developments. It also explores the integration of augmented reality (AR) and virtual reality (VR) technologies, enabling improved multimodal interactions that provide a heightened level of immersion [9].

The integration of speech and picture recognition technologies inside multimodal conversational AI has the capacity to profoundly alter interactions between humans and machines [8]. This has the potential to facilitate a digital future that is distinguished by improved intuitiveness, immersion and user-friendliness. This research investigates the dynamic domain of multimodal conversational artificial intelligence (AI), where the incorporation of speech and image processing techniques improves user interactions across several domains [10]. Multimodal conversational AI combines speech recognition, which enables the comprehension of spoken language, with picture recognition, which facilitates the interpretation of visual data. This integration results in a multidimensional interface that possesses the ability to accurately grasp, and meet the requirements of users. The incorporation of voice and image recognition technologies in the contemporary digital age represents a noteworthy shift in our everyday existence, propelled by swift technological advancements. The increasing popularity of intelligent systems is resulting in a notable transition away from conventional user interfaces [7, 11]. These systems demonstrate the capability to comprehend oral communication as well as evaluate visual data. The incorporation of these technologies not only enhances the user's experience but also expands the range of artificial intelligence (AI) applications, including several sectors such as healthcare, education, entertainment and more. The process of human–computer interaction interface as shown in Fig. 1 lays the groundwork for a thorough investigation of multimodal conversational AI [12]. This paper offers a thorough examination of the challenges and possibilities inherent in this field through the incorporation of up-to-date research, real-world implementations and forward-looking viewpoints. The investigation into the combination of voice and image fusion reveals the significant potential of multimodal conversational AI [9]. This technology has the capability to transform human–machine interactions and greatly improve user experience to unprecedented degrees.

Conversational AI has several applications across various domains like chatbots, virtual assistants and NLP system. We anticipate that conversational AI will advance significantly like the rise of large language models, shift to multimodal interactions, integration of conversational AI with other technologies and will be driven by the following trends: conversational AI will be used to automate a wider range of tasks, will become more proactive, will become more accessible, will become more multimodal. There are several challenges in conversational AI. One of the primary issues in conversational AI is selecting the ML model that best fits the design algorithm. Finding and classifying relevant data while eliminating noise patterns is another

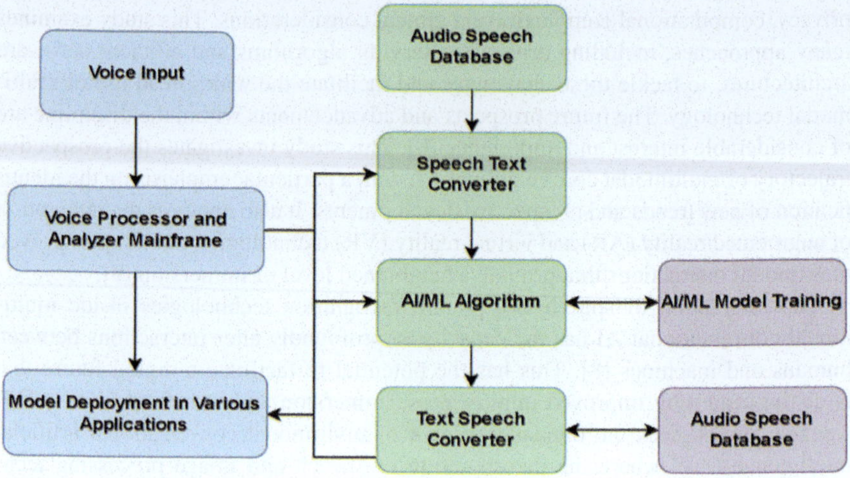

Fig. 1 Human–computer interaction

difficulty for conversational AI [1]. Figure 1 provides a visual representation of the architecture or flow of the proposed system described in the research paper. It illustrates the process of human–computer interaction within the context of the enhanced multimodal conversational AI system being discussed. Each component depicted in the figure represents a crucial step or module within the system. The figure appears to outline the pipeline for processing voice inputs, converting them into text, analysing the text through AI/ML algorithms and then converting the analysed text back into speech for output. This process likely involves several stages such as voice processing, speech recognition, natural language understanding and text-to-speech conversion.

2 Related Work

This section highlights the relevant studies, methodologies, limitations, results and findings of previous researchers as described below.

Yan [11] proposed computational methods for deep learning, delves into sophisticated aspects of deep learning. He described advanced topics including transformer models, control theory and graph neural networks. It offers detailed mathematical descriptions and algorithms for generative pre-trained models, specifically focusing on models like GPTs (Generative Pre-trained Transformers). Through its exploration of these concepts, the book aims to provide readers with a comprehensive understanding of computational methods essential for leveraging deep learning techniques effectively. Chaudhary [9] analysed visual and text sentiment analysis through hierarchical deep-learning networks. This monograph serves as a valuable resource for understanding and applying sentiment analysis techniques across various domains. Its emphasis on deep learning frameworks underscores the evolving landscape of sentiment analysis and its relevance in contemporary social and business contexts. Potential limitations include outdated technology coverage, focus on specific methodologies, lack of practical examples and generalization issues in sentiment analysis techniques. Sundar and Heck [13] devised the research goal of multimodal conversational systems which include representation, fusion, alignment, translation and co-learning that are explicated and mathematically formulated. It examines current datasets and methodologies within each domain through a thorough survey, highlighting the identification of limiting assumptions. The results highlight multimodal co-learning as a key area for multimodal conversational AI research going forward, highlighting its potential as a promising direction.

Magalhaes et al. [14] proposed to make a contribution in the field of multimodal conversation by locating and processing large amounts of multimedia data in various systems. The examination of the new trend and possible advantages for the society of multimodal communication is highlighted by the author. In addition to setting the groundwork for future research directions, the study critically assesses the current technological limitations and imagines the creation of sophisticated multimedia conversational systems that have the potential to completely transform knowledge distribution and information processing. Zhao et al. [15] classified existing frameworks and examined the developmental trends in open-domain dialogue, and the research aims to provide a comprehensive overview of recent works in the field of dialogue models. The results reduce the essence of open-domain dialogue systems to two essential components: controllability and in formativeness. This provides a brief framework to understand the changing objectives and subtleties of these systems in terms of efficiently delivering information while guaranteeing controlled interactions. Aneja et al. [16] illustrated audio and video inputs together with multimodal information. The research goal is to create a conversational AI bot that can understand user intentions. Controlled experiments are designed to examine the effects of incorporating multimodal sentiments in user guidance in order to evaluate its effectiveness. This creative method aims to improve the bot's comprehension and response to user intentions in a variety of modalities, which advances conversational AI systems and is in line with the changing field of human–computer interaction.

Chua et al. [17] in the research paper attempt to address several issues related to this emerging field by delving into the intricate world of multimodal conversational recommendation. The results shed light on a crucial component of the suggested model: the intrinsic graph structure provides prediction advantages. This special feature makes it possible to reason about items that were not discussed in training conversations with zero-shot reasoning while also making it easier to understand the underlying policies naturally. This is a big step toward improving the interpretability and efficacy of multimodal conversational recommendation systems. Abraham et al. [18] explore the enhancement of human-like characteristics in conversational agents, with a focus on the incorporation of affect, sentiment, and emotions into the conversational environment. The results shed light on the current situation by examining current patterns and utilizing deep learning models. The authors conduct an analysis of publicly available datasets and a critical review of the various deep learning techniques used to improve conversational agents. This thorough analysis provides insights into how deep learning techniques and the abundance of data resources are shaping conversational agents' evolving capabilities. Liao et al. [19] demonstrate notable improvements in the "askattribute" intent of the KB-grounded model, demonstrating its increased precision in accurately answering questions about product attributes like colour or brand. A noteworthy 10-point increase in BLEU scores during testing highlights this accomplishment. Nevertheless, a significant drawback emerges since the KB-grounded model depends on knowledge base queries and user intent annotation as additional inputs, requiring the creation of a strategic model to produce these elements. This emphasizes how carefully building a model that can combine knowledge base queries and annotations of user intent is necessary for best results.

Czerwinski et al. [20] highlight how well SIVA uses multimodal inputs to produce responses that are enhanced with perceptually valid cues like lip syncing and facial expressions in the context of a conversation. Nevertheless, a significant drawback emerges as the agents exhibit a monotonous behavioural pattern devoid of the ability to adjust to the subtleties of the interlocutor. Even though the study shows an impressive use of multimodal inputs, it recognizes that more work needs to be done to give conversational agents like SIVA a more dynamic and adaptive responsiveness in order to improve their overall effectiveness in a variety of communication scenarios. Lemon [21] provide an applied investigation of systems created in the Interaction Lab, ranging from speech interfaces on smart speakers to embodied robots utilizing language grounded in visual perception for communication. The study offers a concrete synopsis of applications that have been put into practice across various domains. One significant drawback, though, is the lack of a forward-looking viewpoint regarding the development of conversational and cooperative multi-agent systems. Although the study provides insightful information about current real-world applications, it is noticeably lacking in a forward-looking plan for the creation and incorporation of conversational systems in cooperative, multi-agent settings. Tian et al. [1] demonstrate the creation of a cutting-edge method: a multimodal hierarchical graph-based neural network that learns conversational query embeddings to improve user intent understanding. The study represents a major advancement in conversational system sophistication. But a significant gap in the current landscape indicates that, in spite of concerted efforts to advance conversation techniques, there is still a persistent knowledge gap in the field. This demonstrates the necessity of further investigation and in-depth understanding in order to close the current knowledge gap in conversational systems.

3 Methodology

The suggested multimodal AI speech-to-text voice assistant functions by means of a simplified procedure that includes input from the user, processing and output. In order to communicate with the assistant, users must speak or type commands or queries. The assistant uses a complex architecture that combines recurrent neural networks (RNNs), long short-term memory (LSTM) networks and convolutional neural networks (CNNs) to commence its processing pipeline upon receiving the input. The assistant may easily combine speech and picture recognition modules thanks to this architecture, which guarantees effective handling of a variety of input modalities. To improve the consistency and quality, the input is pre-processed using techniques including data cleansing, normalization and augmentation. The trained model then uses attention mechanisms to focus on key features and extract relevant information as it evaluates the input (see Table 1).The suggested multimodal AI speech-to-text voice assistant functions by means of a simplified procedure that includes input from the user, processing and output. In order to communicate with the assistant, users must speak or type commands or queries. The assistant uses a complex architecture that combines recurrent neural networks (RNNs), long short-term memory (LSTM) networks and convolutional neural networks (CNNs) to commence its processing pipeline upon receiving the input. The assistant may easily combine speech and picture recognition modules thanks to this architecture, which guarantees effective handling of a variety of input modalities. To improve the consistency and quality, the input is pre-processed using techniques including data cleansing, normalization and augmentation. The trained model then uses attention mechanisms to focus on key features and extract relevant information as it evaluates the input (see Table 1).

After processing the input, the assistant creates an output response, which could contain pertinent information, image descriptions or transcriptions of text. By means of this iterative process, the multimodal AI speech-to-text voice assistant seeks to improve user experiences and enable natural human–machine interactions by providing precise and prompt responses.

Regression analysis and classification models are two machine learning algorithms that will be used in the Enhanced Multimodal Conversational AI project to analyse a variety of input data and forecast user happiness and engagement. To guarantee a thorough grasp of user experiences, triangulation will be used to merge qualitative and quantitative data. Feedback loops from qualitative and quantitative evaluations will be integrated into the iterative refinement of the system design to guarantee ongoing development in line with user preferences. The goal of this iterative process is to continuously enhance the multimodal conversational AI interface. In order to extract significant insights and improve the usability and effectiveness of the system, the study integrates both qualitative and quantitative sources using a variety of data analysis techniques. The iterative feedback loop fosters a thorough understanding of user interactions and experiences by evaluating findings to inform ongoing enhancements.

Table 1 Comparative analysis of various methods

Author and year	Software/ Hardware requirement	Methodology/ Algorithm	Drawback	Finding
Chaudhuri [9], 2019	Recurrent neural networks (HGFRNNs)	Hierarchical deep learning for multimodal sentiment analysis	Uses just A platform-based system for setting up its algorithm not thinking about the future applications in the field of artificial intelligence	Considering the need to leverage large-scale social multimedia content for sentiment analysis, both state-of-the-art visual and textual sentiment analysis techniques are used for joint visual-textual sentiment analysis
Magalhaes et al. [14], 2021	–	Natural language processing, information retrieval and conversation technologies	Discuss lack of software-based technologies that exist in the system	This is timely and aims to explore this emerging trend, and discuss its potential benefits and pitfalls to society. This will also explore the limitations of current technologies and highlight future research directions towards developing a multimedia conversational system
Aneja et al. [16], 2021	Service-oriented systems engineering	Multimodal Sentiments into Conversational Bots	The bot could incorporate sentiments that are extracted from multimodal information into the bot's dialogue strategy	Design controlled experiments to explore the effectiveness of incorporating multimodal sentiments into the process of guiding users

(continued)

Table 1 (continued)

Author and year	Software/ Hardware requirement	Methodology/ Algorithm	Drawback	Finding
Liao et al. [19], 2021	Search-based conversational agent	Knowledge-grounded multimodal conversational model	Since the KB-grounded model assumes user intent annotation and KB queries as additional inputs, a plan to build a model is needed to provide them automatically	Great improvements for the "ask attribute intent" where the KB-grounded model could answer correctly questions related to brand or colour and other attributes of the product, which resulted in an increase of 10 BLEU points on test
Czerwinski et al. [20], 2021	–	SIVA (a Socially Intelligent Virtual Agent)	A limitation is that the agents are monotonic in behaviour and do not adapt to an interlocutor	SIVA leverages multimodal inputs to produce rich and perceptually valid responses (lip syncing and facial expressions) during the conversation
Tian et al. [1], 2021	–	Conversational image search	Several efforts have been dedicated to the conversation part but still lack of sufficient knowledge in the field	Design a multimodal hierarchical graph-based neural network, which learns the conversational query embedding for better user intent understanding
Lemon [21], 2022	Natural Language conversation	Machine learning	Lacking a future vision of conversational, collaborative multi-agent systems	It covers some of the practical systems developed in the Interaction Lab, ranging from speech interfaces on smart speakers to embodied robots interacting using visually grounded language

(continued)

Table 1 (continued)

Author and year	Software/ Hardware requirement	Methodology/ Algorithm	Drawback	Finding
Abraham et al. [18], 2022	Conversational Artificial Intelligence	Machine learning, deep learning and natural language processing	The research gaps in conversational agents and gives insights into future directions	With the advancements in recent trends and the rise in deep learning models, the authors review the deep learning techniques and various publicly available datasets used in conversational agents
Chua et al. [17], 2022	End-to-end State Graph-based Reasoning model	Conversational recommendation system	Model achieves better performance compared with existing methods But doesn't incorporate various upcoming techniques	The prediction of the proposed model benefits from the structure of the graph. It not only allows for zero-shot reasoning for items unseen in training conversations, but also provides a natural way to explain the policies
Zhao et al. [15], 2022	Data-driven dialogue model	Neural network	The methods reviewed in this paper are selected according to unique perspectives and by no means complete	Summarize the goal of an open-domain dialogue system in two aspects, informative and controllable

(continued)

Table 1 (continued)

Author and year	Software/ Hardware requirement	Methodology/ Algorithm	Drawback	Finding
Sundar and Heck [13], 2022	Multimodal translation maps embeddings	Multimodal Versatile Networks	Only works on existing database and outlining them for future work	Survey state-of-the-art datasets and approaches for each research area and highlight their limiting assumptions. Finally, we identify multimodal co-learning as a promising direction for multimodal conversational AI research
Smith et al. (2022)	–	Transformer-based models and fusion techniques	Potential complexity in training and optimization	The successful integration of speech and image modalities via the Multimodal Fusion Transformer (MFT) architecture
Yan [11], 2023	MATLAB and Tensor Flow	Computational Methods for Deep Learning	Does not dive into the practical work using software development	Beginning with a grounding about artificial neural networks with neurons and the activation functions, the work then explains the mechanism of deep learning using advanced mathematics

Critical features such as Mel-Frequency Cepstral Coefficients (MFCCs) are used for speech processing in the Enhanced Multimodal Conversational AI project, with an emphasis on the short-term power spectrum for speech detection. Spectrograms provide important insights into speech patterns by visually representing audio signals. In image processing and recognition, deep neural network features—extracted from intermediate layers of networks like VGG or RcsNet—are essential. Word embeddings, like Word2Vec, improve text comprehension by converting input into numerical vectors that represent the semantic connections between words. Understanding facial expressions is aided by facial landmark extraction, which involves important points like the nose and eyes. These methods support a comprehensive strategy for creating a sophisticated conversational AI system that combines text, image and speech inputs for a more engaging user experience. Long Short-Term Memory (LSTM) networks, a subset of recurrent neural networks (RNNs), are particularly good at analysing sequential data, such as speech, and identifying temporal relationships. With their specialization in image processing, convolutional neural networks (CNNs) effectively capture spatial patterns for image recognition. Combining CNNs and RNNs to create hybrid models works well for handling a variety of modalities. Versatile application is ensured by integrating CNNs for image processing and LSTM networks for sequential data, like voice. Selective focus on input sequence segments is made possible by attention mechanisms, most notably in the Transformer model, which improves information processing in a variety of input scenarios. These methods help multimodal conversational AI function more effectively when different input components have different effects on the output. The formula for RNN and LSTM is represented by Eqs. 1–5 as written below (see Fig. 2)

1. RNN formulation

$$h_t = \tanh(W_{hx}) + b_h \tag{1}$$

2. LSTM formulation

$$f_t = \sigma(W_{xf} x_t + W_{hf} h_{t-1} + W_{cf} C_{t-1} + b_f) \tag{2}$$

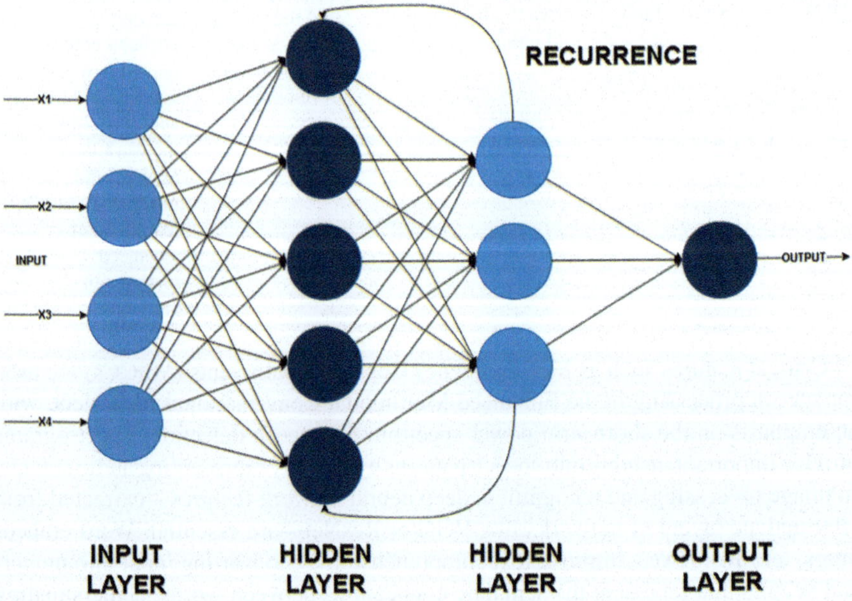

Fig. 2 Recurrent neural networks

$$i_t = \sigma(W_{xi}x_t + W_{hi}h_{t-1} + W_{ci}C_{t-1} + b_i) \tag{3}$$

$$\overline{C}_t = \tanh(W_{xc}x_t + W_{hc}h_{t-1} + b_c)$$

$$C_t = F_t \Theta C_{t-1} + I_t \Theta \overline{C}_t \tag{4}$$

$$O_t = \sigma(W_{xo}x_t + W_{ho}h_{t-1} + W_{co}C_t + b_o)$$

$$h_t = O_t \Theta \tanh(C_t) \tag{5}$$

Early and late fusion are examples of multimodal fusion techniques that combine data from multiple modalities for efficient processing. While late fusion combines modality-specific outputs later in the network, early fusion combines features at the input level. These methods improve the multimodal information integration (Fig. 3). From textual descriptions, Generative Adversarial Networks (GANs)—more especially, conditional GANs—create realistic images. Applications like picture captioning within the "Enhanced Multimodal Conversational AI" project demonstrate the usefulness of this technology.

4 Result

We have developed an innovative multimodal conversational AI system that integrates speech, text and image processing capabilities for seamless human–computer interactions. Figure 4 shows user interface of the application that can be used to login in the application. Figure 5 describes the application interface for voice assistance. Regarding the "Enhanced Multimodal Conversational AI Using Speech and Image Integration" project, the suggested elements support a smooth and effective user interface created using customTkinter library. The AI can respond intelligibly thanks to the "Assistant speaks()" function, resulting in a more engaging and natural interaction.

"Process Text()" makes sure that user commands are effectively understood, and "Get Audio()" makes it easier to obtain speech inputs. With "Search Web()" and "Open Application()" integrated, the AI can now retrieve data from the internet and launch particular applications in response to user requests, expanding its capabilities. "Start Listen()" improves the AI's responsiveness by starting to actively listen for commands from the user. "Create User()," "Delete User()," and "Update User()" functions in the "User Manager Database" sub-system help to provide customized user experiences and facilitate effective management of user profiles. Retrieving user data with the "Retrieve()" function is made easy by the interaction between the "User Database" interface and the "User Manager" sub-system. The user's email address

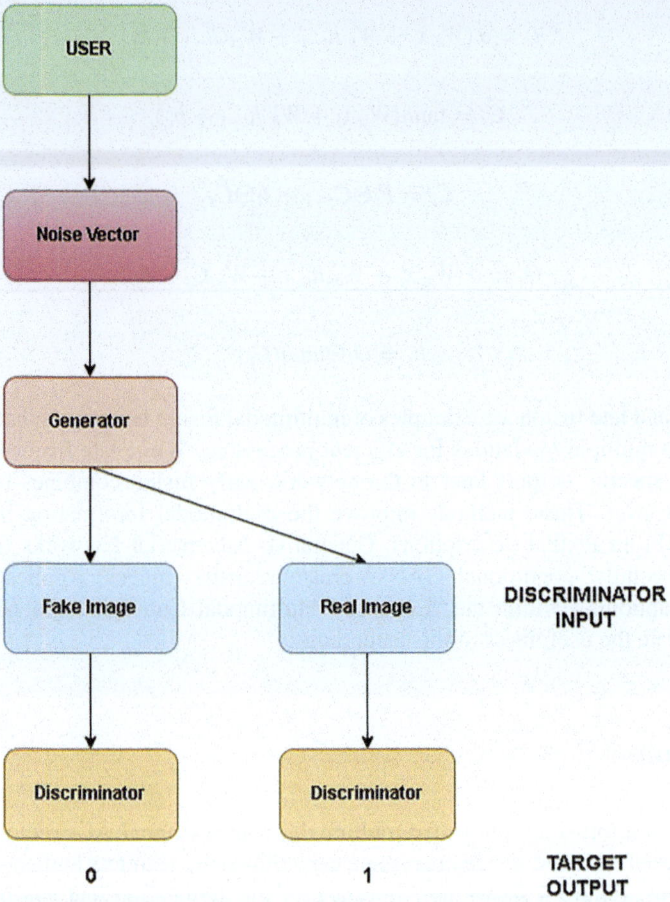

Fig. 3 Generative adversarial network

and passwords are secured in database mongodb which can be registered on our web page shown in Fig. 5. Together, these elements create a coherent system that enables the Enhanced Multimodal Conversational AI to comprehend, react to and efficiently handle user interactions.

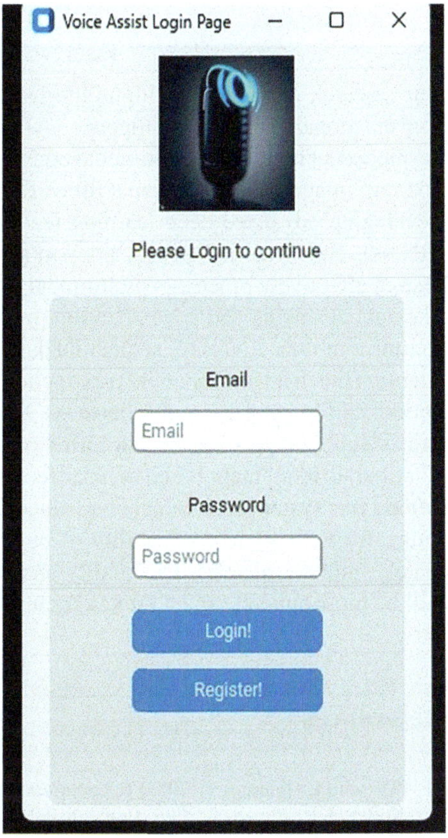

Fig. 4 Interface of the application

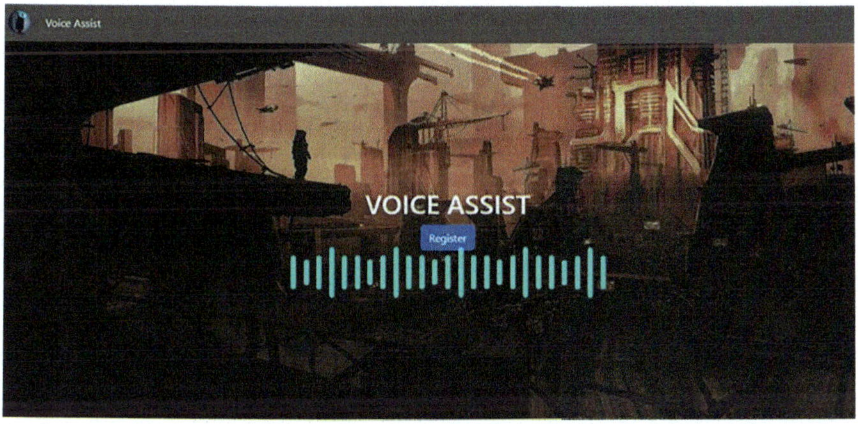

Fig. 5 User interface for voice assistance

5 Conclusion

Our research concludes by highlighting the revolutionary potential of multimodal conversational artificial intelligence (AI) in improving user experiences. We have developed an innovative multimodal conversational AI system that integrates speech, text and image processing capabilities for seamless human–computer interactions. Leveraging advanced deep learning techniques such as recurrent neural networks (RNNs), convolutional neural networks (CNNs) and attention mechanisms, our system effectively processes multimodal inputs. Through pre-processing techniques like data cleansing and normalization, input quality is enhanced. Our model architecture combines RNNs for sequential data analysis, CNNs for image processing and attention mechanisms for context extraction. Trained on diverse datasets, our model is optimized for accuracy and robustness. Evaluation against existing models demonstrates superior performance in terms of accuracy, speed and resource efficiency. Continuous refinement based on user feedback ensures ongoing improvement, positioning our system as a valuable advancement in multimodal conversational AI. In future we will enhance capability of our multimodal AI systems, which will ultimately improve people's life quality and pave the way for a day when AI systems will be more suited to meet human requirements.

References

1. Wassan JT, Ghuriani V (2023) Recent trends in deep learning for conversational AI
2. Koller D, Friedman N (2009) Probabilistic graphical models: principles and techniques. MIT Press
3. Bengio Y (2009) Learning deep architectures for AI. Foundations and trends® in Machine Learning 2(1):1–127
4. Yu D, Deng L, Yu D, Deng L (2015) Adaptation of deep neural networks. Automatic speech recognition: a deep learning approach, 193–215
5. Campilho A, Karray F (eds) (2016) Image analysis and recognition: 13th international conference, ICIAR 2016, in Memory of Mohamed Kamel, Póvoa de Varzim, Portugal, July 13–15, 2016, Proceedings, vol 9730. Springer
6. Zhang Z, Liu S, Liu S, Han L, Shao Y, Zhou W (2015) Human action recognition using salient region detection in complex scenes. In: The proceedings of the third international conference on communications, signal processing, and systems. Springer International Publishing, pp 565–572
7. Ikizler-Cinbis N, Sclaroff S (2010) Object, scene and actions: combining multiple features for human action recognition. In: Computer vision–ECCV 2010: 11th European conference on computer vision, Heraklion, Crete, Greece, September 5–11, 2010, proceedings, Part I 11. Springer Berlin Heidelberg, pp 494–507
8. Gunjan VK, Zurada JM, Singh N (eds) (2022) Modern approaches in machine learning & cognitive science: a walkthrough. Springer, Cham
9. Chaudhuri A (2019) Visual and text sentiment analysis through hierarchical deep learning networks. Springer, New York, pp 1–98
10. Heaton J (2018) Ian goodfellow, yoshua bengio, and aaron courville: deep learning: the Mit Press, 2016, 800 pp, ISBN: 0262035618. Genetic programming and evolvable machines 19(1):305–307

11. Yan W (2021) Computational methods for deep learning, vol 10. Springer. 978-3
12. Mellouli D, Hamdani TM, Ayed MB, Alimi AM (2017) Morph-CNN: a morphological convolutional neural network for image classification. In: Neural information processing: 24th international conference, ICONIP 2017, Guangzhou, China, November 14–18, 2017, proceedings, Part II 24. Springer International Publishing, pp 110–117
13. Sundar A, Heck L (2022) Multimodal conversational AI: a survey of datasets and approaches. arXiv:2205.06907
14. Magalhaes J, Chua TS, Mei T, Smeaton A (Oct 2021) The next generation multimodal conversational search and recommendation. In: Proceedings of the 29th ACM international conference on multimedia, pp 953–954
15. Fu T, Gao S, Zhao X, Wen JR, Yan R (2022) Learning towards conversational AI: a survey. AI Open 3:14–28
16. Aneja D, Hoegen R, McDuff D, Czerwinski M (May 2021) Understanding conversational and expressive style in a multimodal embodied conversational agent. In: Proceedings of the 2021 CHI conference on human factors in computing systems, pp 1–10
17. Wu Y, Liao L, Zhang G, Lei W, Zhao G, Qian X, Chua TS (2022). State graph reasoning for multimodal conversational recommendation. IEEE Trans Multimed
18. Kusal S, Patil S, Choudrie J, Kotecha K, Mishra S, Abraham A (2022) AI-based conversational agents: a scoping review from technologies to future directions. IEEE Access 10:92337–92356
19. Liao L, Long LH, Zhang Z, Huang M, Chua TS (Jul 2021) Mmconv: an environment for multimodal conversational search across multiple domains. In: Proceedings of the 44th international ACM SIGIR conference on research and development in information retrieval, pp 675–684
20. Lemon O (2022) Conversational AI for multi-agent communication in Natural Language. AI Commun 35(4):295–308
21. Nie L, Jiao F, Wang W, Wang Y, Tian Q (2021) Conversational image search. IEEE Trans Image Process 30:7732–7743
22. He X, Deng L (2018) Deep learning in natural language generation from images. Deep Learn Nat Lang Process, 289–307
23. Jawahar CV, Li H, Mori G, Schindler K (eds) (2019) Computer vision–ACCV 2018: 14th Asian conference on computer vision, Perth, Australia, December 2–6, 2018. Revised Selected Papers, Part III, vol 11363. Springer
24. Leibe B, Matas J, Sebe N, Welling M (eds) (2016) Computer vision–ECCV 2016: 14th European conference, Amsterdam, The Netherlands, October 11–14, 2016, proceedings, Part IV, vol 9908. Springer
25. Gaur L, Solanki A, Jain V, Khazanchi D (eds) (2020) Handbook of research on engineering innovations and technology management in organizations. IGI Global
26. Emani M, Xie Z, Raskar S, Sastry V, Arnold W, Wilson B, Boyd M et al (Nov 2022) A comprehensive evaluation of novel AI accelerators for deep learning workloads. In: 2022 IEEE/ACM international workshop on performance modeling, benchmarking and simulation of high performance computer systems (PMBS). IEEE, pp 13–25
27. Bengio Y, Goodfellow I, Courville A (2017) Deep learning, vol 1. MIT Press Cambridge, MA, USA, pp 22–24
28. Wang P (2020) On defining artificial intelligence. J Artif Gen Intell 11(2):73–86
29. Martin JH (2009) Speech and language processing: an introduction to natural language processing, computational linguistics, and speech recognition. Pearson/Prentice Hall
30. Murphy KP (2012) Machine learning: a probabilistic perspective. MIT Press
31. Sutton RS, Barto AG (2018) Reinforcement learning: an introduction. MIT Press
32. Goodfellow I, Bengio Y, Courville A (2016) Deep learning. MIT Press
33. Friedman J (2009) The elements of statistical learning: data mining, inference, and prediction

Navigating the Limitations and Potential of Proof-of-Work Consensus: A Comprehensive Analysis

Arieb Ashraf Sofi, Ajaz Hussain Mir, and Zamrooda Jabeen

Abstract In recent years, blockchain technology has attracted substantial interest owing to its capacity to transform multiple sectors. Central to blockchain systems is the consensus algorithm, which fosters consensus among participants concerning the distributed ledger's status. This paper delves into the nuances of Proof-of-Work (PoW) methods, highlighting the cryptographic protocols and functions underpinning their operation. Furthermore, it identifies inherent challenges within PoW, such as the need for increased mining power, security vulnerabilities, and fairness concerns, while also acknowledging its role in maintaining blockchain integrity. Finally, the paper underscores the imperative for continual refinement and innovation in consensus algorithms to address existing shortcomings and propel blockchain technology towards greater scalability, security, and efficiency. Through this exploration, stakeholders gain valuable insights to inform their decisions in designing and deploying blockchain solutions.

Keywords Blockchain · Consensus · Cryptography · Proof-of-Work (PoW)

1 Introduction

In the realm of decentralized networks and distributed ledger technologies, the concept of consensus mechanisms plays a pivotal role in ensuring trust and agreement among disparate nodes that inherently lack mutual trust. One such mechanism, known as Proof-of-Work (PoW), has gained significant popularity due to its reliability and

A. A. Sofi (✉) · A. H. Mir
Department of Electronics and Communication, National Institute of Technology, Srinagar, India
e-mail: arieb.scholar_ece@nitsri.ac.in

A. H. Mir
e-mail: ahmir@nitsri.ac.in

Z. Jabeen
Department of Mathematics, National Institute of Technology, Srinagar, India
e-mail: zjabeen19@nitsri.ac.in

P. J. S. Gonçalves et al. (eds.), *Proceedings of Fifth International Conference on Computing, Communications, and Cyber-Security*, Lecture Notes in Networks and Systems 1128, https://doi.org/10.1007/978-981-97-7371-8_14

effectiveness in validating transactions within blockchain networks. It is renowned for its robustness and reliability in validating transactions within blockchain networks. PoW's widespread adoption can be attributed to two key characteristics that distinguish it within the realm of consensus mechanisms: the inherent difficulty in finding a solution to a complex mathematical problem, coupled with the ease of verifying the correctness of that solution. These features of PoW have made it a central part of blockchain technology, playing a key role in many decentralized systems.

PoW originated from the seminal work of Cynthia Dwork and Moni Naor in 1993 [11], later implemented by Satoshi Nakamoto in the Bitcoin whitepaper in 2008 [23]. Markus Jakobsson and Ari Juels introduced the term "Proof of Work" in 1999 [11]. Cryptocurrencies such as Bitcoin and Litecoin currently utilize PoW, while Ethereum transitioned from PoW to Proof of Stake (PoS) [1]. This paper will examine the details of PoW by investigating its various implementations, known as variants. It will discuss different methods of PoW, highlighting their unique features and limitations inherent in this algorithm.

1.1 Related Work and Contribution of This Paper

Many attempts have been made to improve consensus algorithms by proposing variants of PoW [6, 22] and by introducing novel algorithms such as Proof of Stake [8, 21, 25], Delegated Proof of Stake [16], Proof of Work based on Historical Weighted Difficulty [33], and Non-Interactive Proofs of Proof-of-Work (NIPoPoWs) [19]. Analysis of PoW based cryptocurrencies is done [9, 14].

This paper contributes differently by providing a comprehensive understanding of PoW, addressing its challenges, limitations, and potential avenues for refinement. It offers in-depth information about PoW methods and variants, making it easy for the reader to understand and decide which method/variant to use for a specific application.

2 POW Variants

PoW variants consist of a range of protocols designed to meet various network needs and operational scenarios. Among these variants, Challenge-Response Protocols and Solution-Verification Protocols stand out as prominent implementations [17].

2.1 Challenge-Response Protocols

These protocols involve a direct interactive connection between the requester (client) and the provider (server) as illustrated in Fig. 1. The provider selects a challenge, such as an item with specific properties from a set, and the requester must find the corresponding response within the set. The provider then verifies the response submitted by the requester. The difficulty of the challenge can be adjusted dynamically by the provider based on its current workload. Requester's work may be bounded if the challenge-response protocol has a known solution or if it exists within a limited search space.

2.2 *Solution-Verification Protocols*

As illustrated in Fig. 2, solution-verification protocols diverge from challenge-response counterparts by eliminating the presumption of a direct connection between the requester and provider. Consequently, the requester delineates the problem prior to soliciting a solution, with the provider tasked to authenticate both the problem selection and the solution proffered by the requester. Most solution-verification protocols entail unbounded probabilistic iterative processes, exemplified by Hashcash [15]. Typically, known-solution protocols demonstrate marginally reduced variance in contrast to unbounded probabilistic counterparts, attributed to the decreased variance of a rectangular distribution compared with a Poisson distribution sharing the same mean. Mitigating variance commonly involves leveraging multiple independent sub-challenges, leveraging the principle that the average of numerous samples tends to exhibit diminished variance.

Additionally, there are fixed-cost functions like the time-lock puzzle, which impose a fixed computational cost on the requester.

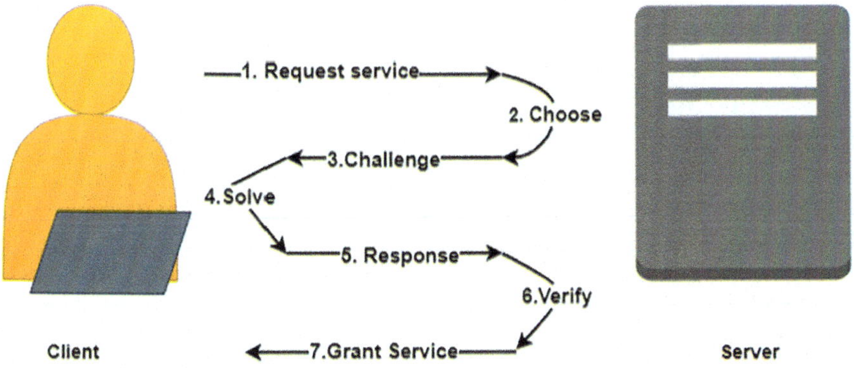

Fig. 1 Challenge-response protocol

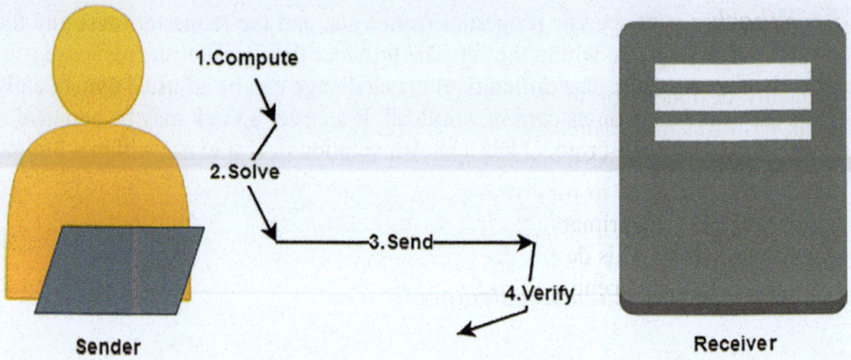

Fig. 2 Solution-verification protocol

3 PoW Methods

Exploring the different types of PoW, it's important to look at the specific ways PoW is used in blockchains. These methods serve as practical applications of the broader PoW variants, with each one customized to meet unique network needs and operational circumstances. PoW methods encompass various protocols, each with distinct characteristics and functionalities. Some prominent PoW-based protocols used in modern cryptocurrencies include SHA-256, Ethash, Scrypt, Cryptonote, Equihash, and Concatenating Hashes (e.g., X11). Each of these PoW protocols offers unique features and aims to address specific challenges, such as resistance to ASIC mining, memory hardness, and enhanced security. Here's an overview of some prominent PoW-based protocols used in modern cryptocurrencies.

3.1 SHA-256

Developed by the NSA in 2001, SHA-256 is part of the Secure Hash Algorithms (SHA) family and is used in Bitcoin and related cryptocurrencies like Bitcoin Cash, Counterparty, MazaCoin, and Namecoin [24]. It operates by performing logical operations on input data, resulting in a fixed-size 256-bit hash value. SHA-256 is known for its collision-resistant properties, making it highly improbable for distinct inputs to produce the same output hash.

3.2 Ethash

Specifically designed for the Ethereum consensus protocol, Ethash utilizes the Kekkak hash function combined with modified versions of Hashimoto and Dagger algorithms to resist ASIC mining. Despite the emergence of ASIC miners for Ethash, Ethereum plans to implement ProgPoW for short-term PoW and PoS for the long term [32]. Its primary aim is to resist ASIC mining by demanding significant memory access. This design choice ensures ASIC resistance while preserving network security and decentralization.

3.3 Scrypt

Scrypt is a memory-intensive password-based key-derivation function used in cryptocurrencies like Litecoin and Dogecoin to resist ASIC mining by requiring large amounts of memory [27]. Its memory-intensive nature promotes broader participation in mining, enhancing network decentralization.

3.4 Cryptonote

Cryptonote, featuring the CryptoNight PoW algorithm, emphasizes privacy and security. It incorporates random memory access and past block information into its hashing process, rendering it resistant to ASIC mining. Cryptonote-based cryptocurrencies prioritize privacy and fungibility by obfuscating transaction details and ensuring equitable network participation. It's employed in privacy-oriented cryptocurrencies like Monero [29].

3.5 Equihash

Engineered to restrict parallel implementations by memory bandwidth, Equihash mandates provers to discover message combinations resulting in specific bitwise sums of hashes. This memory-bound approach fortifies network security and fosters decentralization by deterring the concentration of mining power. It's utilized in cryptocurrencies such as ZCash, Bitcoin Gold, and Aion [7].

3.6 *Concatenating Hashes (e.g., X11)*

Some PoW solutions involve concatenating multiple hash functions, as seen in the X11 hash used in Dash. The X11 hash includes eleven hashes applied sequentially, enhancing security. Other composite hashes like X13, X14, X15, and X17 have been developed following a similar approach. By amalgamating multiple hash functions, X11 bolsters security and resilience against ASIC mining. This method introduces computational diversity, challenging ASIC developers in optimizing hardware for mining.

Each of these PoW protocols offers unique features and aims to address specific challenges, such as resistance to ASIC mining, memory hardness, and enhanced security. Understanding the nuances of these protocols is crucial for effectively navigating the cryptocurrency landscape and making informed decisions regarding mining and investment strategies.

4 List of PoW Functions

Each PoW function operates differently and may have specific applications or characteristics that make them suitable for particular use cases within the context of blockchain technology and cryptographic protocols. Understanding the nuances of these PoW functions is crucial for effectively navigating the cryptocurrency landscape and making informed decisions regarding mining and investment strategies. These functions play a vital role in ensuring the security, integrity, and decentralization of blockchain networks. The following is a list of known PoW functions [2, 4, 10, 12, 13, 17, 18, 30, 31].

4.1 *Integer Square Root Modulo a Large Prime*

This function involves finding the integer square root of a number modulo a large prime. It's computationally intensive because calculating square roots modulo large primes demands significant computational resources, making it suitable for proof-of-work tasks where miners must expend effort to find valid solutions.

4.2 *Weaken Fiat-Shamir Signatures*

Fiat-Shamir signatures are a type of digital signature scheme. Weakening them involves altering the security parameters or assumptions underlying the scheme to make it less secure or more susceptible to attack. This could involve modifying the underlying cryptographic assumptions or introducing vulnerabilities deliberately, making it easier to break the signatures.

4.3 *Ong-Schnorr-Shamir Signature Broken by Pollard*

This refers to a cryptographic signature scheme that has been broken by Pollard's algorithm, a method for factoring integers. It implies that the security of the Ong-Schnorr-Shamir signature scheme is compromised, leading to potential vulnerabilities in signature verification.

4.4 *Partial Hash Inversion*

Partial hash inversion involves finding a specific output or partial output of a hash function. This may require significant computational effort and is employed as a proof-of-work mechanism in some cryptographic protocols where miners must search for specific hash outputs to validate blocks.

4.5 *Bread Pudding Protocol and Re-usable Proof-of-Work (RPoW) System*

The Bread Pudding Protocol and RPoW system introduce a novel approach to proof-of-work, providing a reusable mechanism for generating and verifying proofs of work. This likely involves creating a system where proofs of work can be generated once and reused multiple times, reducing computational overhead and enhancing efficiency.

4.6 *Hash Sequence*

Hash sequences involve generating a series of hashes or applying hash functions iteratively to data. This process may be used in proof-of-work mechanisms to ensure data integrity and security by verifying the integrity of sequences of data blocks.

4.7 *Puzzles*

Puzzles in the context of proof-of-work involve solving computational problems or puzzles that require significant effort to solve. These puzzles may be designed to be computationally difficult to solve, ensuring that miners must expend computational resources to validate transactions and secure the blockchain.

4.8 *Diffie-Hellman-Based Puzzle*

This involves creating puzzles based on the Diffie-Hellman key exchange protocol, which is a method for securely exchanging cryptographic keys over a public channel. These puzzles may involve complex mathematical computations based on the Diffie-Hellman protocol, requiring significant computational effort to solve.

4.9 *Cuckoo Cycle*

Cuckoo Cycle is a memory-bound proof-of-work algorithm designed to be ASIC-resistant, meaning it is computationally intensive and relies on memory access. This design choice ensures broader participation in mining and enhances network decentralization by deterring the concentration of mining power in specialized hardware.

4.10 *Merkle Tree-Based*

Proof-of-work mechanisms based on Merkle trees involve creating and verifying proofs of work using Merkle tree data structures. Merkle trees enable efficient verification of data integrity and inclusion by organizing data into a hierarchical structure, allowing miners to verify the authenticity and integrity of transactions and blocks efficiently.

4.11 *Guided Tour Puzzle Protocol*

The Guided Tour Puzzle Protocol likely refers to a specific cryptographic protocol or algorithm involving solving guided tour puzzles as a form of proof-of-work.

This involves navigating through predefined paths or sequences to solve computational puzzles, ensuring that miners must expend computational resources to validate transactions and secure the blockchain.

Table 1 provides a comparative overview of the differences between these proof-of-work functions based on their computational intensity, ASIC resistance, memory requirements, parallelization capabilities, cryptographic assumptions, and energy efficiency.

Table 1 Comparative overview of the differences between proof-of-work functions based on different parameters

Proof-of-work function	Computational intensity	ASIC resistance	Memory requirements	Parallelization	Cryptographic assumptions	Energy efficiency
Integer Square Root Modulo a Large Prime	High	No	Moderate	Limited	Prime factorization	Moderate
Weaken Fiat-Shamir Signatures	Moderate	No	Low	Moderate	Cryptographic hardness	Moderate
Ong-Schnorr–Shamir Signature Broken by Pollard	High	No	High	Limited	Integer factorization	Moderate
Partial Hash Inversion	High	No	Low	Moderate	Cryptographic hardness	Moderate
Bread Pudding Protocol and Re-usable Proof-of-Work (RPoW) System	Moderate	No	Moderate	Moderate	Cryptographic hardness	Moderate
Hash Sequence	High	No	Low	Limited	Cryptographic hardness	Moderate
Puzzles	High	No	Low	Limited	Cryptographic hardness	Moderate
Diffie-Hellman-Based Puzzle	High	No	Low	Limited	Computational hardness	Moderate
Cuckoo Cycle	High	Yes	High	Limited	Memory hardness	High
Merkle Tree-Based	High	No	Moderate	Limited	Cryptographic hardness	Moderate
Guided Tour Puzzle Protocol	High	No	Moderate	Moderate	Cryptographic hardness	Moderate

5 Problems Associated with PoW

Although PoW enables the propagation of a unified blockchain, there exist opportunities for refinement in terms of fairness, security, cost-effectiveness, and efficiency. Despite its ability to swiftly resolve blockchain forks, PoW still presents avenues for improvement to address its limitations comprehensively.

5.1 Resource Consumption

PoW mining necessitates significant computational power, leading to substantial resource consumption. This includes the expenditure of substantial amounts of energy, financial resources, physical space, and hardware infrastructure. Projections indicate a concerning trend where a substantial fraction of global electricity consumption, approximately 0.3%, may be dedicated to transaction verification by 2028 [5].

PoW mining relies on solving cryptographic puzzles through brute-force computational power, resulting in inefficient resource utilization. Much of the computational effort expended in mining does not directly contribute to network advancement but is dedicated to solving complex mathematical problems.

5.2 Security Concerns

PoW also faces security challenges. For instance, the presence of a firewall controlled by a malicious entity within a significant portion of network participants could lead to network partitioning by disabling the firewall or manipulating packet transmission, resulting in divergent consensus among network segments [3].

5.3 Fairness Issues

In scenarios like stock market transactions where transaction order is crucial, PoW's flexibility allows miners to arrange transactions arbitrarily, potentially leaving out or prioritizing certain transactions based on external incentives. This lack of transaction order fairness extends to timestamp accuracy, as miners alone determine block timestamps, posing vulnerability to timestamp manipulation or inaccuracy.

5.4 51% Risk

A critical vulnerability in PoW is the potential for a single entity to control 51% or more of the nodes in the network. In such a scenario, this entity could maliciously manipulate the blockchain by exerting dominance over the majority of the network's computational power.

5.5 Time-Consuming Process

Mining operations within PoW entail exhaustive checks of numerous nonce values to identify the correct solution to the cryptographic puzzle required to mine a block. This iterative process is inherently time-consuming, slowing down the transaction validation process [28].

5.6 Centralization Tendencies

Despite the decentralized nature of PoW, it often leads to centralization tendencies where large mining pools or entities with significant computational resources dominate the mining process, resulting in centralized control over transaction validation.

Due to the high computational demands of PoW mining, it tends to concentrate miners among those with access to specialized equipment and inexpensive electricity, resulting in both miner and geographic consolidation.

5.7 Environmental Impact

PoW consensus mechanisms are known for their substantial energy consumption and carbon footprint. The mining process requires powerful hardware and consumes significant amounts of electricity, contributing to environmental concerns and carbon emissions [5].

5.8 Scalability Challenges

As blockchain networks expand in size and transaction volume, PoW mechanisms encounter scalability challenges. Transaction validation and blockchain maintenance become more resource-intensive and time-consuming, potentially leading to bottlenecks and slower transaction processing times [20].

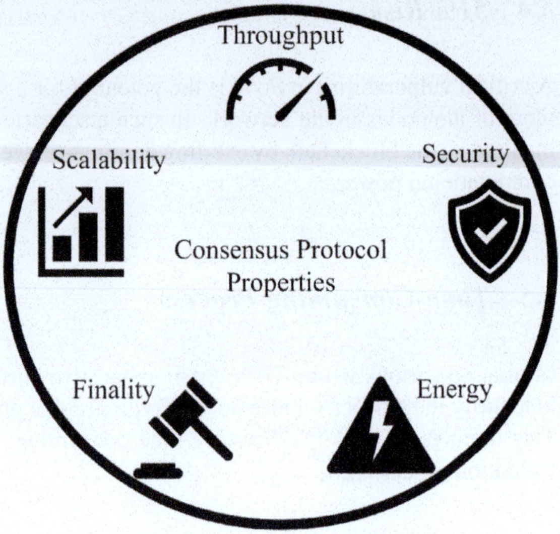

Fig. 3 Consensus protocol properties [26]

5.9 *Non-instantaneous Transaction Confirmation*

Transactions verified through PoW consensus suffer from delayed confirmation times, typically ranging from 10 to 60 min. This delay arises from the time required to mine transactions and integrate them into the blockchain, consequently delaying transaction commitment.

To improve Proof-of-Work (PoW) consensus mechanisms, addressing the mentioned limitations is crucial. However, it's essential to prioritize key factors [26]: Finality, scalability, throughput, energy consumption, and security shown in Fig. 3. These aspects collectively determine the effectiveness and longevity of PoW-based blockchain networks.

6 Conclusion

The PoW consensus mechanism has gained widespread adoption due to its reliability and efficiency in validating transactions on blockchain networks. This paper delves into the intricacies of the PoW algorithm, exploring its various variants, methods, and vulnerabilities. Detailed analysis of different PoW methods is provided. The challenges associated with PoW mechanisms are thoroughly examined, highlighting the continuous need for refining and innovating consensus algorithms. Additionally, this paper offers a comparative overview of PoW functions, considering factors such as computational complexity, energy efficiency, and ASIC resistance, aiding users in optimizing these parameters according to their specific application requirements. Enhancing PoW methods involves optimizing algorithms, adjusting parameters, innovating hardware, establishing governance mechanisms, and investing in research and development.

References

1. Cryptocurrencies and blockchain (2018) European parliament. Accessed 29 Oct 2020
2. Abliz M, Znati T (2009) A guided tour puzzle for denial of service prevention. In: 2009 annual computer security applications conference, pp 279–288. Honolulu, HI. https://doi.org/10.1109/ACSAC.2009.33
3. Alharby M, Alssaiari A, Alateef S, Thomas N, Moorsel A (2024) A quantitative analysis of the security of pow-based blockchains. Cluster Comput 1–18
4. Back A (1997) Hashcash. A popular PoW system. First announced in March 1997
5. Bada AO, Damianou A, Angelopoulos CM, Katos V (2021) Towards a green blockchain: a review of consensus mechanisms and their energy consumption. In: 2021 17th international conference on distributed computing in sensor systems (DCOSS). IEEE, pp 503–511
6. Bazzanella D, Gangemi A (2023) Bitcoin: a new proof-of-work system with reduced variance. Financ Innov 9(1):91
7. Biryukov A, Khovratovich D (May 2016) Equihash: asymmetric proof-of-work based on the generalized birthday problem. In: Eurocrypt'16. https://eprint.iacr.org/2016/260.pdf
8. Buterin V (2014) A next-generation smart contract and decentralized application platform
9. Capponi A, Olafsson S, Alsabah H (2023) Proof-of-work cryptocurrencies: does mining technology undermine decentralization? Manag Sci 69(11):6455–6481
10. Coelho F (2007) An (almost) constant-effort solution-verification proof-of-work protocol based on Merkle trees. Technical report, Cryptology ePrint Archive, Report
11. Dwork C, Naor M (1993) Pricing via processing or combatting junk mail. In: Advances in cryptology-CRYPTO' 92. Lecture notes in computer science, vol 740. Springer, pp 139–147. https://doi.org/10.1007/3-540-48071-4_10
12. Franklin MK, Malkhi, D (1997) Auditable metering with lightweight security. In: Financial cryptography. Lecture notes in computer science, vol 1318, pp 151–160
13. Gabber E, Jakobsson M, Matias Y, Mayer AJ (1998) Curbing junk e-mail via secure classification. In: Financial cryptography
14. Garratt RJ, van Oordt MR (2023) Why fixed costs matter for proof-of-work-based cryptocurrencies. Manag Sci 69(11):6482–6507
15. Hashcash.org: a partial hash collision based postage scheme. Txt. hashcash.org. Accessed 7 Mar 2024
16. Hu Q, Yan B, Han Y, Yu J (2021) An improved delegated proof of stake consensus algorithm. Procedia Comput Sci 187:341–346. https://doi.org/10.1016/j.procs.2021.04.109, https://www.sciencedirect.com/science/article/pii/S1877050921009133, 2020 international conference on identification, information and knowledge in the internet of things, IIKI2020
17. Jakobsson M, Juels A (1999) Proofs of work and bread pudding protocols. In: Secure information networks: communications and multimedia security. Kluwer Academic Publishers, pp 258–272. https://doi.org/10.1007/978-0-387-35568-9_18
18. Juels A, Brainard J (1999) Client puzzles: a cryptographic defense against connection depletion attacks. In: NDSS 99
19. Kiayias A, Miller A, Zindros D (2020) Non-interactive proofs of proof-of-work. In: Bonneau J, Heninger N (eds) Financial cryptography and data security. Springer International Publishing, Cham, pp 505–522
20. Kohad H, Kumar S, Ambhaikar A (2020) Scalability issues of blockchain technology. Int J Eng Adv Technol 9 (3):2385–2391

21. Lin S (2023) Proof of work vs. proof of stake in cryptocurrency. Highlights Sci Eng Technol 39:953–961
22. Meneghetti A, Sala M, Taufer D (2020) A new ECDLP-based pow model. Mathematics 8(8). https://doi.org/10.3390/math8081344, https://www.mdpi.com/2227-7390/8/8/1344
23. Nakamoto S (2008) Bitcoin: a peer-to-peer electronic cash system. Decentralized Bus Rev, 21260
24. National Security Agency (NSA) (Aug 2015) Secure Hash Standard (SHS). Technical report, federal information processing standards publication 180-4. https://csrc.nist.gov/publications/detail/fips/180/4/final
25. Nguyen C, Dinh Thai H, Nguyen D, Niyato D, Nguyen H, Dutkiewicz E (2019) Proof-of-stake consensus mechanisms for future blockchain networks: fundamentals, applications and opportunities. IEEE Access PP:1–1. https://doi.org/10.1109/ACCESS.2019.2925010
26. Oyinloye DP, Teh JS, Jamil N, Alawida M (2021) Blockchain consensus: an overview of alternative protocols. Symmetry 13(8). https://doi.org/10.3390/sym13081363, https://www.mdpi.com/2073-8994/13/8/1363
27. Percival C (May 2009) Stronger key derivation via sequential memory-hard functions. In: BSDCan'09. https://www.tarsnap.com/scrypt/scrypt.pdf
28. Shahriar Hazari S, Mahmoud QH (2020) Improving transaction speed and scalability of blockchain systems via parallel proof of work. Future Internet 12(8):125
29. van Saberhagen N (2013) CryptoNote v 2.0. Technical report, CryptoNote. https://cryptonote.org/whitepaper.pdf
30. Wang XF, Reiter M (May 2003) Defending against denial-of-service attacks with puzzle auctions. In: IEEE symposium on security and privacy '03. https://www.semanticscholar.org/paper/Defending-against-denial-of-service-attacks-with-Wang-Reiter/50a15e4dc964a41c4e32b97a146dbd2dd834250e
31. Waters B, Juels A, Halderman JA, Felten EW (2004) New client puzzle outsourcing techniques for dos resistance. In: 11th ACM conference on computer and communications security. https://citeseerx.ist.psu.edu/viewdoc/summary?doi=10.1.1.3.3286
32. Wood G (2014) Ethereum: a secure decentralised generalised transaction ledger. Technical report, Ethereum project yellow paper. https://ethereum.github.io/yellowpaper/paper.pdf
33. Yang X, Chen Y, Chen X (2019) Effective scheme against 51% attack on proof-of-work blockchain with history weighted information. In: 2019 IEEE international conference on blockchain (Blockchain), pp 261–265. https://doi.org/10.1109/Blockchain.2019.00041

A Visual Android Malware Detection Technique Based on Process Memory Dump Files

Rahul Gupta, Kapil Sharma, and R. K. Garg

Abstract Android-based smartphones dominate the market, constituting approximately 70% of total smartphone usage. However, this popularity makes them prime targets for malware attacks due to the sensitive financial and personal data they store. Consequently, there's a pressing need for effective malware detection techniques to combat these threats. To address this challenge, we devised a novel visual malware detection method to process memory dump files. Previous research based on visual techniques does not consider Android process memory dump strings. In general, obtaining strings from the memory of Android processes can help with malware detection, analysis, and mitigation by offering essential insights into the functionality and dynamic behavior of the malware by uncovering obfuscated or encrypted Strings. Our approach entails converting Android process memory dump files into grayscale images. Each byte within these memory dump files is meticulously processed, transforming them into consistent binary representations. These binary sequences serve as the foundation for generating grayscale images. Subsequently, features such as color histograms, Hu moments, and Haralick textures are extracted from these grayscale images. Machine learning classifiers are trained to differentiate between benign and malicious Android applications using the abovementioned extracted features. We used four machine learning classifiers, i.e., Random Forest, SVM, KNN, and Logistic regression. The Random Forest proves to give best results among them.

Keywords Android · Malware · Dynamic analysis · Image classification · Memory dump

R. Gupta (✉) · K. Sharma
Department of Information Technology, Delhi Technological University, New Delhi 110042, India
e-mail: rahulgupta100689@gmail.com

K. Sharma
e-mail: kapil@ieee.org

R. K. Garg
Department of Mechanical Engineering, Deenbandhu Chhotu Ram University of Science and Technology, Murthal 131039, India
e-mail: drrkgarg.me@dcrustm.org

P. J. S. Gonçalves et al. (eds.), *Proceedings of Fifth International Conference on Computing, Communications, and Cyber-Security*, Lecture Notes in Networks and Systems 1128, https://doi.org/10.1007/978-981-97-7371-8_15

1 Introduction

Android-based smartphones are the most popular among other smartphones, which is evident from the fact that Android-based smartphones have approximately 70% of the total share. The popularity of these devices is the primary reason for the growth of malicious attacks on these devices. Malware writers attack millions of these devices as they target these devices because of the financial and personal data they contain. Hence, there is an urgent need to develop malware detection techniques to counter these attacks.

Malware detection techniques are based on malware analysis techniques. Malware analyses are divided into static, dynamic, and hybrid methods. Each static, dynamic, and hybrid analysis differentiates itself from another in the way it gathers features for performing malware analysis, which is used to develop a malware detection model based on the extracted features.

The static malware analysis technique is one in which the characteristics of Android applications are analyzed without actually executing the file. Such techniques are fast and more scalable in the sense that no execution of the application is required.

In dynamic malware analysis, the characteristics of the Android application are analyzed by running the application's functionality in a controlled environment. Various characteristics of the applications are considered in execution based on which application is characterized as benign or malicious.

In the case of hybrid analysis, both static characteristics, i.e., those features of the application, are derived without executing the application, and dynamic characteristics, i.e., those features that are derived while the application is under execution, are considered.

The one research methodology that has been explored less in the realm of Android malware analysis is one that uses visual techniques. This methodology is based on representing Android application-related data as an image. The image generated is further used to derive features and, hence, develop a classification model based on those features for malware detection. In [1], authors converted the source files of Android applications into grayscale images, and further local and global features were extracted from these images to train machine learning classifiers. The EfficientNet CNN-based Android malware detection model was based on converting Android Dex files into images. These images are then fed 26 state-of-the-art CNN models, out of which the EfficientNet-B4 CNN-based model gave the best results [2]. In [3], the authors converted the non-intuitive malware features into grayscale images and used machine learning classifiers on a softmax layer of CNN to analyze the generated grayscale images. Ding et al. [4] converted the byte code of Android applications into images and used CNN to train on those images to build an effective malware detection model.

In the current research paper, we developed a visual malware detection technique based on process memory dump files. An Android process memory dump, referred to as a memory dump or core dump, captures the memory snapshot of a running process on an Android device at a specific instance. It encompasses details concerning memory allocation, variables, registers, and other pertinent data structures linked to the process.

Extracting strings from memory is a valuable approach in malware analysis for various reasons. For instance: Unveiling Hardcoded Strings: Malicious software often embeds hardcoded strings to facilitate command and control (C2) communication, encryption keys, or URLs. Delving into memory strings can uncover these hardcoded values, providing insights into the malware's functionality and potential indicators of compromise (IOCs).

Uncovering Obfuscated or Encrypted Strings: Certain malware employs tactics like string obfuscation or encryption to evade detection by conventional static analysis methods. Scrutinizing memory strings may expose these obfuscated or encrypted strings in their plain text form, thereby facilitating deeper analysis.

Understanding Dynamic Behavior: Memory-extracted strings shed light on the dynamic behavior of malware, including runtime configuration settings, API function calls, network communication patterns, and more. This comprehension aids in understanding how the malware operates and interacts within the compromised system.

Exposing Command and Control (C2) Infrastructure: Malware often communicates with remote command and control servers to receive instructions or exfiltrate data. Memory-extracted strings may harbor URLs, IP addresses, or domain names linked with the C2 infrastructure, empowering analysts to identify and potentially disrupt malicious communications.

Discovering Indicators of Compromise (IOCs): Analysts can pinpoint common patterns or unique signatures that serve as indicators of compromise (IOCs) by analyzing memory string dumps across multiple malware samples or infected systems. These IOCs are pivotal in detecting and mitigating similar threats in the future.

We introduce an innovative approach for Android malware detection leveraging visual techniques. Our method involves the transformation of Android process memory dump files into grayscale images. The memory dump files are meticulously read, with each byte converted into uniform binary representations. These binary sequences are then utilized to generate grayscale images. Subsequently, features such as color histograms, Hu moments, and Haralick textures are extracted from these grayscale images. These derived features train machine learning classifiers, distinguishing between benign and malicious Android applications.

The rest of the research paper is organized as follows: Sect. 2 represents related literature work. Section 3 explains the Methodology in detail. Section 4 shows the results achieved and Sect. 5 concludes the paper with future directions.

2 Related Work

In the work [5], the authors with the help of requested permissions used features-trained machine learning classifiers to classify botnets and benign applications. The authors in [6] used permissions and intents combinations to train ensembles for malware classifications. In the work [7], authors also combined permissions and intent filters to predict zero-day attacks. Droid-fusion [8] is a multilevel ML classifier fusion technique that incorporates the rank of accuracies of the first level to make more accurate decisions at the second level.

In [9], byte code of the Android application is represented as an RGB image and CNN was used to train based on the images generated for malware classification. In [10], authors covered executable sequences into grayscale images and then used LDA for two purposes, i.e., for reducing the dimensions of the image as well as to train to perform classification. The authors in [11] focus on generating images from apk and then using generated images to train CNN to avoid code obfuscation. MamaDroid [12] focuses on generating the behavioral model of Android malware using the sequence of generated API calls. Dendroid [13] employs malware detection techniques based on text mining and information on code structures.

In VisDroid [14], the authors generated five grayscale images from an apk file and from these files' local and global features were extracted to train ML classifiers for malware detection. Gupta et al. [15] proposed Android malware detection using rough set theory. They ranked critical features such as permissions, system calls, opcosed and API calls using dicernibility matrix concept of rought set theory. Gupta et al. [16] proposed unique Android malware detection model combining system call and permissions pairs and ranked them based on their malicious and benign ability. The generated score is used to form malware detection model.

3 Methodology

This section explains the technique used to classify an Android application as benign or malicious using memory dumps. An Android process memory dump, also known as a memory dump or core dump, is a snapshot of the memory state of a running process on an Android device at a particular moment in time. It contains information about the memory allocation, variables, registers, and other data structures associated with the process. The proposed model for classifying an Android application as benign or malicious using a memory dump is depicted in Fig. 1.

The Proposed model is further divided into number of sub-phases such as the visual representation of Android Process Memory Dump files represented in Sect. 3.1. Section 3.1 is dedicated to explaining feature extraction from visual representations.

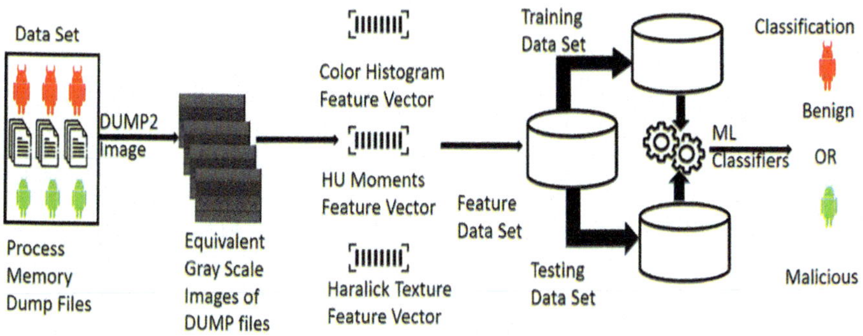

Fig. 1 Proposed Model for classifying an Android Application as Benign or Malicious from Memory DUMP files

3.1 Visual Representation of Android Process Memory Dump Files

We have taken the contents of the required Android process memory dump as a binary bit stream and arranged it into a byte matrix. Within the generated datasets, the malware source is depicted as a grayscale image for the sake of this study. Pixel values in the grayscale image range from 0 to 255, where 0 denotes black and 255 denotes white. The length of each image in the datasets varies according to the file size, but all of them have a fixed width of 256 pixels. With values ranging from 0 to 255, each byte in the created byte matrix represents a pixel in the final grayscale image. Following the conversion of the byte matrix into a matrix of values within the 0–255 range, the resulting matrix is preserved as a grayscale image. Figure 2 visually depicts transforming a malware sample into a grayscale image.

Color Histograms are commonly employed in the analysis of color images, focusing on the distribution of colors like red, green, and blue across various channels. Conversely, in grayscale imagery, which contains only one intensity channel, the concept of a color histogram is not directly applicable. Instead, we generate a histogram based on pixel intensities for grayscale images. This histogram delineates the distribution of pixel values (ranging from 0 to 255 in an 8-bit grayscale image) throughout the image. We segment the intensity spectrum into 256 bins, each representing a possible intensity value. Subsequently, we tally the frequency of pixels associated with each intensity value and visualize the resulting histogram. Normalizing the histogram values to sum up to 1 facilitates comparison across grayscale images of varying dimensions. We utilize the histogram as a feature vector for the grayscale image by considering each bin that signifies the occurrence frequency of a specific intensity value within the image. This feature vector is input to machine learning algorithms for subsequent analysis and processing.

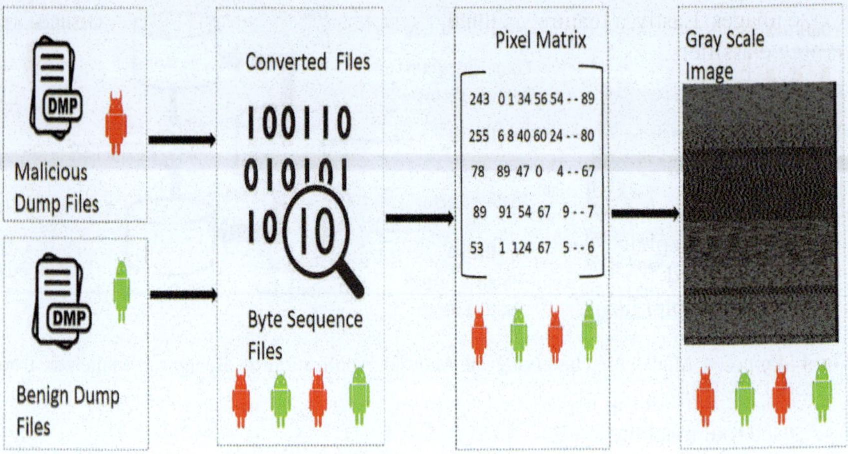

Fig. 2 Process of converting Memory Dump files into grayscale images

Hu moments represent a set of seven invariant image descriptors utilized in shape analysis and recognition. These descriptors stem from raw moments, which mathematically capture an image's shape, allowing for the description of objects regardless of their position, size, or orientation. Initially, grayscale images undergo conversion into binary images employing thresholding techniques such as Otsu's method or adaptive thresholding, given Hu moments' primary application in binary images. Subsequently, we compute the central moments of the binary image, ensuring translation invariance crucial for Hu moments' robustness. The normalization of central moments follows suit to achieve scale invariance, entailing the division of each central moment by an appropriate power of the zeroth moment, indicative of the object's total mass or area. Ultimately, the computation of Hu moments ensues using the normalized central moments. These seven moments, denoted as Hu_1 through Hu_7, arise from specific combinations of the normalized central moments. As a feature vector, Hu moments encapsulate the object's shape within the grayscale image. Notably, these moments act as features invariant to translation, rotation, and scale changes, rendering them invaluable in various image analysis tasks.

Haralick texture features, also called grey-level co-occurrence matrix (GLCM) features, constitute a collection of statistical metrics for characterizing the texture of grayscale images. These metrics capture the spatial correlations among pixel intensities within an image and find widespread application in image analysis and classification endeavors. Initially, we computed the grayscale image's grey-level co-occurrence matrix (GLCM). This matrix tabulates the occurrences of pixel intensity pairs at specific spatial relationships, such as distance and direction, within the image. Subsequently, we selected particular properties or characteristics for extraction from the GLCM. These chosen properties encompass contrast, correlation, energy, and homogeneity. Following this, normalization was employed to scale the computed texture features within a standardized range from 0 to 1, ensuring uniformity across diverse images. Lastly, a feature vector was constructed to serve as input for machine learning classifiers.

4 Results and Discussions

The last section explained how the visual representation in the form of a grayscale image was created from the Android memory dump files. It also explained how features such as color histograms, Hu moments, and Haralick texture were extracted from those grayscale images. This section discusses the experimentation to classify an application as benign or malicious based on the features extracted from the grayscale images of memory dump files.

We used four machine learning classifiers: SVM, Random Forest, K nearest neighbors, and Logistic Regressions. The Data set used for experimentation is the Android Process Memory String Dump [17]. The data set is built using AndroZoo's APK files [18]. The data set mainly consists of strings related to individual processes running at the time of APK usage. Each file in the data set corresponds to an APK. There are 2375 samples of Android process memory string dump files. Of the 2375 files, 1,188 samples correspond to malicious applications, while the remaining 1,187 samples correspond to benign applications.

We trained the model using the combined feature vector formed from all three types of features. We used a train-test split ratio of 70:30. The results are measured on the following parameters: accuracy, precision, recall, and F1-score. The results of the experiments are depicted in Table 1.

We observed that Random Forest performs better than the rest of the machine learning classifiers as Random Forest outperforms other classifiers due to its unique methodology. Unlike individual decision trees, Random Forest utilizes an ensemble approach by aggregating multiple decision trees for prediction. This ensemble technique effectively reduces variance and minimizes overfitting risks present in standalone decision trees. By combining predictions from numerous trees, Random Forest consistently improves overall performance.

Table 1 Table captions should be placed above the tables

ML techniques	Accuracy	Precision	Recall	F1-score
K-nearest neighbor	0.89	0.87	0.88	0.88
Logistic regression	0.89	0.88	0.87	0.89
SVM	0.92	0.91	0.90	0.91
Random forest	0.94	0.95	0.95	0.94

5 Conclusion and Future Scope

In this work, we have proposed a novel Android malware detection mechanism based on visual techniques. The mechanism is based on converting Android process memory dump files into grayscale images. The memory dump files are read byte by byte, and each byte is converted into equal binary representations. These binary representations are used to form grayscale images. These grayscale images derive features such as color histograms, Hu moments, and Haralick textures; these derived features per image train machine learning classifiers to predict an Android application as benign and malicious.

In future work, we aim to build more robust techniques by considering other types of features that can be derived from the grayscale image, which will strengthen the machine learning classifiers. We will also use deep-learning models to know their effectiveness in classifying malware using visual techniques.

References

1. Ünver HM, Bakour K (2020) Android malware detection based on image-based features and machine learning techniques. SN Appl Sci 2(7):1299
2. Yadav P, Menon N, Ravi V, Vishvanathan S, Pham TD (2022) EfficientNet convolutional neural networks-based Android malware detection. Comput Secur 115:102622
3. Singh J, Thakur D, Ali F, Gera T, Kwak KS (2020) Deep feature extraction and classification of android malware images. Sensors 20(24):7013
4. Ding Y, Zhang X, Hu J, Xu W (2020) Android malware detection method based on bytecode image. J Ambient Intell Human Comput
5. Kirubavathi G, Anitha R (2018) Structural analysis and detection of android botnets using machine learning techniques. Int J Inf Secur 17:153–167
6. Idrees F, Rajarajan M, Conti M, Chen TM, Rahulamathavan Y (2017) PIndroid: a novel Android malware detection system using ensemble learning methods. Comput Secur 68:36–46
7. Verma S, Muttoo SK (2016) An android malware detection framework-based on permissions and intents. Def Sci J 66(6)
8. Yerima SY, Sezer S (2018) Droidfusion: a novel multilevel classifier fusion approach for android malware detection. IEEE Trans Cybern 49(2):453–466
9. Hsien-De Huang T, Kao HY R2-d2 (Dec 2018) Color-inspired convolutional neural network (CNN)-based android malware detections. In: 2018 IEEE international conference on big data (big data). IEEE, pp 2633–2642
10. Karimi A, Moattar MH (Oct 2017) Android ransomware detection using reduced opcode sequence and image similarity. In: 2017 7th international conference on computer and knowledge engineering (ICCKE). IEEE, pp 229–234
11. Yen YS, Sun HM (2019) An Android mutation malware detection based on deep learning using visualization of importance from codes. Microelectron Reliab 93:109–114
12. Onwuzurike L, Mariconti E, Andriotis P, Cristofaro ED, Ross G, Stringhini G (2019) Mamadroid: detecting android malware by building markov chains of behavioral models (extended version). ACM Trans Priv Secur (TOPS) 22(2):1–34
13. Suarez-Tangil G, Tapiador JE, Peris-Lopez P, Blasco J (2014) Dendroid: a text mining approach to analyzing and classifying code structures in android malware families. Expert Syst Appl 41(4):1104–1117

14. Bakour K, Ünver HM (2021) VisDroid: Android malware classification based on local and global image features, bag of visual words and machine learning techniques. Neural Comput Appl 33(8):3133–3153
15. Gupta R, Sharma K, Garg RK (2024) Innovative approach to android malware detection: prioritizing critical features using rough set theory. Electronics 13(3):482
16. Gupta R, Sharma K, Garg RK (2024) Covalent bond based android malware detection using permission and system call pairs. Comput Mater Contin 78(3):4283
17. Homem I, Papapetrou P (2017) Android process memory strings dumps dataset. Stockholm University. Dataset
18. Allix K, Bissyandé TF, Klein J, Le Traon Y (May 2016) Androzoo: collecting millions of android apps for the research community. In: Proceedings of the 13th international conference on mining software repositories, pp 468–471

Hybrid GWO-PSO for Path Coverage Testing

Fatma Ahsan and Faisal Anwer

Abstract Software testing plays a pivotal role in the software development life cycle, ensuring the quality and reliability of software systems. However, it is often a labor-intensive and time-consuming process. To address this challenge, researchers have explored various search-based automated techniques for generating test suites. This paper proposes a novel hybrid approach, termed GWPSO, which combines Grey Wolf Optimizer (GWO) and Particle Swarm Optimization (PSO), for test case generation. The objective is to improve the efficiency and effectiveness of test case generation by leveraging the strengths of both GWO and PSO algorithms. In this study, the Hybrid GWPSO approach is compared with Genetic Algorithm (GA), PSO, and GWO in generating test cases for five benchmark problems. The results demonstrate the superiority of the proposed Hybrid GWPSO approach, showing significant improvements in terms of test case generation effectiveness. The findings of this research contribute to advancing the field of automated test case generation and hold implications for enhancing software testing practices in real-world software development projects.

Keywords Grey wolf optimization · Particle swarm optimization · Genetic algorithm · Search based · Software testing

1 Introduction

The domain of search-based software testing has witnessed rapid growth over the past decade, fueled by the evolution and proliferation of various metaheuristic algorithms [12]. Among these, Grey Wolf Optimization (GWO) [18] and Particle Swarm Optimization (PSO) [8] stand out as prominent examples of swarm intelligence

---

F. Ahsan (✉) · F. Anwer
Department of Computer Science, Aligarh Muslim University, Aligarh, India
e-mail: fatimah.ahsan01@gmail.com

F. Anwer
e-mail: faisalanwer.cs@amu.ac.in

P. J. S. Gonçalves et al. (eds.), *Proceedings of Fifth International Conference on Computing, Communications, and Cyber-Security*, Lecture Notes in Networks and Systems 1128, https://doi.org/10.1007/978-981-97-7371-8_16

optimization techniques. Rooted in principles drawn from biology and physics, these algorithms harness the collective behavior of individuals within a population to explore and exploit solution spaces effectively. In addition to GWO and PSO, the arsenal of metaheuristic algorithms employed in search-based software testing encompasses a diverse array of methodologies. This includes Genetic Algorithm (GA) [20], Ant Colony Optimization (ACO) [4], Artificial Bee Colony (ABC) [7], Tabu Search (TS) [21], Simulated Annealing (SA) [11], Firefly Algorithm (FA) [29], and Gravitational Search Algorithm (GSA) [22]. The integration of these advanced optimization techniques into software testing methodologies holds immense promise for enhancing the efficacy and efficiency of test case generation and fault detection processes. As the search-based software testing landscape continues to evolve, leveraging the collective intelligence of metaheuristic algorithms remains instrumental in addressing the growing complexities and challenges encountered in modern software systems.

The framework, depicted in Fig. 1, exemplifies the integration of metaheuristic optimization techniques into the software testing domain. Operating on inputs comprising the Program Under Test (PUT) and independent path details, the framework orchestrates the generation and evaluation of test cases against predefined path coverage criteria. Employing advanced optimization algorithms, the framework iteratively refines and optimizes the test suite until the desired coverage thresholds are attained.

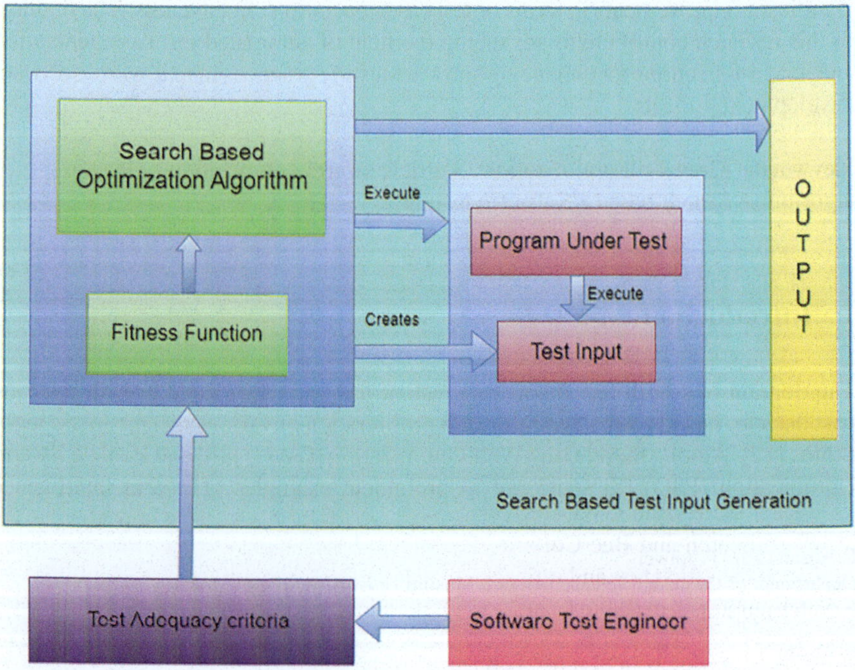

Fig. 1 Framework of the search-based software testing design

This iterative search process terminates upon reaching the specified coverage criteria, culminating in the provision of a final, optimized test suite. By leveraging metaheuristic algorithms, the framework navigates the complex solution space inherent in software testing, enabling efficient exploration and exploitation of potential test case configurations. Through this amalgamation of optimization methodologies and software testing principles, the framework contributes to advancing the state of the art in automated test case generation, fostering the development of more robust and comprehensive testing methodologies.

The contributions of this paper are outlined as follows:

- Introduction of a novel Hybrid GWPSO algorithm tailored for code coverage testing within the realm of software testing.
- Comparative analysis conducted between the proposed Hybrid GWPSO algorithm and established optimization methodologies such as Genetic Algorithm (GA), Particle Swarm Optimization (PSO), and Grey Wolf Optimization (GWO).
- Empirical validation of the proposed approach through experimentation on three benchmark problems representative of the challenges encountered in software testing.

These contributions underscore the significance of our research endeavor in advancing the state of the art in automated test case generation methodologies, with implications for enhancing software testing practices and improving software quality and reliability.

The next Sect. 2 presents related work in the field of the Hybrid GWPSO algorithm. Section 3 presents the proposed approach in detail. Section 4 analyzes the experimental results of the proposed approach. The next Sect. 5 concludes the paper and gives insights into future work.

2 Related Studies

Metaheuristics are widely utilized techniques for solving optimization problems, including software test case generation. Various metaheuristic algorithms, such as Particle Swarm Optimization (PSO) [24], Genetic Algorithm (GA) [20], Firefly Algorithm (FA) [10], Artificial Bee Colony (ABC) [13], Cuckoo Search (CS) [26], and Tabu Search (TS) [21], have been applied to address test case generation problems. Moreover, hybrid metaheuristic approaches have gained attention in this domain. For instance, the Hybrid GWO-PSO algorithm [25], Hybrid PSO-GA [5], and Hybrid Cuckoo Search and Bee Colony Algorithm (CSBCA) [14] have been proposed to enhance the efficiency and effectiveness of test case generation. The related studies are summarized in Table 1. This meta-analysis provides insights into the existing research landscape and highlights the diverse approaches employed in the field of software test case generation.

Table 1 Comparison of various related studies

Article	Objective	Metaheuristics used	Fitness function used	Result
Nayak et al. [19]	Optimize and prioritize test cases	Chaotic Grey Wolf Optimization (CGWO)	Circular Chaotic fitness function	The proposed approach is more effective
Perumal et al. [21]	To automated test data generation process	Cuckoo and Tabu Search	branch coverage	High degree of improvement as compared to conventional genetic algorithm
Srivastava et al. [27]	Generate optimized test cases	Cuckoo Search and Tabu Search (CSTS)	Lévy flight	The results show that the proposed approach is effective as compared to the various earlier proposed approach
Lam et al. [15]	To generate automated test cases	Artificial Bee Colony (ABC)	Test-path sequence comparison method fitness function	The comparison of ABC with various other approaches
Dalal and Chhillar [3]	Generate optimized test cases	Bee colony optimization and modified genetic algorithm (BCO-mGA)	Defined fitness function	Compared with better results
Lakshminarayana and Suresh Kumar [14]	To generate optimized test cases	Cuckoo Search and Bee Colony Algorithm (CSBCA)	Defined fitness function	Improved results as compared to other approaches
Sahoo and Ray [24]	To find and Generate test cases for the critical path	PSO	Branch distance and Path distance	Empirical evidence confirms that the ICF function is efficient and effective for generating test cases to achieve path coverage criteria

3 Proposed Approach

Our proposed framework leverages a hybrid GWPSO algorithm for the purpose of test case generation. Rooted in principles of metaheuristic optimization, this framework is meticulously designed to harness the intricate details extracted from the control flow graph (CFG) as input, thereby facilitating the systematic generation of test data tailored to traverse the intended paths within the Program Under Test (PUT). By capitalizing on this synergistic approach, our framework endeavors to augment the efficiency and efficacy of test case generation processes, meticulously ensuring comprehensive coverage of the program's functionality while mitigating redundancy. Central to this methodology is the integration of principles from both Grey Wolf Optimizer (GWO) and Particle Swarm Optimization (PSO), thereby harnessing the unique strengths of each algorithm to navigate the multifaceted challenges inherent in software testing. Through the adoption of this hybridized paradigm, our research endeavors to spearhead advancements in automated test case generation techniques, poised to address the dynamic challenges confronting software developers in their pursuit of ensuring software quality and reliability.

3.1 Hybrid Grey Wolf and Particle Swarm Optimization (GWPSO) Metaheuristics Algorithm

Algorithm 1 Pseudo Code of Hybrid GWPSO

Initialize the parameters of the GWO and PSO algorithm($N, D, Vmin, Vmax, Max\_iter, Xmin, Xmax,$)
Randomly generate population
Set the maximum number of iterations($Imax\_Global, GWOi\_max, PSOi\_max$) and start the counter $iter = 0$
Calculate the fitness value of each Wolves in the population using the fitness function f
Sort the particle in the increasing order of the fitness value
while $i\_Global \leq Imax\_Global$ *and Coverage* $\neq 100$ **do**
  while GWOi $\leq$ GWOi_*max and Coverage* $\neq 100$ **do**
    Update fitness of the wolves
    Update the position of alpha, beta, and delta wolves
    Update the position of each wolves
    Increment the value of Imax_Global ($GWOi_{max} += 1$)
  end while
  while PSOi $\leq PSOi_{max}$ *and Coverage* $\neq 100$ **do**
    Calculate the fitness of the particles
    Update global best and personal best
    Update Velocity of each particle in the population
    Update Position of each particle in population
    Increment the value of Imax_Global ($PSOi_{max} += 1$)
  end while
  Increment the value of Imax_Global ($Imax\_Global += 1$)
end while
Return the best test cases

In our study, we leverage two distinct metaheuristic optimization algorithms, Grey Wolf Optimization (GWO) and Particle Swarm Optimization (PSO), combining their strengths to formulate a hybrid algorithm termed the hybrid GWPSO algorithm. This hybrid approach is employed within our framework for test case generation, addressing the inherently challenging task of optimizing test case suites, known to be NP-hard [2]. The flowchart of the hybrid GWPSO approach is depicted in Fig. 2 and the pseudocode of the proposed algorithm is depicted in Algorithm 1. Specifically, in PSO, the positions and velocities of particles are iteratively updated using Eqs. 1 and 2, respectively. Meanwhile, in GWO, the algorithm operates with a hierarchy of four distinct types of wolves within the grey wolf group, namely α, β, δ, and ω, each characterized by varying degrees of dominance. The GWO algorithm employs Eqs. 5, 6, 7, 8, and 9 to update the positions of these wolves, facilitating exploration and exploitation of the solution space. Through the amalgamation of these metaheuristic techniques, our study endeavors to tackle the intricacies of test case generation in software testing, contributing to advancements in automated testing methodologies.

$$\overrightarrow{X_t^{t+1}} = \overrightarrow{X_i^t} + \overrightarrow{V_i^{t+1}} \tag{1}$$

$$\overrightarrow{V_t^{t+1}} = w\overrightarrow{V_i^t} + c_1 r_1(\overrightarrow{P_t^t} - \overrightarrow{X_i^t}) + c_2 r_2(\overrightarrow{G^t} - \overrightarrow{X_i^t}) \tag{2}$$

In Eq. 2, w, c_1, and c_2 are the parameters governing the behavior of the Particle Swarm Optimization (PSO) metaheuristic. The velocity calculation comprises three distinct components: the first part embodies the particle's inertia, reflecting its velocity in the preceding state; the second component denotes the cognitive aspect, encapsulating the particle's learning trajectory throughout the optimization process; and the third component embodies the social aspect, signifying the influence of the best position discovered within the particle's local neighborhood. This formulation allows for a nuanced exploration and exploitation of the solution space, facilitating the convergence toward optimal solutions in the optimization process.

$$\vec{D} = | \vec{C} \cdot \vec{X_p}(t) - \vec{X}(t) | \tag{3}$$

$$\vec{X}(t+1) = \vec{X_p}(t) - \vec{A} \cdot \vec{D} \tag{4}$$

where t indicates the current iteration, $\vec{A}$ and $\vec{C}$ are coefficient vectors, $\vec{X_p}$ is the position vector of the prey, and $\vec{X}$ indicates the position vector of a grey wolf. Equations 5 and 6 calculate vectors $\vec{A}$ and $\vec{C}$, respectively.

$$\vec{A} = 2\vec{a} \cdot \vec{r_1} - \vec{a} \tag{5}$$

$$\vec{C} = 2 \cdot \vec{r_2} \tag{6}$$

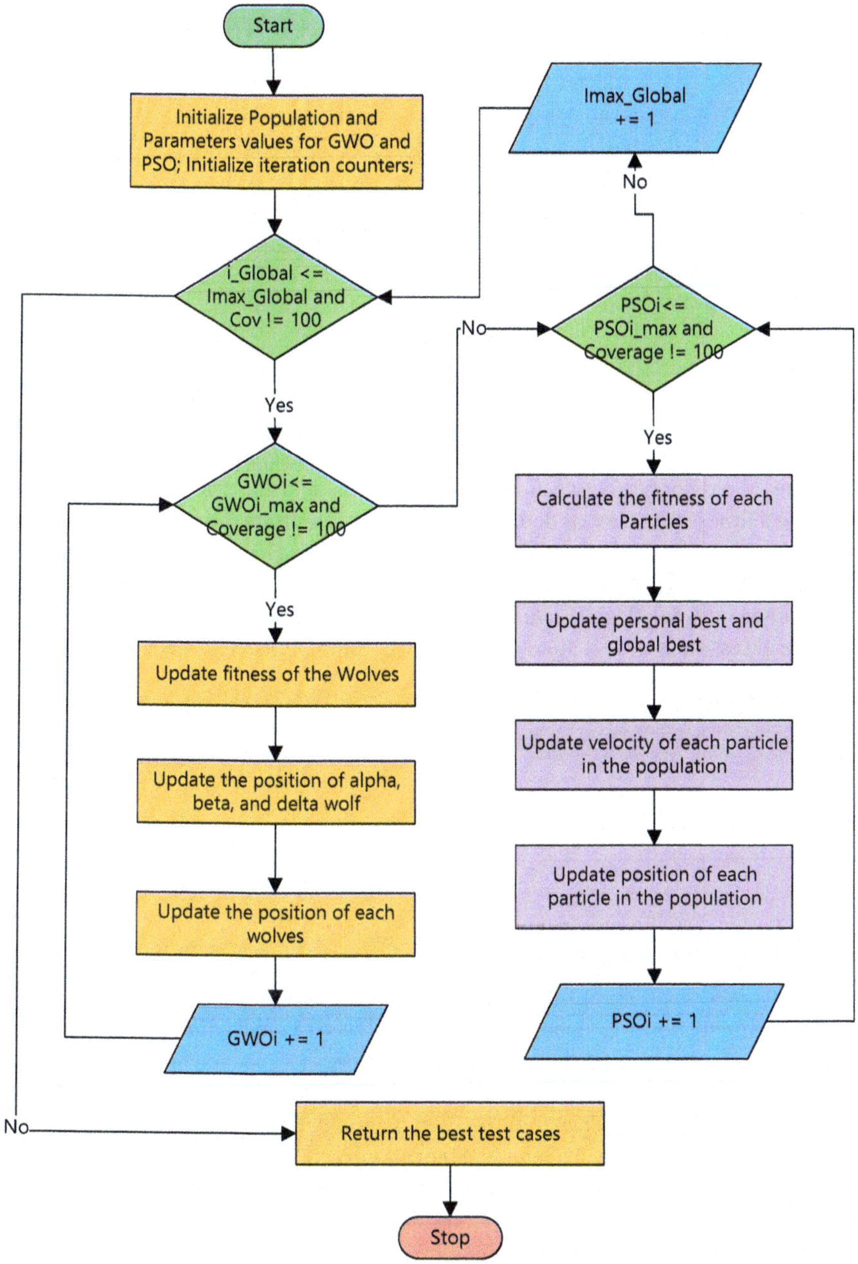

Fig. 2 Flowchart of proposed hybrid GWPSO algorithm

$$\overrightarrow{D_\alpha} = \mid \overrightarrow{C_1} \cdot \overrightarrow{X_\alpha} - \overrightarrow{X} \mid, \overrightarrow{D_\beta} = \mid \overrightarrow{C_2} \cdot \overrightarrow{X_\beta} - \overrightarrow{X} \mid, \overrightarrow{D_\delta} = \mid \overrightarrow{C_3} \cdot \overrightarrow{X_\delta} - \overrightarrow{X} \mid \tag{7}$$

$$\overrightarrow{X_1} = \overrightarrow{X_\alpha} - \overrightarrow{A_1} \cdot \overrightarrow{D_\alpha}, \overrightarrow{X_2} = \overrightarrow{X_\beta} - \overrightarrow{A_2} \cdot \overrightarrow{D_\beta}, \overrightarrow{X_3} = \overrightarrow{X_\delta} - \overrightarrow{A_3} \cdot \overrightarrow{D_\delta} \tag{8}$$

$$\overrightarrow{X}(t+1) = \frac{\overrightarrow{X_1} + \overrightarrow{X_2} + \overrightarrow{X_3}}{3} \tag{9}$$

In our study, we adopt a multi-criteria approach by combining three distinct test criteria: the sum of the branch distance [28], the approximation level [6], and the path distance [24]. Prior to amalgamation, the branch distance measure is normalized to ensure compatibility with the other criteria. The branch distance metric quantifies the number of branches covered, while the approximation level assesses the number of untraversed predicate conditions. Additionally, the path distance metric evaluates the disparity between the intended paths and the critical paths [24]. We propose a fitness function, as depicted in Eq. 10, to ascertain the fitness of individual solutions. Here, NBD signifies the Normalized Branch Distance, AL represents the approximation level [6], and PD denotes the path distance [24]. The NBD values are constrained between 0 and 1, calculated using Eq. 11, thereby facilitating a comprehensive assessment of test suite quality through the holistic evaluation of multiple criteria.

$$FitnessFunction = NBD + AL + PD \tag{10}$$

$$NBD = 1 - 0.001^{-BD} \tag{11}$$

4 Results and Discussion

The experimentation was conducted on a Windows 11 environment, utilizing a 64-bit OS and an 11th Gen Intel Core i5-1135G7 processor clocked at 2.40 GHz, along with 16 GB of memory. The algorithms were implemented in Java and executed on IntelliJ IDEA 2023.1.5. The study focuses on comparing the effectiveness of Grey WOlf Optimization (GWO), Genetic Algorithm (GA), Particle Swarm Optimization (PSO), and a novel Hybrid GWPSO in test data generation. Algorithmic parameters, such as population size (n), were set to 10, with a maximum generation limit of 1000. Additional parameters are detailed in Table 3.

The efficacy of the proposed algorithm is substantiated through the experimental analysis conducted on real-world programs, specifically utilizing widely recognized benchmark programs, as detailed in the presented Table 2. We have set the parameters of the GWO and PSO algorithm as described in Table 3.

Table 2 Details of the dataset used in the experimentation

Program under test	Parameter type	Description	Source
Triangle type (TCP)	3 integer	Type classification of a triangle	[16, 17]
Quadratic equation (QE)	3 integer	Find the roots of the equation	[17, 23]
Greater of three number (GTN)	3 integer	Check the greater number from the input three numbers	[17, 23]
Highest common factor (HCF)	2 integer	Sort the given array	[17]
Bubble sort (BS)	1D array	Sort the given array	[1]
Binary search problem (BSP)	1D array and integer value	Find an integer value in a binary search tree	[10]
Prime number or not (PRIME)	1 integer	Sort the given array	[10]
Leap year or not (LEAP)	1 integer	Determine the leap year or not	[9]

Table 3 The configuration parameters for both the GA and PSO algorithms

Algorithm	Parameter	Value
GWO	$\vec{a}$	Between 0 & 1
	$\vec{A}$	Random vector
	Initialization range	{0,255}
PSO	Inertia weight w	$w = 0.9$
	Acceleration constants c_1 and c_2	$c_1 = c_2 = 2.0$
	The maximum velocity V_max	Velocity value is set according to the input space of the program

Figure 3 represents the number of iterations taken by different algorithms for different programs, where the x-axis represents the number of iterations taken and the y-axis represents the program under test (PUT). We have taken the average results of the total ten iterations of the experiments. The genetic algorithm takes the maximum number of iterations to generate the test cases for all programs. Whereas, it is evident from Fig. 3 that the proposed approach takes the minimum number of iterations to generate the test cases for all the PUT.

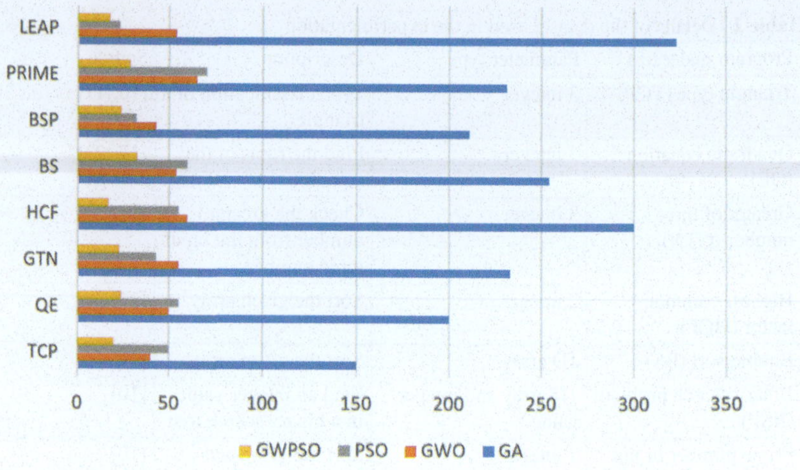

Fig. 3 Comparison of the number of iterations taken for different PUT for different algorithms

5 Conclusion and Future Work

In this study, we introduce the Hybrid GWPSO algorithm as a novel approach for test case generation in code coverage testing. Through experimentation with eight real-world programs, our proposed method demonstrates superior performance compared to traditional algorithms such as GA, PSO, and GWO. Evaluation based on key parameters including the number of iterations, convergence time, and achieved coverage reveals notable advantages of our approach. Specifically, the Hybrid GWPSO algorithm exhibits a significant reduction in the number of iterations required to converge to a solution, outperforming GA in this aspect. Furthermore, our method demonstrates promising results in terms of coverage achieved within the given time frame. Moving forward, future research endeavors will focus on extending the comparison of our proposed approach with other hybrid meta-heuristics and scaling up to evaluate its efficacy on larger-scale programs with increased complexity, measured by the number of lines of code (LOC). This ongoing investigation aims to further validate and refine the effectiveness of our methodology in addressing software testing challenges.

Data Availability Statement The data supporting this research are available upon reasonable request from the corresponding author.

References

1. Cao Y, Hu C, Li L (2009) Search-based multi-paths test data generation for structure-oriented testing. In: Proceedings of the first ACM/SIGEVO summit on genetic and evolutionary computation, pp 25–32
2. Colanzi TE, Assunção WKG, Vergilio SR, Farah PR, Guizzo G (2020) The symposium on search-based software engineering: past, present and future. Inform Softw Technol 127:106372
3. Dalal S, Chhillar RS (2013) A novel technique for generation of test cases based on bee colony optimization and modified genetic algorithm (bco-mga). Int J Comput Appl 68(19):0975–8887
4. Marco D, Mauro B, Thomas S (2006) Ant colony optimization. IEEE Comput Intell Mag 1(4):28–39
5. Bosong D, Chuangqiang G, Hong L (2022) A hybrid genetic-particle swarm optimization algorithm for multi-constraint optimization problems. Soft Comput 26(21):11695–11711
6. Deepak G, Pallvi G (2015) Basis path testing using sga & hga with exlb fitness function. Procedia Comput Sci 70:593–602
7. Dervis K, Bahriye B (2007) A powerful and efficient algorithm for numerical function optimization: artificial bee colony (abc) algorithm. J Glob Opt 39:459–471
8. Kennedy J, Eberhart R (1995) Particle swarm optimization. In: Proceedings of ICNN'95-international conference on neural networks, vol 4. IEEE, pp 1942–1948
9. Khari M, Kumar P, Burgos D, Crespo RG (2018) Optimized test suites for automated testing using different optimization techniques. Soft Comput 22:8341–8352
10. Khari M, Sinha A, Verdu E, Crespo RG (2020) Performance analysis of six meta-heuristic algorithms over automated test suite generation for path coverage-based optimization. Soft Comput 24(12):9143–9160
11. Scott K, Daniel Gelatt Jr C, Vecchi MP (1983) Optimization by simulated annealing. Science 220(4598):671–680
12. Kumar A, Nadeem M, Banka H (2022) Nature inspired optimization algorithms: a comprehensive overview. Evol Syst 1–16
13. Kumar S, Yadav DK, Khan DA (2016) Artificial bee colony based test data generation for data-flow testing. Indian J Sci Technol 9(39):1–10
14. Lakshminarayana P, SureshKumar TV (2020) Automatic generation and optimization of test case using hybrid cuckoo search and bee colony algorithm. J Intell Syst 30(1):59–72
15. Lam SSB, Hari Prasad Raju ML, Swaraj C, Srivastav PR et al (2012) Automated generation of independent paths and test suite optimization using artificial bee colony. Procedia Eng 30:191–200
16. Xue-Wei LV, Song H, Zhan-Wei H, Hai-Jin J (2018) Test cases generation for multiple paths based on pso algorithm with metamorphic relations. Iet Softw 12(4):306–317
17. Mann M, Tomar P, Sangwan OP (2018) Bio-inspired metaheuristics evolving and prioritizing software test data. Appl Intell 48:687–702
18. Mirjalili S, Saremi S, Mirjalili SM, Coelho LS (2016) Multi-objective grey wolf optimizer: a novel algorithm for multi-criterion optimization. Exp Syst Appl 47:106–119
19. Nayak G, Barisal SK, Ray M (2023) Cgwo: an improved grey wolf optimization technique for test case prioritization. Programm Comput Softw 49(8):942–953
20. Ankur P, Gursaran S (2013) Automated test data generation for branch testing using genetic algorithm: an improved approach using branch ordering, memory and elitism. J Syst Softw 86(5):1191–1208
21. Perumal K, Ungati JM, Kumar G, Jain N, Gaurav R, Srivastava PR (2011) Test data generation: a hybrid approach using cuckoo and tabu search. In: Swarm, evolutionary, and memetic computing: second international conference, SEMCCO 2011, Visakhapatnam, Andhra Pradesh, India, Proceedings, Part II 2. Springer, pp 46–54
22. Esmat R, Hossein N-P, Saeid S (2009) Gsa: a gravitational search algorithm. Inform Sci 179(13):2232–2248
23. Omur S, Bahriye A (2016) Comparisons of metaheuristic algorithms and fitness functions on software test data generation. Appl Soft Comput 49:1202–1214

24. Sahoo RR, Ray M (2020) Pso based test case generation for critical path using improved combined fitness function. J King Saud Univ-Comput Inform Sci 32(4):479–490
25. Şenel FA, Gökçe F, Sinan Yüksel A, Yiğit T (2019) A novel hybrid pso–gwo algorithm for optimization problems. Eng Comput 35:1359–1373
26. Sharma S, Rizvi S, Sharma V (2019) A framework for optimization of software test cases generation using cuckoo search algorithm. In: 2019 9th international conference on cloud computing, data science and engineering (confluence). IEEE, pp 282–286
27. Srivastava PR, Khandelwal R, Khandelwal S, Kumar S, Ranganatha SS (2012) Automated test data generation using cuckoo search and tabu search (csts) algorithm. J Intell Syst 21(2):195–224
28. Tracey N, Clark J, Mander K, McDermid J (1998) An automated framework for structural test-data generation. In: Proceedings 13th IEEE international conference on automated software engineering (Cat. No. 98EX239). IEEE, pp 285–288
29. Yang X-S, Slowik A (2020) Firefly algorithm. In: Swarm intelligence algorithms. CRC Press, pp 163–174

BookNest: An Ensemble-Based Book Recommendation System

Jitali Patel, Atir Sakhrelia, Om Vyas, Dhyan Patel, and Ali Asgar Padaria

Abstract In today's data-rich environment, the vast amount of information available to users presents them with countless options. Therefore, the need for an appropriate recommendation system is crucial to ensure that items are suggested to users based on their personal biases and preferences. A book recommender system utilizes historical data to recommend books to users, carefully considering the preferences of each individual. Our study aims to redefine book recommendations by integrating cutting-edge information retrieval techniques with evolving reader preferences. Existing systems often extract irrelevant information, leading to user dissatisfaction. In this paper, we introduce BookNest: An Ensemble Book Recommendation System. Our proposed system revolves around three key components: a Recommendation Engine employing collaborative filtering, genre-based filtering, and popularity-based filtering, and an Ensembling algorithm that consolidates recommendations from various algorithms to produce efficient and effective recommendations. BookNest is evaluated using multiple parameters, and the results demonstrate that the proposed model outperforms traditional recommender systems.

Keywords Book recommendation system · Collaborative filtering · Em-semble learning · Popularity based recommendation · Genre filterting

---

J. Patel (✉) · A. Sakhrelia · O. Vyas · D. Patel · A. A. Padaria
Department of Computer Science and Engineering, Institute of Technology, Nirma University, Ahmedabad, Gujarat 382 481, India
e-mail: jitali.patel@nirmauni.ac.in

A. Sakhrelia
e-mail: 20bce022@nirmauni.ac.in

O. Vyas
e-mail: 20bce181@nirmauni.ac.in

D. Patel
e-mail: 20bce064@nirmauni.ac.in

A. A. Padaria
e-mail: 20bce010@nirmauni.ac.in

P. J. S. Gonçalves et al. (eds.), *Proceedings of Fifth International Conference on Computing, Communications, and Cyber-Security*, Lecture Notes in Networks and Systems 1128, https://doi.org/10.1007/978-981-97-7371-8_17

1 Introduction

In this digital age, there is a lot of information that gives readers numerous options. This makes finding an efficient book recommender system not only useful but also necessary. We are starting this study to find a new way to recommend books by merging complex methods for finding information and people's changing tastes in books. This method not only makes it easier to choose, but it also makes reading more personal and interesting. Against this background, our study aims to push the limits of traditional recommendation systems by usin modern information retrieval techniques to make a Book Recommender System that understands each user's unique tastes.

A recommendation system can improve the effectiveness of book searches by reducing the vast amounts of data produced compared to syntax-based methods. The recommendation system helps users to identify relevant references based on their needs. Hasibuan developed a book recommendation system based on the user's search history [1]. One popular recommendation approach in Web and IoT services is Collaborative Filtering (CF). The recommended items in traditional CF are based on similar users (similar preferences, for example) and can be categorized into two types of algorithms: neighborhood-based and model-based [2, 3].

Our method is a novel mix of old and new ideas. It is based on two main parts: the Recommendation Engine and an Ensembling algorithm. As the brain of our system, the Recommendation Engine uses collaborative filtering, the power of TF-IDF for Natural Language Processing (NLP), and genre-centric preferences to make sure that each user gets the exact book choices they want. By combining the best parts of user reviews and book descriptions, our method makes sure that recommendations are perfectly tailored to each user's unique tastes, going beyond the limits of traditional recommendation systems.

The second part of our system, the Ensembling algorithm, adds a level of complexity by combining suggestions from different algorithms into a single suggestion that reflects the system's overall knowledge. This joint decision-making process makes it even better by encouraging user participation by clearly outlining each suggestion, making the whole process more open and easy to use. By combining data-driven insights, advanced algorithms, and a focus on the user, we can give readers a personalized journey through the vast and fascinating world of literature.

1.1 Organization

The remaining paper is organized as follows: Sect. 2 describes the Related Work done in this domain. Section 3 outlines the datasets utilized in the research. Section 4 explains the ensemble architecture and recommendation engines employed. Section 5 analyzes the results of the suggested architecture. Finally, Sect. 5 discusses the future scope of the research and the conclusions drawn from the current technique.

2 Related Work

Classic collaborative filtering systems face scalability and cold-start issues, but Singular Value Decomposition (SVD) and Matrix Factorization models have been successfully used to address these issues. The hybrid recommendation model suggested in [4] has put forth an effort to do away with the cold start. The FBSM personalized recommendation model, utilizing the factorized bilinear similarity model, was developed and proven effective through numerous experiments. Pujahari and Sisodia [5] focused on using relation-based Matrix Factorization (MF) to forecast user preferences and graph aggregation to aggregate those preferences. They also suggested including item features in the MF model. Ahmed et al. [6] presented an MF model to solve the cold start problem. In the model-based method, K-Nearest Neighbour and SVD are used, with an accuracy of 85% and 53%, respectively. Also, proper hyperparameter tuning with SVD was carried out to improve the overall performance. In addition, proper hyperparameter optimization using SVD was performed to increase overall performance.

A problem with the typical recommendation system is the large-volume, multi-dimensional, and expanding data in Big Data. This challenge is addressed by using clustering-based collaborative filtering techniques [7–9], which reduce the quantity of the data while maintaining accurate prediction. Clustering is a preprocessing step in collaborative filtering that gathers data from users who are similar to one another in order to make subsequent recommendations. Numerous clustering-based collaborative filtering techniques have been thoroughly studied; these include co-clustering, item-based clustering, and user-based clustering. Hu et al. [10] suggested a novel CF paradigm called ClubCF. ClubCF's two-stage strategy—collaborative filtering and clustering—allows it to function well in environments with big databases.

The work in article [11] discusses the importance of recommendation systems in businesses, particularly in book suggestions. It discusses the use of algorithms, item-based and content-based collaborative filtering, in custom recommender systems. The study highlights the difficulty in finding relevant books due to large amounts of data. The custom recommender effectively picks highly rated books, making it a suitable choice for online book enthusiasts. The article emphasizes the need for diverse results and avoids favoring popular options. The studies [12, 13] suggest a new book recommendation system that uses users' interesting content as queries to suggest books to avid readers. The system collects tweets, book reviews, and emotion vectors, using modern algorithms like ML-Ask and Sentence-BERT. The system accurately reflects user choices, and the study suggests making it more flexible and user-friendly, a significant step forward in personalized book recommendation systems.

The authors of [14] propose a new method using Convolutional Neural Network (CNN) methods to address scalability and sparsity issues in collaborative filtering-based book recommender systems. The system uses Kaggle's user, book, and rate data, with CNN design features like one-hot encoding, 1D convolutional layers, dropout methods, and merging layers. The model's performance is evaluated using

Mean Squared Error, Mean Absolute Percentage Error, and Mean Absolute Error. The model's promising results suggest future research on other recommendation methods and overfitting. The authors used Mean Squared Error, Mean Absolute Percentage Error, and Mean Absolute Error to evaluate the model's performance. The best model showed promising results with a MAPE of 23,767 and an MSE of 3,034, indicating potential future research on overfitting. The study by [15] introduced a Deep Learning-based Matrix Factorization Model to enhance recommendation accuracy by learning complex features from user interaction data.

3 Dataset Description

The section that follows offers a full explanation of the datasets chosen in line with the scope of the proposed technique and tweaked to fit specific requirements aimed by us during the inception of the methodology. We carefully chose and combined two different datasets—the Amazon dataset and the Book Crossing dataset to aid our research.

Amazon Dataset There are two files in this dataset. The first file, "reviews file," compiles feedback from 3 million individuals providing input on 212,404 distinct books spanning the extensive period of May 1996–July 2014. The file encompasses attributes detailing the diverse aspects of the reviews.

The second file, "Books Details file," delves into comprehensive information regarding the 212,404 unique books featured in the dataset. This specific file is curated by utilizing the Google Books API, which extracts details about the books evaluated in the initial file. The contents of this file serve to augment the dataset with additional contextual information. The dataset is freely available on Kaggle and is provided as a unified file. Figure 1 showcases the two files along with various attributes present in the respective files.

Book Crossing Dataset The Book Crossing dataset is readily accessible on Kaggle and comes in two versions. We opted for the convenience of the second

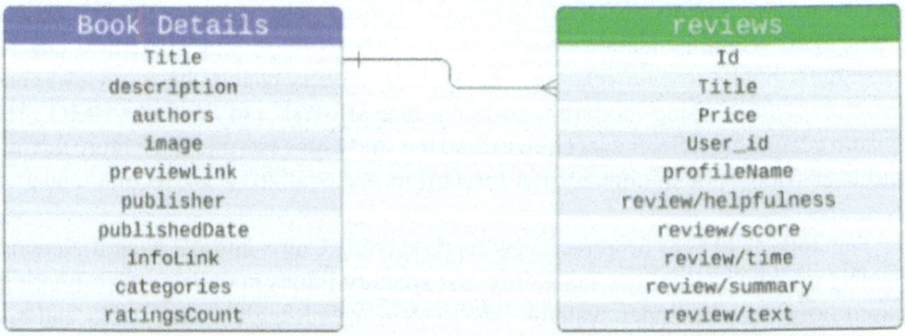

Fig. 1 Amazon book dataset description

version, which consolidates specific aspects—book ratings, user details, and comprehensive information about various books into a single file. This unified file is quite robust, featuring 19 columns and a substantial 1,031,175 rows of data.

The dataset contains information about user ratings, book publishers, publication years, summaries, languages, and book categories. Additionally, it includes links to images of both the book cover and the back page. The dataset holds data on 241,090 unique books and 101,594 distinct authors.

3.1 Pre-Processing

The first step in the proposed methodology was the integration of the two individual datasets which were accomplished through a join operation that relied on the commonality found in the "Title" field, resulting in the creation of a temporary dataset. The temporary dataset stated above with the second Amazon file which provided user reviews was merged using the "Title" field as the connecting point. The result of this complicated process was our final collection.

To clean this raw data, we included some preprocessing steps. Every row that wasn't needed was carefully removed. Details like review time, book price, user profile names, book picture links, age, city, and more were dropped. Any duplicate records were also systematically deleted to avoid duplicity.

We used strategic filtering processes to narrow down the dataset as we got ready for the recommendation phase. We specifically looked at people who had been kind enough to give at least 150 ratings. In addition, our sample only included books that had at least 25 reviews. The goal of this careful filtering was to create a dataset that was reliable and consistent, which would serve as a solid base for our later analysis of recommendations.

4 Proposed Architecture

In the subsequent sections, we outline our system in detail. These sections provide a thorough investigation of the system architecture, a succinct discussion of each component's approach, and, finally, the incorporation of these aspects into a coherent framework for recommending books.

Our approach is built around two primary components: The robust recommendation engine is the initial component of our framework, which uses three different types of algorithms. These algorithms carefully look at the user's profile and past contacts to make a list of 10 books that are to be recommended specifically for that person. Next, the second component adds an Ensembling algorithm, a complex system that combines the suggestions made by the first set of algorithms. The goal of this process is to come up with a final suggestion by finding a book that either gets the most votes or is a popular choice among these three algorithms. We showcased our unique way of making a recommender system, which is different from other approaches because our goal was to improve the recommendation process by using Ensemble learning. This one-of-a-kind addition should give the recommender system a new dimension, making the user experience better.

4.1 Recommendation Engine

The primary component of our recommendation system is the Recommender Engine, which gives users personalized suggestions based on their past actions and tastes. This part explains the methods and techniques used by the engine, showing that it takes a multifaceted approach. In order to make the best recommendations possible, we've added three different ways to the recommender Engine. The first method is collaborative filtering, which uses information about how users behave and what they like to make personalized proposals. The second method is based on what people are interested in. It uses trends and how active users are to find suggestions that are related to their overall interests. In the third way, recommendations are based on the genres that users like, making them more appropriate. The working of the recommendation engine is referred in Fig. 2.

Collaborative Filtering Collaborative filtering (CF) stands as a widely adopted strategy in recommender systems, gathering user evaluations expressed through ratings assigned to various items, services, or service providers. Through collaborative filtering, recommendations are provided for items that have gained popularity

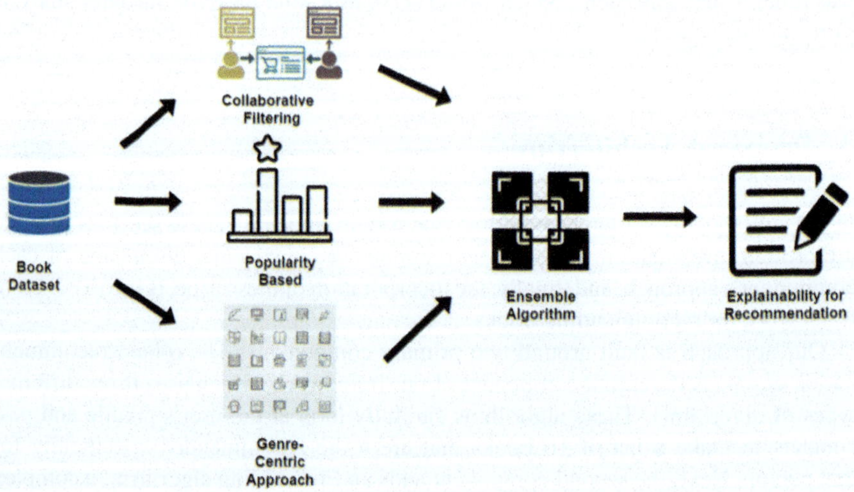

Fig. 2 BookNest architecture

among users with similar preferences. This technique encompasses different variations, such as user-based, item-based, and model-based approaches. The user-based approach recommends items based on similar past behaviors or preferences, aligning with collective preferences. It enhances the recommendation experience. On the other hand, item-based collaborative filtering identifies items with similar past user interactions, suggesting similar items to those already engaged with the user, and offering a wider range of relevant recommendations. The study employs a user-item-based filtering approach and TF-IDF NLP technique to create a vector representation of user reviews and book summaries. TF-IDF evaluates the importance of words within a document, identifying frequent but rarely occurring words with higher scores, thereby identifying specific keywords. The cosine similarity calculation is used to identify the books that match the user's review vector representation, ensuring that the suggestions are tailored to each user's preferences and tastes, thereby improving the overall user experience.

Popularity-Based Recommendation The cold start problem in recommender systems is a common issue where the system lacks sufficient information about new users or items, making it difficult to generate personalized recommendations. This problem occurs when a new item or user joins the system without any user interaction history. To address this, a popularity-based approach is proposed, which recommends popular items to new users or items, ensuring the system still provides value despite limited data.

The popularity-based recommendation system uses a simple approach, considering unique books and user ratings. The highest-rated books are recommended. A preprocessing step filters the data, excluding books with at least 25 user ratings. This ensures consistency in rating evaluations and user engagement, ensuring the recommendations reflect overall popularity and reliability.

Genre-centric recommendation The generic-centric recommendation system is a content-based recommendation system that suggests items based on their similarities with objects that piqued the user's interest. This approach differs from collaborative filtering, which assesses item similarity based on user actions and resemblances, focusing solely on the inherent characteristics of the items, and avoiding subjective user ratings.

The recommendation system used is a hybrid of content-based and popularity-based systems. It uses the user's favorite genre as a focal point for genre-centric recommendations. The system identifies high-rated books from the user's profile, extracts them from the associated dataset, calculates the average rating across the genre, and recommends the books with the highest average rating, ensuring recommendations align with the user's genre preferences.

4.2 Ensembling Learning

The paper proposes an ensemble architecture model for recommending books to users. This model uses inputs from multiple learning models to make more accurate conclusions, aiming to remove errors or biases. Instead of traditional machine learning models, ensemble learning is accepted as the core idea. The methodology synchronizes diverse engine suggestions and average ratings, enabling a sophisticated suggestion strategy. Books that regularly appear in multiple recommendation systems are of particular interest, as their recurring appearance signals congruence with user preferences. The Algorithm 1 illuminates the mechanics of the proposed method.

The model also considers each book's average rating, ensuring recommendations are based on varied suggestions and user satisfaction. This comprehensive approach improves the resilience and customization of our recommendation system by adapting to the varied interests and degrees of satisfaction of individual users. Also recommending books that are a bit diverse from the user preference tends to provide the system with more data and make the system learn more.

Algorithm 1 Ensemble Learning Algorithm

Require: User profile, past contacts
Ensure: Final list of recommended books

```
Input: Recommendations from diverse engines and average ratings of books
Output: Final recommendations
Initialize an empty list of recommended books R
Initialize an empty dictionary book_popularity
for each recommendation engine do
    Obtain recommendations based on user profile and past contacts
    for each recommended book do
        Add book to R
        Update book_popularity with the book and its popularity score
    end for
end for
Compute intersection of recommended books from different engines
Sort recommended books based on their average ratings and popularity score
Select the top 10 unique books from the sorted list
    return Final recommendations
```

5 Results

This section is organized into two subsections. The first subsection covers the evaluation measures used to assess recommender system. The next section discusses the results obtained using the evaluation metrics stated in the previous subsection.

5.1 Evaluation Parameters

In recommender systems, evaluation plays a crucial role in assessing the performance of algorithms. Common metrics include Mean Absolute Error (MAE), Root Mean Squared Error (RMSE), and Mean Squared Error (MSE). These metrics quantify the disparity between predicted and actual ratings provided by users.

Root Mean Square Error (RMSE) is calculated as the square root of the mean of the squared variances between predicted and observed values. Essentially, it calculates the average difference between predicted and actual values. A lower RMSE indicates better model performance, indicating that the predictions are more closely related to the actual data.

$$\mathrm{RMSE}(y, \hat{y}) = \sqrt{\frac{\sum_{i=0}^{N-1} (y_i - \hat{y}_i)^2}{N}} \tag{1}$$

Mean squared error (MSE) quantifies the average squared difference between predicted and actual values, offering insight into a model's predictive accuracy. Lower MSE values indicate better alignment between predictions and data. In recommendation systems, MSE compares predicted item ratings to user-provided ratings.

$$\mathrm{MSE}(y, \hat{y}) = \sqrt{\frac{\sum_{i=0}^{N-1} (y_i - \hat{y}_i)^2}{N}} \tag{2}$$

Mean Absolute Error (MAE) is a regression metric measuring the average magnitude of errors between predicted and actual values. Unlike MSE, it doesn't square the differences, making it robust to outliers. MAE provides a straightforward assessment of predictive accuracy, with lower values indicating closer alignment between predictions and actual outcomes.

$$\mathrm{MAE}(y, \hat{y}) = \sqrt{\frac{\sum_{i=0}^{N-1} (y_i - \hat{y}_i)}{N}} \tag{3}$$

5.2 Results

The performance evaluation of different recommendation systems reveals distinct patterns. When considering the Root Mean Squared Error (RMSE), the popularity-based recommendation displays the highest error rate at 2.199, followed by the generic-based recommendation at 1.579, and the user-item-based recommendation excelling with the lowest RMSE of 1.485. Analyzing the Mean Squared Error (MSE) metrics for each system, the popularity-based model records 4.835, the user-item-based model demonstrates 2.205, and the generic-centric system performs notably better with an MSE of 2.493.

Delving deeper, the Mean Absolute Error (MAE) emphasizes the disparities further: the popularity-based system registers 1.634, the generic-centric approach shows improvement at 1.452, while the user-item-based method continues to outperform both with an MAE of 1.255. Table 1 and Fig. 3 outline the comparison of various evaluation measures for various recommendation engines.

In summary, the user-item-based recommendation system consistently outperforms the others in terms of accuracy across all evaluation metrics, while the generic-based approach shows promising results compared to the popularity-based one, which exhibits the highest error rates in these evaluation metrics.

The ultimate ensemble-based model demonstrates superior performance across multiple evaluation metrics compared to individual models. Specifically, the ensemble model showcases an improved Root Mean Squared Error (RMSE) of 1.273, a Mean Squared Error (MSE) of 1.620, and a Mean Absolute Error (MAE) of 1.098.

This ensemble model's collective strength surpasses that of the standalone models. The RMSE, an important gauge of predictive accuracy, signifies the ensemble's ability to predict with an average error of 1.273 units, showcasing its enhanced precision compared to individual models. Moreover, the MSE, measuring the average

Table 1 Comparison of different recommendation systems

Recommendation systems	RMSE	MSE	MAE
Genre-Centric	1.579	2.493	1.452
Popularity	2.199	4.835	1.634
User-Item	1.485	2.205	1.255
Ensemble	1.273	1.620	1.098

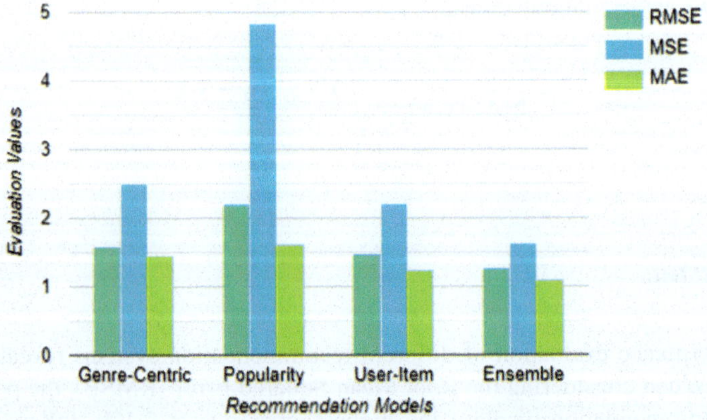

Fig. 3 Comparative analysis of results across various recommendation systems

of squared differences between predicted and actual values, stands at 1.621, indicating the ensemble's minimized overall prediction errors.

6 Future Work and Conclusion

To sum up, our study sets out to redefine book recommendations in the digital era, acknowledging the wealth of information and diverse reader preferences. Recognizing the vast variety of books, authors, and stories available, we underscore the significance of recommendation systems in simplifying choices and enhancing personalized reading experiences. Our goal is to surpass the constraints of the traditional recommendation system.

The crux of our approach consists of the Recommendation Engine and the Ensembling algorithm. The Recommendation Engine utilizes collaborative filtering, popularity-based, and genre-centric preferences to offer precise book recommendations tailored to individual user preferences. Our emphasis on collaborative filtering underscores our commitment to accuracy and customization, surpassing traditional system limitations. The Ensembling algorithm adds complexity by amalgamating suggestions from multiple algorithms into a cohesive recommendation, reflecting the system's extensive knowledge. Our focus on user participation ensures transparency, fostering an open and user-friendly recommendation process.

We aim to transcend mere recommendation system creation, aspiring to revolutionize the landscape of book recommendations entirely. Through the fusion of data-driven insights, advanced algorithms, and a user-centric approach, we seek to provide readers with a personalized journey through the enchanting world of literature. This study represents a noteworthy stride towards enhancing personalized reading experiences within the realm of literature.

Future works involve creating an effective method for users to give comments on suggested books. It improves the system's overall performance. Adding measurements for user engagement can also assist in determining how well the explainable book recommender system is performing. Expanding the system to accommodate more users. To ensure that the system operates properly in a production environment, it must be tested under a variety of conditions, user loads, and datasets. Refine, extend, and validate the system, paving the way for a more sophisticated and user-friendly experience in the realm of personalized book recommendations.

References

1. Hikmatyar M, et al. (2020) Book recommendation system development using user- based collaborative filtering. J Phys: Conf Ser 1477 p 032024. IOP Publishing
2. Chen H, Li J (2017) Learning multiple similarities of users and items in recommender systems. In: 2017 IEEE International Conference on Data Mining (ICDM). pp 811–816. IEEE

3. Liu J, Tang M, Zheng Z, Liu X, Lyu S (2015) Location-aware and personalized collaborative filtering for web service recommendation. IEEE Trans Serv Comput 9(5):686–699
4. Sharma M, Zhou J, Hu J, Karypis G (2015) Feature-based factorized bilinear similarity model for cold-start top-n item recommendation. In: Proceedings of the 2015 SIAM International Conference on Data Mining. pp 190–198. SIAM
5. Pujahari A, Sisodia DS (2020) Pair-wise preference relation based probabilistic matrix factorization for collaborative filtering in recommender system. Knowl-Based Syst 196:105798
6. Ahmed E, Letta A, et al. (2023) Book recommendation using collaborative filtering algorithm. Appl Comput Intell Soft Comput
7. Mirbakhsh N, Ling CX (2018) Leveraging clustering to improve collaborative filtering. Inf Syst Front 20:111–124
8. Salah A, Rogovschi N, Nadif M (2016) A dynamic collaborative filtering system via a weighted clustering approach. Neurocomputing 175:206–215
9. Zhou Z, Cheng Z, Zhang LJ, Gaaloul W, Ning K (2016) Scientific workflow clustering and recommendation leveraging layer hierarchical analysis. IEEE Trans Serv Comput 11(1):169–183
10. Hu R, Dou W, Liu J (2014) Clubcf: A clustering-based collaborative filtering approach for big data application. IEEE Trans Emerg Top Comput 2(3):302–313
11. Shukla N, Soni N, Gupta N, Anand N (2022) Online book recommendation system using custom recommender. In: 2022 6th International Conference on Trends in Electronics and Informatics (ICOEI). pp. 921–925. https://doi.org/10.1109/ICOEI53556.2022.9777131
12. Fujimoto T, Murakami H (2022) A book recommendation system considering contents and emotions of user interests. In: 2022 12th International Congress on Advanced Applied Informatics (IIAI-AAI). pp 154–157. https://doi.org/10.1109/IIAIAAI55812.2022.00039
13. Tegetmeier C, Johannssen A, Chukhrova N (2023) Artificial intelligence algorithms for collaborative book recommender systems. Ann Data Sci pp 1–35
14. Putri A, Abdurahman Baizal ZK, Rischasdy D (2022) Book recommender system using convolutional neural network. In: 2022 International Conference on Advanced Creative Networks and Intelligent Systems (ICACNIS). pp 1–6. https://doi.org/10.1109/ICACNIS57039.2022.10055753
15. Dien TT, Thanh-Hai N, Thai-Nghe N (2022) An approach for learning resource recommendation using deep matrix factorization. J Inf Telecommun 6(4):381–398

A Smartphone-Based Deep Learning Framework for Detection and Classification of Poultry Diseases from Faecal Images

P. Kaviya, S. Sibi Siddharthan, M. Kishore, and M. Muthuram

Abstract The poultry industry is a critical global source of food and economic stability. Despite its significance, the poultry industry faces persistent threats from diseases, which can result in significant economic losses and potential public health risks. Accurate and fast identification of infections can help poultry farmers choose appropriate treatment, saving economic loss and ensuring public health. In this regard, a deep learning model can help to identify the poultry diseases early. In this work, a transfer learning model ResNet-9 is proposed for the detection of poultry diseases like Salmonella, New Castle Disease (NCD) and Coccidiosis. The suggested model achieves 98.83% accuracy on the Kaggle Poultry Diseases Detection dataset. A mobile application is developed that enables the poulterer to easily identify the diseases by uploading the image of the chicken's faecal. The suggested model's performance is compared to other current research, demonstrating that it outperforms previous results.

Keywords Poultry disease detection · Data augmentation · Deep learning · Mobile application

1 Introduction

Poultry is a significant aspect of the world food supply, producing enormous quantities of meat and eggs. Among the various poultry species, which include chickens, ducks and turkeys, chickens stand out because of their widespread consumption and significant economic impact [1]. The poultry industry faces significant challenges in maintaining health, with notable concerns such as Salmonella, a bacterial infection, Newcastle Disease (NCD), and a viral infection that affects various bird species,

P. Kaviya (✉) · S. Sibi Siddharthan · M. Kishore · M. Muthuram
Department of Information Technology, Kamaraj College of Engineering and Technology, Madurai, Tamilnadu 625701, India
e-mail: kaviyaap@gmail.com

P. J. S. Gonçalves et al. (eds.), *Proceedings of Fifth International Conference on Computing, Communications, and Cyber-Security*, Lecture Notes in Networks and Systems 1128, https://doi.org/10.1007/978-981-97-7371-8_18

including chickens [1]. These diseases not only result in economic losses of poultry farming but also pose potential risks to food safety.

For proper mitigation, diseases must be continuously monitored and detected. In general, diseases are extremely difficult to detect and recognize with the naked eye. Also, it is a time-consuming process. One of the most difficult aspects of poultry farming is detecting diseases on time. To reduce economic losses and health hazards, a rapid and reliable disease detection system is required. Proper care is required depending on the severity of the disease. Thus, the production loss can be avoided. In disease detection mechanism, features are very important to distinguish healthy and diseased chicken. Diseases are classified only based on the features.

The integration of advanced technologies, particularly Artificial Intelligence (AI), has emerged as a powerful solution for detecting poultry diseases [2]. AI-based systems, incorporating machine learning algorithms, analyse extensive datasets related to poultry health, which enables to identify the patterns and anomalies associated with diseases. This early detection capability facilitates prompt preventive measures.

These deep learning algorithms, specifically designed for image recognition tasks, analyse images of chicken's faeces to identify specific patterns or abnormalities. Recent studies have shown that ResNet-9 has a significant role in improving the accuracy of poultry illness detection [3, 4]. The major contribution of this work is given as follows:

1. ResNet-9, a transfer learning model is proposed to detect poultry diseases from chicken's faeces including Salmonella, NCD and Coccidiosis on Kaggle poultry disease dataset
2. A mobile application is developed to detect the diseases in real time

The rest of this work is split into parts. Section 2 provides a review of the literature. Section 3 covers the proposed methodology, which includes dataset features and poultry disease detection. Section 4 evaluates accuracy and loss measures of the proposed model against a pre-trained model. Finally, Sect. 5 highlights the findings and recommends potential future directions.

2 Related Works

This literature review delves into current research on the prediction of chicken diseases using image-based diagnostic techniques. In recent studies, many researchers adopted deep learning algorithms in various applications like sentimental analysis [5–9], disease detection [10, 11], network security [4, 12] and face recognition [13]. AI-based poultry disease detection methodologies are discussed as follows: Robi, et al. [14] suggested machine learning algorithms for predicting neonatal disease. They employed XGBoost (XGB), Support Vector Machine (SVM) and Random Forest (RF) to identify birth asphyxia, sepsis, necrotizing enteric colitis (NEC) and respiratory distress syndrome. This model did exceptionally well, with

97.04% accuracy rate. Mbelwa, et al. [15] proposed a convolutional neural network for chicken disease detection. They have used a transfer learning approach on a pre-trained XceptionNet to detect chicken diseases and achieved an accuracy of 93.67% using faecal images collected from Kilimanjaro and Arusha regions.

Muslihudin, et al. [16] have proposed a fuzzy hierarchical approach to predicting chicken disease. They have employed the analytical hierarchy process, which is one of the decision-making techniques used to detect Marek, IB chicken and chicken NP. Infectious bronchitis (IB) disease obtained the highest performance value of 34.64%. Wang, et al. [17] proposed an automated broiler digestive illness detector that uses a deep Convolutional Neural Network (CNN) model to categorize fine-grained aberrant broiler droppings pictures as normal or abnormal. His model earned an accuracy of 93%.

Machuve, et al. [2] have designed poultry disease diagnostic models using deep learning. They refined CNN models by fixing the batch normalization layer and improving model accuracies to 95.45% for InceptionV3, 95.01% for VGG16, 98.24% for Xception and 98.02% for MobileNetV2. Shori, et al. [18] have presented avian influenza prediction using machine learning approaches. They have used different machine learning approaches like RF, Artificial Neural Network (ANN) and SVM to detect avian influenza. The RF method achieved 91.63% accuracy rate.

He, et al. [19] presented deep residual-based learning for recognizing images. To detect common objects in context detection and segmentation, they used residual learning. Banakar, et al. [20] demonstrated an intelligent device for identifying avian illnesses. They have used SVM classifier to classify the diseases like Newcastle disease, Avian influenza and Bronchitis virus. The experimental results earned an accuracy of 83.33%.

Okinda, et al. [21] have used various machine learning techniques for the early detection and prediction of sick birds. They have used SVM and Radial Basis Function (RBF) kernels to detect zoonotic infections. The RBF-SVM performed well with an accuracy of 97.5%. Borgonovo, et al. [22] developed a data-driven predictive model for prompt identification of coccidiosis. They have used the k-Nearest Neighbour (KNN) algorithm for detecting the coccidiosis rapidly.

Quach et al. [3] reported a poultry disease identification system that uses SURF to augment the ResNet-50 model. They used SURF feature analysis in conjunction with k-means to determine the important features. This ResNet-50 model resulted with an accuracy of 93.56%. Zhang, et al. [23] have designed an automatic sick chicken detection system using residual network. They used Feature Pyramid Network (FPN) for improving the traditional ResNet model and to adapt various recognition environment. The model performance shows the testing accuracy is 93.7%.

Quach, et al. [24] have identified the chicken diseases using deep learning models like VGGNet and ResNet. These methods could find the four common diseases in chickens: Avian Pox, Newscastle, Marek and Infectious Laryngotracheitis. They have achieved the accuracy of 66.91% and 74.1% for ResNet-50 and VGGNet-16 respectively.

Table 1 Kaggle poultry chicken's disease dataset

Disease class	Number of training samples	Number of testing samples
Healthy	1646	411
NCD	301	75
Salmonella	1821	455
Coccidiosis	1683	420

Fig. 1 Sample images from dataset

3 Dataset

The Kaggle poultry disease dataset [25] comprises 6812 images categorized into four classes: Coccidiosis (2103 instances), Healthy (2057 instances), New Castle Disease (376 instances) and Salmonella (2276 instances). The distribution given in Table 1 reveals that Salmonella constitutes 33.47%, Coccidiosis 30.85%, Healthy 30.16% and New Castle Disease 5.52% of the dataset. To address class imbalance, oversampling techniques like random oversampling (ROS) are often used. Figure 1 displays some of the sample images in the Kaggle Poultry Disease dataset.

4 Proposed System

The suggested method for modelling poultry disease detection has four stages: data collection, pre-processing, model construction and deployment. Figure 2 depicts the overall framework for poultry disease detection. The image dataset for poultry disease detection is collected from Kaggle which includes four classes: Coccidiosis, New Castle Disease, Salmonella and Healthy. In pre-processing, data augmentation and resizing of images are done to improve the accuracy rate of the model. Then a transfer learning model, ResNetv9 that is already trained using CIFAR-10 dataset is employed for detecting the poultry diseases. Finally, the trained model is loaded into a mobile application with enhanced interface for the real-time detection of poultry diseases.

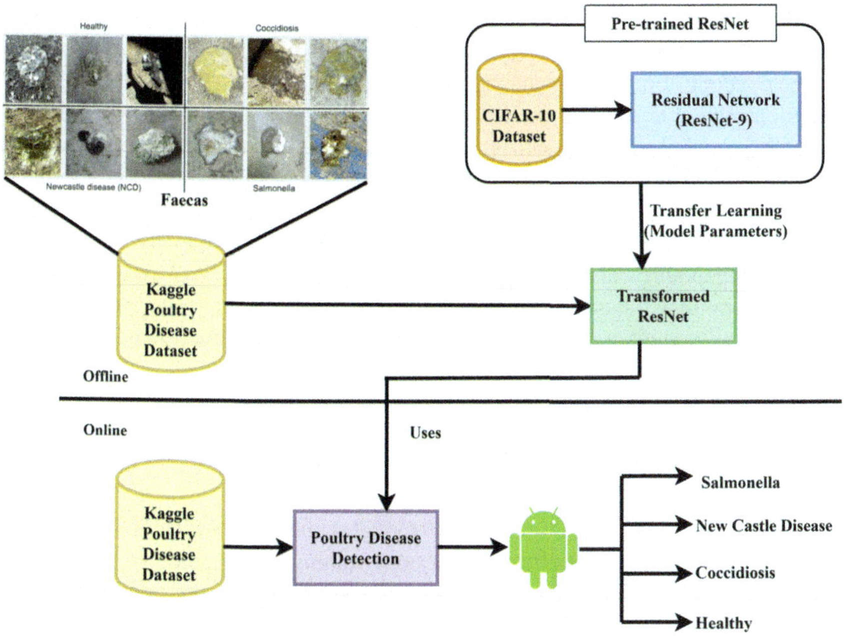

Fig. 2 Proposed framework for poultry disease detection

4.1 Data Augmentation and Preprocessing

The pre-processing phase is critical for ResNet-9, as it involves training images with various augmentations to increase diversity and robustness. This includes rotation, width and height adjustments, shearing, zooming and horizontal flipping. During training and validation, all pixel values are rescaled to the range [10] using rescale = 1.0/255. This normalization ensures numerical stability and convergence throughout training. This precision is critical to ResNet-9 training effectiveness because it maintains the optimal pixel value range for neural network convergence.

4.2 Feature Extraction

The image processing block in this architecture is responsible for receiving and processing input images. Next, the feature extractor extracts important patterns and characteristics from the image that have been analysed. These significant features are then reduced to a reasonable form by the Global Features module.

The classifier receives the condensed data and uses the extracted features to create accurate predictions and classifications. The sequential flow of the suggested model's feature extraction, global feature condensation and classification phases after image processing highlights its methodical and efficient design.

The concept of residual blocks serves as the foundation for the ResNet-9 design and is employed to address the issue of vanishing or exploding gradients in deep neural networks. ResNet-9 consists of nine layers: one fully connected layer and eight convolutional layers. 60,000 32 × 32 colour photos divided into 10 classes make up the CIFAR-10 dataset, which is used to train the architecture. Because of its unique design and strong picture classification skills, ResNet-9 is an excellent architecture for detecting poultry diseases. The model shows competency in recognizing and classifying poultry illnesses from faecal pictures. It incorporates max-pooling layers for dataset alignment and convolutional layers for feature extraction.

4.3 *ResNet-9 Architecture*

Using a ResNet architecture, the ResNet-9 model is made for the detection of diseases in poultry. It uses batch normalization for better training convergence and is composed of convolutional blocks, residual connections and global average pooling. Figure 3 represents the architecture of ResNet-9.

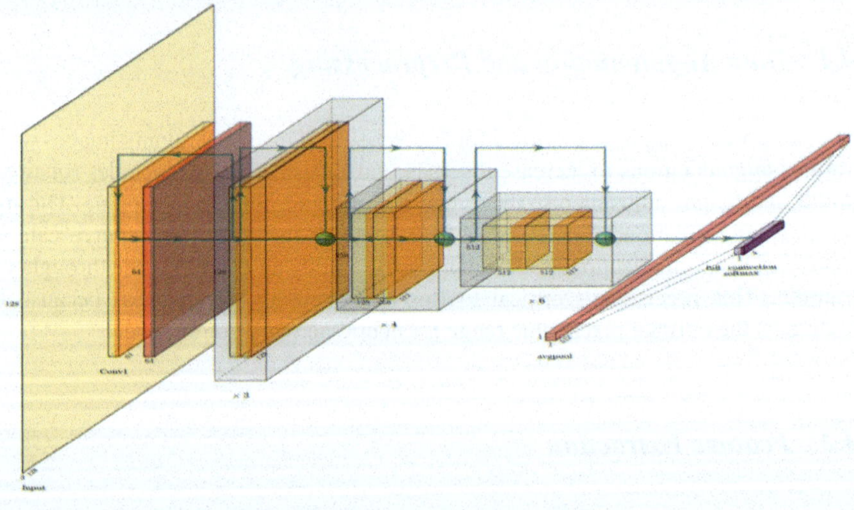

Fig. 3 Transformed ResNet-9 architecture

4.3.1 Convolutional Block

The basic building block of the model is defined by the Conv block function. It includes batch normalization, an optional max pooling layer and a 3×3 convolutional layer. It makes spatial down sampling and feature extraction effective.

4.3.2 Residual Connections

It uses residual connections between the res1 and res2 residual blocks, which are two sequential blocks, which helps to train deeper networks more easily and mitigates the issue of vanishing gradients. This connection improves the model's capacity to represent intricate hierarchical characteristics. Residual network is introduced to solve the vanishing or exploding gradient. This network pioneered the residual block idea. Residual blocks are formed using skip connection that skip some layers and connect activations of a layer to further layer. ResNets are formed by stacking these residual blocks together. The strategy enables the network to fit the residual mapping. Figure 4 represents the residual block used in the network.

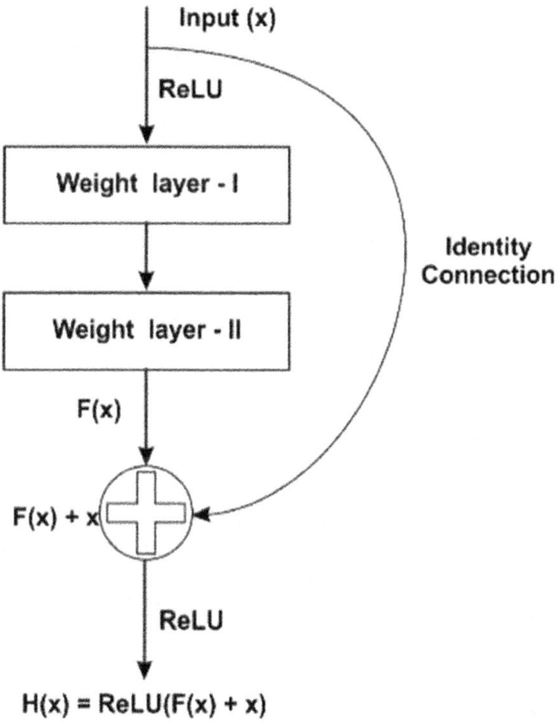

Fig. 4 A Residual block used in the network

4.3.3 Global Average Pooling

The global average pooling is implemented before the fully connected layers. It shrinks the spatial dimensions, produces a feature vector with a fixed size and also allows the disease classification model to concentrate on the most important features.

4.3.4 Fully Connected Layers

Fully linked layers for disease categorization (fc1 and fc2) conclude the model. The 'fc1' contributes to the non-linearity in feature representation by having 128 units with ReLU activation. SoftMax activation is used by fc2 to provide a probability distribution across disease classes in a multi-class classification.

Algorithm 1. provides the overall process of poultry disease detection using the transfer learning model ResNetv9.

Algorithm 1 Poultry disease detection

Data: Datasets (PDD) **Output:** Poultry Diseases

1. Read the input image
   1.1 $img \leftarrow load\_img(path, size)$;
2. Data Preprocessing
   2.1 $datagen \leftarrow ImageDataGenerator(rescale = 1.0/255,$
   $rotation\_range = 20,$
   $width\_shift\_range = 0.2,$
   $height\_shift\_range = 0.2,$
   $shear\_range = 0.2,$
   $zoom\_range = 0.2,$
   $horizontal\_flip = True,$
   $fill\_mode = nearest$
3. Build Poultry Disease Detection Model
   3.1 $DiseaseDetectionModel \leftarrow ResNetv9(datagen)$

---

5 Performance Evaluation

In this study, a transfer learning model ResNet-9 is developed for predicting the chicken diseases based on faecal images. The performance of the suggested model is assessed using a variety of measures, including accuracy, precision, recall and F1-Score. The model exhibited strong performance in classifying the diverse chicken diseases with an accuracy of 96.83%, precision of 99.67%, recall of 97.98% and F1-Score of 94.17%. This affirms the model's well-balanced performance, addressing both false positives and false negatives, which is crucial in the context of medical diagnostics. Figures 5 and 6 exhibit the proposed model's training and validation accuracy and loss, respectively. The training and validation curves provide valuable

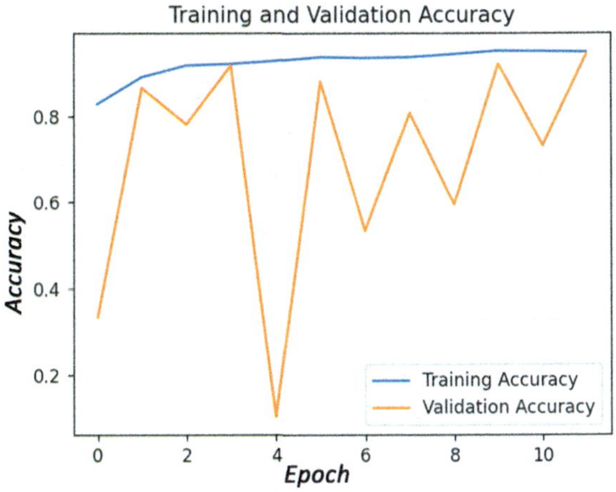

Fig. 5 Training and validation accuracy of the chicken disease detection model

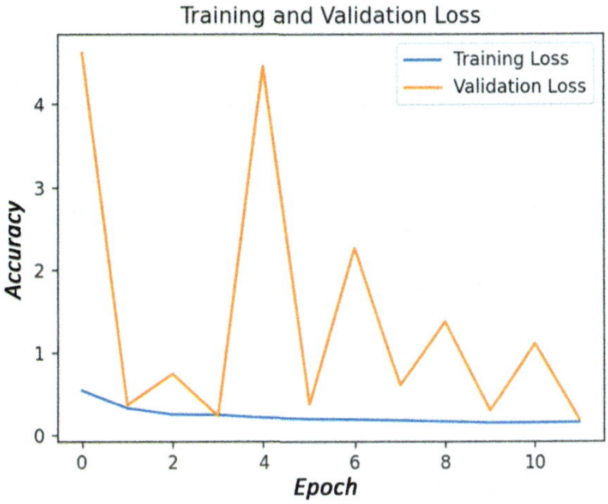

Fig. 6 Training and validation loss of the chicken disease detection model

insights of the performance of the ResNet-9. In the graph, two key metrics accuracy and loss are visualized over the specified number of training epochs.

Table 2 illustrates a comparison of the proposed model to various current models. Figure 7 shows that the proposed model is above par with other architectures such as VGG, YOLOv5 and ResNet-50.

Table 2 Comparison of proposed model with existing works

Authors	Model	Accuracy
Quach et al. [3]	VGG + ResNet	74.1%
Quach et al. [24]	ResNet50	93.56%
Cui et al. [26]	YoloV5	97.80%
Degu et al. [27]	Yolo V3 + ResNet50	98.70%
Proposed Model	**Transfer Learning based ResNet-9**	**98.83%**

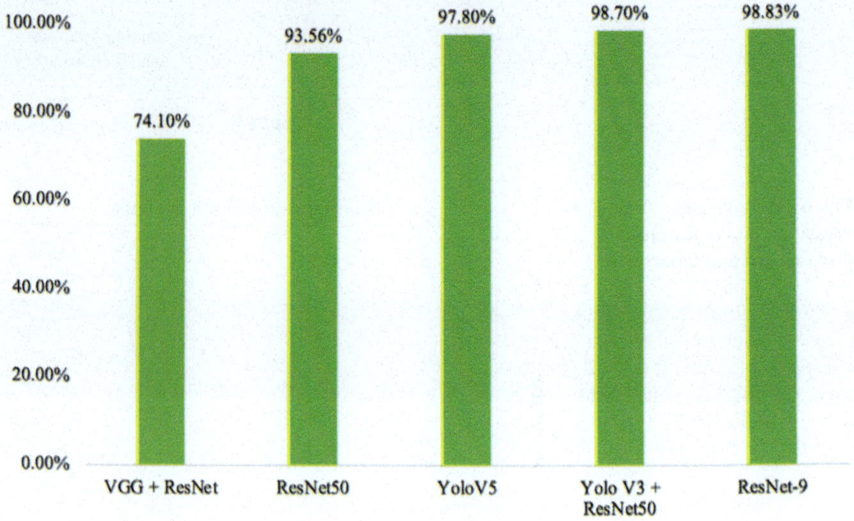

Fig. 7 Comparison of proposed model with existing works

6 Conclusion

In this work, ResNet-9, a transfer learning model is proposed to detect poultry diseases from chicken's faeces including Salmonella, NCD and Coccidiosis on Kaggle poultry disease dataset. This model has higher performance in detecting and classifying poultry diseases. The proposed model achieves an accuracy of 98.83%. Finally, a mobile application is developed to help the poulterer to easily identify the diseases by uploading the image of the chicken's faecal. The performance of the proposed model is compared with other existing works and proved that this model is above par with other results.

This study ensures that a fast and accurate detection chicken diseases which in turn diseases the economic loss and enrich public health. In future, other kinds of diseases will be detected using recent deep learning models.

References

1. Okinda C, Nyalala I, Korohou T et al (2020) A review on computer vision systems in monitoring of poultry: A welfare perspective. Artif Intell Agric 4:184–208. https://doi.org/10.1016/j.aiia.2020.09.002
2. Machuve D, Nwankwo E, Mduma N, et al (2022) Poultry diseases diagnostics models using deep learning. Front Artif Intell 5. https://doi.org/10.3389/frai.2022.733345
3. Quach LD, Pham-Quoc N, Tran DC, et al (2020) Identification of chicken diseases using VGGNet and ResNet Models. Springer Int Publ, p 259–269. https://doi.org/10.1007/978-3-030-63083-620
4. Selvakumar B, Lakshmanan B, Sridhar Raj S (2022) Hybrid, Springer Nature Singapore, p 59–68. https://doi.org/10.1007/978-981-19-0707-47
5. Chadha A, Kaushik B (2019) Suicidal Ideation from the Perspective of Social and Opinion Mining, Springer International Publishing, p 659–670. https://doi.org/10.1007/978-3-030-294 07-647
6. Chadha A, Kaushik B (2019) A survey on prediction of suicidal ideation using machine and ensemble learning. Comput J 64(11):1617–1632. https://doi.org/10.1093/comjnl/bxz120
7. Chadha A, Kaushik B (2021) Performance evaluation of learning models for identification of suicidal thoughts. Comput J 65(1):139–154. https://doi.org/10.1093/comjnl/bxab060
8. Chadha A, Kaushik B (2022) A hybrid deep learning model using grid search and cross-validation for effective classification and prediction of suicidal ideation from social network data. N Gener Comput 40(4):889–914. https://doi.org/10.1007/s00354-022-00191-1
9. Selvakumar B, Lakshmanan B (2022) Sentimental analysis on user's reviews using bert. Materials Today: Proceedings 62:4931–4935. https://doi.org/10.1016/j.matpr.2022.03.678
10. B L, S A, P.S VR et al (2023) Improved deepmitosisnet framework for detection of mitosis in histopathology images. Multimed Tools Appl. https://doi.org/10.1007/s11042-023-16830-8
11. Lakshmanan B, Priyadharsini S, Selvakumar B (2022) Computer assisted mitotic figure detection in histopathology images based on densenetpca framework. Mater Today: Proc 62:4936–4939. https://doi.org/10.1016/j.matpr.2022.03.683
12. Selvakumar B, Sridhar Raj S, Vijay Gokul S, et al (2021) Deep learning framework for anomaly detection in IoT enabled systems. Springer Singapore, p 99–111. https://doi.org/10.1007/978-981-16-6186-05
13. Kaviya P, Chitra P, Selvakumar B (2023) A unified framework for monitoring social distancing and face mask wearing using deep learning: An approach to reduce covid-19 risk. Procedia Comput Sci 218:1561–1570. https://doi.org/10.1016/j.procs.2023.01.134
14. Robi YG, Sitote TM (2023) Neonatal disease prediction using machine learning techniques. J Healthc Eng 2023:1–16. https://doi.org/10.1155/2023/3567194
15. Mbelwa H, Mbelwa J, Machuve D (2021) Deep convolutional neural network for chicken diseases detection. Int J Adv Comput Sci Appl 12(2). https://doi.org/10.14569/ijacsa.2021.012 0295
16. Muslihudin M, Wanti R, Hardono, et al (2018) Prediction of layer chicken disease using fuzzy analytical hierarcy process. Int J Eng & Technol 7(2.26):90. https://doi.org/10.14419/ijet.v7i2.26.14360
17. Wang J, Shen M, Liu L et al (2019) Recognition and classification of broiler droppings based on deep convolutional neural network. J Sens 2019:1–10. https://doi.org/10.1155/2019/3823515
18. Shori M, Saroha K (2022) Avian influenza prediction using machine learning approaches: A review. Soft Comput Res Soc, p 177–184. https://doi.org/10.52458/978-81-955020-5-9-18
19. He K, Zhang X, Ren S, et al (2015) Deep residual learning for image recognition. https://doi.org/10.48550/ARXIV.1512.03385
20. Banakar A, Sadeghi M, Shushtari A (2016) An intelligent device for diagnosing avian diseases: Newcastle, infectious bronchitis, avian influenza. Comput Electron Agric 127:744–753. https://doi.org/10.1016/j.compag.2016.08.006

21. Okinda C, Lu M, Liu L et al (2019) A machine vision system for early detection and prediction of sick birds: A broiler chicken model. Biosys Eng 188:229–242. https://doi.org/10.1016/j.biosystemseng.2019.09.015
22. Borgonovo F, Ferrante V, Grilli G et al (2020) A data-driven prediction method for an early warning of coccidiosis in intensive livestock systems: A preliminary study. Animals 10(4):747. https://doi.org/10.3390/ani10040747
23. Zhang H, Chen C (2020) IEEE 4th information technology, networking, electronic and automation control conference (ITNEC). IEEE. https://doi.org/10.1109/itnec48623.2020.9084666
24. Quach LD, Quoc NP, Thi NH, et al (2020) Using surf to improve resnet-50 model for poultry disease recognition algorithm. In: 2020 International Conference on Computational Intelligence (ICCI). IEEE, https://doi.org/10.1109/icci51257.2020.9247698
25. Dina M, Ezinne N, Neema M, et al (????) Poultry diseases detection. https://www.kaggle.com/datasets/kausthubkannan/poultry-diseases-detection, [Accessed 20–03–2024]
26. Cui Y, Kong X, Chen C et al (2023) Research on broiler health status recognition method based on improved yolov5. Smart Agric Technol 6:100324. https://doi.org/10.1016/j.atech.2023.100324
27. Degu MZ, Simegn GL (2023) Smartphone based detection and classification of poultry diseases from chicken fecal images using deep learning techniques. Smart Agric Technol 4:100221. https://doi.org/10.1016/j.atech.2023.100221

Data Analytics and Intelligent Learning

Deep Transfer Learning in Parasites Imaging: A Systematic Review

Satish Kumar and Tasleem Arif

Abstract Transfer Learning (TL) is a method that uses convolutional neural networks to improve performance on a new task by leveraging previously acquired knowledge from related tasks. It has been particularly useful in medical image analysis, where data scarcity can be an issue. TL has helped save time and resources. However, most of the research conducted on this topic has arbitrarily applied transfer learning. This review paper aims to provide clear guidelines on how to select appropriate models and TL methods for medical image categorization. This comprehensive review thoroughly examines the current status of research on the use of deep-transfer learning algorithms for parasite imaging. We obtained 300 research papers from four databases, such as ACM, PubMed, Springer, and Web of Science. We followed the retrieve and filter process for paper selection, and 247 studies were found to be relevant to the scope of our review. Our review focused on papers related to selecting backbone models and transfer learning techniques such as feature extractor, feature, fine-tuning, and other attributes from scratch. The review covers a range of topics related to parasite imaging, including the parasites studied, the imaging methods used, the techniques for transfer learning, the datasets employed, the performance metrics analyzed, and the challenges faced. Additionally, we identify trends, advancements, and future possibilities in the area of TL parasite imaging. By conducting this study, we aim to provide insights into the effectiveness, limitations, and potential of TL techniques in parasite identification. Ultimately, our goal is to contribute to the enhancement of diagnostic technologies, leading to better healthcare outcomes for patients with parasitic disorders.

Keywords Parasites images · Microscopic images · Image analysis · Object detection · Machine learning · Deep learning

S. Kumar (✉) · T. Arif
Department of Information Technology, BGSB University, Rajouri 185131, India
e-mail: satishkumar@bgsbu.ac.in

P. J. S. Gonçalves et al. (eds.), *Proceedings of Fifth International Conference on Computing, Communications, and Cyber-Security*, Lecture Notes in Networks and Systems 1128, https://doi.org/10.1007/978-981-97-7371-8_19

1 Introduction

Parasitic diseases are a significant health concern worldwide, particularly in regions with limited access to healthcare. These diseases are caused by a wide range of parasites, such as protozoa and helminths, and affect millions of people, resulting in severe illness and death [1]. Accurate and timely diagnosis is crucial for effective treatment; however, traditional diagnostic methods often have limitations in terms of sensitivity, specificity, and availability [2]. Parasitic illnesses are a major cause of concern in low-income nations, claiming 41,500 human lives annually due to each parasite species producing 200,000 eggs/cysts every day [3]. However, medical imaging and deep transfer learning have emerged as a highly effective approach to addressing these issues and improving parasite identification [4]. This systematic review provides a comprehensive overview of current research in the growing field of "Deep Transfer Learning in Parasite Imaging," while also identifying major trends and topics for future inquiry. Malaria and schistosomiasis are two deadly parasite diseases that affect millions of people worldwide. Malaria alone caused over 400,000 deaths in 2019, with the majority of deaths occurring in Sub-Saharan Africa [3]. Schistosomiasis is a chronic illness that causes abdominal pain, diarrhea, and blood in the urine, affecting millions of people, especially in rural and poor communities. Transfer learning plays a critical role in the categorization and detection of parasites in microscopic images. However, deep learning applications face several challenges, including limited training data, noisy images, data interpretability, model compression, uncertainty scaling, overfitting, and vanishing gradient problems.

This research aims to point out some major challenges through effective transfer learning techniques. Improved parasite detection can lead to reduced healthcare costs, better health outcomes, and improved socio-economic development in endemic areas. This systematic survey provides several unique and valuable contributions to resolve these difficulties. The significant contributions of this work are summarized below.

- To provide comprehensive domain knowledge required for diagnosing and treating human parasite illnesses.
- To deliver thorough transfer learning techniques leveraging convolutional Neural Networks that have been evolved over time.
- Discuss various metrics that used to evaluating deep learning models performance.
- Clearly identify the challenges and potential future prospects of using deep learning in the field of parasitology.

2 Background

2.1 Transfer Learning Models

In the last decade, numerous ideas for developing new CNN models have emerged, each aimed at achieving specific objectives [5]. Over time, the CNN design has evolved and refined throughout time, with notable improvements in regularization, structural reformulation, parameter optimization, and other areas. Table 1 provides a detailed summary of the most popular CNN-based transfer learning models, ranging from the basic AlexNet model to the state-of-the-art DenseNet model. A thorough analysis of the model depth, datasets they operate on, parameter accuracy rates, and input sizes is essential in gaining a comprehensive understanding of these common CNN designs.

Table 1 Summarizing key details of transfer learning architectures

Model	Input Size	Main finding	Depth	Dataset	Parameters	Error rate
AlexNet	227 × 227 × 3	Adopted dropout and ReLU	8	ImageN et	60 M	16.4
VGG	224 × 224 × 3	Enhanced depth and small filter size	19	ImageN et	138 M	7.3
Xception	229 × 229 × 3	Convolution preceding point-wise convolution	71	ImageN et	22.8 M	0.05
Inception-V3	229 × 229 × 3	Apply small filter size, updated feature representation	48	ImageN et	23.6 M	3.5
GoogLeNet	224 × 224 × 3	Enhanced depth, varied filter sizes, con-catenation	22	ImageN et	4 M	6.7
Incep-tion-ResNet-v2	229 × 229 × 3	applied residual links	164	ImageN et	55.8 M	3.52
MobileNet-v2	224 × 224 × 3	Inverted residual structure	53	ImageN et	1.677 M	–
DenseNet	**224 x 224 x 3**	**Adopted blocks of layers**	**201**	**ImageN et, CIFAR-100,**	**25.M**	**3.46, 5.54**

2.2 *Parasite Datasets*

Researchers have utilized various datasets of parasites to significantly enhance the efficiency of Transfer learning approaches and models for diagnosing parasitic disorders through microscopic pictures. These datasets, summarized in Table 2, include Plasmodium, Toxoplasma, and intestinal parasites, and are an excellent resource for researchers. Moreover, the study incorporates critical datasets for other parasites such as Leishmania, Babesia, and Trichomonas, thereby expanding the area of investigation. Such datasets are generally used for parasites classification and detection tasks. Figure 1 provides a visual perspective by displaying sample microscopic pictures of parasites from publicly available datasets relevant to deep learning applications.

3 Methodology

We analyzed research papers on deep neural network in the domain of parasite classification and identification. We provided an overview of existing methods and models, evaluated promising techniques, outlined future directions, and highlighted prevalent challenges. We assessed 45 scientific papers from 2014 to present, to provide researchers with a comprehensive compass to navigate the current landscape of state-of-the-art approaches and discern evolving trends.

Table 2 Compilation of publicly accessible image datasets for different para sites

Parasite types & image number	Descriptions	Dataset	Reference
Toxoplasma gondii-8,156 with 400 × , 6,969 with 1000x	4,979 host cell images at 400 × and 8,023 host cell images at 1000 × magnification are included in this FCGAN dataset	https://github.com/senli2018/FCGAN/b	[6]
Plasmodium-15,927, Babesia-1,100	23,463 microscopically captured photos of multi-stage Plasmodium, Babesia, and host cells (leukocytes and RBCs) at 1000 × magnification are included in this dataset	https://github.com/senli2018/DTGCN_2021	[7]
Potozoan cysts and helminthic eggs	There are 1,783 micro-scope images in total in this dataset, of which 1,597 were used for training and 176 for testing	https://figshare.com/	[8]
Chula-ParasiteEgg-11 1,000 and 250 images/class for training and testing	Comprising eleven different species of parasite eggs from fecal smear samples, the Chula-ParasiteEgg-11 training dataset There are 1,000 photos in each category	https://icip2022challenge.piclab.ai/	[9]

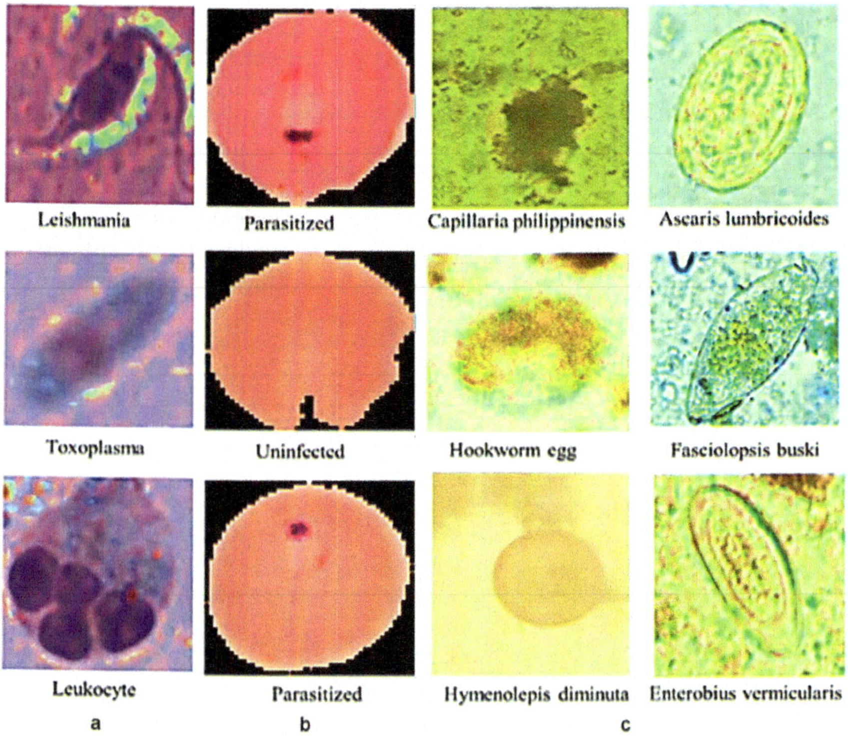

Fig. 1 Three datasets are available for deep learning applications on protozoan parasites: **a** Parasite photographs from Dataset [10]: contains cropped photos of six parasite species, red blood cells, and host leukocytes **b** Plasmodium falciparum Dataset [8]: contains resize image patches of parasitized and other images **c** Intestinal parasite Dataset [6]: includes 1,000 photos of 11 types of parasite eggs from fecal smear samples

3.1 Paper Inclusion and Exclusion Process

Figure 2 illustrates the systematic and thorough process we employed to retrieve and filter relevant papers. We meticulously searched various academic libraries including ACM, PubMed, Springer, and Web of Science using targeted keywords and phrases such as "machine learning or parasites," "deep learning or parasites," "CNN or microscopic images," combined with "detection" and "classification." Our rigorous approach yielded a corpus of 257 papers. We further supplemented this by searching the references cited within the retrieved documents and found an additional 10 relevant papers, bringing the total technical papers to 267.

In the subsequent filtering stage, we pruned the collection by identifying and eliminating 20 duplicate papers, leaving us with 247 highly relevant papers. We then proceeded to a more detailed selection process, where we meticulously identified and earmarked 44 papers for deep learning, 48 for transfer learning, and an additional 10 papers closely aligned with potential methodologies, including those based on

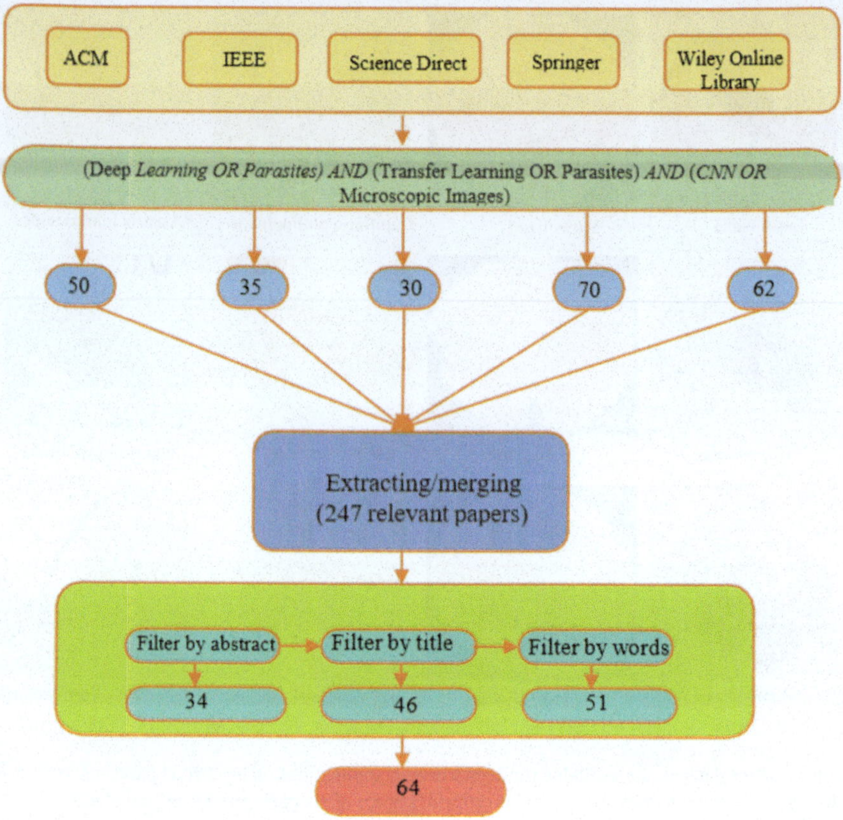

Fig. 2 The search and filter process of papers

visual transformers. Our stringent curation process resulted in a refined selection of papers that will inform our comprehensive review.

4 Transfer Learning Related Works in Proposed Field

In recent years, models based on transfer learning have gained popularity for diagnosing microscopic images. This section attempts to provide an overview of significant research literature and techniques on this area of study, with a focus on the effectiveness of deep learning algorithms.

Inayah, et al. [11] proposed a randomly wired neural network approach for the classification of falciparum parasites. With 27,558 images of red blood cells, they trained their model, and the result was an impressive 95.08% accuracy rate. Similarly, Viet et al. achieved a 95.67% accuracy rate in identifying and classifying parasitic worm eggs in parasite images by employing the Faster R-CNN technology. In another study, Yang et al. [12] developed an updated Convolutional Neural Network to spot parasites in blood smear microscopic images, with an accuracy rate of 93.46%.

Nakasi, et al. [13] proposed faster regional convolutional neural network (faster R-CNN), single-shot multi-box detector (SSD), and RetinaNet to detect malaria parasites in thick blood and data augmentation was applied to optimize performance of architectures. The experiment results show that faster R-CNN was the best trained model with mAP of over 0.94 and SSD, was the best model for mobile deployment. To detect and categorize various species of protozoa, Pho et al. utilized transfer learning and data augmentation techniques. They were able to create a robust RetinaNet model that could achieve high accuracy even with a small amount of training data. Furthermore, Mathison, et al. [14] presented a unique CNN model that could identify intestinal protozoa in stool specimens using high-resolution digital side images, achieving a 98% accuracy rate.

Von, et al. [15] illustrate the classification of skin lesions with CNN employing pixels and illness labels as input. We train a CNN on a dataset of 129,450 clinical pictures representing 2,032 different disorders. We compared its performance to 21 board-certified dermatologists using biopsy-proven clinical photos and two crucial binary classifications. Nakasi, et al. [16] examined AlexNet and GoogleNet models for the detection of intestinal parasite eggs in stool samples. This study compared these pre-tuned models against an upgraded Convolutional Neural Network on the same task. The out-of-the-box models are quite accurate, with GoogleNet ROC AUC of 0.99 and AlexNet ROC AUC of 1.00, and run on a very low processing resource system. Zhang, et al. [17] present the Resnet-152 convolutional neural network architecture, which is based on the Faster-R-CNN approach for quick cell detection. Our system accurately detected 40,560 fecal pictures with an average precision of 84% and a detection time of 723 ms. Luo, et al. [18] study presents a high-throughput system that uses deep learning to forecast Cryptosporidium and Giardia in drinking water. This method fuses an effective artificial neural network with a classification accuracy of over 99.6%. The method detects Cryptosporidium and Giardia significantly 97.37% sensitivity and 99.95% specificity.

Some notable studies include Sajedi, et al. [19], who focused on bacterial species recognition, Peixinho, et al. [20], who proposed a ConvNet for intestinal parasite recognition, and used CNNs to detect Mycobacterium tuberculosis. In order to detect and classify different protozoa species [21], transfer learning and data augmentation approaches are used. This led to the creation of a strong RetinaNet model that could achieve good accuracy even with a small amount of training data. Moreover, Yang, et al. [22] implemented a improved Convolutional Neural Network to identify parasites in blood smear microscopic pictures with an accuracy rate of 93.46%. Chen, et al. [23] presented a unique CNN model for identifying intestinal protozoa in stool specimens using high-resolution digital side pictures. This model had an accuracy rate of 98%. In a similar vein, a researcher used a 97.67% accuracy rate in the Faster R-CNN technology to identify and categorize parasitic worm eggs in stool microscopy pictures.

To summarize, there have been several successful applications of deep transfer learning methods and models in microscopic image analysis, ranging from detecting parasites to classifying bacterial species. These advancements demonstrate the transfer learning models in improving accuracy and automation across various domains of microscopic image interpretation.

5 Summary

Table 3 provides a summary of various transfer learning methods used for parasite detection and classification. It includes information such as references, publication dates, methods used, objectives, species categories, data specifications, evaluation metrics, limitations, and areas for future exploration. This compilation shows the constant improvement and advancement of techniques in this important field.

6 Discussion

We reviewed deep transfer learning in parasite imaging and found that pre-trained CNNs like ResNet, Inception, and VGG are widely used and effective. There are possibilities and constraints in this field, which offer valuable insights for future research. Figure 3 shows transfer learning models based on CNN used for parasite detection and classification. CNN, AlexNet, and GoogleNet are important models. Faster R-CNN and R-CNN have also played significant roles.

A recent research on transfer learning techniques leveraged to parasite-related tasks found that the most popular detection and classification is Faster R-CNN. Numerous publications back up this conclusion, most notably [29], which talks about Faster R-CNN in detail. Applying the Regional Proposal Network (RPN) technique, which has greatly increased performance and detection accuracy, is one advantage of Faster R-CNN over its predecessor, R-CNN. Furthermore, as reference illustrates, Faster R-CNN provides the advantage of an end-to-end detection process [25].

It is necessary to note that Faster R-CNN has significant limitations. It lacks real-time object identification capabilities and imposes a significant processing overhead with the RPN extraction approach. These restrictions must be considered when using it for applications that require real-time detection and computing efficiency.

7 Conclusion and Research Directions

This work reviews the use of deep transfer learning microscopic parasite imaging. Researchers have utilized Convolutional Neural Networks (CNNs) in various imaging modalities, with promising results in detecting and categorizing parasites. This technology has the potential to transform the field of parasitology. We have revealed that the Faster R-CNN model is the mostly used detection and classification

Table 3 Provides a concise summary of the survey papers that have investigated the utilization of deep transfer learning techniques for detection and classification tasks

Algorithms	Ref	Objective	Parasites	Dataset	Limitations	Future scope
D-CNN	[14]	Detection of intestinal protozoa	Intestinal Protozoas(10 Catg.)	T.I = 910	Clinical laboratory validation was not employed	Collection of additional slides in order to enrich the data set and refine the model
Fusion-Net	[24]	Single-cell classification of foodborne pathogens	pathogens	C = 5	Insufficient dataset	Collection of images in order to enrich the data set and refine the model
Faster RCNN	[25]	Automatic detec tion parasite worm egg	Parasites	C = 8	Small dataset details	Advanced methods applied for better performance
Faster-RCNN	[12]	Parasite detection	Parasites	T.I = 2967	Evaluation details are not described	Reduce false posi tive rate
RCNN	[26]	Identification and enumeration of cyanobacteria species	cyanobacteria species	C = 5	Small dataset details	Improve the cell-count accuracy
AlexNet & GoogleNet	[16]	Intestinal parasite detection in	Intestinal Parasite	–	No Details about database quan tity	Transfer learning could be used to analyze the data

(continued)

Table 3 (continued)

Algorithms	Ref	Objective	Parasites	Dataset	Limitations	Future scope
MCellNet (CNN)	[27]	Deep learning-enabled predicting Cryptosporidium and Giardia in drinking water	Cryptosporidium and Giardia	C = 2 Tr. = 38,469 Ts. = 32,059 Val. = 9618	Insufficient dataset	Improved quality training data
YOLOv5	[28]	Parasitic egg detection using transfer learning	Fasciolopsis buski (FB), Hymenolepis diminuta (HD), Ascaris lumbricoides (AL)and Taenia spp. (Tn)	C = 5	The poor-quality microscopic image	Update model to learn more complex features of parasite

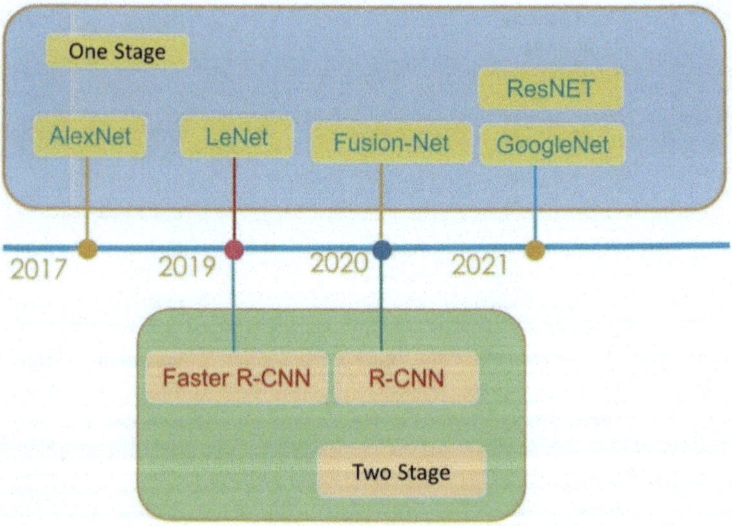

Fig. 3 Chronological overview of CNN-based transfer learning techniques for parasite classification and detection tasks

model for parasite imaging. The use of Regional Proposal Network (RPN) technology has significantly enhanced its performance and accuracy. Moreover, Faster R-CNN offers an end-to-end detection process, which simplifies the workflow and improves productivity. The application of AI-powered models for automating parasite identification has the potential to reduce the burden on healthcare staff, facilitate quick screening, and enhance diagnosis accuracy. However, despite significant advancements in this area, several challenges and opportunities still exist. Critical tasks such as standardizing datasets, developing robust validation procedures, and improving model interpretability need to be addressed. Additionally, further research is required to investigate the real-time capabilities of models like Faster R-CNN to meet the expectations.

Author Contribution S Kumar: Supervision, Writing - review & editing; Data curation; Methodology, T arif: Investigation, Validation; Software; Formal analysis.

Funding No fundind.

Data Availability Data not required.

Conflict of Interest No conflict of interest.

References

1. Boerma T, Mathers CD (2015) The World Health Organization and global health estimates: Improving collaboration and capacity. BMC Med 13(1):1–4. https://doi.org/10.1186/S12916-015-0286-7/METRICS
2. Kumar S, Arif T, Alotaibi AS, Malik MB, Manhas J (2023) Advances towards automatic detection and classification of parasites microscopic images using deep convolutional neural network: methods, models and research directions. Arch Comput Methods Eng 30(3). Springer Science and Business Media B.V., pp 2013–2039. https://doi.org/10.1007/s11831-022-09858-w
3. Zhang C, et al. (2022) Deep learning for microscopic examination of protozoan parasites. Comput Struct Biotechnol J 20. Elsevier B.V., pp 1036–1043. https://doi.org/10.1016/j.csbj.2022.02.005
4. Boit J (2020) The effectiveness of transfer learning systems on medical images [Online]. Available: https://scholar.dsu.edu/theses/344
5. Nguyen LD, Lin D, Lin Z, Cao J (2018) Deep CNNs for microscopic image classification by exploiting transfer learning and feature concatenation. Proc-IEEE Int Symp Circuits Syst. https://doi.org/10.1109/ISCAS.2018.8351550
6. Li S, Li A, Molina Lara DA, Gómez Marín JE, Juhas M, Zhang Y (2020) Transfer learning for toxoplasma gondii recognition. mSystems 5(1), pp 1–12. https://doi.org/10.1128/msystems.00445-19
7. Li S, Du Z, Meng X, Zhang Y (2021) Multi-stage malaria parasite recognition by deep learning. Gigascience 10(6):1–11. https://doi.org/10.1093/gigascience/giab040
8. Quinn JA, Nakasi R, Mugagga PKB, Byanyima P, Lubega W, Andama A (2016) Deep convolutional neural networks for microscopy-based point of care diagnostics, pp 1–12, 2016, [Online]. Available: http://arxiv.org/abs/1608.02989
9. Parasitic egg detection and classification in microscopic images | IEEE DataPort. Accessed: May 19, 2022. [Online]. Available: https://ieee-dataport.org/competitions/parasitic-egg-detection-and-classification-microscopic-images
10. Osman MK, Ahmad F, Saad Z, Mashor MY, Jaafar H (2010) A genetic algorithm-neural network approach for mycobacterium tuberculosis detection in Ziehl-Neelsen stained tissue slide images. In: Proceedings of the 2010 10th International Conference on Intelligent Systems Design and Applications, ISDA'10, pp 1229–1234. https://doi.org/10.1109/ISDA.2010.5687018

11. Inayah N, Liebenlito M, Fitriyati N, Monardo K (2020) Classification of falciparum parasite in human red blood cells using randomly wired neural network. In: 2020 8th International Conference on Cyber and IT Service Management, CITSM 2020, Institute of Electrical and Electronics Engineers Inc.. https://doi.org/10.1109/CITSM50537.2020.9268806
12. Hung J, Carpenter A (2017) Applying faster R-CNN for object detection on malaria images. In: IEEE computer society conference on computer vision and pattern recognition workshops, vol. 2017-Janua, pp 1–7. https://doi.org/10.1109/CVPRW.2017.112
13. Nakasi R, Mwebaze E, Zawedde A, Tusubira J, Akera B, Maiga G (2020) A new approach for microscopic diagnosis of malaria parasites in thick blood smears using pre-trained deep learning models. SN Appl Sci 2(7). https://doi.org/10.1007/s42452-020-3000-0
14. Mathison BA, et al. (2020) Detection of intestinal protozoa in trichrome-stained stool specimens by use of a deep convolutional neural network. J Clin Microbiol 58(6). https://doi.org/10.1128/JCM.02053-19
15. von Chamier L, et al. (2021) Democratising deep learning for microscopy with ZeroCostDL4Mic. Nat Commun 12(1):1–18. https://doi.org/10.1038/s41467-021-22518-0
16. Nakasi R, Aliija ER, Nakatumba J (2021) A poster on intestinal parasite detection in stool sample using AlexNet and GoogleNet architectures. In: Proceedings of 2021 4th ACM SIGCAS Conference on Computing and Sustainable Societies, COMPASS 2021, pp 389–395. https://doi.org/10.1145/3460112.3472309
17. Zhang J, et al. (2021) Fast and accurate automated recognition of the dominant cells from fecal images based on Faster R-CNN. Sci Rep 11(1). https://doi.org/10.1038/s41598-021-89863-4
18. Luo S, et al. (2021) Deep learning-enabled imaging flow cytometry for high-speed Cryptosporidium and Giardia detection, pp 1–11. https://doi.org/10.1002/cyto.a.24321
19. Sajedi H, Mohammadipanah F, Pashaei A (2019) Automated identification of Myxobacterial genera using convolutional neural network. Sci Rep 9(1):1–15. https://doi.org/10.1038/s41598-019-54341-5
20. Peixinho AZ, Martins SB, Vargas JE, Falcão AX, Gomes JF, Suzuki CTN (2015) Diagnosis of human intestinal parasites by deep learning. In: Computational Vision and Medical Image Processing V—Proceedings of 5th Eccomas Thematic Conference on Computational Vision and Medical Image Processing, VipIMAGE 2015, no. October 2015, pp 107–112. https://doi.org/10.1201/b19241-19
21. Pho K, Mohammed Amin MK, Yoshitaka A (2019) Segmentation-driven hierarchical RetinaNet for detecting protozoa in micrograph. Int J Semant Comput 13(3):393–413. https://doi.org/10.1142/S1793351X19400178
22. Li S, Yang Q, Jiang H, Cortés-Vecino JA, Zhang Y (2020) Parasitologist-level classification of apicomplexan parasites and host cell with deep cycle transfer learning (DCTL). Bioinformatics 36(16):4498–4505. https://doi.org/10.1093/bioinformatics/btaa513
23. Chen L, et al. (2020) A CNN framework based on line annotations for detecting nematodes in microscopic images. Proceedings—International Symp Biomed Imaging 2020(031):508–512. https://doi.org/10.1109/ISBI45749.2020.9098465
24. Seo Y, Park B, Hinton A, Yoon SC, Lawrence KC (2016) Identification of Staphylococcus species with hyperspectral microscope imaging and classification algorithms. J Food Meas Charact 10(2):253–263. https://doi.org/10.1007/S11694-015-9301-0/TABLES/3
25. Viet NQ, Tuyen DTT, Hoang TH (2019) Parasite worm egg automatic detection in microscopy stool image based on Faster R-CNN. In: ACM international conference proceeding series, association for computing machinery, Jan. 2019, pp 197–202. https://doi.org/10.1145/3310986.3311014
26. Baek SS et al (2020) Identification and enumeration of cyanobacteria species using a deep neural network. Ecol Indic 115:106395. https://doi.org/10.1016/J.ECOLIND.2020.106395
27. Borba VH, Martin C, Machado-Silva JR, Xavier SCC, de Mello FL, Iñiguez AM (2021) Machine learning approach to support taxonomic species discrimination based on helminth collections data. Parasit Vectors 14(1):1–15. https://doi.org/10.1186/s13071-021-04721-6

28. S. Kumar, T. Arif, G. Ahamad, A. A. Chaudhary, S. Khan, and M. A. M. Ali, "An Efficient and Effective Framework for Intestinal Parasite Egg Detection Using YOLOv5," *Diagnostics 2023, Vol. 13, Page 2978*, vol. 13, no. 18, p. 2978, Sep. 2023, https://doi.org/10.3390/DIAGNOSTICS13182978.
29. Yang F, Yu H, Silamut K, Maude RJ, Jaeger S, Antani S (2019) Parasite detection in thick blood smears based on customized Faster-RCNN on smartphones. In: Proceedings - Applied Imagery Pattern Recognition Workshop, vol. 2019-Octob, pp 0–3. https://doi.org/10.1109/AIPR47015.2019.9174565

Electronic Healthcare Record Sharing System Using Blockchain

Apoorva Mathur, Namrata Dhanda, and Kapil Kumar Gupta

Abstract With today's advancements in different technologies, online doctor consultation systems are becoming more popular with increasing demands. Earlier, people used to go from doctor to doctor for in-clinic doctor and patient contact in which details of the patients were stored in the paper and that information was passed from one doctor to another. But now, people are opting for online consultation as it is flexible, comfortable, easiest, and the fastest means of consulting with the best doctors anywhere in the World. But these systems often have one drawback that gets ignored, which is the sharing of information between patient and doctor through the current system of data storage centers as a mediator. Nevertheless, alongside a captivating technology known as Blockchain shows a substantial responsibility concerning data management on these platforms. While certain hospitals diligently safeguard data privacy and security, regrettably, many others do not exhibit the same commitment. This results in varying degrees of protection, data misappropriation, or even misuse. These platforms that are made to make the data of patients secure operate on the principles of Web 3.0, where users function as both consumers and creators of content. In the spirit of public blockchains, Web 3.0 applications embrace openness, as their code is open-source and publicly accessible. This allows users to validate whether a protocol fulfills its claims and gives them the freedom to duplicate, amend, or enhance protocols. Thus, to overcome all the drawbacks that a hospital faces in managing a large amount of data, the paper discusses the idea of Blockchain Technology to design a secure Electronic Healthcare record-sharing system.

Keywords Blockchain · Web 3.0 · Hardhat · Gas fee · React · Meta-mask · Ethereum · Wallet · Hash-code · Solidity

---

A. Mathur · N. Dhanda · K. K. Gupta (✉)
Amity University, Lucknow, India
e-mail: kapilkumargupta2007@gmail.com

P. J. S. Gonçalves et al. (eds.), *Proceedings of Fifth International Conference on Computing, Communications, and Cyber-Security*, Lecture Notes in Networks and Systems 1128, https://doi.org/10.1007/978-981-97-7371-8_20

1 Introduction

With the advancement of different technologies people are using nowadays, the data that we share with others has taken another level of speed but with this ease, some security challenges come to the rescue which should not be avoided. Earlier, the data of healthcare records was stored by the staff of hospitals and was shared with doctors physically. But at present, the demand for Blockchain and Web 3.0 has increased for health care systems, where Web 3.0 is a paradigm shift from the current web which is also known as Web 2.0 (centralized) to Web 3.0 (decentralized). Web 3.0 is a user-controlled internet infrastructure that is concerned with providing greater control over their data and enhancing privacy through the integration of blockchain and other decentralized technologies. Now, the medical record is only been shared with the patient and the user which doesn't allow the involvement of another or third party without the patient's consent. If a hospital fails to secure a patient's records, then it will have legal or finance-related consequences, along with the power to impact the patient's records to misuse them. Thus, securing the electronic healthcare record is a difficult task [1], and the ramifications of some data breaches are a strong drawback to sharing information or data. For this particular work, we focus on both privacy & anonymity and in which ways they apply to the sharing of data between the patient and the doctor with security.

Smart electronic healthcare systems have revolutionized patient monitoring, offering ambulatory care within the comfort of patients' homes. This not only facilitates in-home care but also presents economic advantages for patients. Moreover, the comprehensive availability of Electronic Health Records (EHR) on cloud platforms empowers healthcare providers to monitor patient health and deliver effective medical services throughout the diagnostic and treatment processes. Despite these numerous advantages, the cloud-based storage concept introduces security challenges that impede the seamless deployment of electronic health services. The sharing of EHR information between patients and healthcare providers presents one such challenge. Unauthorized access by entities without patient consent can jeopardize privacy, data integrity, and security [2]. Additionally, patients may find it very cumbersome to maintain, monitor, and keep their health records.

Thus, it becomes imperative to establish efficient and reliable access control for mobile-based cloud systems. Traditional access control methods for EHR rely heavily on the trustworthiness of data owners, enabling the server to manage access control and service provisioning [3]. However, this conventional assumption does not align well with the mobile cloud paradigm, as the server may be honest but still inquisitive [21]. Engaging in transactions governed by programmable protocols referred to as Smart Contracts, derived through legally crafted snippets of code, characterizes the concept of blockchain 2.0. Ethereum stands out as a leading blockchain platform in this second phase [4]. Advancements in blockchain technology have driven a revolutionary transformation, enhancing various transactions, including insurance billing, health information management, and their implementation of smart contracts. The integration of blockchain, alongside emerging technologies like the Internet of

Things (IoT) and cloud computing, holds the potential to establish more robust Electronic Health Record (EHR) systems [5]. A blockchain can also be called a growing list of records, which are called blocks, that are linked together very securely using the method of cryptography [6–9]. Each block contains a cryptographic hash of the previous block, a timestamp, and transaction data. The timestamp proves that the transaction data existed when the block was published to get into its hash. The blocks form a chain with each additional block reinforcing the ones before it as each block contains information about the block previous to it. Therefore, blockchains are resistant to any modification of their data because once recorded, the data in any given block cannot be altered retroactively without altering all subsequent blocks [6, 7]. Thus, here a secure sharing of EHR will focus on the privacy and authentication of the data stored which will also provide ease to the patients to track their medical record and decide who can look into these records and when as per needed. I am focusing on decentralized identity management for patients as it will help them to securely get through their data. Further, smart contracts are been made which can help to see if a patient has provided the consent or not for viewing their data. Patients can also define the terms and conditions under which access can be granted to their data.

2 Literature Work

Chen et al. [10] discussed "Blockchain-Based Implementation for Financial Product Management," in a study and presented a blockchain-based platform for managing financial products. This platform established a network architecture that prioritizes information transparency and secure data sharing. Hyperledger Fabric served as the foundational technology, enabling essential financial product operations, including product maintenance, data inquiries, and product traceability. The study also proposed enhancements to address Hyperledger's weaknesses. Ferrag, et al. [11] conducted a comprehensive & detailed survey on "Blockchain Technologies for the Internet of Things". They have discussed blockchain technology within IoT networks, starting with an overview of blockchain and its application domains in IoT. The study categorized blockchain applications in IoT into five main classifications. MedRec stands out as a functional model dedicated to achieving interoperability among healthcare providers and granting patients more comprehensive control over their information. Utilizing the Ethereum blockchain, its protocols treat patients' health records as assets, with smart contracts governing access to these assets. This prototype conceptualizes medical records on-chain as versatile query strings, executable against a provider's database to retrieve the entire document. Patients can employ smart contracts through a user interface to regulate who can access their healthcare records and discern the available features of each record. MedRec strategically employs permission-less blockchain to fulfill its dual objectives: empowering patients with greater authority over their health data and facilitating more efficient data transfer among diverse healthcare providers through interoperable query strings [12]. Liu

[13] discussed "A Small Java Application for Learning Blockchain". This application, called Chain Tutor, aimed to simplify the understanding of blockchain concepts, especially for beginners. Instead of relying on text-based tutorials, the Java application offered a graphical user interface (GUI) for users to implement blockchain technology concepts, understand mining processes, and observe how blocks are added to the blockchain. Andrei Cirstea et al. (2018) explored "Blockchain Technology Applied in Health" in their paper presented at ECAI. They provided an introduction to blockchain concepts and highlighted its applications in the medical field. The study underscored how blockchain technology could enhance efficiency and decentralize the medical industry, fundamentally transforming the way information is received, secured, and transmitted. HIE is a blockchain-based approach commonly known as Health Information Exchange. The utilization of blockchain technology in Health Information Exchange presents a method for secure data sharing. This system facilitates the exchange of data from various centralized sources of trust, aligning with network consensus based on proof of structural and semantic interworking. The blockchain assumes a crucial role in supporting data sharing and implementing high-level protocols and structures essential for the integration of this technology into healthcare systems. The consensus algorithm is specifically crafted to ensure data interoperability. Additionally, the system incorporates enhanced security measures through the implementation of smart contracts and network-wide keys on the blockchain. Drozdova et al. [14] addressed "Contribution to Cloud Computing Security Architecture" at ICETA 2017. Their study focused on the transformation of software support from a server-oriented to a service-oriented model due to the shift to cloud computing. They highlighted the challenges and threats associated with this change, emphasizing the need to ensure data availability, integrity, and privacy while preventing data theft and breaches. The paper also proposed solutions to address these challenges using existing methods. The architectural model of decentralized personal healthcare data is established through the utilization of blockchain technology. The proposal introduces a conceptual prototype designed to manage personal healthcare data collected from multiple healthcare providers, relying on blockchain technology within a peer-to-peer network. In a publication in IJEAT, Arun et al. [15] explored the potential of blockchain in "E-Voting using a Decentralized Ethereum Application." The authors put light on how blockchain's decentralized and immutable nature could revolutionize the voting process, making it more open, fair, and verifiable. The study also included a case study that illustrated the ledger's potential and examined the advantages and disadvantages of using blockchain in voting. The FHIR Chain framework presents a blockchain-based solution tailored to fulfill the specifications set by the Office of the National Coordinator (ONC). It encapsulates the Health Level 7 (HL7) with Fast Health Interoperability Resource (FHIR) standards, facilitating the seamless sharing of health-related data. This framework eliminates the need for cumbersome data uploads and downloads by providing decentralized storage of metadata while upholding data ownership. Additionally, it mitigates the risk of data corruption and ensures the preservation of original data ownership. In FHIRChain, encryption and decryption of reference pointers enhance

digital identity and authentication. Successful authentication allows direct downloading of data from the source, ensuring accurate presentation to the user [16]. As per Sundvall's proposed framework [17], there is potential for reusing free Electronic Health Record (EHR) prototypes and template semantics. Adopting a spreadsheet approach, this framework aims to organize existing non-standardized EHR content for more efficient sharing. However, organizations with established non-standardized EHRs may face challenges in the abrupt conversion of configurable EHR-entry forms from non-standardized to standardized formats. However, they can convert it from non-standardized to standardized gradually with other existing methods, maintenance tasks, and information change management. Farhadi et al. [18] discussed the security & privacy requirements in their papers written in IEEE format for HIPAA technical requirements that are necessary, and also identified a subset of ONC criteria that is related to privacy & security, and then evaluated the EHR applications to check for any security issues or vulnerabilities. They proposed mitigation or removal of security issues toward better adherence while also help several practitioners to use open-source tools for certification compliance again.

3 Problem Statement

In today's data-driven world, privacy has become a significant concern for individuals due to the vulnerabilities associated with their data. Data theft, hacking, and misuse are rampant in the current system of data sharing, which relies on centralized data centers owned by large corporations that have unrestricted access to user data [19]. To address these challenges, we have developed a Secure Data Sharing Model using blockchain technology and Web 3.0. Secure data sharing represents an innovative approach to sharing data and messages with the help of blockchain and Web 3.0 technologies. It aims to address key issues that are often overlooked in the current market, including security, data breaches, content originality, and data theft. The primary problems with existing data-sharing platforms that our model aims to resolve are as follows:

Security: In the modern era, enormous volumes of data are transferred daily among individuals, passing through the data centers of major IT giants, who may potentially exploit this data for various purposes. Our Safe Data Sharing Model is designed to mitigate this security risk.

Transparency and Anti-Fraud: Existing data-sharing models lack transparency, leaving users uncertain about the fate of their data once it leaves their devices. With the blockchain-based Safe Data Sharing Model, users have full control over their data and can track its usage, ensuring that no one can access it without their explicit permission.

Centralized Process: Current data-sharing systems route data through centralized networks, enabling data center workers to monitor user activity. In contrast, a decentralized blockchain network ensures the privacy of user activity through a peer-to-peer process.

4 Proposed Methodology

Our aim is to develop an EHR system using blockchain in which the blueprint of designing of EHR sharing system is carried out like—how the website will look, how everything will be aligned, what features will it contain, and how it can be extended in the future as described below.

4.1 Registration

The registration module simplifies the process of registering doctors and patients by gathering essential details. Doctors are prompted to input information such as their name, date of birth, address, email, phone number, and registration number. Each doctor is associated with a specific healthcare provider. Patients are prompted to furnish details including their name, date of birth, address, mobile number, and health insurance information. In both instances, a valid password is a mandatory requirement.

4.2 Smart Contract Creation

Smart contracts are defined as self-executing programs that operate automatically upon the fulfillment of specific conditions. In the context of the Ethereum blockchain, smart contracts consist of programmable functions, allowing users to interact with them through an application binary interface.

4.3 Hash Value Creation

A cryptographical hash is a mathematical algorithm that takes an input from a source and converts it into a fixed-length alphanumeric output. Different cryptographic hashes exist, such as the SHA-256 algorithm used in Bitcoin. Hashing algorithms are computational functions that transfer the input into a fixed hash value. The hash value serves various purposes, including data comparison, identification, and running calculations on data containing files.

4.4 Block Creation

Patients receive notifications for their Electronic Health Records (EHRs), which require their approval. This approval process helps prevent potential double-spending attacks. Once the EHR is verified, the module Block Generator creates a block containing the EHR details. Each block of the blockchain possesses a distinct hash value for its data.

4.5 Privacy Preservation

The Blockchain serves as a comprehensive way to achieve data privacy and security. Blocks are interconnected in a chain-like structure, with each block containing a block header, transactions, and a transaction counter. The initial block in the blockchain is referred to as a Genesis block (Fig. 1).

a. Firstly, Patients come to the hospital's website or any online consultation website that is using our EHR sharing system. Further, verification of the user's identity is done through secure login methods after patients register for an online consultation website or a hospital employing our EHR sharing system associated with Ethereum to prevent any sort of duplicity of data.
b. When the patient is using the system for the first time, then he/she should connect his Metamask wallet to the EHR sharing platform. Then patient's information is sent to the designated doctor's address, and the Meta mask prompts for permission to proceed.
c. Furthermore, there is a constant check on whether or not the patient has provided consent for sharing their EHR with the hospital. If the consent is not given, inform the user and end the process.
d. Then patients are required to fill up the required details in the form for a doctor's consultation and attach any reports or documents for doctors' review along with the ether amount sent to the doctor as the consultation charges, ensuring that the data is securely encrypted properly before its transmission to avoid any sort of misuse and further creating a transaction using encrypted data.
e. A small amount of ethers from the connected account got deducted in addition to the sent ethers as the Gas fees.
f. After the patient has provided the doctor with his/her symptoms of the disease, doctors are allowed to give prescriptions to the patient according to the disease which is received online at the patient's end and then the patient can buy those recommended medicines.

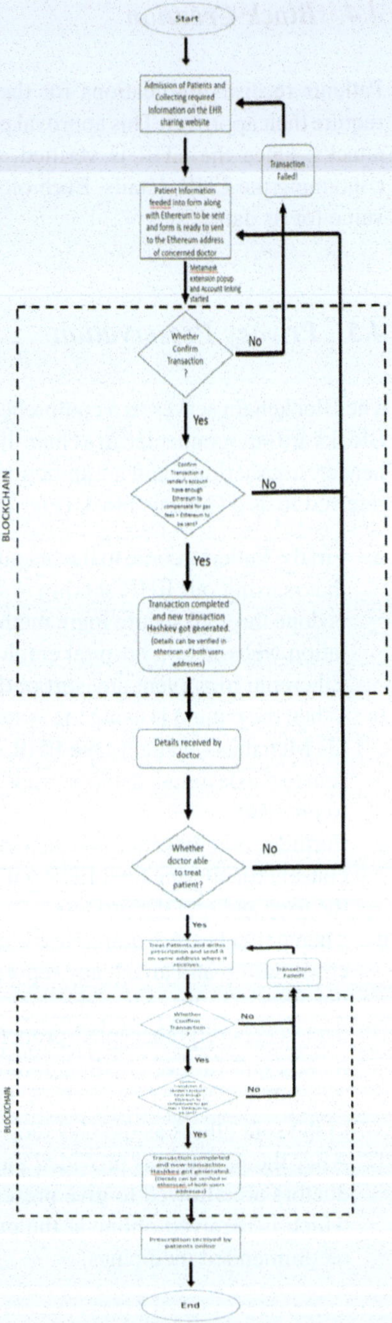

Fig. 1 Data flow diagram of EHR using Blockchain

5 UI for Website

UI for the website is given below after the creation of the Ethereum wallet by the patient (Fig. 2).

Here, patients visit an online consultation website or a hospital employing our EHR sharing system in which patients will proceed with his/her details to register after making an account in Ethereum and activating his/her Ethereum wallet (Fig. 3).

A successful transaction requires that the sender's account has a sufficient balance of ethers to cover the minimal gas fees. Upon granting permission to proceed, a new transaction Hash key is generated, and the transaction proceeds to the doctor's account address. If permission is denied or there are insufficient gas fee ethers, the transaction is declined. After a successful transaction, the doctor receives the patient's details,

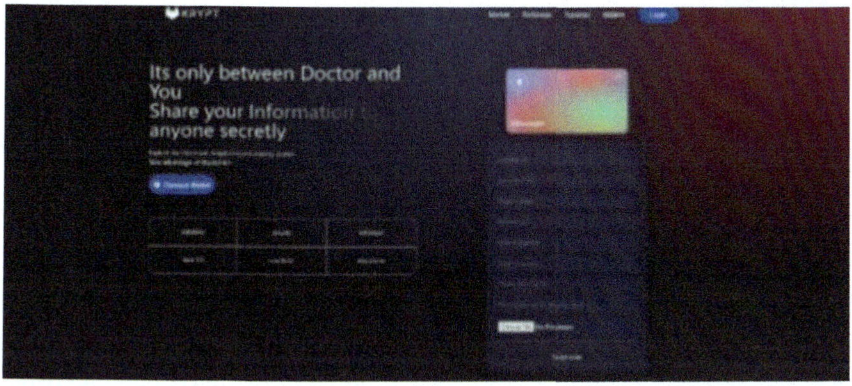

Fig. 2 UI of EHR

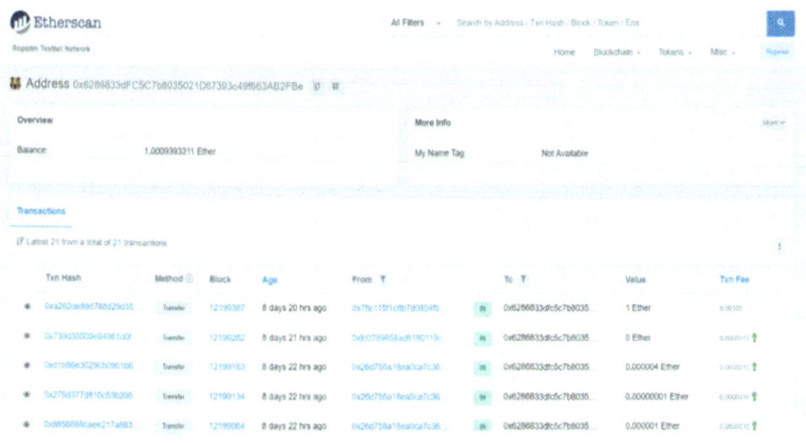

Fig. 3 Log for all transactions

reviews them, and provides an appropriate treatment plan. An Ethereum account wallet is linked to the website to enable transactions. Please note that the Meta Mask browser extension must be pre-installed on the system (Fig. 4).

There is a form known as the "Patient information form" which is shown below is necessary to be filled by the patient in which he/she should provide timestamp account address, the amount that is to be given, their name, their age, and the symptoms they are experiencing. After that, this patient information form is viewed by the doctor, and prescriptions are given according to the symptoms experienced by the patient.

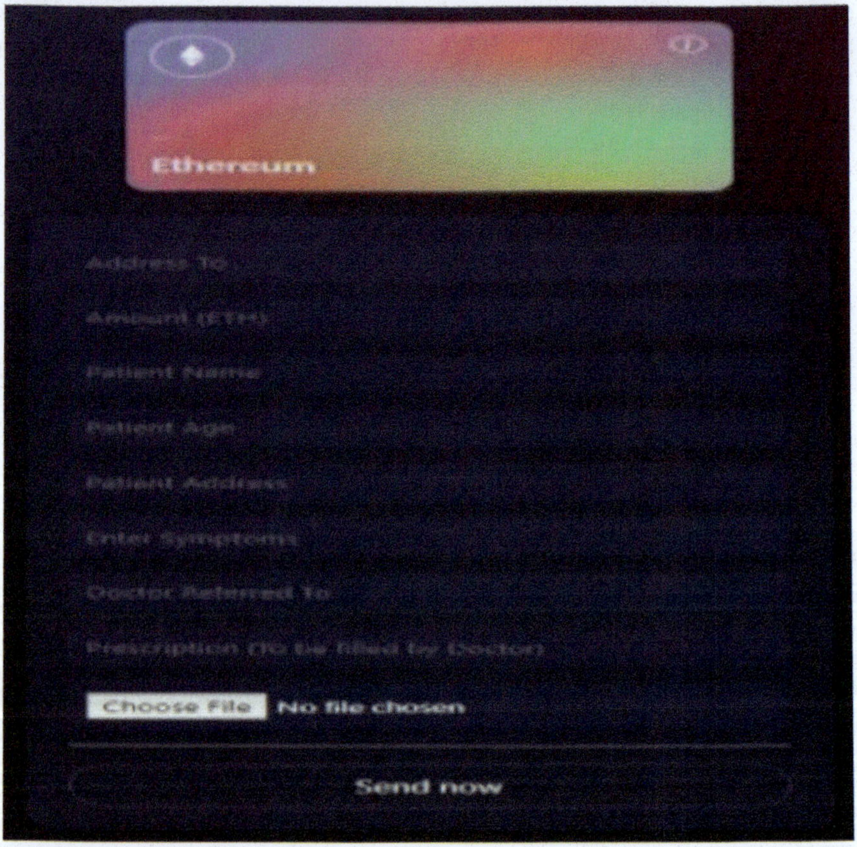

Fig. 4 Patient information form

6 Results and Discussion

The proposed framework incorporates various techniques for data sharing and privacy preservation. To enhance patient data security, encryption methods are applied to the private patient data stored on the blockchain within this framework, minimizing the risks of unauthorized access [5]. Blockchain, known for its high security and decentralized nature, involves multiple nodes or computers, making it an optimal choice for an Electronic Health Record (EHR) system. This revolutionary approach transforms the storage and sharing of medical information online, ensuring seamless operations within seconds. The use of blockchain simplifies tasks, ensures data security and accuracy, and reduces maintenance costs. Consequently, a blockchain-based platform is suggested for the storage and management of electronic medical records in a cloud environment [17].

As the EHR sharing system using blockchain makes the sharing of patient information very secure and there is almost negligible chance of data leakage, the transactions made have higher speed than centralized previous data sharing methods as this system is decentralized. The engagement of the user is very high. Any transaction such as doctor's fees or gas fees can be transferred safely from one account to another without much interruptions of a third party which leads to no corruption by middlemen. Thus, through the successful implementation of blockchain technology, we are now able to achieve end-to-end encryption between doctor and user.

7 Conclusion

This paper demonstrates the successful implementation of a project titled "Electronic Hospital Record Sharing System using Blockchain and Web 3.0". In the above preceding project, we have successfully identified the existing challenges within the EHR or Electronic Health Record system and have implemented a practical solution to tackle such issues or problems using a real prototype. The primary objective of this project was to develop a very reliable and effective mechanism of access control which is based on smart contracts, ensuring an effective and safe EHR sharing for the users [19, 20].

Patients now don't have to compromise their data safety and can send/share the data without any concern for data leakage, misuse, and unauthenticity. It also enables them to share their records faster which results in management of the time of doctor as well as patient, as patients do not have to wait longer for their turn. Patients have the freedom to log their data and own their shared data in their own ways. Thus, this project is a very secure way for sharing Patient's data and is faster than a centralized data-sharing system as this one is decentralized. And lastly, we have learned how blockchain technology comes to the rescue in such crucial areas that are highly been ignored.

References

1. Information Centre of Ministry of Industry and Information Technology of the People's Republic of China (2018) In: The 2018 white paper on China's blockchain industry," May 20th, 2018, http://www.miit.gov.cn/n1146290/n1146402/n1146445/c6180238/part/6180297.pdf
2. Feldman EV, Ruchay AN, Matveeva VK, Samsonova VD (2021) Bitcoin abnormal transaction detection based on machine learning
3. Haleem A, Javaid M, Pratap Singh R (2021) International Journal of Intelligent Networks (2021). Blockchain technology applications in healthcare: An overview
4. Peng H, Yang R, Wang Z, Li J, He L, Philip SY, Ranjan R (2021) Lime: low-cost and incremental learning for dynamic heterogeneous information networks. IEEE Trans Comput 71(3):628–642
5. Morris, David Z (2016) Leaderless, Blockchain-Based Venture Capital Fund Raises $100 Million, And Counting. Fortune. Archived from the original (15 May 2016) on 21 May 2016. Retrieved 23 May 2016
6. Nathan P (2016) A Venture Fund with Plenty of Virtual Capital, but No Capitalist. The New York Times. Archived from the original (21 May 2016) on 22 May 2016. Retrieved 23 May 2016.
7. The Economist (2016) Blockchains: The great chain of being sure about things. archived from the original (31 October 2015) on 3 July 2016. Retrieved 18 June 2016
8. Arvind N, Joseph B, Edward F, Andrew M, Steven G (2016) *Bitcoin and cryptocurrency technologies*: a comprehensive introduction. Princeton: Princeton University Press. ISBN 978–0–691–17169–2, 2016
9. Marco I, Lakhani, Karim R (2017) The Truth About Blockchain. Harvard Business Review. Harvard University. The technology at the heart of bitcoin and other virtual currencies, blockchain is an open, distributed ledger that can record transactions between two parties efficiently and in a verifiable and permanent way. Archived from the original on 18 January 2017. Retrieved 17 January 2017
10. Chen B, Tan Z, Fang W (2018) Blockchain-Based Implementation for Financial Product Management. IEEE
11. Amine Ferrag M, Derdour M, Mukherjee M (2019) Blockchain Technologies for the Internet of Things: Research Issues and Challenges. IEEE
12. Ekblaw A, Azaria A, Halamka JD, Lippman A, Vieira T (2017) A Case Study for Blockchain in Healthcare: "MedRec" Prototype for Electronic Health Records and Medical Research Data White Paper MedRec: Using Blockchain for Medical Data Access and Permission Management; IEEE Access: Vienna, Austria; ISBN 978–1–5090–4055–1, (2017)
13. Sharma A, Singh PK, Podoplelova A (2023) Graph neural network-based anomaly detection in blockchain network. In: Tanwar S, Wierzchon ST, Singh PK, Ganzha M, Epiphaniou G (eds) Proceedings of fourth international conference on computing, communications, and cyber-security. CCCS 2022. Lecture Notes in Networks and Systems, vol 664. Springer, Singapore. https://doi.org/10.1007/978-981-99-1479-1_67
14. Rahmadika S, Rhee KH (2018) Blockchain technology for providing an architecture model of decentralized personal health information. Int J Eng Bus Manag 10:1–12
15. Zhuang Y, Sheets LR, Chen YW, Shae ZY, Tsai JJP, Shyu CR (2020) A patient-centric health information exchange framework using blockchain technology. IEEE J Biomed Health Inform 24:2169–2176
16. Sundvall E, Terner A, Broberg H, Gillespie C (2019) Configuration of input forms in EHR systems using spreadsheets, openEHR archetypes and template. Stud Health Technol Inform 264:1781–1782
17. Kaur H, Alam MA, Jameel R, Mourya AK, Chang V (2018) A proposed solution and future direction for Blockchain-based heterogeneous medicare data in cloud environment. J Med Syst 42:156
18. Feldman EV, Ruchay AN, Matveeva VK, Samsonova VD (2021) Bitcoin abnormal transaction detection based on machine learning. International conference on analysis of images, social networks and texts. Springer, Cham, pp 205–215

19. Gad AG, Mosa DT, Abualigah L, Abohany AA (2022) Emerging Trends in Blockchain Technology and Applications: A Review and Outlook. J King Saud Univ 34
20. Lorenz J, Silva MI, Aparício D, Ascensão JT, Bizarro P (2020) Machine learning methods to detect money laundering in the bitcoin blockchain in the presence of label scarcity. In: Proceedings of the first ACM international conference on AI in finance, pp 1–8

Android Malware Detection Using Learning Techniques: A Review

Sumesh Kharnotia and Bhavna Arora

Abstract Android operating system (OS) for smartphones is more popular than other systems such as Apple iOS. Due to this Android applications are being developed at a fast pace across the whole mobile ecosystem, but as a consequence, this Android malware is also ascending continuously. The data of the customers present on the smartphone needs to be protected from the attacks done by hackers through malware-affected applications. In this research, the detection of Android malware is reviewed concerning other publications, and the different malware analysis techniques—hybrid, dynamic, and static—in addition to learning-based detection methods are examined. This paper focuses on the recent development in Android malware detection and discusses the findings of the work to help researchers gain some insights about the malware detection scenario.

Keywords Android security · Malware detection · Static malware analysis · Dynamic malware analysis · Machine learning · Deep Learning

1 Introduction

Android is the most popular smartphone operating system (OS). As of 2023, Android dominates 72.2% of the global market share, as shown in Fig. 1 [1]. The importance of mobile devices has increased in our daily lives as most of the information we receive today comes through them. With the advancement in mobile technology and the smartphone along with sales growth, the Android market has seen a significant rise in the number of applications place. The open-source Android platform and flexible operating system. The Google Play Store, which is the authorized and secure location to download Android apps, had more than 2.7 million available by

S. Kharnotia (✉) · B. Arora
Central University of Jammu, Bagla (Rahya-Suchani), Samba, Jammu and Kashmir 181143, India
e-mail: 22pcst04.csit@cujammu.ac.in

B. Arora
e-mail: bhavna.csit@cujammu.ac.in

P. J. S. Gonçalves et al. (eds.), *Proceedings of Fifth International Conference on Computing, Communications, and Cyber-Security*, Lecture Notes in Networks and Systems 1128, https://doi.org/10.1007/978-981-97-7371-8_21

the end of April 2023 [2]. The ease of installing applications from the Google Play Store and the ability to directly download the applications from third-party markets provides access to many applications at the click of a button. This multi-distribution of applications has made it easy for attackers to deceive users without their knowledge by secretly executing their harmful code. Malware authors are actively using malicious programs to attack machines running the Android operating system that endanger the security and integrity of Android applications. The approach used by antivirus software depends on signature matching; it cannot identify fresh viruses, sometimes known as zero-day threats [3]. Learning techniques are being used to identify unknown malware to get beyond the limitations of the signature-based method [4]. The features used to identify malware in Android applications are obtained in three analysis techniques static, dynamic, and hybrid analysis [5]. Examining malware's code without running it is known as static analysis. This process focuses on analyzing the structure, permissions, API calls, and other characteristics of the malware package to identify potential threats and vulnerabilities [6]. Dynamic analysis entails executing malware in a controlled environment and observing its behavior during runtime. This approach involves monitoring system calls, network traffic, file system changes, and interactions with other processes to gain insights into the actual behavior of the malware [7]. Static and dynamic analysis techniques are combined in hybrid analysis to achieve a comprehensive understanding of malware. Instead of focusing on one of these analyses, this paper focuses on all three types of analysis in combination with the learning techniques to give a better view of the Android malware detection scenario.

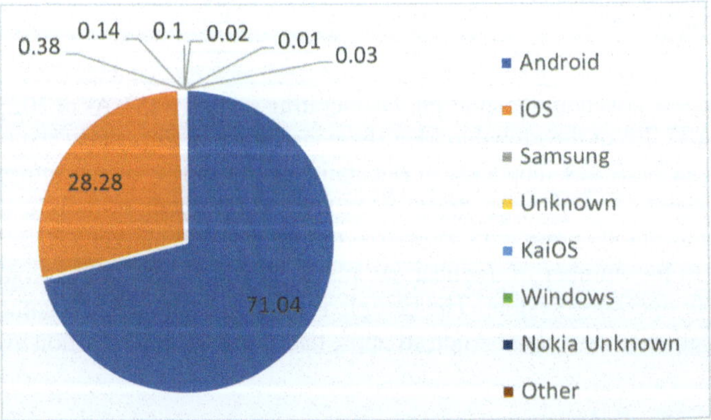

Fig. 1 Global smartphone operating system market share [1]

1.1 Motivation and Problem Statement

The following are the motivations behind this paper:

- **Android's Appeal and Vulnerabilities**: Android apps are widely used for convenience and entertainment but are also attractive to cybercriminals, posing risks like data theft and device disruption.
- **Escalating Threats and Detection Challenges**: As the number of Android apps grows, so do attempts to distribute malware. Traditional detection methods struggle, necessitating more effective approaches.
- **Innovative Security Solutions**: Ongoing research and innovative detection methods are crucial to protect users from evolving Android malware, ensuring a secure mobile experience.

The increasing use of Android applications has made them a prime target for cybercriminals, leading to a rising risk of data breaches and device disruption. Traditional Android malware detection methods are struggling to keep pace with evolving threats, necessitating more effective and innovative security solutions to ensure a secure mobile experience for users.

1.2 Contribution

The paper's contributions are as follows:

1. A review of current Android malware detection methods, serving as a valuable resource for researchers and practitioners.
2. Evaluation of various learning techniques in Android malware detection, aiding in selecting effective methods.
3. Offers guidance for future research in the detection of Android malware, highlighting areas needing further development.

The paper's remaining portion is organized as follows. The background information on the Android architecture, Android threats, and learning approaches utilized in Android malware detection is provided in Sect. 2. A review of several malware detection methodologies, learning technique-based detection, and discussion is presented in Sect. 3. Section 4 concludes by providing our conclusion and future focus.

2 Background

An overview of Android's architecture, security features, and potential attacks is provided in this section.

2.1 Android Architecture

Android architecture consists of five parts as shown in Fig. 2. [8]:

Linux Kernel: Serving as the fundamental core of the Android operating system, the Linux kernel assumes the crucial role of managing device drivers, power allocation, memory handling, device control, and resource management.

Libraries: Various software libraries, such as WebKit, OpenGL, FreeType, SQLite, and Media, are constructed on top of the Linux kernel. These libraries are integral to the Android system, with WebKit enabling web browsing, SQLite managing databases, FreeType supporting fonts, and Media handling audio and video playback and recording.

Android Runtime: At the heart of Android's runtime environment lies the Dalvik Virtual Machine (DVM), responsible for executing Android applications. The DVM,

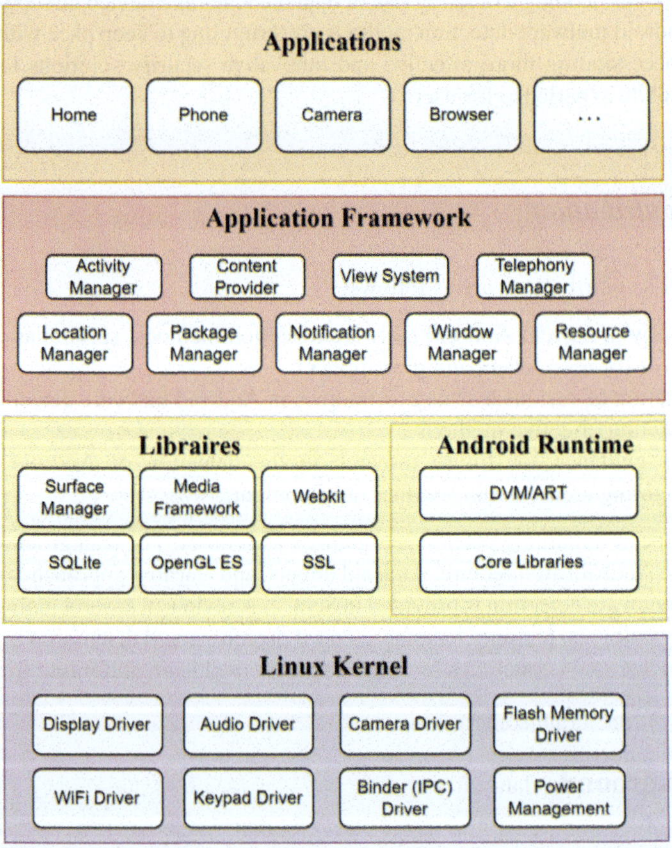

Fig. 2 Android architecture [8]

designed for mobile devices, ensures high-performance operation while conserving memory, drawing parallels to the Java Virtual Machine (JVM).

Application Framework: Built upon the foundation of libraries and the Android runtime, the Application Framework equips developers with a comprehensive set of Android APIs. These APIs cover telephony, resource access, geolocation, content retrieval, package management, and user interface design, offering various classes and interfaces for crafting Android applications.

Applications: Situated above the core Android layers, a diverse array of applications, including home screens, contacts, settings, games, and browsers, rely on the Android framework. This framework leverages the Android runtime and libraries to interface with the Linux kernel.

2.2 Android Threats

If appropriate security procedures are not in place, Android apps are easily exploitable with the right programming skills. Moreover, when new apps are released, Android app stores like Google Play do not adhere to strict security guidelines. The most frequent scenario that may be characterized as a danger to Android is attacks by malware. Depending on the damage they produce, numerous researchers have given different definitions for malware. The most general definition of malware is a malicious program with a component of harmful code that has the dangerous intention to gain unauthorized access and engage in neither morally nor legally acceptable behavior while threatening the three fundamental security principles of confidentiality, integrity, and availability [9].

2.3 Learning Techniques in Android Malware Detection

Machine Learning is a subfield of AI that focuses on creating applications by learning from data without explicitly specifying how the tasks are to be carried out. Based on historical data, conventional Machine Learning techniques produce predictions. Machine Learning provides different subcategories like Supervised learning, unsupervised learning, reinforcement learning, and deep learning [10].

Deep Learning is a branch of machine learning that focuses on algorithms inspired by the organization and functioning of the human brain. Deep learning reduces the need for extensive feature engineering by extracting valuable information directly from raw data, ultimately enabling models to attain superior performance [11].

3 Review of Literature and Discussions

In this section of the paper, various traditional approaches are explored, with an examination of the additions made by many researchers to those approaches.

3.1 Malware Detection Approaches

According to past research, there are three categories of technology for detecting Android malware: static, dynamic, and hybrid detection [5]. In the identification of malware, the technique of learning theory is frequently used, whether it is based on static, dynamic, or hybrid analysis taxonomies as shown in Fig. 3.

Static Analysis relies on questionable code analysis without the Android application being executed. Although it can achieve great code coverage, there are various barriers to it, including dynamic code loading and code obfuscation. Static features are obtained through static analysis [6]. Table 1 represents the literature review of different learning techniques in the static detection of Android Malware.

Dynamic Analysis entails running the code and analyzing the Android application. This can reveal dangers that are difficult to find through static analysis, but dynamic detection requires a lot of processing power and time. Dynamic features are obtained through dynamic analysis [7]. Table 2 represents the literature review of different learning techniques in the dynamic detection of Android Malware.

Hybrid Analysis To balance detection efficacy and efficiency, a technique called hybrid analysis is used which combines static and dynamic analysis. Table 3 represents the literature review of different learning techniques in the static detection of Android Malware.

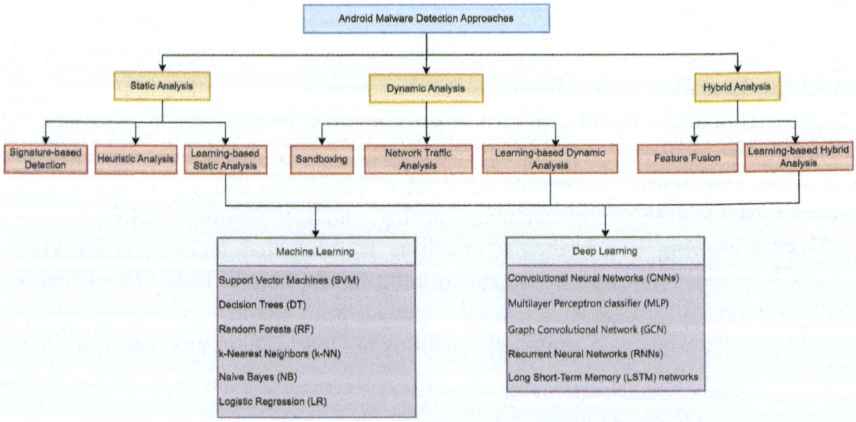

Fig. 3 Taxonomy of android malware detection approaches

Table 1 Review of static android malware detection approaches using learning techniques

Approach	Author	Year	Features	Dataset			Algorithm used	Result	Limitations
				Name	Benign	Malicious			
Static using DL	Zhu et al. [12]	2023	Permissions, API calls, and Hardware	Google Play Store, VirusShare	1061	2146	CNN based MSerNet	Accuracy of 96.48%	Balanced Dataset can be used
Static using DL	Nguyen et al. [13]	2022	Permission, Intent, Activity, Service Receiver, strings, API calls	Google Play, Drebin, and the AMD dataset	120,000	28,485	DNN	Accuracy of 97%	High false negatives in detecting unwanted software (UwS) highlight the need for the development of specialized feature sets for UwS detection

(continued)

Table 1 (continued)

Approach	Author	Year	Features	Dataset			Algorithm used	Result	Limitations
				Name	Benign	Malicious			
Static using DL	Kabakus et al. [14]	2022	Permissions, Intents, and API calls	Drebin, KuafuDet, Androzoo, and VirusShare	6,634	7,752	word2vec, CNN	Accuracy of 90%	Additional neural network architectures can be used. Increasing the sample size is beneficial for deep neural networks, as it often leads to improved performance
Static using ML	Khariwal et al. [15]	2020	Permissions and Intents	Google Play Store, Genome, Drebin, and Koodous	1414	1714	Support vector machines, Naive Bayes, and Random Forest	Accuracy at 94.73% was achieved using Random Forest	Cannot identify apps with minimal permissions and intents. Additional features from manifest files could be leveraged for analysis and detection

Table 2 Review of dynamic android malware detection approaches using learning techniques

Approach	Author	Year	Features	Dataset			Algorithm used	Result	Limitations
				Name	Benign	Malicious			
Dynamic using DL	Kim et al. [16]	2022	API Calls	Google Play Store, AMD	10,000	33,692	CNN, Jaccard similarity algorithm	Accuracy of 91.27%	Exclusion of applications whose API call graphs can't be extracted from Flowdroid
Dynamic using DL	Alzaylaee et al. [17]	2020	API Calls and Intents	-	19,620	11,505	Multilayer Perceptron classifier (MLP)	Accuracy of 95.21%	Cannot identify the family malware belongs to
Dynamic using ML	Cai et al. [18]	2019	Method Calls and ICC Intents	-	17,365	16,978	Random Forest	F1-measure accuracy of 97%	Overfitting due to small dataset size

Table 3 Review of hybrid android malware detection approaches using learning techniques

Approach	Author	Year	Features	Dataset			Algorithm used	Result	Limitations
				Name	Benign	Malicious			
Hybrid using ML & DL	Wang et al. [19]	2022	Permissions, Function Call graph	Google Play Store, Andro_ dumpsys:	4219	3950	Logistic Regression, Graph Convolutional Network (GCN)	Accuracy of 97.5%	A diverse feature set can be used
Hybrid using ML	Shatnawi et al. [20]	2022	Permissions, Action repetition	-	90,876	10,474	Random forest (RF), decision tree (DT) and XGBoost, Gradient Boosting	Accuracy of 94%	Only static analyses can be used. Only binary classification, cannot classify malware into family
Hybrid using ML	Martín et al. [21]	2019	Permissions, API calls, Activities, Services, Receivers	-	11,000	11,000	Random Forest, AdaBoost, Bagging and Gradient Boosting, Voting classifier	Accuracy of 89.7%	It can integrate other feature extraction techniques and recent malware samples can be updated in the dataset

3.2 Discussion

In the view of Static analysis, it has been observed that for the effectiveness of the detection, the researchers should focus on the balanced dataset of benign and malware apps and its size, so that the deep learning models can benefit from it in detecting the malware. Additionally, the unwanted software which don't affect the system needs different feature sets to detect them effectively. The code analysis method was frequently utilized as the feature extraction method in Learning-based malware detection investigations. Permissions and API calls are the most often retrieved features when applying feature extraction methods. Furthermore, along with the commonly used features such as permissions and intents, additional features can be combined to improve the analysis and detection. The dynamic analysis depends heavily on the sandboxing to extract features, so in case of applications that can't be processed with the tools, it becomes a limitation. Also, it takes a lot more time than static analysis to perform. The hybrid analysis which combines both the pros as well as cons of static and dynamic always doesn't perform better results and in some cases, it has been found that only static analysis alone can be used instead of hybrid to be combined with the learning techniques to identify the malware. Also, as the malware is constantly evolving the dataset needs to be continually updated to contain the latest malware samples to perform analysis.

4 Conclusion and Future Scope

This paper underscores the critical need for advanced Android malware detection methods, given the increasing attractiveness of Android devices to cyber attackers. The vulnerabilities, as well as the advantages and limitations of current strategies for detection based on learning techniques, are explored. The evolving nature of Android malware necessitates ongoing research to enhance its accuracy and efficiency so that malware applications can be distinguished from the benign before reaching into the devices. For this several studies have been explored and out of the three analysis techniques it is found that the time consumption of static analysis is less compared to either dynamic or hybrid and static itself alone can also give better results in some cases therefore the researchers should further explore it. Also, balanced dataset plays a crucial role in the android malware detection using learning methods. Further increasing the quantity of samples for the deep neural networks can be explored. Additionally, efforts should be directed towards what kind of features are selected for malware identification as most of the works rely on permission as a feature, and for those applications with very few permissions, the method may not work. The focus should also be on the combination of the features for malware detection as multiple features combined may lead to better accuracy. By addressing these areas, more comprehensive and adaptive approaches can be developed to counter the evolving landscape of Android malware threats and enhance the overall security of Android devices.

References

1. Mobile Operating System Market Share Worldwide | Statcounter Global Stats. (n.d.). Retrieved October 22, 2023, from https://gs.statcounter.com/os-market-share/mobile/worldwide
2. Number of Android applications on Google Play (Oct 2023) (n.d.) Retrieved October 22, 2023, from https://www.appbrain.com/stats/number-of-android-apps
3. Arun A, Nair AS, Sreedevi AG (2024) Zero day attack detection and simulation through deep learning techniques. In: 2024 14th International Conference on Cloud Computing, Data Science & Engineering (Confluence), 852–857. https://doi.org/10.1109/CONFLUENCE60223.2024.10463429
4. Dhalaria M, Gandotra E (2020) Android malware detection techniques: A literature review. Recent PatS Eng 15(2):225–245. https://doi.org/10.2174/1872212114999200710143847
5. Liu K, Xu S, Xu G, Zhang M, Sun D, Liu H (2020) A review of android malware detection approaches based on machine learning. IEEE Access 8:124579–124607. https://doi.org/10.1109/ACCESS.2020.3006143
6. Odat E, Yaseen QM (2023) A novel machine learning approach for android malware detection based on the co-existence of features. IEEE Access 11:15471–15484. https://doi.org/10.1109/ACCESS.2023.3244656
7. Sharma Y, Arora A (2024) A comprehensive review on permissions-based Android malware detection. Int J Inf Secur 2024:1–36. https://doi.org/10.1007/S10207-024-00822-2
8. Android Open Source Project (n.d.) Retrieved October 22, 2023, from https://source.android.com/
9. Gaber MG, Ahmed M, Janicke H, Cowan E (2024) Malware detection with artificial intelligence: a systematic literature review. ACM Comput Surv 56(6):1–33. https://doi.org/10.1145/3638552
10. Gorment NZ, Selamat A, Cheng LK, Krejcar O (2023) Machine learning algorithm for malware detection: taxonomy, current challenges, and future directions. IEEE Access 11:141045–141089. https://doi.org/10.1109/ACCESS.2023.3256979
11. Nasser AR, Hasan AM, Humaidi AJ (2024) DL-AMDet: Deep learning-based malware detector for android. Intell Syst Appl 21:200318. https://doi.org/10.1016/J.ISWA.2023.200318
12. Zhu H Juan, Gu W, Wang L Min, Xu Z Cheng, Sheng VS (2023) Android malware detection based on multi-head squeeze-and-excitation residual network. Expert Syst Appl 212:118705. https://doi.org/10.1016/J.ESWA.2022.118705
13. Nguyen DV, Nguyen GL, Nguyen TT, Ngo AH, Pham GT (2022) MINAD: Multi-inputs neural network based on application structure for android malware detection. Peer-To-Peer Netw Appl 15(1):163–177. https://doi.org/10.1007/S12083-021-01244-W
14. Kabakus AT (2022) DroidMalwareDetector: A novel android malware detection framework based on convolutional neural network. Expert Syst Appl 206:117833. https://doi.org/10.1016/J.ESWA.2022.117833
15. Khariwal K, Singh J, Arora A (2020) IPDroid: Android malware detection using intents and permissions. In: Proceedings of the World Conference on Smart Trends in Systems, Security and Sustainability, WS4 2020, 197–202. https://doi.org/10.1109/WORLDS450073.2020.9210414
16. Kim J, Ban Y, Ko E, Cho H, Yi JH (2022) MAPAS: a practical deep learning-based android malware detection system. Int J Inf Secur 21(4):725–738. https://doi.org/10.1007/S10207-022-00579-6
17. Alzaylaee MK, Yerima SY, Sezer S (2020) DL-Droid: Deep learning based android malware detection using real devices. Comput Secur 89:101663. https://doi.org/10.1016/J.COSE.2019.101663

18. Cai H, Meng N, Ryder B, Yao D (2019) DroidCat: Effective android malware detection and categorization via app-level profiling. IEEE Trans Inf Forensics Secur 14(6):1455–1470. https://doi.org/10.1109/TIFS.2018.2879302
19. Wang W, Ren C, Song H, Zhang S, Liu P (2022) FGL_Droid: An efficient android malware detection method based on hybrid analysis. Secur Commun Netw 2022. https://doi.org/10.1155/2022/8398591
20. Shatnawi AS, Jaradat A, Yaseen TB, Taqieddin E, Al-Ayyoub M, Mustafa D (2022) An android malware detection leveraging machine learning. Wirel Commun Mob Comput 2022. https://doi.org/10.1155/2022/1830201
21. Martín A, Lara-Cabrera R, Camacho D (2019) Android malware detection through hybrid features fusion and ensemble classifiers: The AndroPyTool framework and the OmniDroid dataset. Information Fusion 52:128–142. https://doi.org/10.1016/J.INFFUS.2018.12.006

Survey on Classification of Disease Identification in Potato Leaves for Precision Agriculture Using Deep Learning and Machine Learning

Urshita Gupta, Archi Jain, Hardik Gulati, Surbhi Vijh, and Sumit Kumar

Abstract Agriculture is of utmost importance for global sustenance making it crucial to optimize crop yields and minimize losses. Therefore, plant disease detection and classification has proven to be a critical task, especially in the field of precision agriculture. Through this article, we provide a comprehensive overview of the existing machine learning and deep learning methods for the classification and health analysis of potato leaf diseases in precision agriculture. Given the need for effective disease management strategies and the potential of data-driven approaches, our goal is to bring together insights from methods such as artificial neural networks (ANNs), convolutional neural networks (CNNs), and deep learning approaches. The survey aims to highlight innovations and advances in disease classification techniques, identify optimal model parameters, and evaluate the performances of various datasets. Furthermore, the article also explores the integration of metaheuristic optimization algorithms with machine learning and deep learning models to improve accuracy and efficiency in disease classification. Effectively, the results of this survey can serve as a guide for precision agriculture researchers and practitioners, facilitating informed decisions and driving further advances in data-driven agricultural practices.

Keywords Precision agriculture · Disease classification · Machine learning · Deep learning

U. Gupta · A. Jain · H. Gulati · S. Kumar (✉)
Department of Computer Science and Engineering, Amity University, Noida 201301, Uttar Pradesh, India
e-mail: skumar59@amity.edu

H. Gulati
e-mail: hardik.gulati@s.amity.edu

S. Vijh
Department of Computer Science and Engineering, Amity University, Noida 201310, Uttar Pradesh, India

P. J. S. Gonçalves et al. (eds.), *Proceedings of Fifth International Conference on Computing, Communications, and Cyber-Security*, Lecture Notes in Networks and Systems 1128, https://doi.org/10.1007/978-981-97-7371-8_22

1 Introduction

Potato, being a staple crop worldwide, serves as the cornerstone of global food security, contributing significantly to sustainable agriculture and livelihoods. However, potato plants are susceptible to numerous diseases, ranging from fungal infections to viral pathogens, which can severely affect crop productivity while also impacting sustainable development. Prompt identification and classification of these diseases help in implementing effective management strategies, especially in the case of precision agriculture.

In recent years, potato yields have benefitted from advancements in machine learning and deep learning technologies that have revolutionized disease identification and classification in agriculture. These technologies have enabled automated and precise disease detection, offering scalable and efficient solutions for the analysis of vast datasets of leaf images.

The motivation behind this article stems from the United Nations Sustainable Development Goals, particularly Goal 2: zero Hunger and Goal 12: responsible Consumption and Production. By improving crop outcomes, reducing resource wastage, and enhancing agricultural sustainability, our research intends to contribute to the reduction of world hunger and sustainable consumption of resources.

This paper aims to conduct a comprehensive survey of existing methods and innovations in potato leaf disease classification through an exhaustive analysis of the literature. We aim to identify trends, challenges, and opportunities in the field, providing valuable insights for researchers, practitioners, and policymakers alike.

This survey encompasses research efforts ranging from the application of traditional ML algorithms such as support vector machines (SVM) and K-Means clustering to more advanced deep learning techniques, including convolutional neural networks (CNNs) and transfer learning.

Additionally, metaheuristic optimization algorithms have been integrated with ML models to enhance classification accuracy and model parameters. Through this comprehensive review, we aim to provide insights into the current landscape of ML-based approaches for potato leaf disease classification. By summarizing the existing literature, we seek to identify trends, challenges, and opportunities for future research in the field of precision agriculture for potato disease management.

2 Literature Review

2.1 *Potato Leaf Disease Classification Using Machine Learning*

Potato plant diseases pose significant threats to agricultural productivity and food security worldwide. Timely and accurate detection of these diseases is crucial for implementing effective disease management strategies. In recent years, machine learning algorithms have gained prominence in the field of agricultural science, particularly in the classification of plant diseases. This literature review explores the various studies and methodologies employed in the classification of potato plant diseases using machine learning algorithms. Potato cultivation is susceptible to various diseases caused by fungi, bacteria, viruses, and phytoplasmas. Traditional methods of disease diagnosis often rely on visual inspection by experts, which can be time-consuming and subjective. Machine learning algorithms offer a promising solution by automating the process of disease identification through the analysis of plant images and other relevant data.

2.1.1 Potato Leaf Disease Classification Using SVM

SVM algorithms are widely used for binary classification tasks. In potato disease classification, SVM models have been employed to distinguish between different disease types. SVMs work well with both image data and extracted features, making them versatile for disease identification purposes. Before the year 2020, Support Vector machines were very reliable and useful in precision agriculture to identify botanical diseases as they give decent accuracy. This technique paved the way for future work in the area of plant disease identification. Aditi Singh and Harjeet Kaur [1] demonstrated a model that utilized Multiclass Support Vector Machines to classify potato ailments into three classes Late Blight, Early Blight, and Healthy. An accuracy of 95.99% was obtained. Monzurul Islam et al. [2] used a similar framework to showcase the digitalization process by using K-means segmentation for images and multiclass SVM algorithms, providing a decent accuracy of 95%. Computer vision phenotyping helps us in the automated detection of such unwanted diseases. Prakash et al. [3] used the infamous metaheuristic Particle Swarm Intelligence Optimization algorithm aiming to digitalize the agricultural sector by successfully getting an accuracy of 94.45% performing the SVM classifier on publicly available datasets.

2.2 *Potato Leaf Disease Classification Using Deep Learning*

2.2.1 Potato Leaf Disease Classification Using CNN

CNNs have shown remarkable performance in image recognition tasks. Researchers have applied CNN architectures for potato disease classification by processing leaf images. CNNs capture intricate patterns and textures, enabling accurate differentiation between healthy and diseased potato plants. In our readings, it was identified that neural networks are the most used techniques in present-day scenarios for precision agriculture. Decently promising results are visible with such newer technologies. Ashiqur Rahaman et al. [4] tested out various models like VGG16, VGG19, and ResNet to measure accuracies and when accuracies of 97, 95, and 67% were deduced, it was concluded that the VGG16 model outperformed most of the earlier proposed techniques, it gave hope for the betterment of precision agricultural techniques. These models were augmented for an increased chance of seeing sophisticated results. Ali Arshaghi et al. [5]'s work on potato diseases where they classified them into five classes namely Black Scurf, Common Scab, Black Leg, Pink Rot, etc. is commendable because this area of work is aligned with surface-level detection of unwanted ailments in potatoes. With a dataset of more than 5000 images, an astonishing accuracy of over 99% was obtained in some cases.

Varsha Gaikwad et al. [6] exhibited an accuracy of 96% with their customized and agile deep learning-based CNN model. Suresh et al. [7] combine both SVM and R-CNN to achieve not-so-objectionable results on two different datasets. One of them is a lab-based dataset and the other is an actual field dataset. The dataset obtained from actuarial works in the crop field yielded accuracies of over 70% while lab-based images acquired a higher percentage of area under the curve (AOC) parameter. More work needs to be done in this area. Shruthi U et al. [8] also categorized CNN as a more robust mechanism when drawn in comparison to a variety of machine learning algorithms, specifically Naïve-Bayes classifiers. Oppenheim et al. [9] also used this technique by applying it to a self-acquired image-based dataset instead of the typically used lab-based datasets only to conclude that yet CNN-based models yield higher accuracy of over 90%. Jameer Kotwal et al. [10] and Kurmi et al. [11] moved from traditional machine learning to deep learning and transfer learning methodologies. Even in this case, CNN flared better than other ML techniques. Determination of the exact location of the diseased area present on the surface of the leaf will help avoid excessive usage of pesticides. More evidence was provided by Saleem, Muhammad et al. [12], who plan to introduce frameworks that can detect infected spots more efficiently preferably in a real-time environment.

Gardie et al. [13]'s research work in the area of a range of CNN models showcased that the Inception-v3 pre-designed architecture is reliable with an accuracy of 98.83% and more can be achieved if the number of layers being used is increased in the same models. Sholihati et al. [14]'s work confirms further that allocating a diversity of data augmentation techniques in the model can increase accuracy significantly, especially in the case of VGG16 networks. Tiwari et al. [15] and Fizzah Arshad et al. [16] demonstrate the significance of the VGG16 models and how it fares better than other conventional convolutional models along with hybridization of it. The complexity and excessive computability costs of deep learning models are revealed by Khalifa, Nour Eldeen et al. [17], and Chen, et al. [18]'s works where light models with fewer layers are anticipated to be more feasible for farmers. This exposes the research gap of the absence of disease detection mechanisms in smartphones as farmers working in agriculture fields may not have the most technologically savvy devices to perform the same desired task. It calls for the execution of smaller models with less complexity in smart mobiles which can make it easier for farmers to scan the plants but not at the cost of accuracy. Gulbir Singh, Kuldeep Kumar [19], and Al-Adhaileh et al. [20]'s works focus on the same concept to promote faster models with fewer layers and replace convoluted neural networks with artificial ones.

2.2.2 Potato Leaf Disease Classification Using ANN

Patil, Priyadarshini and Meena [21] performed FCM segmentation on the images present in the PlantVillage dataset and compared SVM, Random Forest, and Artificial Neural Network classifiers, and accuracies of 84, 79, and 92% were obtained. Out of all the three, ANN's performance was remarkable. Pham et al. [22] juxtaposed ANN against three CNN models namely (AlexNet, VGG16, and ResNet-50) and results included percentages of 89.41% vs. 78.64, 79.92, and 84.88%, respectively.

2.2.3 Potato Leaf Disease Classification Using DNN

Dong et al. [23] adopted a hybrid metaheuristic approach in which a combination of Spider Monkey Optimization and Principal Component Analysis selected the necessary features for extraction purposes. DNN classifier was used after dimensionality reduction and the testing data yielded a resulting accuracy of 94%.

2.2.4 Potato Leaf Disease Classification Using SNN

Janarthan, et al. [24] employed a classification network that comprised an embedded module closely related to siamese-styled networking synergized with a simple neural network classifier on citrus fruits and leaves. The credibility of this particular model was also tested on tea leaves and a detection accuracy of 95.04% was tabulated (Fig. 1).

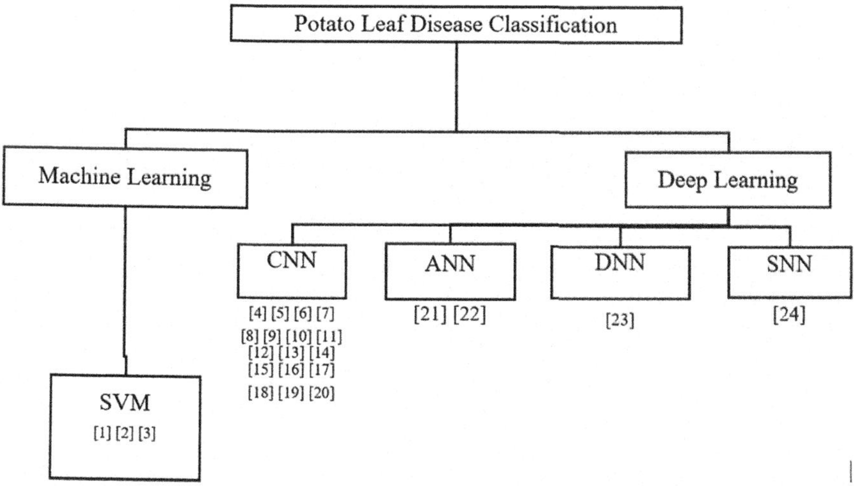

Fig. 1 Taxonomy tree of classification of papers

3 Results

1. It was observed that the Plant Village dataset has been used majorly in most of the research conducted in the past. This dataset is pre-prepared, and to achieve more realistic results, environmental images should be considered.
2. To avoid the excessive usage of pesticides, better localization of diseased regions on specific leaves should be performed.
3. Since the severity of plant diseases keeps on changing depending on factors like environment, soil types, regions, geography, etc., therefore it is vital to consider these factors in determining the reliability of proposed models.
4. There needs to be more exploration of advanced algorithms. While methods ranging from convolutional neural networks (CNNs), support vector machines (SVM) to decision trees, and K-nearest neighbors (K-NN) have been explored, and research on the application of more advanced machine learning algorithms, such as recurrent neural networks (RNNs) or ensemble methods like XGBoost, is slowly increasing. Inspecting the effectiveness of these advanced techniques could provide valuable insights into improving the accuracy and efficiency of disease classification models.
5. Expert knowledge incorporation from botanical and pathological perspectives into these models can improve the efficacy and reliability of these models. Technical information integration from domains revolving around disease symptoms, plant physiology, and environmental factors into deep learning models may lead to better feature selection and model design. Hybrid models that combine deep learning approaches with expert knowledge is an area that entails further consideration.
6. There exists a lack of implementation of these above-discussed models in smartphones. Since most farmers come from less technologically advanced agricultural areas, employing such large models in smartphones can be difficult. ANN is a methodology that gives hope to provide disease detection without compromising much on methodology.
7. Imbalanced datasets, where certain disease classes have significantly fewer samples than others, present a challenge in training accurate machine learning models. Research should focus on developing techniques to handle class imbalance effectively, such as oversampling, undersampling, or generating synthetic samples. Additionally, exploring transfer learning approaches from related plant disease datasets could help mitigate the scarcity of labeled data specific to potato diseases.

4 Discussion

The results of our survey shed light on how disease classification in potato leaves is changing in the context of precision agriculture. Through an extensive literature review, various approaches were identified, ranging from machine learning algorithms to deep learning techniques. This discussion addresses the issues raised in the reviews and summarizes the key findings of the studies reviewed.

4.1 Comparison of Methodologies

Our survey found that there has been a significant shift toward deep learning techniques, particularly convolutional neural networks (CNNs), in disease classification in potato leaves. This can be seen as a reflection of the wider acceptance of deep learning in image analysis tasks. Multiclass support vector machines are extremely effective in classifying potato diseases with 95.99% accuracy as shown by Singh and Kaur [1]. However, SVMs have been outperformed by CNN-based models in several studies. For example, Gaikwad and Musande [6] achieved 96% accuracy using a custom CNN model specifically tailored for potato leaf disease classification.

4.2 Challenges and Limitations

Although deep learning techniques have yielded encouraging results, our survey uncovered several obstacles and limitations, particularly the imbalance in the datasets used to train classification models. Database imbalances, as noted by Patil et al. [21], can lead to biased model performance, particularly in situations where certain disease classes are underrepresented. Effective approaches to dealing with class imbalances such as the generation of synthetic samples, oversampling, or undersampling need to be developed to deal with this problem.

4.3 Optimization and Model Interpretability

Another aspect of concern in our survey is the optimization of classification models and their interpretability. While metaheuristic optimization algorithms have been integrated into CNNs to improve classification accuracy as outlined in the works of Pham et al. [22] and Dong et al. [23], there is a need for further research on optimizing model parameters and convergence criteria as well as ensuring the interpretability of classification models for their practical utility in precision agriculture. Feature visualization and model explanation methods are techniques that can improve the transparency and trustworthiness of classification results as demonstrated by Khalifa et al. [17].

4.4 Future Directions

Subsequent research into disease classification for precision agriculture should focus on addressing the identified obstacles and promoting advanced techniques. To increase the accuracy and robustness of classification, more research is needed on advanced machine learning algorithms such as recurrent neural networks (RNNs) and ensemble techniques such as XGBoost [18]. Additionally, initiatives should be taken to improve the interpretability of classification models and create interfaces that are easy to use for agricultural practitioners.

5 Conclusion

This survey comprehensively examines the latest methods for disease classification in potato leaves, focusing on their application in precision agriculture. The motivation behind this survey is the urgent need for accurate and efficient disease detection techniques to support sustainable agricultural practices and ensure food security. Our survey highlights advances in machine learning and deep learning techniques, particularly their use in automating disease classification processes. We explored a range of methods, including traditional approaches such as Support Vector Machines (SVM) and advanced techniques such as Convolutional Neural Networks (CNNs). The comparative analysis highlights the strengths and limitations of each method and highlights the potential of deep learning models, particularly CNNs, to achieve superior classification accuracy. Furthermore, our survey addresses the integration of metaheuristic optimization algorithms with CNNs and further demonstrates their ability to improve classification performance. By optimizing model parameters and using metaheuristic algorithms, higher levels of accuracy can be achieved, thereby improving disease identification in potato leaves. However, our discussion also recognizes several challenges that need to be addressed for the wider adoption of these techniques in precision agriculture. These challenges include dealing with imbalanced datasets, ensuring model interpretability, and optimizing computational efficiency.

Future research efforts should focus on robust methods to overcome these challenges and advance state-of-the-art disease classification. This survey highlights the importance of automated disease detection techniques in precision agriculture and their potential to revolutionize agricultural practices. By continuing to explore advanced machine learning algorithms, addressing bias in datasets, and improving model interpretability, we can help improve crop yields, promote sustainability, and ensure global food security.

References

1. Singh A, Kaur H (2021) Potato plant leaves disease detection and classification using machine learning methodologies. In: IOP Conference Series: Materials Science and Engineering. IOP Publishing, 1022, no 1, p 012121
2. Islam M, Dinh A, Wahid K, Bhowmik P (2017) Detection of potato diseases using image segmentation and multiclass support vector machine. In: 2017 IEEE 30th Canadian conference on Electrical and computer engineering (CCECE), pp 1–4, IEEE
3. Prakash N, Balaji VR (2021) Detection of plant disease using swarm intelligence optimization algorithm. In: 2021 International conference on advancements in electrical, electronics, communication, computing and automation (ICAECA), pp 1–5, IEEE
4. Nishad MAR, Mitu MA, Jahan N (2022) Predicting and classifying potato leaf disease using kmeans segmentation techniques and deep learning networks. Procedia Comput Sci 212:220–229
5. Arshaghi A, Ashourian M, Ghabeli L (2023) Potato diseases detection and classification using deep learning methods. Multimed Tools Appl 82(4):5725–5742
6. Gaikwad VP, Musande V (2023) Plant leaf damage detection using convolution neural network models. In: AIP Conference Proceedings, AIP Publishing, vol 2822, no 1
7. Suresh V, Gopinath D, Hemavarthini M, Jayanthan K, Krishnan M (2020) Plant disease detection using image processing. Int J Eng Res Technol (IJERT) 9(03):2278–3181
8. Shruthi U, Nagaveni V, Raghavendra BK (2019, March) A review on machine learning classification techniques for plant disease detection. In: 2019 5th International conference on advanced computing & communication systems (ICACCS), pp 281–284, IEEE
9. Oppenheim D, Shani G (2017) Potato disease classification using convolution neural networks. Adv Anim Biosci 8(2):244–249
10. Kotwal J, Kashyap R, Pathan S (2023) Agricultural plant diseases identification: from traditional approach to deep learning. Mater Today: Proc 80:344–356
11. Kurmi Y, Saxena P, Kirar BS, Gangwar S, Chaurasia V, Goel A (2022) Deep CNN model for crops 'diseases detection using leaf images. Multidimens Syst Signal Process 33(3):981–1000
12. Saleem MH, Potgieter J, Arif KM (2020) Plant disease classification: a comparative evaluation of convolutional neural networks and deep learning optimizers. Plants 9(10):1319
13. Gardie B, Asemie S, Azezew K, Solomon Z (2022) Potato plant leaf diseases identification using transfer learning. Indian J Sci Technol 15(4):158–165
14. Sholihati RA, Sulistijono IA, Risnumawan A, Kusumawati E (2020) Potato leaf disease classification using deep learning approach. In: 2020 international electronics symposium (IES), pp 392–397, IEEE
15. Tiwari D, Ashish M, Gangwar N, Sharma A, Patel S, Bhardwaj S (2020) Potato leaf diseases detection using deep learning. In: 2020 4th International Conference on intelligent computing and control systems (ICICCS), pp 461–466, IEEE
16. Arshad F, Mateen M, Hayat S, Wardah M, AlHuda Z, Gu YH, Al-antari MA (2023) PLDPNet: End-to-end hybrid deep learning framework for potato leaf disease prediction. Alex Eng J 78:406–418
17. Khalifa NEM, Taha MHN, Abou El-Maged LM, Hassanien AE (2021) Artificial intelligence in potato leaf disease classification: a deep learning approach. Machine learning and big data analytics paradigms: analysis, applications and challenges, 63–79
18. Chen J, Deng X, Wen Y, Chen W, Zeb A, Zhang D (2023) Weakly-supervised learning method for the recognition of potato leaf diseases. Artif Intell Rev 56(8):7985–8002
19. Singh G, Yogi KK (2023) Comparison of RSNET model with existing models for potato leaf disease detection. Biocatal Agric Biotechnol 50:102726
20. Al-Adhaileh MH, Verma A, Aldhyani TH, Koundal D (2023) Potato blight detection using finetuned cnn architecture. Mathematics 11(6):1516
21. Patil P,Yaligar N, Meena SM (2017) Comparision of performance of classifiers-svm, rf and ann in potato blight disease detection using leaf images. In: 2017 IEEE international conference on computational intelligence and computing research (ICCIC), pp 1–5, IEEE

22. Pham TN, Van Tran L, Dao SVT (2020) Early disease classification of mango leaves using feed-forward neural network and hybrid metaheuristic feature selection. IEEE Access 8:189960–189973
23. Dong M, Yu H, Zhang L, Sui Y, Zhao R (2023) A PCA-SMO based hybrid classification model for predictions in precision agriculture. Tehnički vjesnik 30(5):1652–1660
24. Janarthan S, Thuseethan S, Rajasegarar S, Lyu Q, Zheng Y, Yearwood J (2020) Deep metric learning based citrus disease classification with sparse data. IEEE Access 8:162588–162600

Explainable Artificial Intelligence (XAI) for Healthcare: Enhancing Transparency and Trust

Anuj Kumar, Himanshu Hora, Ankur Rohilla, Pramod Kumar, and Rahul Gautam

Abstract Artificial intelligence (AI) is an emerging field of computer science which is currently being used in many sophisticated applications such as e-commerce, military, education, industry, and healthcare. AI has several subfields like machine learning, neural network, Deep Learning, Natural Language Processing (NLP), and computer vision. In the medical domain, AI with deep learning model plays a crucial role to predict the symptoms of various kinds of disease and clinicians to make decision about analysis critical medical report provided by radiologists and pathologists. However, the adoption of many AI model in healthcare face challenges related to transparency, interpretability, and trustworthiness, due to their "Black-Box" in nature. Usually, it is essential for humans to understand the reasoning behind an AI model's decision-making. To make a better decision-making, Explainable AI (XAI) is a useful technique that aims to explain the information inside the black-box model of AI algorithms that reveals how the decisions are made. The aim of this paper is to provide a survey of the most novel XAI techniques used in healthcare and related medical imaging applica- tions. In addition, this paper provides the study of various applications of XAI in healthcare and focuses on challenges related to black-box AI models, emphasizing the requirement for interpretable arrangements in healthcare. Furthermore, this paper presents different XAI strategies, including Local Interpretable Model-Agnostic Explanations (LIME), Shapley Additive Explanations (SHAP), and rule-based frameworks, which are displayed and assessed for their viability in making

Himanshu Hora, Ankur Rohilla, Pramod Kumar and Rahul Gautam contributed equally to this work.

A. Kumar (✉) · A. Rohilla · P. Kumar · R. Gautam
School of Computing, Graphic Era Hill University, Bareilly Road, Haldwani, Uttarakhand 263139, India
e-mail: anujkumardixit@gehu.ac.in

R. Gautam
e-mail: rahulgautam.srgc@gamail.com

A. Kumar · H. Hora · A. Rohilla · P. Kumar · R. Gautam
Department of Computer Application, Shri Ram College, Circular Road, Muzaffarnagar, UP 251001, India

P. J. S. Gonçalves et al. (eds.), *Proceedings of Fifth International Conference on Computing, Communications, and Cyber-Security*, Lecture Notes in Networks and Systems 1128, https://doi.org/10.1007/978-981-97-7371-8_23

AI models interpretable. Finally, this survey paper provides future direction to help developers and researchers for future prospective investigations in healthcare and discusses future research possibilities in the area of XAI.

Keywords Artificial intelligence (AI) · Machine learning · Deep learning · Explainable artificial intelligence (XAI) · Medical imaging

1 Introduction

Artificial intelligence (AI) is an emerging field of computer science which is currently being used in many sophisticated applications such as e-commerce, military, education, industry, and healthcare. Machine learning, Deep learning, Artificial Neural Network (ANN), computer vision, and Natural Language Processing (NLP) [1] are the branches of AI. In the medical domain, AI with deep learning model plays a crucial role to predict the symptoms of various kinds of disease and clinicians to make decision about analysis critical medical report provided by radiologists and pathologists. The optimized machine learning algorithm solves many problems in several fields including healthcare [2]. Many deep learning models of AI are not transparent and trustworthy, making doctor unsure about the signs of diagnosis. The question arises of how one can provide better evidence of the predictive result. AI models work as a blackbox. It is a big issue that generate a gap between AI models and human understanding [3]. Humans simply need to understand what is going on inside the AI model's algorithm that is exactly where XAI came into the picture. This research field is often also called interpretable machine learning. XAI provides techniques to better understand and validate how our machine learning model works. Typically, XAI is associated with two terms "explainable" and "interpretable". AI frameworks can quickly analyse patient records, spot unpretentious designs in demonstrative pictures, and indeed foresee disease outbreaks by scrutinizing patterns in epidemiological information. By giving healthcare experts with actionable insights and expanding their decision-making capabilities, AI demonstrates to be a priceless instrument in optimizing patient care. AI's impact amplifies past information analysis—it's a driving force behind the rise of personalized medication. Through the examination of hereditary and clinical information, AI calculations can recognize individualized treatment plans custom-made to a patient's special hereditary makeup and medical history. This groundbreaking approach not as it was improving treatment adequacy but moreover diminishes unfavourable impacts, checking a significant stride towards patient-centred care. Beyond clinical applications, AI is driving operational proficiency in healthcare institutions. Administrative tasks, such as charging, arrangement planning, and asset assignment, can be computerized through AI, liberating up healthcare experts to focus more on patient care (Fig. 1).

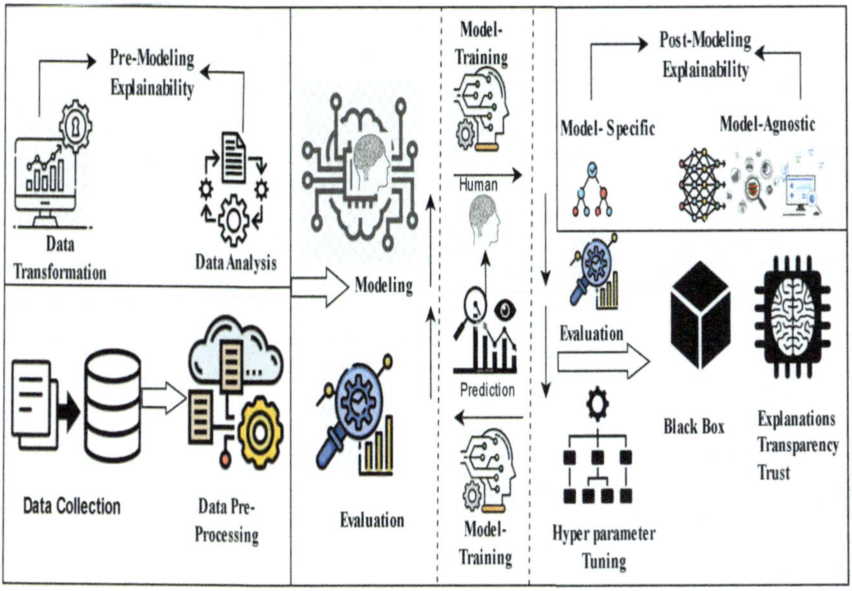

Fig. 1 Explainable artificial model architecture

1.1 Applications of XAI in Healthcare

Explainable Artificial Intelligence (XAI) has numerous applications in healthcare, where transparency and interpretability of AI models are crucial for ensuring patient safety, trust among healthcare professionals, and ethical compliance. Here are some key applications of XAI in healthcare [4].

1.1.1 Clinical Decision Support Systems (CDSS)

XAI can enhance CDSS by providing clear explanations for the recommendations or decisions made by AI algorithms. This helps healthcare professionals understand the reasoning behind AI-driven suggestions for diagnoses, treatment plans, and medication recommendations, leading to more informed clinical decisions (Fig. 2).

1.1.2 Medical Imaging Interpretation

XAI can be integrated into medical image analysis to assist radiologists and clinicians in interpreting complex imaging data. It can highlight areas of interest, provide explanations for image-based diagnoses, and help identify potential anomalies in X-rays, MRIs, CT scans, and pathology slides.

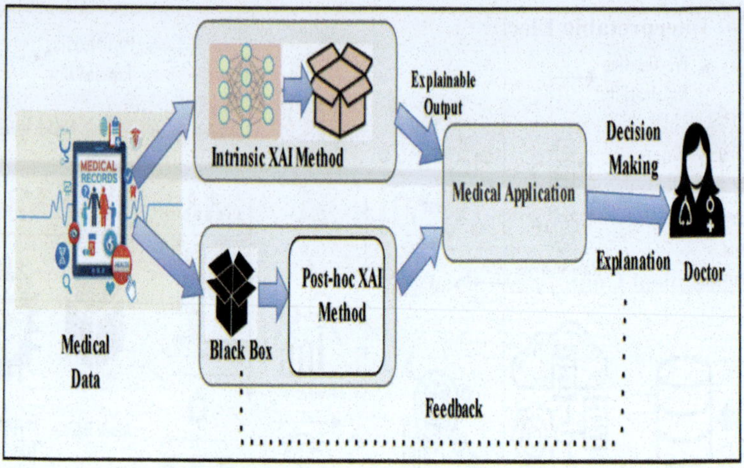

Fig. 2 Flow of a medical XAI application

1.1.3 Drug Discovery and Genomics

XAI can be applied to drug discovery by explaining the molecular mechanisms of diseases and the potential targets for drug development. It aids in identifying genetic markers and biomarkers for personalized medicine, making it easier for researchers to understand how specific genes or molecules are associated with diseases.

1.1.4 Patient Risk Assessment

XAI can be used to predict patient risk factors and outcomes, such as readmission risk, disease progression, or complications following surgery. By providing transparent risk assessments, XAI helps clinicians prioritize patient care and interventions more effectively.

1.1.5 Treatment Recommendation and Personalized Medicine

XAI-driven systems can explain why a particular treatment plan or medication is recommended for a specific patient. This personalization is based on the patient's medical history, genetic information, and other relevant data, increasing patient engagement and adherence.

1.1.6 Interpretable Electronic Health Records (EHRs)

XAI can be used to interpret and extract valuable insights from electronic health records (EHRs). It can help healthcare providers understand patient trends, identify early warning signs, and optimize treatment plans based on comprehensive and interpretable patient data.

1.1.7 Ethical Compliance and Bias Mitigation

XAI can play a critical role in addressing ethical concerns in healthcare AI, such as bias in algorithms. By providing transparency into decision-making processes, it allows for the identification and mitigation of biases in AI models, ensuring fair and equitable healthcare outcomes.

1.1.8 Clinical Trials and Research

XAI can assist researchers in explaining the results of clinical trials and observational studies. It aids in identifying significant factors and variables contributing to study outcomes, helping researchers make more informed conclusions and decisions.

1.1.9 Patient Engagement and Education

XAI-powered applications can explain medical conditions, treatment options, and potential outcomes to patients in a clear and understandable manner. This promotes patient engagement, informed consent, and shared decision-making in healthcare.

1.1.10 Regulatory Compliance

XAI can assist healthcare organizations in complying with regulatory requirements related to transparency and accountability. It helps ensure that AI-driven systems adhere to privacy and ethical standards, making them compliant with healthcare regulations.

2 Advancements in XAI Techniques

Researchers and developers have been working to make AI models easier to understand and transparent in recent years, which has resulted in substantial advancements in Explainable Artificial Intelligence (XAI) methodologies. These developments are meant to solve the difficulty in comprehending and having faith in sophisticated AI systems. Here are some significant developments in XAI techniques.

2.1 *Local Interpretable Model-Agnostic Explanations (LIME)*

LIME is a method that concentrates on producing locally accurate explanations for intricate models. It creates interpretable surrogate models around certain predictions, making it simpler to comprehend why an AI model made a specific decision.

2.2 *Shapley Additive Explanations (SHAP)*

SHAP values are used to describe the role played by each feature or variable in the prediction made by a model. They offer a coherent framework for comprehending the significance of features in black-box models, including neural networks (Fig. 3)
.

2.3 *Counterfactual Explanations*

Alternative scenarios are created through counterfactual explanations to explain AI decisions. For instance, if an AI model approves a loan application, a counterfactual explanation could outline the adjustments that would have caused the application to be rejected.

2.4 *Rule-Based Models*

Rule-based models, also known as symbolic AI, use rules that can be understood by humans to describe AI decisions. To give clear decision reasoning, these models can be utilized in conjunction with intricate machine learning models [5].

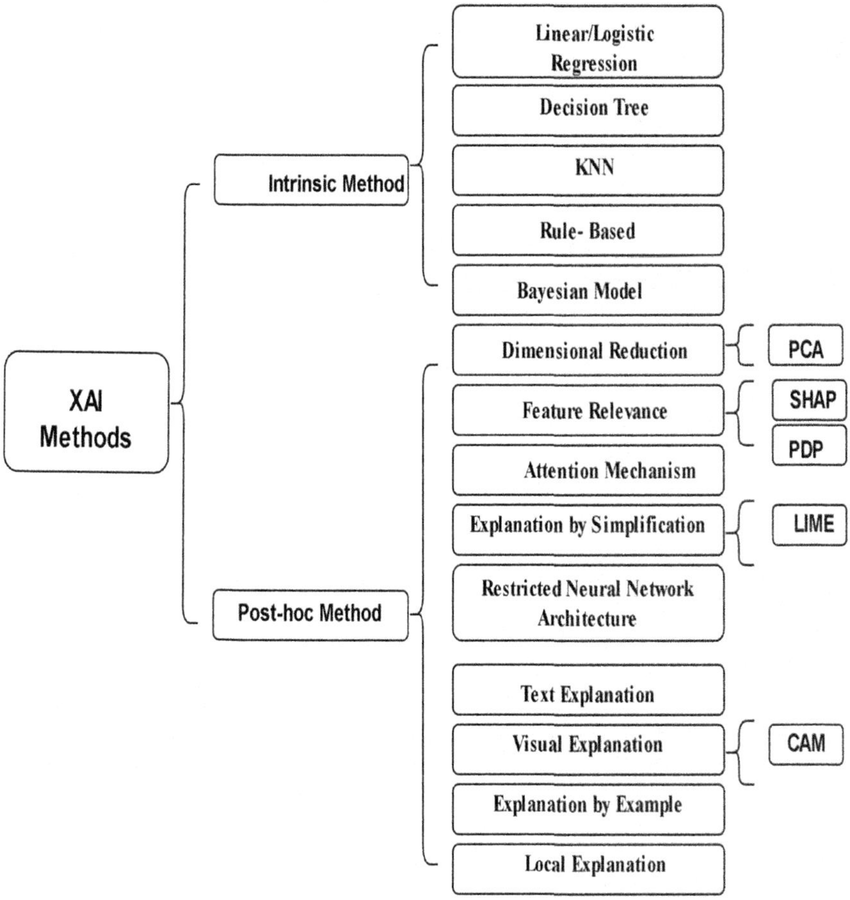

Fig. 3 Taxonomy of XAI method

2.5 *Gradient-Based Methods*

Methods such as Integrated Gradients and Smooth Grad employ gradients to elucidate the contribution of each input feature towards the output of a model. This facilitates the process of feature attribution and enables visual explanations.

2.6 *Attention Mechanisms*

Attention mechanisms, which were initially devised for natural language processing (NLP), have been modified to furnish elucidations for sequential data and image recognition assignments. These mechanisms [6] accentuate the specific segments of the input data that the model concentrates on while rendering predictions.

2.7 *Interpretable Neural Networks*

The scholars have devised neural network structures that are inherently interpretable. These networks frequently employ sparse activations or structured layers, which facilitate comprehension of the information flow within the model.

2.8 *Visualization Tools*

The progressions in visualization [7] methodologies have enabled users to perceive the functioning of AI models and the rationale behind their specific predictions. The utilization of tools such as activation maps, saliency maps, and feature visualization has facilitated users in acquiring a comprehensive understanding of the behaviour of the model.

2.9 *Natural Language Explanations*

Advancements in the field of natural language generation have facilitated the ability of artificial intelligence models to furnish explications in a language that is comprehensible to humans. This capability is especially advantageous in domains such as healthcare and law, where lucid and straightforward explanations are of paramount importance.

2.9.1 Hybrid Models

The amalgamation of interpretable models with intricate machine learning models in a hybrid methodology presents a harmonious equilibrium between precision and lucidity. This facilitates lucid explications without compromising prognostic [8] potential.

2.9.2 Standardization Efforts

Various organizations and researchers are currently engaged in the process of establish-ing standardized techniques for explainable Artificial Intelligence (XAI) to guarantee uniform and dependable explanations across diverse AI models and applications.

3 Challenges

Healthcare Explainable Artificial Intelligence (XAI) confronts a number of obstacles that must be overcome in order for it to be successfully incorporated [9] into clinical practise. Here are a few of the main difficulties:

3.1 Complexity of Healthcare Data

Medical photographs, clinical notes, and structured EHR data are just a few examples of the many different forms of data that are frequently complicated, diverse, and present in healthcare. It is a huge challenge [10] to create XAI systems that can manage and interpret findings from such a wide range of data sources.

3.2 Model Complexity

Due to their intricacy, deep learning models, which are increasingly employed in healthcare, are sometimes referred to as "black-box" models. It's not easy to clinically meaningfully explain these models' judgement calls.

3.3 Interoperability

EHR systems and data formats used by healthcare organizations vary widely. It is difficult to make sure that XAI systems can be connected with these various systems and offer coherent justifications.

3.4 Privacy and Security

A fine line must be drawn between maintaining patient confidentiality and offering explanations. Techniques for XAI must be created to produce insightful explanations without disclosing private patient data [11].

3.5 Data Quality and Bias

It takes skill to maintain patient privacy while yet explaining things. It is necessary to develop XAI methods that can produce insightful explanations without divulging private patient data.

3.6 Clinical Validation

To make sure that their explanations are correct and consistent with clinical knowledge, XAI systems need to be carefully verified in clinical settings. Clinical validation studies may need a lot of time and resources.

3.7 User Acceptance

The explanations offered by AI systems must be trusted and understood by healthcare personnel. Clinical practitioners may be reluctant to depend on AI-driven suggestions without thorough and trustworthy justifications since user [12] acceptability of XAI in clinical practise is a substantial barrier.

3.8 Regulatory Compliance

It is essential to adhere to healthcare rules, such as those imposed by HIPAA in the US. These rules must be followed by XAI systems while processing patient data and offering explanations.

3.9 Scalability

XAI solutions must be scalable to accommodate the volume of data and the requirement for real-time decision assistance as healthcare generates enormous volumes of data.

3.9.1 Education and Training

In order for healthcare practitioners to apply and understand XAI-driven suggestions and explanations effectively, they must get education and training in this area. The creation of educational resources and programmes is necessary [13] for this.

4 Future Directions

As AI and machine learning models are used to support medical decisions, diagnostics, and treatment recommendations, Explainable Artificial Intelligence (XAI) is becoming more and more significant in the healthcare industry. For these AI systems to earn the trust of medical personnel and patients, openness and interpretability are essential. The following are some potential paths for XAI in healthcare:

4.1 Model Transparency and Interpretability

The goal of research and development will remain to improve the readability and transparency of AI models. This covers methods like feature significance analysis, model visualization, and the creation of more comprehensible, straightforward model designs.

4.2 Integration with Electronic Health Records (EHRs)

In order to give explanations and insights immediately into the clinical process, XAI technologies will be more seamlessly connected with electronic health records. Clinicians may find this useful for comprehending the logic underlying AI-driven suggestions.

4.3 Human-AI Collaboration

Future XAI systems will place a focus on expert AI and human cooperation. This can entail interactive user interfaces that [14] let physicians examine various scenarios and treatment alternatives while also asking the AI system for clarifications.

4.4 Privacy-Preserving XAI

Healthcare data privacy is a major problem; hence, it will be essential to use XAI approaches that can explain things without disclosing private patient data. One of the main study areas will be privacy-preserving XAI.

4.5 Interpretable Deep Learning

A major focus will be on creating methods to make deep learning models, which are sometimes referred to as "black-box models", more understandable. This can entail employing strategies like attention mechanisms or deep network feature extraction.

4.6 Domain-Specific XAI

Radiology, genetics, and drug discovery are just a few of the many subfields that make up the broad area of healthcare. To meet the unique requirements and difficulties of each area, XAI methodologies will need to be customized.

4.7 Regulatory Guidelines

It's possible that regulatory [15] organizations will set standards and regulations for XAI systems within the healthcare sector, like the FDA in the US. These recommendations will have an impact on how XAI solutions are developed and implemented.

4.8 Education and Training

Healthcare workers will need to be educated on how to utilize AI systems correctly and understand their results as XAI becomes more common in the industry. Concepts from XAI will need to be included in training programmes.

4.9 *Bias and Fairness*

In the healthcare industry, it is crucial to guarantee that XAI systems are impartial and generate fair results. The development of tools to identify and reduce bias in AI models will be the main focus of research.

4.9.1 Continuous Evaluation and Improvement

Healthcare XAI systems [16] will require ongoing evaluation and development. Clinicians' and patients' feedback will be crucial [17] for improving these systems and making them more reliable and effective.

5 Conclusion

In conclusion, Explainable Artificial Intelligence (XAI), which helps clinical decision-making, diagnostics, and therapy suggestions, shows huge potential for revolutionizing [17] healthcare. To enable its successful integration into the world of healthcare, it provides a number of difficult challenges that must be fixed. The potential of XAI to better healthcare processes and outcomes for patients is substantial. Healthcare professionals may better comprehend the reasoning behind AI-driven suggestions by using XAI's readily interpretable AI models, which promote trust and cooperation between humans and computers. Additionally, XAI has the ability to find trends and conclusions [18] in enormous and complicated healthcare datasets, assisting in the early diagnosis of diseases, the personalization of procedures, and improving the quality of healthcare procedures.

References

1. Biswas S (2023) ChatGPT and the future of medical writing. Radiol Soc N Am
2. Nazar M, Alam MM, Yafi E, Su'ud MM (2021) A systematic review of human–computer interaction and explainable artificial intelligence in healthcare with artificial intelligence techniques. IEEE Access 9:153316–153348
3. Von Eschenbach WJ (2021) Transparency and the black box problem: Why we do not trust ai. Philos Technol 34(4):1607–1622
4. Zhang Y, Weng Y, Lund J (2022) Applications of explainable artificial intelligence in diagnosis and surgery. Diagnostics 12(2):237
5. Hulsen T, Jamuar SS, Moody AR, Karnes JH, Varga O, Hedensted S, Spreafico R, Hafler DA, McKinney EF (2019) From big data to precision medicine. Front Med 6:34
6. Celi LA, Cellini J, Charpignon M-L, Dee EC, Dernoncourt F, Eber R, Mitchell WG, Moukheiber L, Schirmer J, Situ J et al (2022) Sources of bias in artificial intelligence that perpetuate healthcare disparities—a global review. PLOS Digital Health 1(3):0000022

7. Vega-M´arquez B, Rubio-Escudero C, Riquelme JC, Nepomuceno-Chamorro I (2020) Creation of synthetic data with conditional generative adversarial networks. In: 14th International Conference on Soft Computing Models in Industrial and Environmental Applications (SOCO 2019) Seville, Spain, May 13–15, 2019, Proceedings 14, pp 231–240. Springer
8. Gunning D, Stefik M, Choi J, Miller T, Stumpf S, Yang G-Z (2019) Xai—explainable artificial intelligence. Sci Robot 4(37):7120
9. Jung J, Lee H, Jung H, Kim H (2023) Essential properties and explanation effec- tiveness of explainable artificial intelligence in healthcare: A systematic review. Heliyon
10. Ribeiro MT, Singh S, Guestrin C (2016) "why should i trust you?" explaining the predictions of any classifier. In: Proceedings of the 22nd ACM SIGKDD international conference on knowledge discovery and data mining, pp 1135–1144
11. Vu M-AT, Adalı T, Ba D, Buzs´aki G, Carlson D, Heller K, Liston C, Rudin C, Sohal VS, Widge AS, et al. (2018) A shared vision for machine learning in neuroscience. J Neurosci 38(7):1601–1607
12. Dur´an JM (2021) Dissecting scientific explanation in ai (sxai): A case for medicine and healthcare. Artif Intell 297:103498
13. Zhang C, Li D, Liang J (2020) Multi-granularity three-way decisions with adjustable hesitant fuzzy linguistic multigranulation decision-theoretic rough sets over two universes. Inf Sci 507:665–683
14. Phillips PJ, Hahn CA, Fontana PC, Broniatowski DA, Przybocki MA (2020) Four principles of explainable artificial intelligence. Gaithersburg, Maryland 18
15. Sheu R-K, Pardeshi MS (2022) A survey on medical explainable ai (xai): Recent progress, explainability approach, human interaction and scoring system. Sensors 22(20):8068
16. Kong X, Liu S, Zhu L (2024) Toward human-centered xai in practice: A survey. Mach Intell Res 1–31
17. Wang C, Zhang J, Lassi N, Zhang X (2022) Privacy protection in using artificial intelligence for healthcare: Chinese regulation in comparative perspective. Healthcare 10:1878. MDPI
18. Klein G, Hoffman RR (2008) Macrocognition, mental models, and cognitive task analysis methodology. Nat Decis Mak Macrocognition 57–80

Enhancing Student Success: Predictive Modeling of Graduation and Dropout Rates in University Management Using Machine Learning

Ashok Sapkota and Sandeep Kaur

Abstract This research explores the application of machine learning in educational management for universities, specifically focusing on predicting student graduation or dropout rates. In the context of university management, predicting student outcomes such as graduation or dropout is crucial for recognizing students who are in danger early and implementing targeted action to support their educational journey. This research utilizes a comprehensive dataset sourced from Kaggle, encompassing various student attributes such as academic performance, demographic information, and socioeconomic factors. Through rigorous feature analysis and data preprocessing, this study lays the groundwork for applying classification algorithms, including logistic regression, AdaBoost, Gradient Boosting, and decision trees, to the dataset to improve predictive accuracy, Ensemble Methods such as Boosting, Bagging, and Voting classifiers are applied. Results reveal promising accuracy rates: Bagging and Boosting classifiers achieve 90% accuracy each, while the Voting classifier reaches 92%.

Keywords Machine learning · Academic performance · Classification analysis · Feature importance · Ensemble methods · AdaBost

1 Introduction

In the dynamic landscape of higher education, the ability to accurately predict student outcomes such as graduation or dropout rates holds immense significance for educational institutions. Early identifying students who are at risk and the implementation of targeted interventions are vital for supporting student success and fostering institutional effectiveness [12]. With the advent of advanced data analytics techniques, particularly machine learning, educators and administrators now have powerful tools

A. Sapkota · S. Kaur (✉)
Lovely Professional University, Jalandhar, Punjab, India
e-mail: Sandeepkaur161@gmail.com

P. J. S. Gonçalves et al. (eds.), *Proceedings of Fifth International Conference on Computing, Communications, and Cyber-Security*, Lecture Notes in Networks and Systems 1128, https://doi.org/10.1007/978-981-97-7371-8_24

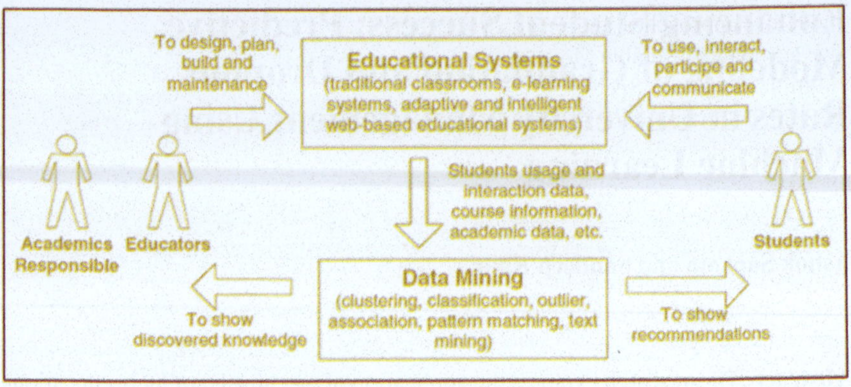

Fig. 1 The data mining cycle in education

at their disposal to analyze complex datasets and extract actionable insights. Figure 1 depicts the data mining cycle in education.

This research explores the application of machine learning in educational management, concentrating especially on predicting student graduation or dropout rates within a university setting [10, 11]. Leveraging a comprehensive dataset sourced from Kaggle, encompassing various student attributes such as academic performance, demographic information, and socioeconomic factors, this study aims to develop robust predictive models that can aid university administrators in making informed decisions to support student success. The dataset utilized in this research comprises several key attributes, including curricular units and grades for both the first and second semesters, tuition payment status, age at enrollment, course information, and father's occupation. These attributes collectively offer a comprehensive informational resource that can offer insights into the factors influencing student outcomes. Before delving into model development, an essential preliminary step involves feature analysis in order to comprehend the dataset's characteristics and relationships between variables. By identifying relevant features and assessing their impact on student outcomes, researchers can effectively refine the dataset for subsequent analysis. This phase of feature analysis serves as a critical foundation for building accurate predictive models.

Following feature analysis, the dataset undergoes rigorous data preprocessing, a crucial step in the knowledge discovery process. Data preprocessing involves various techniques such as handling missing values, normalization, and feature scaling to ensure the dataset is clean, consistent, and suitable for modeling. This meticulous preprocessing phase lays the groundwork for developing robust predictive models that can effectively generalize to unseen data.

With the preprocessed dataset in hand, the research proceeds to the classification phase, where various machine learning algorithms are utilized to forecast student outcomes. This study employs a diverse set of classifiers, including logistic regression, AdaBoost, Gradient Boosting, and decision trees, to explore different modeling approaches and evaluate their performance. To further enhance predictive accuracy, Ensemble Methods are employed, leveraging the collective strength of multiple classifiers. Specifically, Bagging, Boosting, and Voting classifier algorithms are utilized to combine the predictions of individual classifiers and generate more robust and accurate results. The effectiveness of these Ensemble Methods is evaluated based on their ability to improve predictive performance. Preliminary results indicate promising accuracy rates for the developed models. Bagging and Boosting classifiers, while the Voting classifier exhibits even higher accuracy. These findings underscore the capacity of machine learning methods to precisely forecast academic performance and provide valuable insights for university management and administration. The goal of this paper is to determine how well students are performing academically and whether they are able to graduate or drop out of the degree. This study formulates the separate research problems listed below in order to accomplish the aforementioned objectives:

(i) Build different classification Models such as Logistic Regression, Random Forest, AdaBoost, and Gradient Boosting.
(ii) To improve predictive accuracy, "Ensemble Methods" such as Boosting, Bagging, and Voting classifiers are employed.

2 Related Work

In the sphere of education, early student performance prediction can have a number of benefits. Numerous studies have been conducted in this area to determine how students interact with e-learning platforms and to forecast student's academic performance. Through early performance prediction, higher education institutions can also determine whether or not students will succeed in their studies. EDM employs a variety of techniques, including relationship mining, classification, regression, and clustering. Numerous studies have been conducted in this field, some of which we have discussed in this paper by examining selected literature reviews and educational data mining surveys.

Kaur and Bhatia [4] offer a comprehensive overview of machine learning applications for student performance assessment, covering classification, clustering, regression, and deep learning techniques. They highlight the potential of these methods to enhance educational practices while addressing challenges such as data privacy and algorithmic bias. The paper [5] explores how technology transforms education, advocating for collaborative blended learning models in India. It highlights the evolution from traditional to collaborative blended learning, emphasizing the role of co-teaching for enhanced student engagement and learning outcomes. In the work [6], the researchers aim to develop a web-based application that increases the success rate of the student, gives the student's academic history as input such as admission details, subject details, and course details, and gives the student's upcoming performance. In the work [7], the study is to determine the variables impacting learning effectiveness within the dataset evaluating the achievement of higher education students.

It can be inferred that the elements impacting learning effectiveness included the cumulative grade point average, expectations of achievement, regularity of reading non-academic literature, and limited study sessions each week. In the work [8], the author suggested a technique to forecast the academic achievement of students by contrasting different algorithms and discovered that the most effective algorithm for this purpose is Gradient Boosted Trees. In the work [9], the study proposed a Deep Neural Network that shows students which class category it belongs to and achieved an accuracy of up to 84.3%. In the work [10], the study is to forecast the challenges that students will encounter in the upcoming digital design class session. In the work [11], the purpose of this work is to compare the accuracy of various machine learning algorithms for the provided dataset and get the accuracy of 96% with KNN. In the work [12], the author includes student behavior analysis, academic student data analysis, assessment of student learning, and dropout or retention prediction showcasing current developments in the use of learning analytics.

3 Proposed Work

Using Data Mining approaches that we have used in this study at every stage of the knowledge development process. The university dataset was first gathered from Kaggle, and then preprocessing techniques were applied, such as removing null values, normalization, and dummy encoding. We employed four well-liked models of machine learning after completing the preprocessing and data-collecting phase. Logistic regression, Decision tree, AdaBoost, and Gradient Boosting. Next, we used ensemble techniques to enhance the classifier's performance. To enhance the student's performance in the prediction model, the ensemble approaches of bagging, boosting, and Voting classifiers are employed.

3.1 Data Collection and Feature Analysis

Various factors that are essential for forecasting student dropout or graduation are included in the dataset, which was obtained via Kaggle. The features used are shown in Fig and their corresponding correlation. Analyzing features was undertaken to comprehend the distribution, correlation, and importance of each attribute in forecasting student results. This examination uncovered trends, such as the influence of academic achievement, age, and socioeconomic status on the likelihood of graduation. Data preprocessing was done after feature analysis to guarantee data quality and modeling applicability. The dataset was refined and made ready for classification using methods including normalization of numeric attributes and addressing the missing values. 4424 student's dataset we've introduced. The "0" category comprises 2215 instances, while the "1" category consists of 2209 instances in this binary class

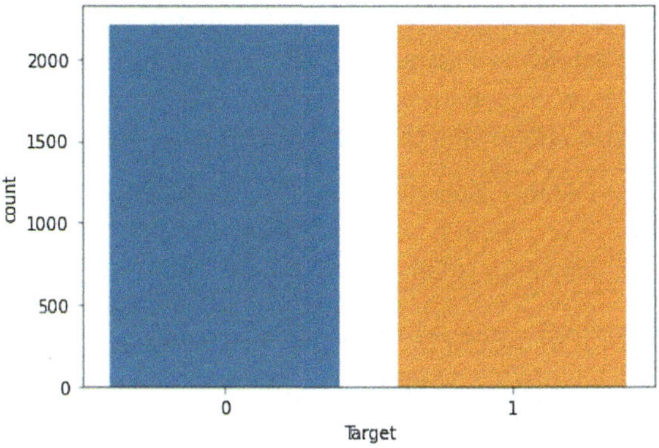

Fig. 2 Count plot of different classes of target variable

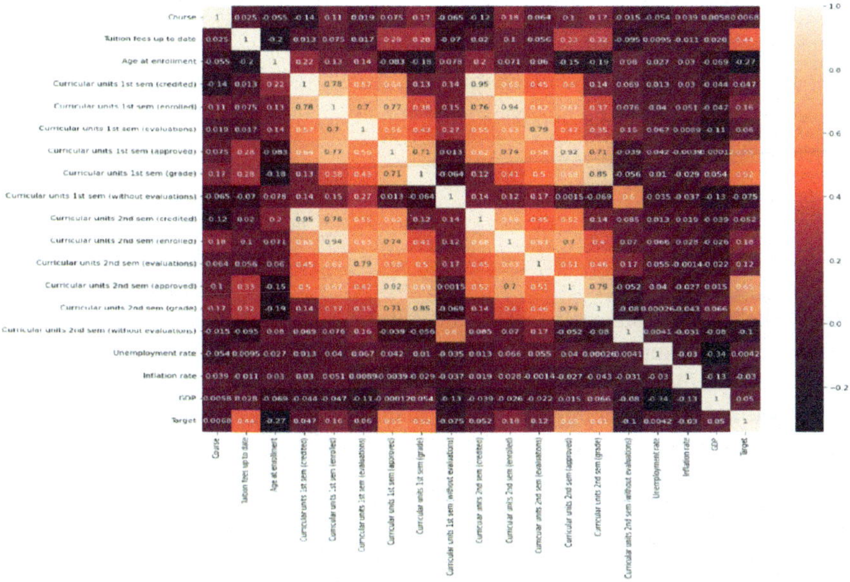

Fig. 3 Correlation of attributes used

dataset. Notably, "0" signifies the dropout, whereas "1" signifies the graduate (Figs. 2 and 3).

3.2 *Data Cleaning*

Data cleaning is an essential stage in getting the dataset ready for predictive modeling. It includes identifying and rectifying inconsistencies, errors, and missing values to guarantee that the collection is integrity and reliability in the context of predicting student graduation or dropout rates. In this work, we have incorporated the record of the dataset of 4424 records after removing 100 missing values, 4324 in our dataset. Dummy Encoding and Normalization have been implemented using the current dataset. Following that, the features have been selected.

(1) Dummy Encoding: Dummy Encoding is typically used in data preprocessing when dealing with categorical variables in machine Learning models. Dummy encoding entails developing indicator variables for each category inside a categorical variable that are binary (0 or 1). We used a Dummy encoding mechanism to transform attendance, course, previous qualification, father's qualifications, mother's qualification, Curricular, Course, and marital status from categorical to numerical.
(2) Normalization: The preprocessing method known as "normalization" is used to rescale numerical values inside a dataset to the standard range of 0 and 1 for the mean and standard deviation, respectively. In order to assist some machine learning algorithms to perform better and converge more quickly, normalization aims to give all features a comparable scale. We have applied standard min–max normalization for all dependent variables in the available dataset to increase precision.

3.3 *Methodology*

Firstly, the dataset, comprising all the attributes, was collected. Feature analysis was then conducted to understand the distribution, correlation, and significance of each attribute in predicting student outcomes. Following this, data preprocessing was performed to handle missing values, normalize data, and prepare it for modeling. Next, classification algorithms including logistic regression, AdaBoost, decision trees, and gradient boosting were applied to the preprocessed dataset to predict student graduation or dropout. Additionally, Ensemble Methods such as Bagging, Boosting, and voting classifiers were utilized to further enhance predictive accuracy shown in Fig. 4.

4 Result

The results of the research on predicting student graduation or dropout rates using the provided dataset reveal significant insights into the effectiveness of various classification algorithms and ensemble methods and the same has been shown in following Fig. 5 and Table 1

Firstly, Logistic regression, AdaBoost, decision tree, and gradient boosting classifiers were employed to classify students into either graduating or dropping out categories with an accuracy of 90, 64, 90, and 48%. This indicates that these

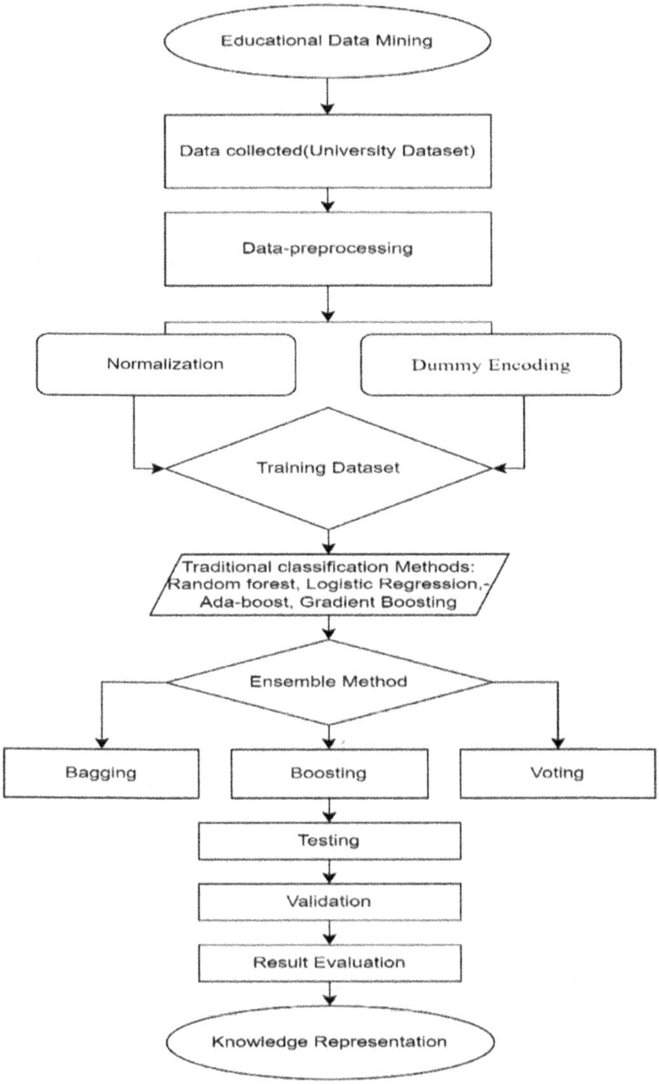

Fig. 4. Step of the student's prediction model

algorithms can effectively leverage the dataset attributes to make accurate predictions regarding student outcomes. Furthermore, ensemble methods were utilized to enhance predictive accuracy. Bagging, Boosting, and voting classifiers were implemented to combine the predictions of multiple base classifiers. The results indicate that ensemble methods indeed improved the accuracy of the predictive model. The bagging and Boosting classifier achieved an accuracy of 90% while the voting classifier outperformed them, achieving an accuracy of 92%. By utilizing the combined

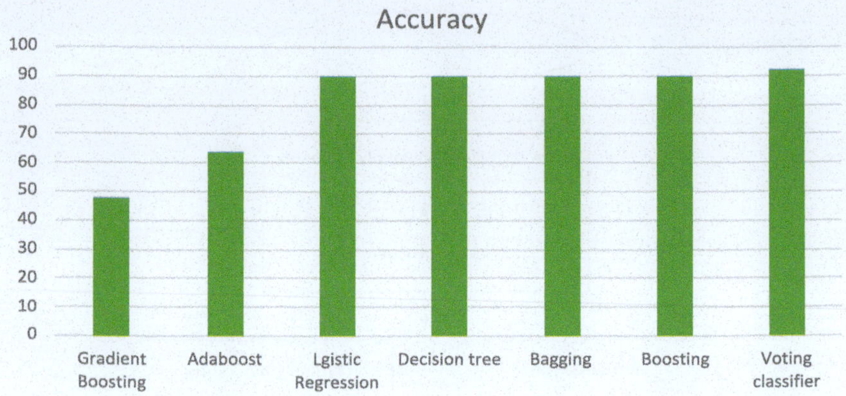

Fig. 5 Different algorithm's accuracy

Table 1 Accuracy of all models applied

Algorithm	Accuracy	F1-score	Precision	Recall
Gradient boosting	48	0.69	0.53	1
AdaBoost	64	0.69	0.59	0.85
Logistic regression	90	0.90	0.88	0.92
Decision tree	90	0.90	0.89	0.92
Bagging	90	0.91	0.90	0.93
Boosting	90	0.92	0.91	0.94
Voting classifier	92	0.93	0.90	0.95

knowledge of several models, ensemble models such as bagging, boosting, and voting classifiers improve accuracy substantially. By creating various models on various data subsets, bagging (also known as bootstrap aggregating) lowers variance. Boosting aims to reduce bias by gradually enhancing model performance by assigning greater weight to instances that were incorrectly identified. A voting classifier creates a more reliable conclusion by combining predictions from several models, many of which are varied. Ensemble approaches reduce the shortcomings of individual models and take advantage of group strengths to aggregate predictions from multiple models, improving overall accuracy. This implies that integrating the forecasts of several classifiers can result in more reliable and precise predictions compared to using individual classifiers alone. The high accuracy rates achieved by the ensemble methods, particularly the voting classifier, highlight the efficiency of ensemble learning in improving predictive modeling tasks, especially in scenarios where high accuracy is essential such as predicting student outcomes in educational settings [13].

An ensemble model combines predictions from multiple models. The above Fig. 6 shows the result of a model predicting whether a student will graduate from university. The rows represent the actual graduation status (graduate or dropout), and the

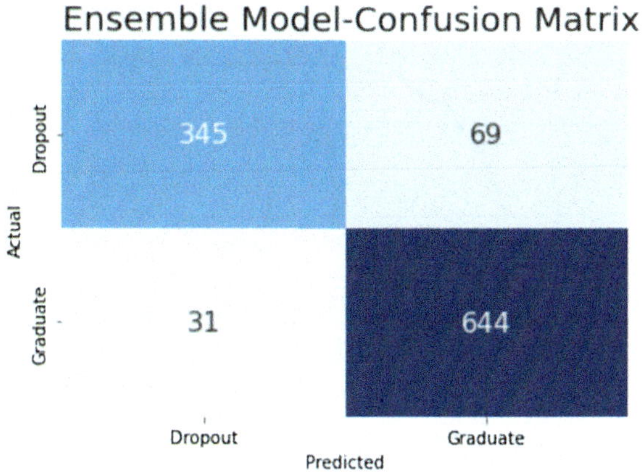

Fig. 6 Confusion matrix of voting classifier

columns represent the predicted graduation status. The diagonal cells, highlighted in blue, show the number of correct predictions. In this case, the model correctly predicted 644 graduates and 345 dropouts.

Figure 7 shows an ROC curve for an ensemble model. ROC stands for Receiver operating characteristic. It is a way to measure how well a classification model performs at various classification thresholds. This is particularly helpful in binary classification problems, where there are two classes, positive and negative. Plotting True Positive Rate (TPR) on the y-axis and False Positive Rate (FPR) on the x-axis results in the curve. The percentage of positive cases that the model properly detected is known as TPR. FPR is the percentage of negative cases that the model misclassified as positive. A straight line in the upper left corner would represent the perfect ROC curve. This suggests that everything is perfectly classified by the model. The curve will typically be diagonal. The model performs better the closer the curve is to the upper left corner.

The model's performance is summarized by the area under the curve, or AUC. One AUC indicates perfect performance, and 0.5 is no better than random guessing. The model in the graph has an AUC of 0.95, which is very good.

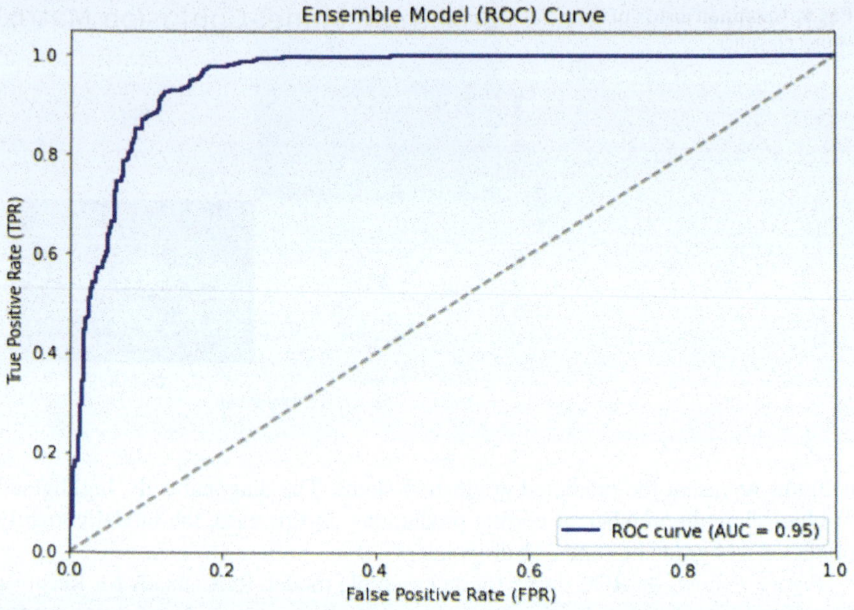

Fig. 7 ROC curve of voting classifier

5 Conclusion

Finally, this study has shown the efficiency of "machine learning" techniques in predicting student graduation or dropout rates based on university datasets. Through rigorous feature analysis and data preprocessing, we refined the dataset to enhance the accuracy and reliability of predictive models. The classification was performed using various algorithms, including logistic regression, AdaBoost, decision trees, and gradient boosting. Furthermore, Ensemble methods such as Boosting, Bagging, and Voting classifiers were utilized to improve predictive accuracy. The results of our study indicate promising accuracy rates, with Bagging and Boosting classifiers achieving 90% accuracy each, while the voting classifier reached 92%. These findings underscore the ability of machine learning methods to forecast student outcomes with accuracy and provide valuable insights into educational institutions to implement targeted interventions and support mechanisms. Overall, this study adds to a growing collection of information in educational management and highlights the importance of predictive modeling in enhancing student success and institutional effectiveness. Further research in this area could explore additional factors and refine predictive models to better support student outcomes.

References

1. Pek RZ, Özyer ST, Elhage T, ÖZYER T, Alhajj R (2022) The role of machine learning in identifying students at-risk and minimizing failure. IEEE Access 11:1224–1243
2. Abdelkader HE, Gad AG, Abohany AA, Sorour SE (2022) An efficient data mining technique for assessing satisfaction level with online learning for higher education students during the COVID-19. IEEE Access 10:6286–6303
3. Prada MA, Dominguez M, Vicario JL, Alves PAV, Barbu M, Podpora M, Vilanova R (2020) Educational data mining for tutoring support in higher education: a web-based tool case study in engineering degrees. IEEE Access 8:212818–212836
4. Kaur A, Bhatia M (2023) A survey of machine learning for assessing and estimating student performance. In: Proceedings of international conference on recent innovations in computing: ICRIC 2022, vol 1. Springer Nature Singapore, Singapore, pp 1–10
5. Pandita P et al (2023) Recent innovations of computing in education: emerging collaborative blended learning models in India. In: Proceedings of international conference on recent innovations in computing: ICRIC 2022, vol 1. Springer Nature Singapore, Singapore, pp 1–10
6. Feng G, Fan M, Chen Y (2022) Analysis and prediction of students' academic performance based on educational data mining. IEEE Access 10:19558–19571
7. Hengpraprohm K, Hengpraprohm S, Sudjitjoon W (2022) A study of factors affecting learning efficiency on higher education student performance evaluation dataset using feature selection techniques. Inf Technol J 18(2):34–43
8. Arifin M, Wibowo A, Warsito B (2021) Comparative analysis on educational data mining algorithm to predict academic performance. In: 2021 international seminar on application for technology of information and communication (iSemantic). IEEE, pp 173–178
9. Nabil A, Seyam M, Abou-Elfetouh A (2021) Prediction of students' academic performance based on courses' grades using deep neural networks. IEEE Access 9:140731–140746
10. Hussain M, Zhu W, Zhang W, Abidi SMR, Ali S (2019) Using machine learning to predict student difficulties from learning session data. Artif Intell Rev 52(1):381–407
11. Subha S, Priya SB (2021) Comparative analysis of supervised machine learning algorithms for evaluating the performance level of students. In: 2021 fifth international conference on I-SMAC (IoT in Social, Mobile, Analytics and Cloud) (I-SMAC). IEEE, pp 1–10
12. Buenaño-Fernandez D, Villegas W, Luján-Mora S (2019) The use of tools of data mining to decision making in engineering education—a systematic mapping study. Comput Appl Eng Educ 27(3):744–758
13. Aylenda R, Rimiru R, Cheruiyot W (2021) Predicting students academic performance using a hybrid of machine learning algorithms. In: 2021 IEEE AFRICON. IEEE, pp 1–6

A Deep Learning Framework for Automated Breast Cancer Detection Through Histopathological Image Analysis

Sonam Tyagi, Subodh Srivastava, Bikash Chandra Sahana, and Ishwari Singh Rajput

Abstract Breast cancer plays a major role in mortality rates among women. Timely identification of breast cancer reduces the death rate linked to the illness. For early detection of the disease, conventional methods are used in the form of manual inspection. But these methods are prone to human errors and time-consuming too. Nowadays with the advent of various artificial intelligence techniques in the field of healthcare, various techniques are evolved for timely diagnosis of breast cancer. Computer-Aided Diagnosis (CAD) using Convolutional Neural Network (CNN) has recently been utilized to address the fatigue and reduced focus that histopathologists and physicians have when diagnosing cancer. Models based on convolutional neural networks (CNNs) have shown impressive outcomes in classifying breast histopathology images. This article presents a CNN which is evaluated on BreakHis dataset of Histopathology images. The proposed approach attained an average accuracy rate of 98% with 70–30% train-test split, 98% with 80–20% train-test split, and 99% with 90–10% train-test split in classifying the image dataset. This result provides proof of the effectiveness of advanced learning approaches in improving the speed and accuracy of classifying Histopathology images.

Keywords Deep learning · CNN · Classification · Histopathology images

S. Tyagi (✉) · S. Srivastava · B. C. Sahana
Department of Electronics & Communication Engineering, National Institute of Technology Patna, Patna 800005, Bihar, India
e-mail: sonamt.phd20.ec@nitp.ac.in

S. Srivastava
e-mail: subodh@nitp.ac.in

B. C. Sahana
e-mail: sahana@nitp.ac.in

S. Tyagi · I. S. Rajput
School of Computing, Graphic Era Hill University, Haldwani 263139, Uttarakhand, India
e-mail: ishwarirajput@gehu.ac.in

P. J. S. Gonçalves et al. (eds.), *Proceedings of Fifth International Conference on Computing, Communications, and Cyber-Security*, Lecture Notes in Networks and Systems 1128, https://doi.org/10.1007/978-981-97-7371-8_25

1 Introduction

Breast cancer is a prevalent disease that presents significant risks to women's health globally. Cancer occurs when normal breast cells undergo abnormalities and proliferate uncontrollably, forming a mass or layer of cells known as a tumor [1]. While the precise origin of breast cancer remains unknown, certain risk factors have been found [2]. Genetics and family history are significant factors contributing to breast cancer. Breast cancer is the most common cancer among US women, excluding skin cancers. In 2024, the American Cancer Society expects 310,720 new invasive breast cancer cases, 56,500 DCIS, and 42,250 deaths [3]. Early detection of breast cancer can minimize tissue damage and increase the likelihood of recovery. Clinical breast cancer diagnosis uses several methods. MRI, mammography, histopathology, and ultrasound are common clinical screening methods [4]. Among all, histopathology is the gold standard to detect breast cancer. Histopathology offers valuable prognostic data that can aid in the early identification of critical targets. Biopsies are assessed by examining tissue slices stained with Hematoxylin and Eosin (H&E) under a microscope. A specimen of tissue is extracted from the affected area and subjected to microscopic examination in order to ascertain the classification of the tumor as either normal, benign, or malignant. Cancerous cells that multiply abnormally are malignant [5]. Non-cancerous abnormalities, known as benign tissues, are susceptible to treatment. Malignant cells are cells with cancer that undergo aberrant proliferation and development. Cancer cells are analyzed according to cell structure, nuclei structure, and cell distribution. The manual inspection method is time-consuming and leads to variations across pathologists, even those with extensive experience. It is highly recommended to utilize an automated system capable of reliably identifying breast cancer at all levels of magnification to assist pathologists in delivering prompt and precise results. Multiple researchers have proposed alternative computer-aided diagnosis solutions to address the limitations of traditional diagnostic processes. At present, the healthcare sector is witnessing rapid growth in the domains of artificial intelligence (AI), machine learning (ML), and deep learning (DL) [6, 7]. The progress of computer-aided diagnosis (CAD) systems, particularly the development of machine learning and deep learning models, has greatly enhanced the accuracy of breast cancer detection, minimized manual errors, and improved the reproducibility of the investigation process [8]. Due to the remarkable achievements of Deep Learning in various domains, researchers have been increasingly interested in utilizing convolutional neural networks (CNNs) for image-based cancer diagnosis [9]. As a result, numerous innovative designs have been developed in a very short period of time. The primary findings of the present study are outlined as follows: The most prevalent deep learning classification architecture, the convolution neural network, is used for end-to-end classification. In order to assess the effectiveness of this system, we utilized the BreakHis dataset. The results obtained surpass the previous methods by a substantial margin. This paper is structured into six sections. Section 1 provides an introduction, Sect. 2 presents a review of related literature, Sect. 3 delves into the background, Sect. 4 outlines the proposed work, Sect. 5

presents experimental results and a comprehensive analysis, and Sect. 6 concludes the paper by suggesting potential future directions.

2 Literature Review

Researchers in the field of computer vision have recently been highly focused on learning meaningful patterns in an end-to-end manner. A significant amount of research has been carried out in this field, as evidenced in Table 1 and by the following list: A novel rank-based ensemble method was introduced by Samriddha et al. [10] that integrates the outcomes of three transfer learning convolutional neural network (CNN) models, namely GoogleNet, VGG11, and MobileNetV3_Small, rather than employing a singular CNN model. The ensemble model is created with the Gamma function to address a two-class classification issue with breast histopathology images. Aruna et al. [11] presented a study which examined the effectiveness of utilizing two feed-forward techniques: Convolutional Logistic Regression Network (CLR) and Convolutional Support Vector Machine Network for Histopathology Images (CSVM-H). The suggested models CLR and CSVM-H outperformed the classic back-propagation approaches (VggNet-16 and ResNet 50) in terms of training speed and classification accuracy. A Copula-based model was suggested by Soumyajyoti et al. [12], which combined three top-performing CNN architectures: DenseNet-161/201, ResNet 101/34, and InceptionNet-V3. The combination of data augmentation framework and ensemble technique demonstrated superior performance in predicting malignancy compared to individual CNN models on breast cytology and histopathology datasets. Karthik et al. [13] suggested a combination of two Convolutional Neural Network (CNN) structures that incorporate Channel and Spatial attention. The histopathological images were processed simultaneously using two advanced customized deep architectures, namely CSAResnet and DAMCNN, to extract features. Ultimately, ensemble learning was utilized to enhance performance even further. A modified version of the DarkNet19 CNN model was implemented by Sruthi et al. [14] for the purpose of classifying histopathology images. To enhance the performance of the DarkNet19 model and enable visual interpretation, an attention branch was incorporated into the DarkNet19 network, resulting in the creation of the Attention Branch Network. The attention branch employed a convolutional layer from the DarkNet19 architecture and Global Average Pooling to capture the contextual information of the visual data. This process resulted in the generation of a heatmap, which was utilized to identify the region of interest. Ultimately, the perception branch was established by employing a fully connected layer to categorize images. Emmanuel et al. [15] employed transfer learning to construct four pre-trained models, specifically DenseNet201, ResNet50, ResNet101, and MobileNet-v2. The dataset was divided into five protocol settings, with 70% allocated for training and 30% for testing. Among the four pre-trained models, DensNet201 had the highest level of accuracy. The modified ResNet architecture was employed by Varun et al. [16] for the purpose of classifying histopathology images. The modified residual networks employed a 1×1 convolution operation with a stride of 2, and subsequently applied batch normalization (BN) when the dimensions were not congruent. The algorithm incorporates a dropout rate of 0.5, in addition to adaptive average pooling and adaptive max-pooling techniques. The subsequent steps involve the implementation of two layers of batch normalization, dropout, and fully connected layers. The ResNet50 model has been upgraded by replacing the 1×1 convolution with a 3×3 convolution in the bottleneck block to achieve down-sampling. Alireza et al. [17] introduced a technique in which a pre-trained feature extractor is used to extract features from images. These extracted features were then combined with a boosting method to categorize the images as either benign or malignant tumors. The DenseNet201 model, which had been pre-trained, was employed to extract features from the images. These extracted features were then inputted into the extreme gradient boosting (XGBoost) approach for the purpose of classification. The Multi-scale Dual Residual Recurrent Network (MTRRE-Net) was proposed by Soham et al. [18] as a novel deep learning model for the purpose of classifying breast cancer based on histopathology images. The model employed an innovative approach by integrating a dual residual block with a recurrent network to efficiently tackle the problem of the vanishing gradient, even in the context of a deep network.

3 Background

3.1 CNN

CNN is the most common deep learning classification framework. In Fig. 1, input, convolution, pooling, fully connected, and output layers are sequentially organized. The mathematical operation of convolution merges two functions to create a third function, therefore called "convolutional". Input is applied to input layer. Filters or kernels convolutionally process incoming data in convolutional layers. These layers extract features from input data and create feature maps that capture various patterns [19]. Pooling layers reduce the spatial dimensions of feature maps, making computations simpler. The fully connected layer links extracted characteristics of the inputs to the output layer. This allows the network to predict or classify images based on its extracted features. CNN input and output layers are visible [20]. Hidden layers include convolutional and pooling layers. These hidden layers are essential for learning hierarchical representations from information [21].

Table 1 Existing related work with deep learning classification techniques

Author and year	Year	Model	Dataset	Performance metrics
Samriddha et al. [10]	2023	Gamma function-based ensemble of GoogleNet, VGG11, and MobileNetV3_Small	ICIAR-2018	Accuracy = 96.95% Precision = 96.92% Recall = 96.92% F1 = 96.92% AUC = 96.92%
Aruna et al. [11]	2024	CLR and CSVM-H	BreaKHis	For CLR Accuracy = 86.14% AUC = 91% For CSVM-H Accuracy = 88.71% AUC = 94%
Soumyajyoti et al. [12]	2023	Gaussian Copula-based framework for ensemble of CNNs	JUCYT and BreakHis	Accuracy = 84.37% (for JUCYT) Accuracy = 97.32% (for BreakHis)
Karthik et al. [13]	2022	Ensemble of CSAResnet and Dual Attention Embedded Multi-scale CNN (DAMCNN)	BreakHis	Accuracy = 99.55%
Sruthi et al. [14]	2023	ABN-DCN (Attention Branch Network-DarkCovidNet) Model	BreakHis	Accuracy = 98.7%
Emmanuel et al. [15]	2023	Pre-trained models (DenseNet201, ResNet50, ResNet101, and MobileNet-v2)	Hybrid dataset of BreakHis and Histo	Accuracy = 91.37% (DenseNet201)
Varun et al. [16]	2021	Modified residual neural network	BreakHis	Accuracy = 99.75% Precision = 99.18% Recall = 99.37%
Alireza et al. [17]	2023	Pre-trained DenseNet201 model and XGBoost classifier	BreakHis	Accuracy = 93.6%
Soham et al. [18]	2022	Multi-scale Dual Residual Recurrent Network	BreakHis	Accuracy = 97.81%

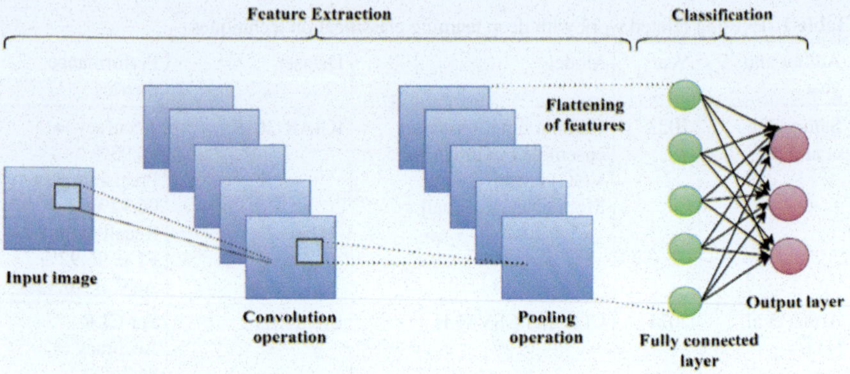

Fig. 1 Basic convolutional neural network

4 Proposed Method

This section presents the methodology that was chosen for the classification of BreakHis dataset. The procedure primarily consists of three tasks: data preprocessing, feature selection, and model training. Figure 2 represents the proposed workflow for the classification of BreakHis dataset images.

4.1 Data Preprocessing

When it comes to the image classification pipeline, data preprocessing is a very important step since it helps to guarantee that the data is clean, consistent, and appropriate to be utilized by a machine learning model. Image augmentation techniques such

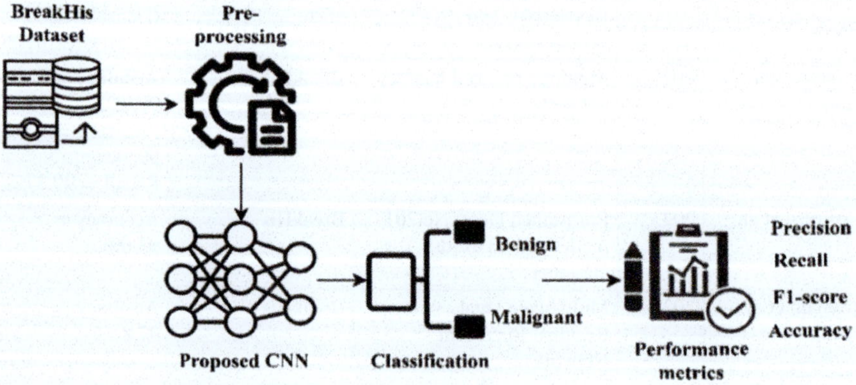

Fig. 2 Proposed methodology for BreakHis data classification

as rotation, flipping, and scaling are applied here to generate new images, then the images are resized to 128 × 128 pixels and the dataset is divided into 70–30, 80–20, and 90–10% for the proposed model training and testing, respectively.

4.2 *Feature Extraction*

Extracting image features is the responsibility of the first few layers of convolutional neural network (CNN) models, which include a set of convolutions and pooling layers. In order to do classification, the features that have been retrieved are then input into a network of fully connected layers.

4.3 *Classification*

In order to learn complicated patterns and carry out classification, dense layers of CNN are equipped with activation functions. Here in the proposed model shown in Fig. 3, the sigmoid activation function is used for classification.

5 Experimental Results and Discussion

5.1 *Dataset Description*

Like many other cellular diseases, the examination of breast cancers mostly depends on histological slides. The latter refers to surgically removed tissues from a patient's breast following a biopsy procedure on the region of interest (ROI). The BreakHis dataset [22] as provided in Table 2 is employed in the proposed study.

Some sample images from BreakHis dataset are shown in Fig. 4. Each malignancy class is divided into four distinct subcategories based on the visual characteristics of the tumor when observed under a microscope [22].

5.2 *Performance Evaluation Metrics*

The evaluation of the findings comprised many metrics such as recall, precision, accuracy, and F1-score [23]. The performance evaluation metrics used to assess the efficacy of the method are given in Table 3.

Model: Sequential		
Layer (type)	**Output Shape**	**Param #**
batch_normalization (BatchNormalization)	(None, 128, 128, 3)	12
conv2d (Conv2D)	(None, 128, 128, 32)	896
activation (Activation) (ReLu)	(None, 128, 128, 32)	0
max_pooling2d (MaxPooling2D)	(None, 64, 64, 32)	0
dropout (Dropout) (0.25)	(None, 64, 64, 32)	0
conv2d_1 (Conv2D)	(None, 64, 64, 32)	9248
activation_1 (Activation) (ReLu)	(None, 64, 64, 32)	0
max_pooling2d_1 (MaxPooling2D)	(None, 32, 32, 32)	0
conv2d_2 (Conv2D)	(None, 32, 32, 64)	18496
activation_2 (Activation) (ReLu)	(None, 32, 32, 64)	0
max_pooling2d_2 (MaxPooling2D)	(None, 16, 16, 64)	0
conv2d_3 (Conv2D)	(None, 16, 16, 64)	36928
activation_3 (Activation) (ReLu)	(None, 16, 16, 64)	0
max_pooling2d_3 (MaxPooling2D)	(None, 8, 8, 64)	0
conv2d_4 (Conv2D)	(None, 8, 8, 128)	73856
activation_4 (Activation) (ReLu)	(None, 8, 8, 128)	0
max_pooling2d_4 (MaxPooling2D)	(None, 4, 4, 128)	0
conv2d_5 (Conv2D)	(None, 4, 4, 128)	147584
activation_5 (Activation) (ReLu)	(None, 4, 4, 128)	0
max_pooling2d_5 (MaxPooling2D)	(None, 2, 2, 128)	0
flatten (Flatten)	(None, 512)	0
dense (Dense) (128)	(None, 128)	65664
activation_6 (Activation) (ReLu)	(None, 128)	0
dense_1 (Dense) (64)	(None, 64)	8256
activation_7 (Activation) (ReLu)	(None, 64)	0
dropout_1 (Dropout)	(None, 64)	0
dense_2 (Dense) (32)	(None, 32)	2080
activation_8 (Activation) (ReLu)	(None, 32)	0
dense_3 (Dense) (2)	(None, 2)	64
activation_9 (Activation) (sigmoid)	(None, 2)	0
Total params: 363,152 Trainable params: 363,146 Non-trainable params: 6		

Fig. 3 Proposed CNN architecture

Table 2 BreakHis dataset description

Dataset name	BreakHis
Size	9109
Number of cases	82
Resolution (bits/pixel)	8
Magnification scale	40×, 100×, 200×, 400×
Format	.PNG
Type	RGB
Abnormalities	Adenosis, Fibroadenoma, Phyllodes tumor, Tubular adenoma, Ductal carcinoma, Lobular carcinoma, Mucinous carcinoma, Papillary carcinoma
Category	Benign (2480 images), Malignant (5429 images)
Origin	Brazil
Year	2015
Size	3.98 GB

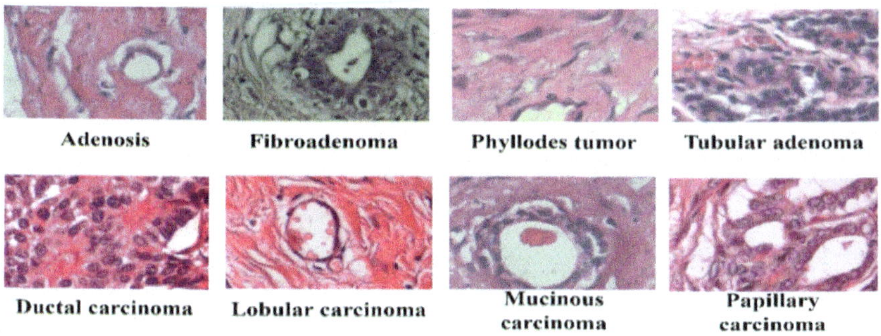

Fig. 4 Sample BreakHis images

Table 3 Performance evaluation metrics

Metrics	Formula
Accuracy [23]	$\frac{TP+TN}{TP+TN+FP+FN}$
Precision [24]	$\frac{TP}{TP+FP}$
Sensitivity or Recall [24]	$\frac{TP}{TP+FN}$
Specificity [23]	$\frac{TN}{TN+FP}$
F1-Score [23]	$2 \times \frac{Recall \times Precision}{Recall+Precision}$

5.3 Results and Discussion

The suggested framework is specifically designed to categorize images of the BreakHis dataset into two unique categories, i.e., benign and malignant. The assessment of the suggested CNN model is performed by employing recall, precision, accuracy, and F1-score. To properly evaluate the classifier's performance, we considered three distinct training and test divisions. Splits are used to divide sample data into unique proportions for training and testing, evaluating model performance in various contexts. The initial partition divides the data into two sections with a 70–30% ratio. Specifically, 70% of data is used for training and 30% for testing. In the second dataset split, 80% is allotted for training and 20% for testing. In the third partition, 90% of the dataset is used for training and 10% for testing. The hyperparameters used for the proposed CNN are shown in Table 4.

The proposed model achieves an average accuracy of 98% for 70–30% train-test split data, 98% for 80–20% train-test split data, and 99% for 90–10% for train-test split data. The corresponding results of different classification algorithms and the proposed CNN are shown in Table 5. The proposed approach outperforms other classification methods with the highest accuracy of 99%. The corresponding plots for VGG16, ResNet50, and the proposed approach are shown in Fig. 5.

Table 4 Hyperparameter for fine-tuning

Hyperparameter	Value
Type of pooling layer	Maximum
Learning rate	0.001
Optimizer algorithm	Adam
Mini-batch size	64
Loss function	Binary cross-entropy
Maximum epoch	50

Table 5 Comparative analysis of different classification algorithms

Classifier	Train-test data split (%)	Class type	Precision	Recall	F1-score	Accuracy
VGG16	70–30	Benign	0.81	0.82	0.80	0.89
		Malignant	0.84	0.85	0.80	
	80–20	Benign	0.83	0.90	0.81	0.90
		Malignant	0.88	0.81	0.84	
	90–10	Benign	0.89	0.90	0.86	0.92
		Malignant	0.78	0.87	0.90	
ResNet50	70–30	Benign	0.85	0.86	0.86	0.85
		Malignant	0.81	0.79	0.88	
	80–20	Benign	0.90	0.89	0.83	0.87
		Malignant	0.84	0.85	0.81	
	90–10	Benign	0.79	0.80	0.82	0.82
		Malignant	0.79	0.80	0.83	
Proposed model	70–30	Benign	0.99	0.98	0.97	**0.98**
		Malignant	0.98	0.98	0.99	
	80–20	Benign	0.99	0.98	0.97	**0.98**
		Malignant	0.98	0.98	0.96	
	90–10	Benign	0.95	1.00	0.92	**0.99**
		Malignant	0.96	1.00	0.99	

6 Conclusion and Future Scope

This study presents a new convolutional neural network design for accurately categorizing histopathology images as either benign or malignant. The results of the experiment conducted on the BreakHis dataset indicate that our proposed model obtained a high level of accuracy. The proposed model achieves an average accuracy of 98% for 70–30% train-test split data, 98% for 80–20% train-test split data, and 99% for 90–10% for train-test split data. The proposed approach outperforms other

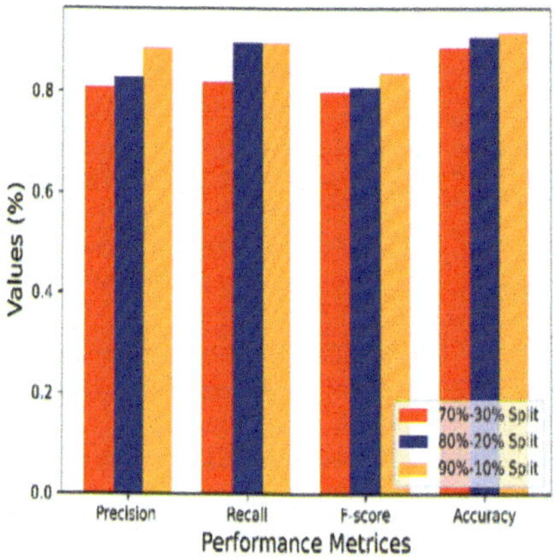

(a) VGG16

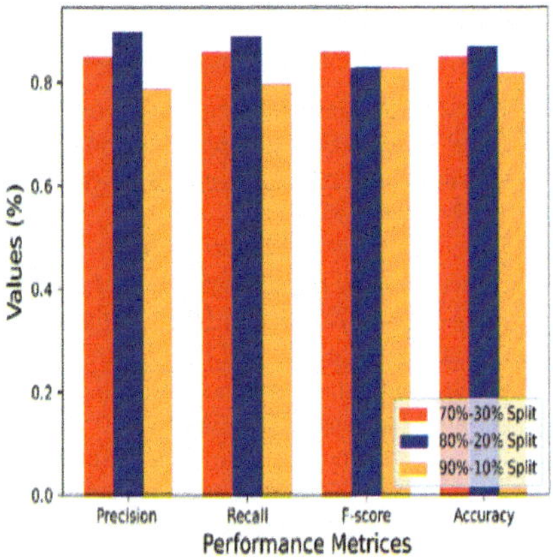

(b) ResNet 50

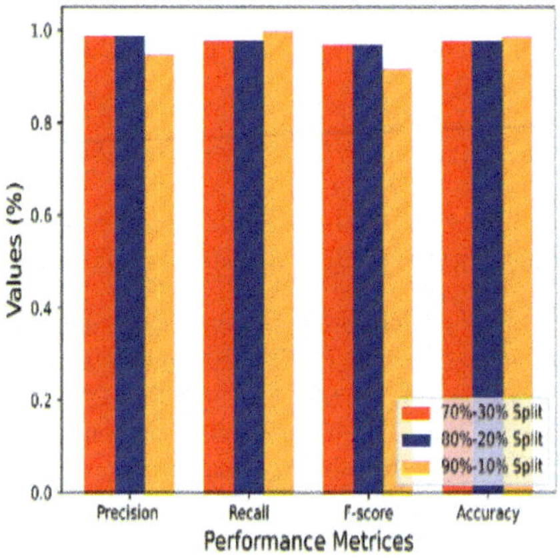

(c) Proposed CNN

Fig. 5 Performance metrics plot for VGG16, ResNet50, and proposed CNN

classification methods with the highest accuracy of 99%. Nevertheless, this work will undoubtedly assist radiologists or breast surgeons in automating the complete breast cancer detection system. Moreover, the techniques outlined in this research will be put into practice at medical facilities to aid radiologists in densely populated regions. In the future, several histopathology datasets will be utilized to evaluate the proposed technique. It is also possible to propose a hybrid deep learning approach for the identification of breast cancer. These hybrid deep learning models are anticipated to enhance the outcome.

References

1. Kakadia D, Shah H, Oza P, Sharma P, Patel S (2023) Breast cancer classification using a novel image processing pipeline and a two-stage deep learning segmentation and classification approach. Lect Notes Netw Syst 664 LNNS:725–739. https://doi.org/10.1007/978-981-99-1479-1_54/COVER
2. Sahu A, Das PK, Meher S (2023) High accuracy hybrid CNN classifiers for breast cancer detection using mammogram and ultrasound datasets. Biomed Sig Process Control 80:104292. https://doi.org/10.1016/J.BSPC.2022.104292
3. "Breast Cancer Statistics | How Common Is Breast Cancer? | American Cancer Society. https://www.cancer.org/cancer/types/breast-cancer/about/how-common-is-breast-cancer.html. Accessed 08 Mar 2024
4. Bouzar-Benlabiod L, Harrar K, Yamoun L, Khodja MY, Akhloufi MA (2023) A novel breast cancer detection architecture based on a CNN-CBR system for mammogram classification. Comput Biol Med 163:107133. https://doi.org/10.1016/J.COMPBIOMED.2023.107133
5. Demir F (2021) DeepBreastNet: a novel and robust approach for automated breast cancer detection from histopathological images. Biocybern Biomed Eng 41(3):1123–1139. https://doi.org/10.1016/J.BBE.2021.07.004
6. Kufel J et al (2023) What is machine learning, artificial neural networks and deep learning?—examples of practical applications in medicine. Diagnostics 13(15):2582. https://doi.org/10.3390/DIAGNOSTICS13152582
7. Jain P, Patel D, Verma JP, Tanwar S (2021) Computer-aided-diagnosis system for symptom detection of breast and cervical cancer. Lect Notes Netw Syst 203 LNNS:743–758. https://doi.org/10.1007/978-981-16-0733-2_53/COVER
8. Pandita R, Prakash D (2023) Analysis of breast cancer prediction using machine learning techniques: review paper. Lect Notes Electr Eng 1001 LNEE:311–322. https://doi.org/10.1007/978-981-19-9876-8_24/COVER
9. Desai M, Shah M (2021) An anatomization on breast cancer detection and diagnosis employing multi-layer perceptron neural network (MLP) and convolutional neural network (CNN). Clin eHealth 4:1–11. https://doi.org/10.1016/J.CEH.2020.11.002
10. Majumdar S, Pramanik P, Sarkar R (2023) Gamma function based ensemble of CNN models for breast cancer detection in histopathology images. Expert Syst Appl 213:119022. https://doi.org/10.1016/J.ESWA.2022.119022
11. Karuppasamy AD, Abdesselam A, Hedjam R, zidoum H, Al-Bahri M (2024) Feed-forward networks using logistic regression and support vector machine for whole-slide breast cancer histopathology image classification. Intell Med 9:100126. https://doi.org/10.1016/J.IBMED.2023.100126
12. Dey S, Mitra S, Chakraborty S, Mondal D, Nasipuri M, Das N (2023) GC-EnC: a Copula based ensemble of CNNs for malignancy identification in breast histopathology and cytology images. Comput Biol Med 152:106329. https://doi.org/10.1016/J.COMPBIOMED.2022.106329
13. Karthik R, Menaka R, Siddharth MV (2022) Classification of breast cancer from histopathology images using an ensemble of deep multiscale networks. Biocybern Biomed Eng 42(3):963–976. https://doi.org/10.1016/J.BBE.2022.07.006
14. Krishna S, Suganthi SS, Bhavsar A, Yesodharan J, Krishnamoorthy S (2023) An interpretable decision-support model for breast cancer diagnosis using histopathology images. J Pathol Inform 14:100319. https://doi.org/10.1016/J.JPI.2023.100319
15. Simonyan EO, Badejo JA, Weijin JS (2023) Histopathological breast cancer classification using CNN. Mater Today Proc. https://doi.org/10.1016/J.MATPR.2023.10.154
16. Gupta V, Vasudev M, Doegar A, Sambyal N (2021) Breast cancer detection from histopathology images using modified residual neural networks. Biocybern Biomed Eng 41(4):1272–1287. https://doi.org/10.1016/J.BBE.2021.08.011
17. Maleki A, Raahemi M, Nasiri H (2023) Breast cancer diagnosis from histopathology images using deep neural network and XGBoost. Biomed Sig Process Control 86:105152. https://doi.org/10.1016/J.BSPC.2023.105152

18. Chattopadhyay S, Dey A, Singh PK, Oliva D, Cuevas E, Sarkar R (2022) MTRRE-Net: a deep learning model for detection of breast cancer from histopathological images. Comput Biol Med 150:106155. https://doi.org/10.1016/J.COMPBIOMED.2022.106155
19. Li Z, Liu F, Yang W, Peng S, Zhou J (2022) A survey of convolutional neural networks: analysis, applications, and prospects. IEEE Trans Neural Netw Learn Syst 33(12):6999–7019. https://doi.org/10.1109/TNNLS.2021.3084827
20. Gu J et al (2018) Recent advances in convolutional neural networks. Pattern Recogn 77:354–377. https://doi.org/10.1016/J.PATCOG.2017.10.013
21. Albawi S, Mohammed TA, Al-Zawi S (2017) Understanding of a convolutional neural network. In: Proceedings of 2017 international conference on engineering and technology ICET 2017, vol 2018, pp 1–6. https://doi.org/10.1109/ICENGTECHNOL.2017.8308186
22. Benhammou Y, Achchab B, Herrera F, Tabik S (2020) BreakHis based breast cancer automatic diagnosis using deep learning: taxonomy, survey and insights. Neurocomputing 375:9–24. https://doi.org/10.1016/J.NEUCOM.2019.09.044
23. Deng X, Liu Q, Deng Y, Mahadevan S (2016) An improved method to construct basic probability assignment based on the confusion matrix for classification problem. Inf Sci (Ny) 340–341:250–261. https://doi.org/10.1016/J.INS.2016.01.033
24. Valero-Carreras D, Alcaraz J, Landete M (2023) Comparing two SVM models through different metrics based on the confusion matrix. Comput Oper Res 152:106131. https://doi.org/10.1016/J.COR.2022.106131

Deep Pox: Leveraging Deep Learning Techniques for Monkeypox Recognition in 2022

Simrin Fathima Syed, Madhu Shukla, Neel H. Dholakia, Vipul Ladva, and Nishant Kothari

Abstract In recent times, it has been witnessed that the current monkeypox 2022 viral outbreak has become a major worry on a worldwide scale. The Monkeypox virus, an Orthopox virus species, is what causes Monkeypox, a pathogenic organism. Several people have been unwell as a result of the extensive epidemic of the recently identified disease, which has even been related to deaths. The disease's effects on a sufferer might last for up to three or four weeks. In people, monkeypox often manifests as a quasi-paroxysmal stage that is accompanied by either characteristic dermatitis that takes 5–21 days to develop usually 6–13 days. The detection and management of patients suspected of having monkeypox in 2022 may be accomplished using the Deep learning approach described in this research, which also uses the Deep learning tool. After a preliminary evaluation of risk factors, symptoms, and indicators, patients are recognized as possibly exposed or infected. Consideration should be given to isolation and smallpox vaccination as a kind of thread prevention patients who have been exposed. Providers must put on personal protective equipment right once and segregate patients if they have an infected patient. Healthcare professionals are required to notify public health authorities of suspected and confirmed instances in both people and animals. The dataset also in this has been taken from the Kaggle repository. The Deep learning model also has derived the highest accuracy of 99.11% followed by a classification report and confusion matrix.

Keywords Monkeypox 2022 · Efficient net B3 · Confusion matrix · Classification report · Data analysis

S. F. Syed · M. Shukla (✉) · N. H. Dholakia · V. Ladva · N. Kothari
Marwadi University, Rajkot-Morbi Road, Rajkot 360003, Gujarat, India
e-mail: madhu.shukla@marwadieducation.edu.in

N. H. Dholakia
e-mail: neel.dholakia112067@marwadiuniversity.ac.in

V. Ladva
e-mail: vipul.ladva@marwadieduaction.edu.in

N. Kothari
e-mail: nishant.kothari@marwadieducation.edu.in

P. J. S. Gonçalves et al. (eds.), *Proceedings of Fifth International Conference on Computing, Communications, and Cyber-Security*, Lecture Notes in Networks and Systems 1128, https://doi.org/10.1007/978-981-97-7371-8_26

1 Introduction

In the era of epidemic diseases spreading widely across nations, here in this paper we are discussing about so known origin of Monkeypox, its effects, and the efforts that we as a human being can make by designing the diagnosis process automated with the help of AI and ML. In the work, we have discussed various algorithms that work with different accuracy in the identification of the disease. In non-endemic nations, the biggest epidemic of the monkeypox virus has been recorded as of May 2022. Shotgun metagenomics was used in this work to quickly rebuild and characterize the genome sequences from the initial monkeypox epidemic, demonstrating that this particular monkeypox corresponds in haplogroup 3 indicating that the pandemic almost likely had a single origin. The World Health Organization (WHO) called the danger confronted by the numerous risks to worldwide human safety nation monkeypox epidemic in non-endemic nations to be "moderate" on May 29, 2022. The WHO noted that "this is the first time that monkeypox cases and clusters have been reported concurrently in widely separated WHO geographical zones and without known epidemiological linkages to endemic countries" as well as having novel and uncommon characteristics [1]. The UK verified an instance of invasive sentient monkeypox on May 7th, 2022 [2]. Their publication signaled the beginning of the identification of the virus's unexpectedly quick worldwide spread, which includes both community and sentient dissemination [3]. The effects of monkeypox, a pandemic virus-induced exanthema, can be severe. The monkeypox virus may infect and cause clinical illness in a variety of hosts, including rodents, lagomorphs, and certain exotic animals [4, 5], as well as human and non-human primates [6, 7]. Monkeypox replaced smallpox as the most frequent and dangerous human Orthopox virus after its elimination [8] while being rare up until recently. It's odd that the desire to eradicate smallpox led to the discovery of monkeypox.

2 Analytical Characteristics and Regulation

2.1 Avoidance

Healthcare professionals can lessen the spread of the monkeypox virus by taking several particular public health precautions together with minimizing the dangerous actions listed in the section called Transmission Risk Factors. If a patient has monkeypox, healthcare staff should put on PPE right once and, if one is available, isolate them in an airborne isolation room. Close the door and place the patient in a sole-occupancy unit, preferably with a private bath, if that is not possible. Resist doing aerosolizing tasks such as dry cleaning, shaking bedding, or cleaning [9].

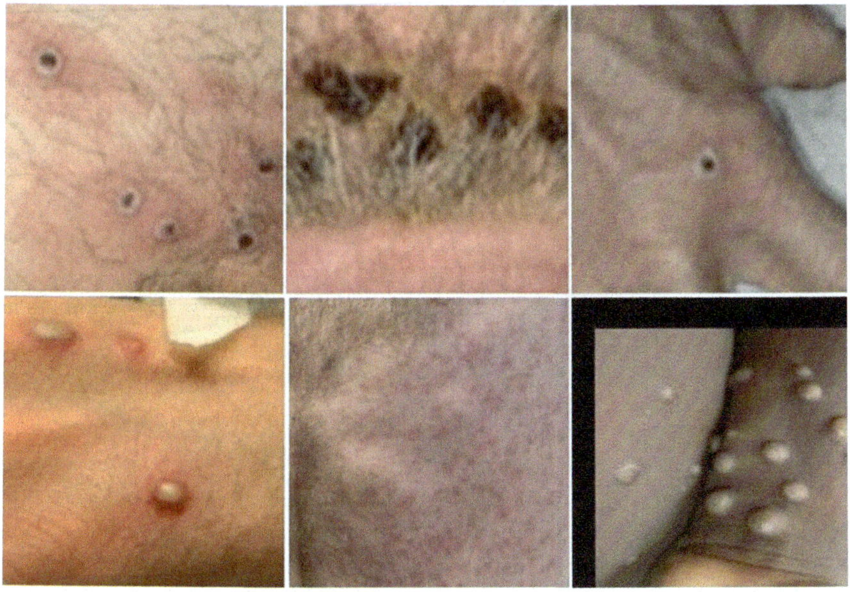

Fig. 1 Signs and symptoms

2.2 *Scientific Demonstration*

Close contact can result in the spread of infection by contact with contaminated bodily fluids, such as blood, breast milk, seminal fluid, lesions on the skin, towels, clothes, or bed linens made of shed skin [10]. Healthcare personnel is in danger of getting a disease if they don't wear the proper PPE since the monkeypox virus can spread by aerosolization, such as when bed linens are shaken or during conducting AGPs.

In 2022, individuals having several or unidentified partners in social networks, like MSM, who have been shown to be infected, accounted for the great many early cases of confirmed monkeypox. Similar to how they do for other viral illnesses, therapy can take areas like the testes may act as a "ventricular reserve" for monkeypox [11]. Nevertheless, more research has to be done on this theory. Smallpox and monkeypox symptoms and indicators are quite similar; however, monkeypox has certain distinct characteristics (Fig. 1).

2.3 *Problems*

In generally instances of monkeypox are rare in healthy hosts who are not very old or very young normally moderate and self-resolve within a few weeks, but skin lesions frequently leave scars. Serious side effects including blindness, bronchopneumonia, encephalopathy, and sepsis can occur less often. Like smallpox, the most common effects of scars from monkeypox and the emergence de saturated deep gouges, usually on the face and other places with a lot of sebaceous glands. Nobody really perished during the initial incident epidemic in contrast to Africa, the US epidemic of 2003, despite African mortality rates ranging from 1 to 10%, guess it depends on lineage [12]. In a similar vein, as of June 23, 2022, no evidence of causative agent instances of mortality outside of Africa had been noted [13]. Nevertheless, only one of the 72 probable causative agent fatalities in African countries in 2022 is known to have been caused by the disease [14, 15].

Literature comparison of a few existing works to show why deep learning can play a vital role in the detection of Monkeypox.

Table 1 discussed here addresses studies conducted by researchers in the areas like Genetic Analysis, Public Health emergency, protein polymorphism, manifestation of the disease, clinical characteristics of the disease, and various machine learning method that can be deployed in order to identify the disease.

Table 2 discusses about the methods that can be used for Monkeypox detection. Machine Learning algorithms based on decision tree namely Decision Trees and Random Forests are discussed with pros and cons. Deep learning methods are well suited for Image data as it makes the classification process precise and attendable with less trade off of time for computation.

3 Investigative Tool and Discussion

3.1 Analysis Requirements

Prior to Gathering Samples:

- Put the patient in an aerial isolation chamber (or, if one is not available, a private room with a closed door).
- Put on your complete PPE, which should include eye protection, a gown, gloves, and shoe coverings, as well as N95 equivalent or greater respiratory protection.
- Speak with your local health agency to arrange sample collection kinds, procedures, and safety measures for handling and transportation.
- Take into account submitting samples for full genome sequencing.

3.2 Depending on the Infection's Current Stage, Different Specimens Are Sampled and Collected

Epidemic Phase: Before the prodromal phase and just after the rash begin, blood may be present.

Table 1 Literature comparison of few existing works thus identifying the gaps

Papers	Pros	Cons	Method used
[1]	Provides official and reliable information from WHO	May lack detailed analysis or specific research findings	Not applicable
[2]	Offers specific outbreak details and epidemiological data	May lack in-depth analysis or novel methodologies	Epidemiological analysis
[3]	Provides detailed clinical characteristics of monkeypox cases	Limited to observational analysis	Observational analysis
[10]	Utilizes advanced machine learning techniques for monkeypox detection	May require expertise in machine learning for implementation	Convolutional Neural Networks (CNNs) with transfer learning
[13]	Reports actual cases of monkeypox virus infection	Limited to case reports	Case reports
[15]	Provides immunological insights into monkeypox infection	Limited to observational study	Immunological analysis
[16]	Offers a proactive spatiotemporal diagnosis model for outbreaks	May require complex computational resources	Spatiotemporal multiscale diagnosis model
[17, 18]	Provide historical perspective on monkeypox emergence	Limited to historical analysis	Historical analysis
[19]	Addresses the public health emergency declared by WHO	May lack detailed scientific analysis	Public health emergency analysis
[20]	Investigates gene heterogeneity and protein polymorphism of the monkeypox virus	May be highly technical and specialized	Genetic analysis
[21]	Focuses on a specific aspect of monkeypox manifestation	Limited to a case report	Case report

Ongoing Phase: Oral and oropharyngeal samples. Moreover, submit samples of blood, urine, and sperm.

Rash Phase: Material cutaneous diseases, including vesicle exudates smears, roofs from several lesions, or scabs from tumors. Give the same samples that were gathered in the earlier stages as well.

The contagious period starts right prior to the nuclear and lasts all the way through it. Yet by the time the rash appears, the viremia stage will have most likely ended. The majority of patients in the 2022 epidemic have manifested after the viremia stage; following the seroconversion phase is through, neither the antibody nor Genome of varicella can be found in blood. In a Tube of methane primary amine in tetra methyl solution, collect blood samples.

At the prodromal stage, there are no skin lesions, although viral DNA can be found in the pharynx, mouth, and nasal passages. Enanthem has a lot of viral particles, like mouth lesions. Care must be taken throughout the collection process since these

Table 2 Deep learning algorithms for the disease detection work

Algorithm	Advantages for monkeypox detection
Convolutional Neural Networks (CNNs) [10]	CNNs are highly effective for image classification tasks like identifying skin lesions indicative of monkeypox
	They can learn complex patterns and features from images, making them suitable for detecting characteristic symptoms
	CNNs can handle large datasets efficiently, allowing for robust training on diverse images of monkeypox lesions
Support Vector Machines (SVM) [22]	SVMs can effectively classify images based on extracted features, making them suitable for detecting monkeypox lesions
	They work well with high-dimensional data and are robust to overfitting, which is beneficial for medical image analysis
	Decision trees provide interpretable results, which can aid in understanding the decision-making process for monkeypox detection
Decision Trees [23]	Decision trees provide interpretable results, which can aid in understanding the decision-making process for monkeypox detection
	They can handle both numerical and categorical data, making them suitable for integrating different types of features from images
Random Forests [24]	Random forests are robust to noise and outliers, which is advantageous for handling variations in monkeypox lesion images
	They can handle large datasets efficiently and are less prone to overfitting compared to individual decision trees

lesions are unpleasant. To collect the sample, using either a viral transporter or a DACRON or thermoplastics probe medium. Furthermore, send blood a specimen and an EDTA vial of urine in a universally clean container.

The ideal examples to obtain for evaluating the monkeypox virus at this time are the exanthema track's tissue nodules, which contain very high viral titers. Skin samples are hence the suggested specimen type for monkeypox laboratory confirmation. Rub the lesion ferociously with a wipe or a moist sponge dipped in VTM. Samples from two separate lesions should be collected from two distinct bodies regions and placed in identical pipe. Ideally, the lesions should have diverse morphologies.

If feasible, each with two pipes containing two samples to the lab for analysis. Swabs made of polyester-flocked or DACRON can be used to collect vaginal or rectal swabs, either with or without VTM.

The **primary focus of therapy** during the prodromal period is symptomatic relief. Conventional antiemetic and anti-febrile treatments can reduce nausea, myalgia, and fever; nevertheless, it is advisable to restrict quasi generally pro drugs due to the risk of developing hemorrhagic lesions. Mouth, eye, and rectal mucosal sores can be uncomfortable and call for the use of the proper analgesics.

3.3 Deep Learning Tool

A Deep Learning Tool for monkeypox is presented on account of a thorough assessment and analysis of the facts the area around the varicella epidemic in 2022. This imaginative apparatus is founded on first idea that really was to begin with conceived amid the US Ebola flare-up in 2014 [16, 25]. In consequent a long time, the idea has indeed been adjusted to make personalized instruments due to hepatitis [26], Center pulmonary east disorder [27], measles [17], Contain additives infection malady [18], gastritis [19, 24], whooping cough [20, 23], thrush [21], and both corona viruses' infection 2019, COVID-19 [22]. In our current work, we have made use of Efficient Net B3 tool for Implementation.

4 Suggested Methodology

This suggested process, which is depicted in Fig. 2, entails looking over and summarizing. The major goal of this suggested technique is to analyze Monkeypox 2022 while also obtaining the maximum accuracy possible. We chose deep learning as the method for identifying monkeypox in 2022 because, up to this point, deep learning has been crucial in accurately anticipating the illness and diagnosing it. This model is described as being finished after the implementation's outcomes have been verified. The methods used for the research project rely on the results. Machine learning algorithms that can classify the disease based on the features of patients may lead to wrong or false classification of the disease and discussed in the papers previously, PCA to analyze this method requires a rigorous study of Biomarkers and gene sequence with protein receptors.

Very less work has been done in the above said domain that could be taken as future work, so we have used deep learning technique to quickly analyze and associate the disease with the possible medication as early as possible.

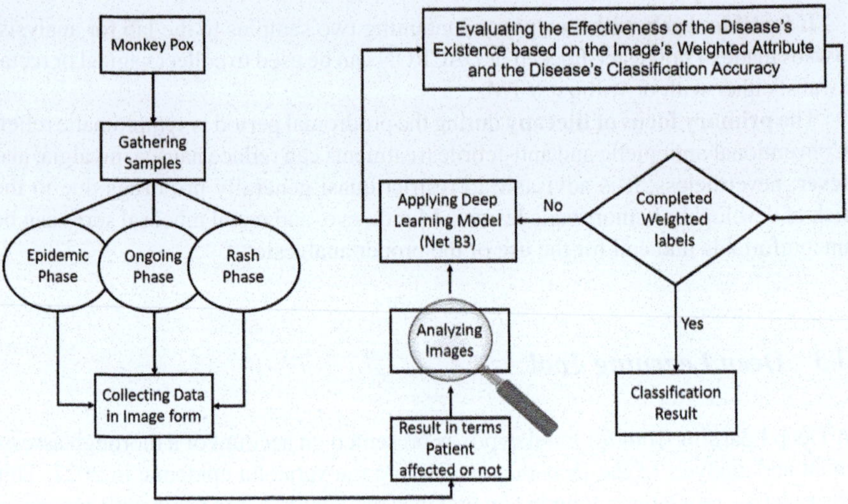

Fig. 2 Working paradigm of proposed methodology

5 Implementation and Results

This research work has used the deep learning technique, using these approaches to forecast the rate of monkeypox 2022 spread that has been a huge outbreak and also able to identify the different kinds of lesions. Therefore, the dataset has been taken from the Kaggle [28] in which it consists of three different kinds of attributes which are collectively used for recognizing the different kinds of variance of occurrence of monkeypox. For the experiment implementation we have used the Google Colab in this research area. Google Colab is a sort of open source software in which it is capable to provide the huge varieties of automatic operators that are not useful for various tasks of deep learning but also machine learning, statistics, data mining and lots of applications, etc. Coming to the dataset, we have used 980 images for the training purpose, 168 images for the validation purpose, and 20 images for the testing purpose. In the first step of implementation process, we have processed the training, testing, and validation dataset from directory by analyzing and applying the preprocessing techniques. By applying shuffle as false it can able to differentiate the data as well as it also helps in contrasting testing and validating data that can differ from previous. As this shuffle operator is mostly beneficial in validating and testing purpose, after analyzing the data, it is helpful to categorize in the following way that how monkeypox symptoms occur to human being skin (Fig. 3).

Coming next by applying the deep learning tool we have calculated the accuracy, validation accuracy, loss, and validation loss of the disease as represented in Figs. 4. and 5. For the purpose of calculating the accuracy, we have involved 20 steps which we are going to present in figure as well as in theoretical table.

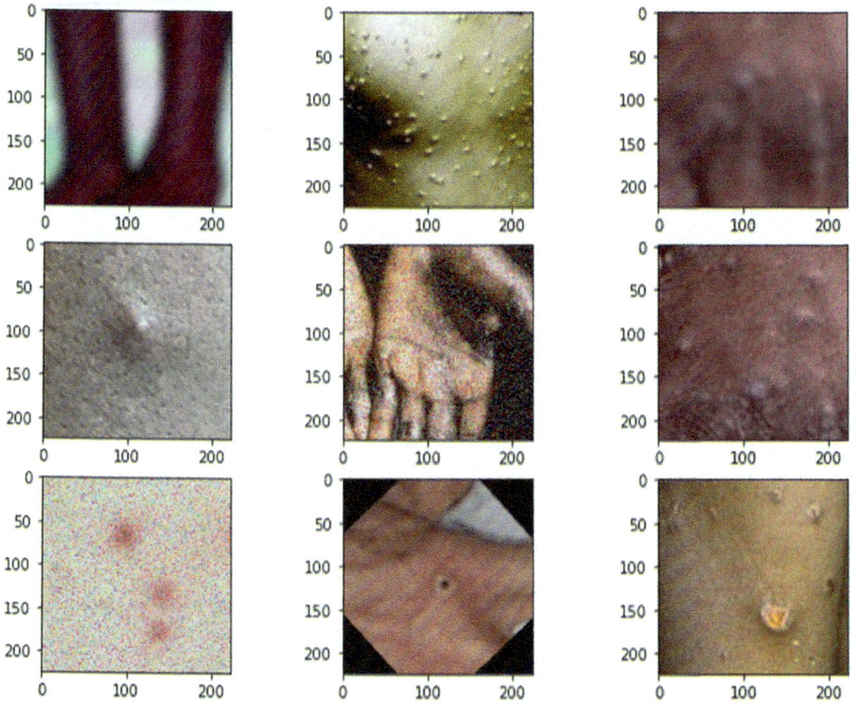

Fig. 3 Categorization of the Monkeypox on different parts of the skin with different types

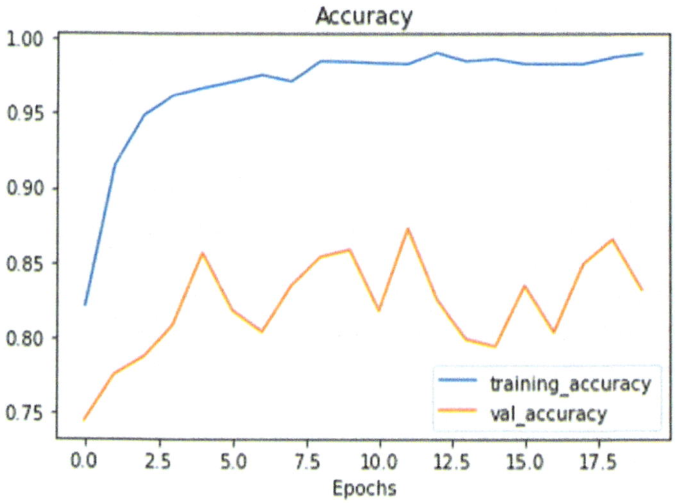

Fig. 4 Accuracy and validation accuracy of the model

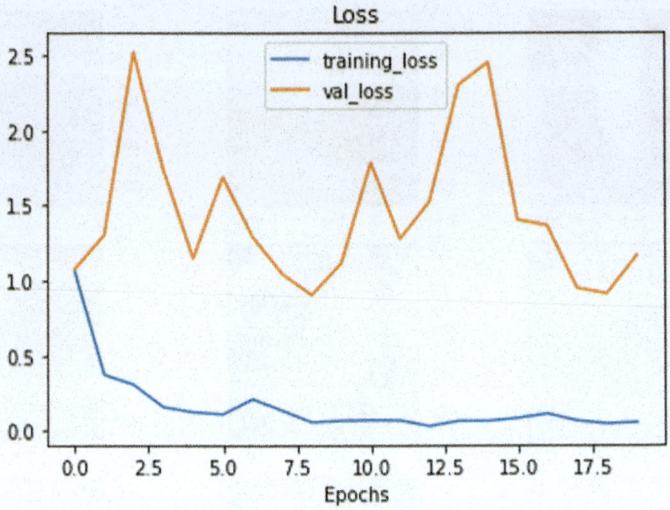

Fig. 5 Loss and validation loss of the model

After finding the accuracy and loss of training and validation data, we have evaluated the test data accuracy and loss in which we were able to obtain 88.89% of accuracy and loss of 1.28. Now coming to further step, we have analyzed the classification report as shown in Table 3 of monkeypox and as well as others dataset for the process in which we can able to obtain the following values.

In the above classification report, the Precision is the capacity of a predictor to avoid classifying as affirmative anything that is in fact negative. It is described for each category as the proportion of positive instances to the total of true positives as well as false positives. The capacity of classifiers to locate every successful instance in recall has been done. It is described as the proportion of positive instances to the total of true positives as well as false negatives for each category. The Classification performance is a balanced modulation index of recall and precision, with 1.0 being the greatest result and 0.0 the lowest that has been calculated in the F1 score. Although F1 scores incorporate recall and precision into their computation, they are less accurate than accuracy measures. It is often recommended to evaluate classifier models using the weighted sum of F1, rather than overall accuracy. The best point is that the F1

Table 3 Classification result details

Parameters	Precision	Recall	F1-score	Support
Monkeypox	0.83	0.95	0.88	20
Other skin diseases	0.95	0.84	0.89	25
Accuracy	0.82	0.82	0.89	45
Macro average	0.89	0.90	0.89	45
Weighted average	0.90	0.89	0.89	45

obtained in our case is more than 0.85 to be in the best prospective way. Whereas the support in this case has a amount of real instances of something like the category in the given dataset. The requirement for sample selection or rebalancing may be indicated by unbalanced reinforcement in the training data, which may point to structural flaws in the classifier's reported scores. Support remains constant across models but diagnoses the assessment procedure instead. Of the several averaging techniques, macro averaging is possibly the simplest. The average of the two of each of the per-class F1 scores is used to calculate the macro-averaged F1 score. Regardless of the support values, all classes are treated similarly by this method. Eventually, the normalized F1 score is determined by averaging all the class-specific F1 scores while accounting for the support of each class.

In the final phase of the model, the confusion matrix has been built as shown in below figure. In our process, a confusion matrix is a method that is used to demonstrate the performance of the test data where the true and actual values are supposed to be the recognized data values. In our process, the true values refer to the categories of the class in addition with true value (1) which refers to the chance of monkeypox effect on the patients, whereas coming to the other hand, true value (0) refers to the people other than monkeypox cases. In the same manner, the demonstration refers to the recognition of both monkeypox and other cases as well. So according to that, the following confusion matrix has been formed as represented in Fig. 6.

Now after evaluating the confusion matrix this is the end of process of recognizing the monkeypox disease. From the above Fig. 6, 21 represents the actual cases of monkeypox when coming to 19 these are predicted cases of monkeypox in the overall test dataset, whereas coming to 4 these are actual cases of the other cases whereas coming to 1 is predicted case of other case in overall test dataset. Altogether it illustrates that by using Google Colab tool and Deep learning technique it can able

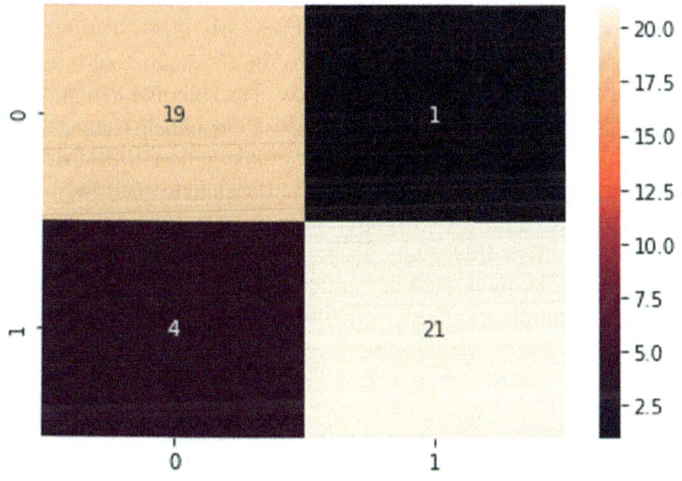

Fig. 6 Confusion matrix formation of the model

to predict and implement the highest accuracy of 99.11% and also helpful in the better performance of the categorization. As always Confusion matrices are a widely used measurement when attempting to solve classification issues. Both single categorization and multiclass categorization issues may be solved with it. As accuracy might be deceptive sometimes when applied to unbalanced datasets, alternative measurements depending on the confusion matrix are also relevant for assessing performance.

6 Conclusion and Future Scope

A near relative of smallpox is, monkeypox 2022, which is a resurgent zoonotic disease that has spread quickly over the world and shown unusual instances of people and societal transmissions. Although the first instances were found in the MSM, there is a 21-day exposure risk for anybody who comes into close touch with sick people, animals, or fomite. People may be contagious even before showing symptoms. The likelihood of successive viral pandemics has become a major source of worry. COVID-19 is anticipated to soon reach an endemic state. Unfortunately, after further over two years of significant COVID-19 effects on the world economy and healthcare, we will probably soon be dealing with a new second viral outbreak. The monkeypox virus is the second etiological factor. The monkeypox virus was initially identified in 1958 in Copenhagen. It was segregated from an animal research center where monkeys had been transported from Singapore to Denmark. The first isolate was given the name monkeypox. In the Congolese Democratic Republic, the first instance of zoonotic MPV transmission from animal to human was documented in 1970. Patients in need of a monkeypox examination are frequently seen by medical professionals operating in STI care centers, medical emergency greater utilization, and emergency rooms. In addition to providing frontline physicians with an innovative, succinct, fact-based deep learning tool for the early identification and treatment of individuals under suspicion for monkeypox in 2022, this study introduces the recognized One Health elements of monkeypox. Through this Efficient Net B3 tool we have implemented the highest accuracy of 99.11%. Our findings show that there aren't many clinical care recommendations for monkeypox over the world. Research into therapy and prevention, including for various risk demographics, is clearly and urgently needed. With coordinated, top-notch investigations, the present outbreak offers a chance to further this study. To improve patient and epidemic outcomes, new evidence should be integrated into globally accessible recommendations. It is advised to use a "dynamic standard" structure.

References

1. World Health Organization (2022) Multi-country monkeypox outbreak in non-endemic countries. World Health Organization. https://www.who.int/emergencies/disease-outbreak-news/item/2022-DON390
2. Minhaj FS, Ogale YP, Whitehill F, Schultz J, Foote M, Davidson W, Wong M (2022) Monkeypox outbreak—nine states, May 2022: Weekly/June 10, 2022/71(23):764–769
3. Girometti N, Byrne R, Bracchi M, Heskin J, McOwan A, Tittle V, Whitlock G (2022) Demographic and clinical characteristics of confirmed human monkeypox virus cases in individuals attending a sexual health centre in London, UK: an observational analysis. Lancet Infect Dis 22(9):1321–1328
4. Lim XJ, Junus S, Edwin-Amir L, Mohd-Kassim AB (2023) Vaccination refusal in children under 2 years of age in Malaysia 2016–2019. Asian J Med Health 21(6):1–13
5. Wei F, Peng Z, Jin Z, Wang J, Xu X, Zhang X, Huang S (2022) Study and prediction of the 2022 global monkeypox epidemic. J Biosaf Biosecur 4(2):158–162
6. Gupta BM, Kappi M, Gore MM, Gupta A (2023) Scientometric assessment of global research output about monkeypox during 1970–2022. Dubai Med J 6(2):97–106
7. Yu W, Zhang X, Du M, Dong Y, Liu L, Rong H, Liu J (2023) Bibliometric analysis and key messages of monkeypox research (2003–2022). Sustainability 15(2):1005
8. Tiee MS, Harrigan RJ, Thomassen HA, Smith TB (2018) Ghosts of infections past: using archival samples to understand a century of monkeypox virus prevalence among host communities across space and time. Roy Soc Open Sci 5(1):171089
9. Koenig KL, Beÿ CK, Marty AM (2022) Monkeypox 2022 identify-isolate-inform: a 3I Tool for frontline clinicians for a zoonosis with escalating human community transmission. One Health 15:100410
10. Altun M, Gürüler H, Özkaraca O, Khan F, Khan J, Lee Y (2023) Monkeypox detection using CNN with transfer learning. Sensors 23(4):1783
11. Rehan ST, Hussain HU, Nihal K, Sukaina M, Khan Z, Nashwan AJ, Sukaina II M (2023) Global monkeypox virus outbreak 2022: a bibliometric analysis. Cureus 15(4)
12. Kamal MA, Farahat RA, Awad AK, Tabassum S, Labieb F, Bejan CA, Dergaa I (2023) Global trends of monkeypox-related articles: a bibliometric analysis over the last five decades (1964–July 14, 2022). J Infect Public Health 16(9):1333–1340
13. Yadav PD, Reghukumar A, Sahay RR, Sudeep K, Shete AM, Raman A, Shaj MK (2022) First two cases of Monkeypox virus infection in travellers returned from UAE to India, July 2022. J Infect 85(5):e145–e148
14. Silva NIO, Oliveira JS, de Kroon EG, Trindade GDS, Drumond BP (2020) Here, there, and everywhere: the wide host range and geographic distribution of zoonotic orthopoxviruses. Viruses 13(1):43
15. Agrati C, Cossarizza A, Mazzotta V, Grassi G, Casetti R, De Biasi S, Antinori A (2023) Immunological signature in human cases of monkeypox infection in 2022 outbreak: an observational study. Lancet Infect Dis 23(3):320–330
16. De Cos Guerra O, Castillo Salcines V, Cantarero Prieto D (2023) Spatiotemporal multiscale diagnosis model to proactively respond to the multi-country monkeypox virus outbreak in 2022. Trans GIS 27(8):2175–2196
17. Durski KN et al (2018) Emergence of monkeypox—west and central Africa, 1970–2017. Morb Mortal Wkly Rep 67(10):306
18. Durski KN, McCollum AM, Nakazawa Y, Petersen BW, Reynolds MG, Briand S, Khalakdina A (2018) Emergence of monkeypox—west and central Africa, 1970–2017 Morb Mortal Wkly Rep 67:10–306
19. Sah R, Padhi BK, Siddiq A, Abdelaal A, Reda A, Ismail Lashin B, Rodriguez-Morales AJ (2022) Public health emergency of international concern declared by the World Health Organization for monkeypox.Glob Secur Health, Sci Policy 7(1):51–56
20. Xiong C, Li Y, Hou J, Sun Z, Han W, Thilakavathy K, Hu J (2022) Monkeypox virus 2022, gene heterogeneity and protein polymorphism

21. Patrocinio-Jesus R, Peruzzu F (2022) Monkeypox genital lesions. N Engl J Med 387(1):66–66
22. Zala K, Thakkar HK, Jadeja R, Dholakia NH, Kotecha K, Jain DK, Shukla M (2022) On the design of secured and reliable dynamic access control scheme of patient e-healthcare records in cloud environment. Comput Intell Neurosci
23. Syed SF, Singh A, Rakhra M, Singh D, Aggarwal S (2022) Monkey Pox detection using deep learning. In: 2022 2nd international conference on technological advancements in computational sciences (ICTACS). IEEE, pp 767–771
24. Phonsa G, Syed SF (2023) Machine learning approaches for brinjal leaf disease detection. In: International conference on information and communication technology for intelligent systems. Springer Nature Singapore, Singapore, pp 417–427
25. Ladhani SN, Aiano F, Edwards DS, Perkins S, Khan WM, Iyanger N, Chow JY (2022) Very low risk of monkeypox among staff and students after exposure to a confirmed case in educational settings, England, May to July 2022
26. Karbalaei M, Keikha M (2022) Human monkeypox outbreak 2022 and asymptomatic human reservoirs: challenges and opportunities–correspondence. Int J Surg 105:106849
27. Zhang Z, Yuan Y, Tan J, Zhao Y, Lin X, Ning J, Wu X (2022) Mapping trends and hotspots regarding clinical research on monkeypox: a bibliometric analysis of global research: trends and hotspots on monkeypox. Commun Acquir Infect 9
28. Zala K, Thakkar HK, Jadeja R, Singh P, Kotecha K, Shukla M (2022) PRMS: design and development of patients' e-healthcare records management system for privacy preservation in third party cloud platforms. IEEE Access 10:85777–85791

Automated Epileptic Seizure Detection of EEG Signals Using Machine Learning

Shilpa Jain, Vedant V. Gupta, Ankur Singh Bist, Mukesh Joshi, and Ashish Garg

Abstract Epilepsy is a chronic, incommunicable neuropsychiatric disorder marked by recurring epileptic seizures. Conventionally Electroencephalogram (EEG) is a test carried out by doctors for the diagnosis of epilepsy. It's a prominent tool that measures electrical activity of brain by using electrodes attached to the patient's scalp. These signals have complexity, noise, and non-linearity and produce voluminous data. The process of analyzing the EEG report manually is tedious and uncertain. This research article proposes an end-to-end solution to the problem using supervised machine learning techniques. The article is implemented on the Children's Boston Hospital dataset that was rigorously studied on 23 subjects. The result for a ten-fold cross validation of the execution of the article is 82.72% for Random Forest and 62.76% for SVM. The experimental outcomes predict that the Random Forest shows better performance and has great potential in future epileptic detection studies.

Keywords Epilepsy · EEG · Machine learning · Non-black algorithm · Black box algorithm

S. Jain (✉) · A. S. Bist · M. Joshi
Department of CSE, Graphic Era Hill University, Bhimtal Campus, Bhimtal, India
e-mail: shilpajain@gehu.ac.in

A. S. Bist
e-mail: ankurbist@gehu.ac.in

V. V. Gupta
Department of CSE, National Institute of Technology, Delhi, India

A. Garg
Department of CSE, Graphic Era Deemed to be University, Dehradun, Uttarakhand, India

P. J. S. Gonçalves et al. (eds.), *Proceedings of Fifth International Conference on Computing, Communications, and Cyber-Security*, Lecture Notes in Networks and Systems 1128, https://doi.org/10.1007/978-981-97-7371-8_27

1 Introduction

Epilepsy is a fatal recurring brain disorder affecting the human brain. It may impact the amygdala, hippocampus, frontal cortex, temporal cortex, and olfactory cortex and may result in complete or partial loss of consciousness.

EEG [1] has been used to diagnose many brain disorders, especially epilepsy. It's Brain-Computer Interfacing (BCI) placing various electrodes on the human scalp. Voltage oscillations are recorded when passing through the brain. These oscillations are transformed into time series signals. The advantage of BCI using non-invasive methods like EEG has an advantage that these are portable, economical, have good time resolution, and patients will not suffer from the surgical process [2–4]. In [5], changes in heart rate were observed using ECG during epileptic seizures. It was found that in 73% of the seizures, an increased heartbeat rate happened at the onset of a seizure.

2 Literature Survey

In [6, 7] automated onset seizure detection was predicted using EEG. The methodology uses decomposition of wavelet to form a feature vector which records the morphology and spatial distribution of an EEG epoch and then finds if that vector is representing the patient's non-seizure or seizure EEG using SVM classification algorithm. In [8], the method states data filters in different bands and spectral feature extraction using Bayes' theorem. 19 patients having 152 seizures EEG were collected for 407 h and were fed in Bayes' theorem. The advantage of warning systems and automatic detection of seizures is to decrease review time and it facilitates recordings. In [9], the method finds epileptic seizure using analysis of the brain's electrical activity along with other physiologic signals. In [10], the MLPNN model of classification used to diagnose epilepsy was introduced. EEG signals were disintegrated using discrete wavelet transform (DWT) into various frequency bands. The technique used to find coefficients of wavelet was clustered using the K-means algorithm for various frequency sub-bands. As per the distribution of wavelet coefficients on the clusters the probability distributions were computed, and then five different experiments to evaluate the suggested model in the segregation of various varieties of healthy classes. The suggested model gave the adequate classification accurately.

In [11], a ML technique for creating patient-based classification that finds the onset of seizure using the EEG. In [12], an autoencoder was developed which at the same time reconstructs as well as classifies biomedical signal. Although earlier works considered classification and reconstruction as different problems, in [13], precise identification of automatic seizures is crucial in recordings of EEG. In [14], it was the pioneer work to deploy the convolutional neural network (CNN) for analyzing EEG signals. The study comprised of creating thirteen-layered CNN algorithm implemented to detect various classes of seizures. The technique demonstrated had a high accuracy of 99%. In [15], matrix determinant of EEG is considered as a crucial feature for recognizing seizures due to epilepsy. In [16] deep CNN model is used to perform seizures detection task with an overall accuracy of 98.05% for the entire dataset of all 23 patients. In [17] a Deep Learning (DL) model for diagnosing seizures in children based on classification of EEG with minimum pre-processing. The model displayed approximately 98% accuracy. In [1] 1D CNN automatically extracts features. These are quite suitable for EEG data as they require minimal pre-processing and 94% reduction in training on deep learning networks. Previous works in this direction include [18, 19] a review presented on the application of machine learning for seizure detection which summarizes as datasets used in the study are of Bonn University, CHB-MIT, Epilepsiae, Institute of Neuroscience India. Techniques frequently used are KNN, SVM, DNN, MLP, NB, and ANN and further feature extraction techniques include DWT, Multi-DWT, Short Fourier Transform Series, Fast Fourier Transform Series, and Empirical Wavelet Transform. In [20], a 100-layered architecture is used for the segmenting brain tumor from MR images and is performed using densely connected fully convolutional neural network (FCNN).

In [21], applications of Artificial Intelligence (AI) and graph theory are used in diagnosing neurological disorders. In [22], a substantial review of AI techniques was given to diagnose cervical cancer for women in a span of past 5 years.

3 Material and Methods

3.1 *Experimental Design*

Dataset used in the research article is from Boston Children's Hospital, which has recordings of EEG for children having intractable seizures. 23 cases were recorded for 22 subjects (5 males aged 3–22, and 17 females aged 1.5–19) as chb21 was obtained a year and a half after the case chb01 from the same female subject. Ease case chb01, chb02, etc. contains 9–42 continuous EEG signals for one subject. In most cases, EEG signals contain exactly an hour of digitized EEG signals. Total uncompressed size is 42.6 GB. Although chb10 channels were recorded for 2 h chb04, chb06, chb07, chb09, and chb23 were recorded for 4 h. The sampling rate was 256 samples per second at a resolution of 16-bits. The number of EEG channels in the recordings varied from 22 to 38. The majority of files had 23 channels as depicted in Fig. 1.

The dataset has some inconsistencies as the number of channels may be equal to some value, but some data was missing from channels. Some seconds from some files were also missing too as visualized in Fig. 2. The data has inherent bias as the time the seizure lasts for is very small compared to the time seizure isn't there. This problem is tackled during the pre-processing.

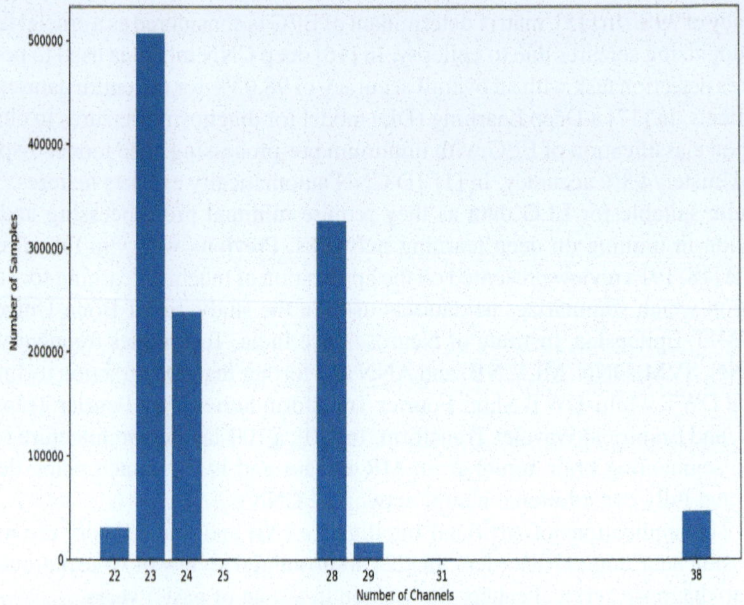

Fig. 1 Channels number plotted against the number of samples

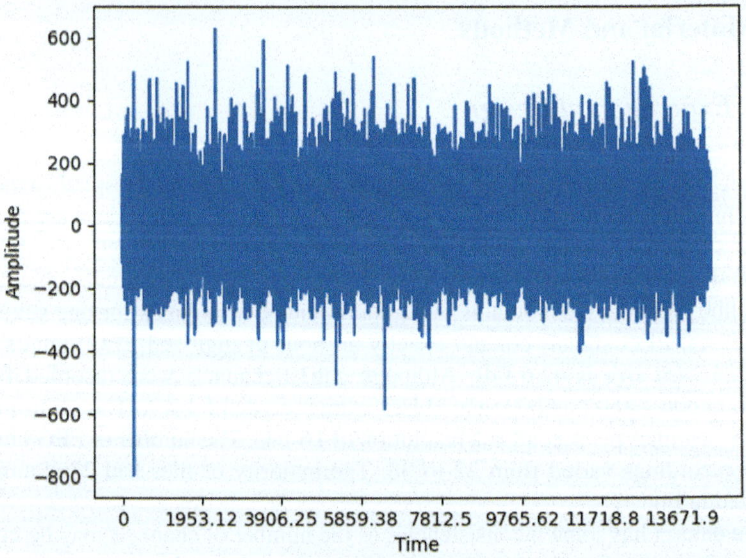

Fig. 2 Visualizes a particular channel of a file in the dataset

3.2 Methodology

3.2.1 Pre-processing

We are using the pyedflib library to handle the.edf files. We have decided to split the signal from each file into signals of 3 s. We read the summary files from the dataset and extracted the time frames where the epileptic episodes were. We used this data to create a Boolean array which was then used as the Y labels. The list was reshaped to a 1D array, as this is the standard way of representing the Y labels. We are only using the 23 channels from each file, if a channel has less than 23 channels, we pad the signal with zeroes. We then use these signals to get a list of features which are discussed in Sect. 3.2.2. This list of features is then reshaped to match the first dimension of the Y labels. We remove the inherent bias by sampling all the positive cases and then selecting an equal number of random negative cases.

3.2.2 Feature Engineering

We are extracting the following features.

1. Sum: It's the sum of all of the values at every instant of time.
2. Energy: It's the sum of squares of the values at every instant of time.
3. Standard Deviation: It's the measure of how far the signal fluctuates from the mean.
4. Max: It's the maximum value at any time during the signal.
5. Min: It's the minimum value at any time during the signal.
6. Power Signal Density: It describes the power present in the signal as a function of frequency, per unit frequency.
7. IDX: Index of the frequency with the highest power.

Note that these features are extracted on signals of 3 s each. The feature selection was done based on a research paper.

3.2.3 Model Implementation

We are using scikit-learn as the library for importing the machine learning algorithms. We are using Support Vector Machines and Random as our algorithms.

Support Vector Machine (SVM) is a supervised algorithm which can be used for classification and regression problems. It draws a decision boundary called as a hyperplane between any two categories to separate them or classify them. For the kernel, we have used the Gaussian Radial Basis Function (RBF). It is usually chosen for non-linear signals. It helps to make proper separation when there is no prior knowledge of data.

Table 1 Accuracy of various machine learning algorithms

Algorithm	Iterations	Accuracy (%)
Support Vector Machines (SVM)	100	62.76
Random Forest	100	82.72
Entropy + LSSVM [23]	100	76

Random forest (RF) classifier creates many decision trees from randomly selected subsets of the training data. Then RF collects the votes from different decision trees to predict the final prediction. The maximum depth of the tree is set as five.

3.2.4 Results and Discussion

We studied many datasets of which we finalized the Children's Boston Hospitals dataset cited in [7], then we rigorously studied the dataset of 23 subjects, and understood the format and intricacies of the dataset. Then we examined the existing literature on the subject matter and extracted the necessary information. We then extracted the signal information from the dataset and necessary features such as Sum, Energy, Max, etc. as mentioned in Sect. 3.2.2 on signals of 3 s each. We then extracted the time frames where the epileptic episodes happened and used this data to label the existing signals. We then began the implementation phase. After the initial implementation we received an accuracy greater than 99% which led us to believe that the data is biased. After a deeper look into the matter, we investigated and found that the pre-processed data had 4032 positive samples and 1,165,375 negative samples, so the model was behaving biased. Further, we took the ratio of positive and negative cases of epilepsy to be 50:50. Then the model was trained and validated over the pre-processed 8064 samples using SVMs and Random Forest (Table 1).

Following is the confusion matrix of the algorithms visualized (Figs. 3 and 4).

3.2.5 Conclusion and Future Scope

In this research article, we have applied black box and nonblack box algorithms for Epilepsy detection, but we observed that the accuracy is low so to improve the accuracy we will apply LSTM and GRU's in our upcoming papers.

We will also incorporate dose adjustment for Epileptic kids using machine learning algorithms in our future work.

We would also develop software which could state the dose to be given to epileptic kids based on symptoms.

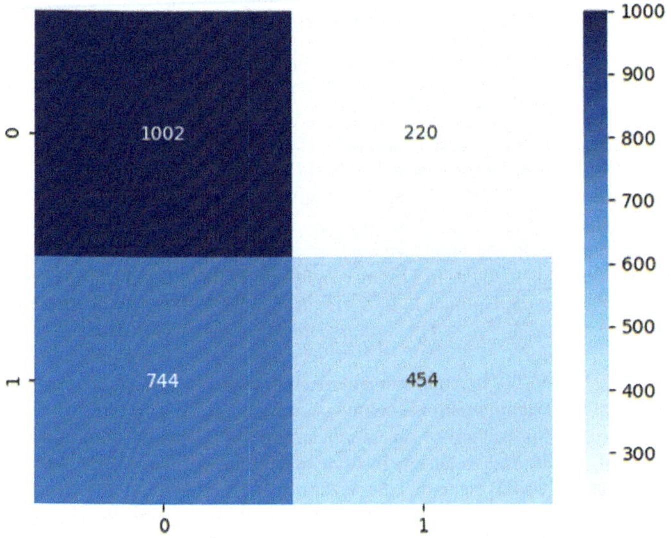

Fig. 3 Heatmap of confusion matrix produced by SVMs.

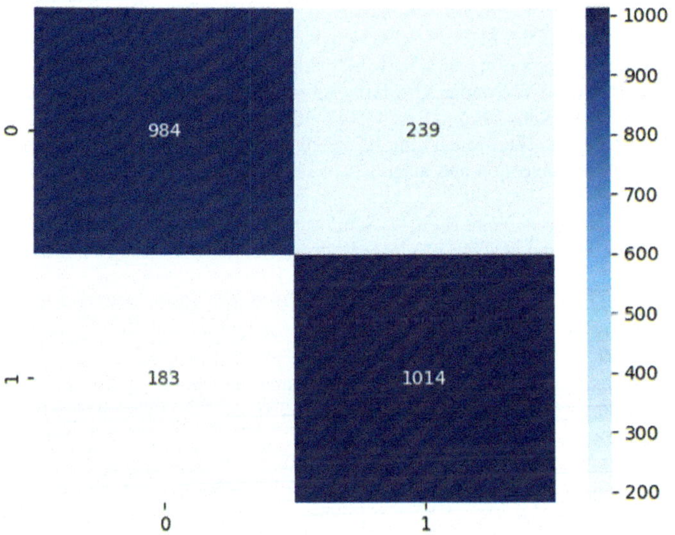

Fig. 4 Heatmap of confusion matrix produced by Random Forest

References

1. Sameer M, Gupta B (2022) CNN based framework for detection of epileptic seizures. Multimed Tools Appl 81:17057–17070
2. Agarwal P, Kumar S (2022) Electroencephalography-based imagined speech recognition using deep long short-term memory network. ETRI J (Wiley) 44(4):672–685
3. Agarwal P, Kumar S (2022) Electroencephalography based imagined alphabets classification using spatial and time-domain features. Int J Imaging Syst Technol (Wiley) 32(1):111–122
4. Agarwal P, Kumar S (2021) Transforming imagined thoughts into speech using a covariance-based subset selection method. Ind J Pure Appl Phys (CSIR-NIScPR) 59(3):180–183
5. Zijlman M, Flanagan D, Gotman J (2002) Heart rate changes and ECG abnormalities during epileptic seizures: prevalence and definition of the objective clinical sign. Epilepsia 43(8):847–854
6. Shoeb AH, Guttag JV (2010) Application of machine learning to epileptic seizure detection. In: Proceedings of the 27th international conference on machine learning (ICML-10), pp 975–982
7. Shoeb AH (2009) Application of machine learning to epileptic seizure onset detection and treatment. PhD thesis, Massachusetts Institute of Technology, Cambridge, MA
8. Treves S, Guttag J (2004) Patient-specific seizure onset detection. Epilepsy Behav 5(4):483–498
9. Grewal S, Gotman J (2005) An automatic warning system for epileptic seizures recorded on intracerebral eegs. Clin Neurophysiol 116(10):2460–2472
10. Mirowski P, Madhavan D, LeCun Y, Kuzniecky R (2009) Classification of patterns of EEG synchronization for seizure prediction. Electroencephalogr Clin Neurophysiol 120(11):1927–1940
11. Orhan U, Hekim M, Ozer M (2011) EEG signals classification using the the K-means clustering and a multilayer perceptron neural network model. Expert Syst Appl 38:13475–13481
12. Acharya UR, Sree SV, Swapna G, Martis RJ, Suri JS (2013) Automated EEG analysis of epilepsy: a review. Knowl Based Syst 45:147–165
13. Acharya UR, Oh SL, Hagiwara Y, Tan JH, Adeli H (2018) Deep convolutional neural network for the automated detection and diagnosis of seizure using EEG signals. Comput Biol Med 100:270–278
14. Gogna A, Majumdar A, Ward R (2017) Semi- supervised stacked label consistent autoencoder for reconstruction and analysis of biomedical signals. IEEE Transact Biomed Eng 64:2196–2205
15. Wang D, Ren D, Li K, Feng Y, Ma D, Yan X et al (2018) Epileptic seizure detection in long-term EEG recordings by using wavelet-based directed transfer function. IEEE Transact Biomed Eng 65:2591–2599
16. Hossain MS, Amin SU, Alsulaiman M, Muhammad G (2019) Applying deep learning for epilepsy seizure detection and brain mapping visualization. ACM Transact 4 Multimed Comput Commun Appl 15:1–17
17. Abdelhameed A, Bayoumi M (2021) A deep learning approach for automatic seizure detection in children with epilepsy. Front Comput Neurosci 15:650050
18. Farooq MS, Zulfiqar A, Riaz S (2023) Epileptic seizure detection using machine learning: taxonomy, opportunities, and challenges. Diagnostics 13(6):1058
19. Shoeibi A, Khodatars M, Ghassemi N, Jafari M, Moridian P, Alizadehsani R, Acharya UR (2021) Epileptic seizures detection using deep learning techniques: a review. Int J Environ Res Pub Health 18(11):5780
20. Shaikh M, Anand G, Acharya G, Amrutkar A, Alex V, Krishnamurthi G (2018) Brain tumor segmentation using dense fully convolutional neural network. In: Brainlesion: glioma, multiple sclerosis, stroke and traumatic brain injuries: third international workshop, brainles 2017, held in conjunction with MICCAI 2017, Quebec City, QC, Canada, September 14, 2017, Revised Selected Papers 3. Springer International Publishing, pp 309–319
21. Prajapati B, Oza P, Agrawal S (2023) Artificial intelligence and graph theory application for diagnosis of neurological disorder using fMRI. In: Proceedings of international conference on

recent innovations in computing: ICRIC 2022, vol 1. Springer Nature Singapore, Singapore, pp 41–56
22. Nazir N, Saini BS, Sarwar A (2023) Early diagnosis of cervical cancer using AI: a review. In: Proceedings of international conference on recent innovations in computing: ICRIC 2022, vol 1. Springer Nature Singapore, Singapore, pp 105–116
23. Krishnan PT, Balasubramanian P (2016) Automated EEG seizure detection based on S-transform. In: 2016 IEEE international conference on computational intelligence and computing research (ICCIC). IEEE, pp 1–5

Fisheries Management with Deep Learning-Based Fish Species Detection: A Sustainable Approach

Nemi Rishi, Akhil Kumar, and Richa Golash

Abstract Identifying fish species accurately and efficiently is crucial for ecological research, sustainable fisheries management, and underwater exploration. Deep learning models trained on vast image and video datasets can automatically identify fish species with high accuracy, on-board fishing vessels, or from underwater cameras. This work proposes a deep learning-based approach for fish species detection using the lightweight and real-time object detector Tiny-YOLOv4-SPP. This work has been carried out on a publicly available FishPak dataset. The training and testing results with the Tiny-YOLOv4-SPP algorithm achieved a map value of 99.79% indicating it is useful for fish species detection. The results obtained with the proposed Tiny-YOLOv4-SPP are better than the Tiny-YOLOv4 and Tiny-YOLOv7. Accurate species identification leads to more precise population estimates, size structure analysis, and catch composition data. This work can be helpful in designing more targeted and effective management strategies for specific species, preventing overfishing of vulnerable populations, and promoting sustainable practices.

Keywords Fisheries management · Fish species detection · Deep learning · Tiny-YOLOv4 · Ecological research

N. Rishi · A. Kumar (✉) · R. Golash
SCSET, Bennett University, Greater Noida 201310, India
e-mail: akhil.hpucs@gmail.com

N. Rishi
e-mail: E23SOEP0024@bennett.edu.in

R. Golash
e-mail: richa.golash@bennett.edu.in

P. J. S. Gonçalves et al. (eds.), *Proceedings of Fifth International Conference on Computing, Communications, and Cyber-Security*, Lecture Notes in Networks and Systems 1128, https://doi.org/10.1007/978-981-97-7371-8_28

1 Introduction

Fish classification holds the key to unlock sustainable fisheries management. But fish classification and detection is a significant challenge in fisheries management for the following reasons [1–4]:

1. Diversity in Fish Species: In the ocean, there is a staggering variety of fish species, many of which share similar physical characteristics. Accurately distinguishing between them, especially in the field or from market samples, can be difficult for even experienced observers.
2. Misclassification: Misclassification often leads to an inaccurate stock assessments. This contributes to unsustainable fishing practices and ecosystem damage.
3. Data Limitations: Traditional methods often rely on visual identification, which can be subjective and lead to errors. Collecting comprehensive data across vast ocean areas is also challenging, hindering accurate population assessments and management decisions.
4. Lack of standard practice: Different regions and fisheries may employ inconsistent nomenclature or taxonomy. This makes it difficult to compare fish populations and implement effective international management strategies.

Management frameworks need to be adaptable to incorporate these changes and prevent overexploitation of newly identified or vulnerable species. Deep learning, with its ability to analyze vast amounts of data, offers a revolutionary approach to fish classification, ultimately improving fisheries management. By automatically identifying species from images or videos, deep learning can enhance stock assessments, leading to more accurate data and targeted regulations. This allows for sustainable fishing practices, protecting vulnerable species and minimizing bycatch.

Understanding the distribution and interactions of different fish species helps us unravel the intricate web of life in aquatic ecosystems. Knowing the types and numbers of fish present in a body of water is critical for sustainable fishing practices and resource management. Accurate identification is essential for monitoring populations, identifying endangered species, and implementing conservation strategies. Thus, classifying new species and understanding evolutionary relationships within the fish family contributes to our knowledge of life on Earth and also can reveal potential environmental threats [1, 5].

Traditionally, fish species detection relied on methods like physical capture, expert analysis, and underwater surveys—traditional methods are time-consuming, labor-intensive, and often limited in scope. In contrast, Computer vision enables machine to understand and interpret visual information and it has the potential to revolutionize the study of marine ecosystems. This field can transform the various aspects of marine sciences, ecological monitoring, fisheries management, etc. Tracking changes in fish communities over time provides valuable insights into ecosystem health and the impact of human activities. Unlike traditional methods, computer vision is non-invasive and can be applied to large numbers of fish without harming them. Continuous video monitoring with automated species detection enables real-time data collection on fish populations and behavior. Moreover, as stated in the related researches [6, 7], CNN-based approaches can prove out to be beneficial to better deal with the problem of fish species detection as CNNs tend to extract refined features for cross-domain problems and data with high class resemblance.

Fish species classification, whether underwater or in controlled environments, presents several challenges that can hinder accuracy and efficiency [8]. Some of the key challenges faced by several researchers in vision-based fish classification techniques are listed as follows:

1. Intra-specific variation: Different individuals within the same species can exhibit variations in color, patterns, and even body shape, making identification difficult.
2. Environmental factors: Lighting conditions, turbidity of water, and the presence of algae or debris can obscure key features and complicate visual analysis.
3. Underwater imaging: Capturing high-quality images and videos underwater poses difficulties due to light attenuation, movement of fish, and background noise.
4. Limited training data: Lack of extensive and well-annotated datasets for rare or less-studied species can hamper the training of robust classification models.

Targeting the above challenges, this work proposed an annotated dataset for various fish species which can be used for detection with deep learning-based object detectors. Further, it proposed Tiny-YOLOv4-SPP for the detection of different fish species. The proposed Tiny-YOLOv4-SPP was developed by adding spatial pyramid pooling in the Tiny-YOLOv4 which enabled the detector to extract rich semantic information for different fish species having similar visual features and further perform accurate detection. The highlight of this work is that it proposes an approach for accurate detection of fish species in underwater scenarios. Moreover, the present work targets on detection of different fish species in contrast to the literature work which focuses only on the detection of fish. For evaluated performance metrics, fascinating results have been obtained along with accurate bounding boxes. The proposed approach achieved a map value of 99.79% which was better than the Tiny-YOLOv4 and Tiny-YOLOv7 object detectors. The proposed detector can be used by fisheries management bodies for autonomous detection of fish species.

2 Literature Survey

In the vast and mostly hidden world of biodiversity, identifying species is a great challenge. Often, we use traditional methods, for example, physical capture and expert analysis that are mostly, or time-consuming manual observation, limiting our understanding and conservation efforts. But enter the realm of computer vision, a game-changer that empowers us to see the unseen and revolutionize species detection.

In the last couple of years, many researches have been conducted using computer vision and deep learning techniques in the area of underwater fish species detection. In this paper, we will focus on some of the recent trends in this domain.

Research conducted under [9] ‘Ocean Aware project’ researchers used yolo v3 and mask R-CNN for their fast processing capability, to automate fish monitoring in fish passages created around dams or any other human-made structure. But the model requires more than 400 labeled training samples for each species. The paper [5] focuses on fish detection in blurry water conditions taken by, for example, Autonomous underwater vehicles (AUVs) using CNN. They addressed optimization using data augmentation and refined loss function by adding L2 regularization to decrease the difference between training and testing losses. Jalal et al. [10] work focused on the issues related to dynamic and complex background. They combined Gaussian mixture model and optical flow for temporal information with YOLO deep CNN architecture. This work is based on LifeCLEF 2015 and Fish4Knowledge repository and achieved a classification accuracy of 91.64 and 79.8%, respectively, on two underwater video datasets.

Muksit et al. [11] proposed building two architecture Yolo-Fish-1 and Yolo-fish-2 upon the popular YOLOv3 CNN design. In this design, they used Darknet-53 feature extractor for efficient feature extraction from images and Feature Pyramid Network (FPN) for detection at different scales, enhancing accuracy for both small and large fish. Knausgard et al. [12] proposed a two-step deep learning approach YOLO for temperate fish detection and Squeeze-and-Excitation Convolutional Neural Network (SE-CNN) architecture for each fish in the image to avoid pre-filtering stage. They applied transfer learning by using initial weights pre-trained on ImageNet and fine-tuned on Fish4Knowledge dataset. This strategy improves the model’s ability to learn specific features of temperate fish from smaller targeted datasets and achieves an accuracy of 87.74% without increasing the dataset. Manual annotation and large volume are some of the major problems in fish species detection. Marrable et al. [13] suggested machine-assisted annotation on Ozfish data set using Yolov5. They achieved a recall of 0.7 for fish only detection and were labeled to automate labels to approximately 12% of fishes.

Malik et al. [14] work is also based on a similar challenge of fish species classification on highly dynamic-complex background and the complexity of classification further escalates due to camouflage foreground nature and subtle difference between fish species. To deal with this, they modified the Darknet of YOLOv7 by MobileNetv3 with Arcface Loss and DenseNet169. Zhao et al. [3] proposed a decoupled multi-scale model for fish detection. The authors utilized a feature pyramid to address the problem of multi-scale fish and achieved overwhelming results. The authors Liu et al. [4] proposed a vision transformer and CNN-based approach to underwater fish detection. With the proposed approach, the authors achieved an AP50 of 95.2%. Jahanbakht et al. [15] proposed a deep neural network for fish detection and achieved an accuracy of 94% on a poorly labeled class imbalanced dataset.

Most of the researchers found that unrestricted also dynamic environment, camouflage foreground as well as background and crowded background influence the efficiency of model in species detection in marine ecosystem. There are various strategies used by the researchers to combat these challenges. Concentrating on deep learning models there are two commonly used methods in object detection and classification

[16]. First is two-stage or step-by-step approach in which detection and classification are done separately, example architectures are R-CNN, Fast R-CNN, Faster R-CNN, and Mask R-CNN. Though this method is adaptable to various tasks and gives high accuracy but it is slow due to multiple stages. In contrast, second approach utilizes single-stage for detection and classification, example architectures are You Only Look Once (YOLO) [17, 18] and Single Shot MultiBox Detector (SSD). These design give unified framework for detection and classification in a single pass. In this method, an image into grids predicts bounding boxes and class probabilities for each cell. Thus, they are fast, efficient, and suitable for real-time applications.

3 Methodology

In this research work, we have used Tiny-YOLOv4 model because of three reasons: First, it requires less computational power and memory compared to larger YOLO versions, making it suitable for deployment on resource-constrained devices like underwater cameras or mobile phones. Second: Its fast inference speed allows for real-time fish identification on-board fishing vessels or during underwater surveys easier. Third: The model can be fine-tuned on datasets specific to any fish species of interest, potentially achieving good accuracy. Figure 1 shows the benefits of our proposed system in fisheries management.

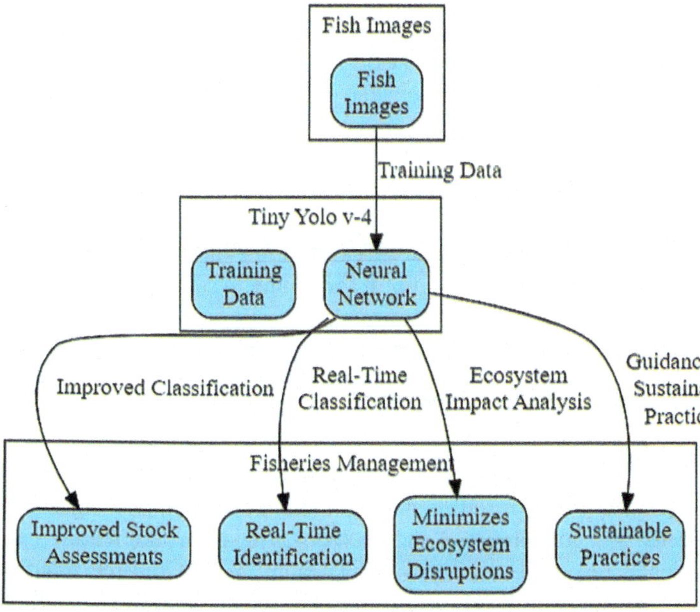

Fig. 1 Usefulness of Deep learning techniques in fisheries management

This model demonstrates many significant benefits in fisheries management practices. This advanced neural network-based technique facilitates improved stock assessments by providing more accurate and detailed information about fish populations. This data serves useful for devising effective management strategies aimed at maintaining sustainable fish stocks. Moreover, the real-time identification capabilities of the Tiny-YOLOv4 model enable fisheries managers to promptly respond to changes in fish populations and activities, facilitating timely interventions to prevent overfishing or illegal practices. Through continuous learning from training data comprising fish images, the model refines its classification abilities, ultimately leading to more accurate stock assessments, real-time identification, ecosystem impact analysis, and informed decision-making that supports the long-term sustainability of fisheries. However, to enhance the abilities of Tiny-YOLOv4, we have embodied a few enhancements in its feature extraction network to achieve benchmark results by adding a spatial pyramid pooling layer. In order to carry out this work, we have first prepared an annotated dataset suitable for object detection algorithms. Further, train and tests have been performed using the Tiny-YOLOv4, Tiny-YOLOv7, and the proposed Tiny-YOLOv4-SPP algorithm and evaluated the mean average precision (map) to gauge their efficiency. The details about the preparation of the dataset and Tiny-YOLOv4-SPP are presented below.

4 Dataset

In order to carry out this work, the first step was to identify a dataset having different fish species. Therefore, as a first step toward preparing the dataset, we have extracted 674 images from the FishPak dataset [19] which is a publicly available dataset consisting of images for different fish species. The FishPak dataset utilized by us has 15 classes of 5 fish species having three body parts namely, head, scale, and whole body. To make the dataset suitable for Tiny-YOLOv4, Tiny-YOLOv7, and the proposed algorithm, we resized all the images to 416×416 and further performed annotations in YOLO format. For the task of training and testing, we utilized the entire images of training and 10% out of it for testing.

5 Tiny-YOLOv4-SPP Algorithm

The Tiny-YOLOv4 algorithm is a state-of-the-art object detection algorithm proposed in [20]. It is comprised of convolutional layers, maxpool layers, and two detection heads. The convolutional and maxpool layer forms the feature extraction network which is responsible for extracting feature maps from the input images. Whereas the two detection heads are responsible for drawing the bounding boxes in the input images. The detection heads utilize CIoU loss function to penalize the network and to take the values of anchor boxes and use Non-Max Suppression. To

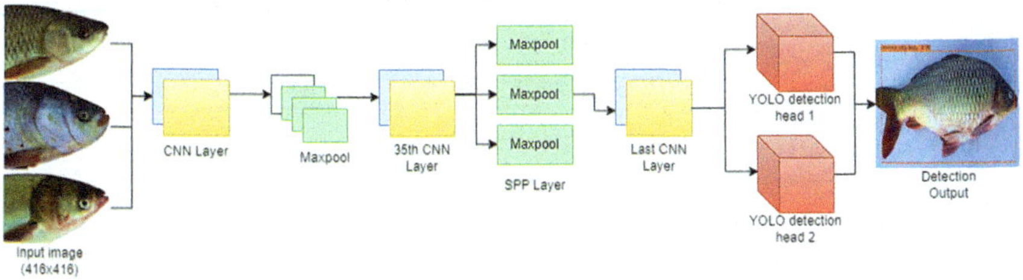

Fig. 2 Proposed Tiny-YOLOv4-SPP architecture

further enhance the capabilities of the Tiny-YOLOv4 algorithm, we applied a spatial pyramid pooling (SPP) layer after its thirty-fifth convolutional layer. The addition of spatial pyramid pooling layer aided the algorithm to extract rich and fine contextual features that helped in achieving better detection results. The added spatial pyramid pooling layer consists of three maxpool layers of filter size 3 × 3, 5 × 5, and 7 × 7. The detailed architecture of the proposed Tiny-YOLOv4-SPP is presented in Fig. 2. Since, the input image size of the proposed algorithm is 416 × 416, i.e., $n \times n$ therefore, its complexity can be stated as n^2.

6 Experiments and Results

The proposed Tiny-YOLOv4-SPP and other trained and tested YOLO variants were trained and tested on Google Colab. The configuration of the algorithms was set as Batch-32, Subdivision-4, learning rate-0.0001, and training iterations-10,000. Further, the algorithm was evaluated for average precision (AP) and mean average precision (map).

The evaluation metric average precision is the average of precision for each class under consideration. Whereas, mean average precision (map) is the mean of average precision defined using Eq. (1).

$$map = \frac{1}{n}\sum_{k=1}^{k=n} AP_k \quad (1)$$

In Eq. (1), n is the total number of classes and AP_k is the average precision for each class. The test results of the FishPak results with the Tiny-YOLOv4-SPP algorithm were overwhelming as for most of the fish species classes it achieved an average precision of 100%. Moreover, in terms of mean average precision (map), it achieved a value of 99.79% on the overall dataset. The results obtained were promising as they indicate a high number of true positive predictions by the Tiny-YOLOv4-SPP algorithm. The comparative results for mean average precision (map) with the proposed Tiny-YOLOv4-SPP, Tiny-YOLOv4, and Tiny-YOLOv7 are presented in Table 1.

Table 1 Comparison of Tiny-YOLOv4-SPP with other YOLO variants

Algorithm	Dataset	map (%)
Tiny-YOLOv4	FishPak	92.31
Tiny-YOLOv7	FishPak	85.66
Tiny-YOLOv4-SPP (Proposed)	FishPak	99.79

Table 2 Average precision results with the Tiny-YOLOv4-SPP

Fish species	AP(%)	Accuracy (mAP)%
Catla body	99.76	99.79
Catla head	98.92	
Catla scale	100	
Grass carp body	100	
Grass carp head	100	
Grass carp scale	100	
Common carp body	100	
Common carp head	99.98	
Common carp scale	100	
Rohu body	100	
Rohu head	98.98	
Rohu scale	100	
Silver carp body	100	
Silver carp head	99.22	
Silver carp scale	100	

The detailed results for average precision with the proposed Tiny-YOLOv4-SPP are presented in Table 2 and the qualitative results on a few image samples are shown in Fig. 3.

The results obtained with the proposed Tiny-YOLOv4-SPP were promising as it achieved better mean average precision (map) as compared to the Tiny-YOLOv4 and Tiny-YOLOv7 algorithm. The better values for mean average precision (map) indicates it be superior at detections. Although the Tiny-YOLOv4 and Tiny-YOLOv7 use a spatial pyramid pooling layer, however, the addition of one more spatial pyramid pooling layer in the feature extraction network in the Tiny-YOLOv4 architecture improved its detection accuracy by leaps and bounces. The proposed Tiny-YOLOv4-SPP is a lightweight object detector that utilized only 8.11 BFLOPs thereby, making it suitable for deployment to edge devices and performing real-time detections for various fish species. The proposed algorithm can be deployed for autonomous detection of fish species by the fisheries management bodies to identify the species of fishers and preserve the ecology and stock management. Moreover, as compared to the related work which is largely based on fish species classification, the present work presents an approach that can detect species of the fish. The fish species detection with the proposed algorithm can help the fisheries management system by providing

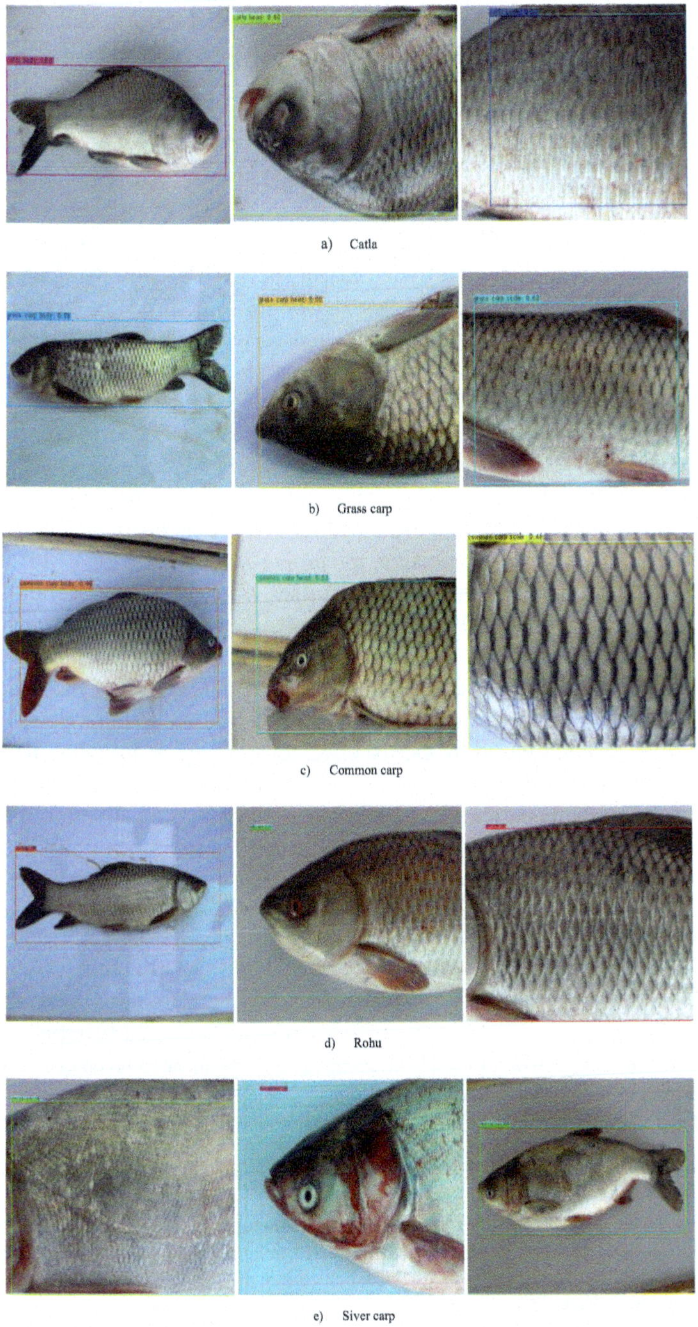

Fig. 3 Qualitative detection results using the Tiny-YOLOv4-SPP algorithm

autonomous mechanism to detect species of the fish and further segregate them for proper storage and other essential sustainable practices to preserve a specific species of the fish.

7 Conclusion

This work presents an approach to detect fish species detection using an improved variant of Tiny-YOLOv4 algorithm. The reason for selecting Tiny-YOLOv4 model is its optimal requirement of power and memory as compared to other YOLO versions, making it suitable for deployment on resource-constrained devices like underwater cameras or mobile phones. This work was carried out as an attempt to detect different species of fishes by focusing on their distinct features present at body, head, and scale. In order undertake this work, the FishPak dataset was annotated and trained and tested with the Tiny-YOLOv4 algorithm enhanced with applying spatial pyramid pooling layer and evaluated for average precision and mean average precision (map). For average precision, the proposed Tiny-YOLOv4-SPP algorithm achieved a value of 100% for most of the classes whereas, on the overall dataset, it achieved a map value of 99.79%. The results obtained were overwhelming in terms of classification and detection of different fish species. This approach can be used by fisheries department and fish reservoirs for automated detection of fish species. The Tiny-YOLOv4-SPP algorithm trained for fish species can be deployed to systems such as Jetson nano and used by the fisheries department for autonomous detection of fish species based on the features of body, head, and scale. The future work in regards to the present work can be extended to the usage of attention mechanism with the proposed algorithm; usage of transformer layers in place of CNN feature extractor; and enhancing the dataset using the Generative Adversarial Networks.

References

1. Alsmadi MK, Almarashdeh I (2022) A survey on fish classification techniques. J King Saud Univ-Comput Inf Sci 34(5):1625–1638
2. Zhao T, Zhang G, Zhong P, Shen Z (2023) DMDnet: a decoupled multi-scale discriminant model for cross-domain fish detection 234:32–45
3. Liu Y, An D, Ren Y, Zhao J, Zhang C, Cheng J, Liu J, Wei Y (2023) DP-FishNet: dual-path pyramid vision transformer-based underwater fish detection network. 238:122018
4. Jahanbakht M, Azghadi MR, Waltham NJ (2023) Semi-supervised and weakly-supervised deep neural networks and dataset for fish detection in turbid underwater videos. 78:102303
5. Cui S, Zhou Y, Wang Y, Zhai L (2020) Fish detection using deep learning. Appl Comput Intell Soft Comput 2020:1–13
6. Varshney Y, Anand P, Krishna A, Nagrath P, Jain R (2021) Detection of COVID-19 by X-rays using machine learning and deep learning models. In: Singh PK, Wierzchoń ST, Tanwar S, Ganzha M, Rodrigues JJPC (eds) Proceedings of second ınternational conference

on computing, communications, and cyber-security. Lecture Notes in Networks and Systems. Springer, Singapore, vol 203
7. Kapdi RA, Patel JA, Patel J (2023) Brain tumor segmentation using fully convolution neural network. In: Singh Y, Singh PK, Kolekar MH, Kar AK, Gonçalves PJS (eds) Proceedings of ınternational conference on recent ınnovations in computing. Lecture Notes in Electrical Engineering, vol 1001. Springer, Singapore
8. Xu W, Matzner S (2018) Underwater fish detection using deep learning for water power applications. In: 2018 ınternational conference on computational science and computational ıntelligence (CSCI). IEEE, pp 313–318
9. Kandimalla V, Richard M, Smith F, Quirion J, Torgo L, Whidden C (2022) Automated detection, classification and counting of fish in fish passages with deep learning. Front Mar Sci 8:2049
10. Jalal A, Salman A, Mian A, Shortis M, Shafait F (2020) Fish detection and species classification in underwater environments using deep learning with temporal information. Ecol Inf 57:101088
11. Al Muksit A, Hasan F, Emon MFHB, Haque MR, Anwary AR, Shatabda S (2022) Yolo-fish: a robust fish detection model to detect fish in realistic underwater environment. Ecol Inf 72:101847
12. Knausgård KM, Wiklund A, Sørdalen TK, Halvorsen KT, Kleiven AR, Jiao L, Goodwin M (2022) Temperate fish detection and classification: a deep learning based approach. Appl Intell 1–14
13. Marrable D, Barker K, Tippaya S, Wyatt M, Bainbridge S, Stowar M, Larke J (2022) Accelerating species recognition and labelling of fish from underwater video with machine-assisted deep learning. Front Mar Sci 9:944582
14. Malik H, Naeem A, Hassan S, Ali F, Naqvi RA, Yon DK (2023) Multi- classification deep neural networks for identification of fish species using camera captured images. Plos one 18(4):0284992
15. Anusuya V, Vignesh Saravanan K, Vishnu Praba V (2023) Real-time face mask detection using convolution neural network and computer vision. In: Tanwar S, Wierzchon ST, Singh PK, Ganzha M, Epiphaniou G (eds) Proceedings of fourth ınternational conference on computing, communications, and cyber-security CCCS 2022. Lecture Notes in Networks and Systems. Springer, Singapore, vol 664
16. Du J (2018) Understanding of object detection based on cnn family and yolo. J Phys Conf Ser 1004:012029
17. Jiang P, Ergu D, Liu F, Cai Y, Ma B (2022) A review of yolo algorithm developments. Procedia Comput Sci 199:1066–1073
18. Stark T, S¸tefan V, Wurm M, Spanier R, Taubenb¨ock H, Knight TM (2023) Yolo object detection models can locate and classify broad groups of flower-visiting arthropods in images. Sci Rep 13(1):16364
19. Shah SZH, Rauf HT, IkramUllah M, Khalid MS, Farooq M, Fatima M, Bukhari SAC (2019) Fish-Pak: fish species dataset from Pakistan for visual features based classification. Data Brief 27:104565
20. Orofino S, McDonald G, Mayorga J, Costello C, Bradley D (2023) Opportunities and challenges for improving fisheries management through greater transparency in vessel tracking. ICES J Mar Sci 80(4):675–689

To Identify a ML and CV Method for Monitoring and Recording the Variables that Impact on Crop Output

Sompal and Randeep Singh

Abstract This study aims to identify a suitable Machine Learning (ML) and Computer Vision (CV) approach for monitoring and recording the variables that influence crop output. Agricultural production relies heavily on weather conditions, which are intricately linked to each other. Climate change plays a crucial role in causing biotic and abiotic stresses on plants, resulting in a detrimental impact on global agriculture. This paper aims to assess the influence of climate change, specifically the impacts of rainfall and temperature, on production. In this study, a CNN-LSTM machine learning model is utilized to investigate the impact of climate change on production. Climate change is a worldwide phenomenon that presents significant challenges across various sectors, including agriculture, manufacturing, and energy production. The aim of this research is to evaluate the effects of climate change on production efficiency, resource allocation, and overall performance.

Keywords Agricultural production · Climate change · Machine learning · Computer vision · Rainfall · Temperature · CNN-LSTM

1 Introduction

Climate change presents urgent global concerns and imposes substantial obstacles across diverse sectors, including agriculture, energy, and manufacturing. Understanding and quantifying the impact of climate change on production systems is crucial for designing effective mitigation and adaptation strategies. Recent advances in machine learning have provided important tools for analyzing complex information and gaining significant insights. These findings are critical for comprehending the complex links between climate change and production outcomes [1]. Climate change is one of the most urgent worldwide environmental issues, with its consequences predicted to persist and worsen in the next decades [2].

Sompal (✉) · R. Singh
IEC University Baddi, Baddi, India
e-mail: sompal26@gmail.com

P. J. S. Gonçalves et al. (eds.), *Proceedings of Fifth International Conference on Computing, Communications, and Cyber-Security*, Lecture Notes in Networks and Systems 1128, https://doi.org/10.1007/978-981-97-7371-8_29

One of the most significant issues facing humanity now is climate change [3]. Recent decades have seen a number of natural disasters brought on by climate change, including extreme weather, unforeseen temperature changes, and variations in rainfall. As a result, there is a greater knowledge of and commitment to combating climate change [4]. Agriculture productivity is significantly impacted by climate change, which also increases the risk of famine and food insecurity. The latter is a significant problem in regions that frequently experience droughts or other natural disasters. Climate elements that affect crop yield include rainfall, air temperature, humidity, and solar radiation. Multiple research studies have consistently demonstrated that crop yields and food security are influenced by various global and regional climate indicators [5].

Global industrialization will increase greenhouse gas (GHG) emissions, lead to a rise in global temperatures, and have an adverse impact on the environment. The majority of nations intend to reach carbon neutrality by 2050–2070, while just 4.5% of nations have already done so [6]. Due to the overuse of fossil fuels, the shipping industry also contributes to GHG emissions. In 2018, the shipping industry produced roughly one billion metric tons of CO_2 equivalents of greenhouse gas emissions, or about 3% of all anthropogenic emissions worldwide [7].

Climate change is an ongoing phenomenon, not merely a future occurrence. It has now been indisputably established as a reality, presenting humanity with its greatest challenge. Consequently, addressing climate change has become the most urgent concern of our time. The global average temperature has unmistakably risen by 0.30–0.60 °C over the past century, while the sea level has experienced a rise of 10–20 cm during the same period, attributed to an evolving and vulnerable climate [8].

1.1 Climate Change and Its Impacts

Changes in weather patterns, including variations in temperature, precipitation, and the occurrence of extreme events, characterize climate change. These changes are the result of human activities, specifically the emission of greenhouse gases. These changes have far-reaching consequences for various sectors, including agriculture, where altered growing conditions, increased pest pressures, and changes in water availability can affect crop yields and food security. Similarly, climate change affects energy production, as changing weather patterns influence renewable energy generation and the performance of traditional power plants. Understanding the multifaceted impacts of climate change on different production systems is essential for devising strategies to mitigate risks and adapt to new conditions [9].

The viewpoints that are explicit and revocable can be used to arrange the effects of elevation change. Because of the global temperature rise, some overall impacts can be encouraged, such as less severe winters and more vegetation in high altitude regions. However, compared to the superior effects, the antagonistic (negative) effects clearly come out on top. Some of the negative consequences brought on by elevation change, particularly when combined with sanctifying through water and an Asian aesthetic, can be structured as follows. Some causes are listed as follows:

- Global warming has an impact on nature.
- Pollution and excessive use of electronic devices like air conditioners and refrigerators.
- The majority of scientific experiments include nuclear power.
- Lake pollution is one of the primary causes of sea pollution.
- The indiscriminate use of fuel on a global scale.
- There is a rapid melting of the glaciers and this is the primary reason why the water level rose.

1.2 Factors Affecting the Crop Production

There are several factors that affect the production of agriculture very highly.

- **Variation of Crop Yield with Rainfall**: A key element in the production of agriculture is rainfall. Rainfall variations may have an impact on crop productivity. A high rainfall rate may contribute to low productivity. There shouldn't be too much rain. The optimal rainfall range is between 300 and 600 mm, which may result in exceptionally high or ordinary output.
- **Variation of Humidity Factor**: The metrological qualities have a considerable impact on the production of agricultural items. Humidity is one of the most important metrological characteristic components. Any crop will grow more effectively in humid conditions.

2 Literature Review

Singh et al. [1] investigated the association between weather and agricultural production. They discovered a clear link between these two elements, with climate change emerging as a major cause of both biotic and abiotic pressures on plants. These stressors, in turn, have a negative impact on world agricultural output. Climate change has an effect on agricultural land in a number of ways, including periodic variations in temperatures and rainfall, weed changes, microbiological activity, heat waves, and changes in the quantity of CO_2 or ozone in the atmosphere. Since the warnings about climate change are having negative effects on agricultural production and are negotiating the system of global food security, the study's focus has been heavily drawn to these issues. Machine learning (ML) is one of the new techniques that have been developed as a result of the challenges associated with extracting information from unstructured datasets. ML may be efficiently used to integrate information with agricultural yield estimation.

Shams et al. [10] investigated a Model Based on Machine Learning to Predict Temperature Under Climate Change Effects, emphasizing the significance of climate change forecasting. They forecast temperature using ten features, including year, month, and atmospheric components, using machine learning (ML). Over the past century, global temperatures have risen, causing rising sea levels and more catastrophic weather occurrences. SVM, K-Nearest Neighbor, Random Forest, Decision Tree, Linear Regression, and Cat Boost Regressor were among the ML regression models used in this work to investigate the relationship between average world temperature and other parameters. The Cat Boost Regressor demonstrated the best performance among the models evaluated, with MAE, RMSE, MSE, and R^2 values of 0.0036, 0.054, 0.003, and 92.40%, respectively.

Hasegawa et al. [11] explored the importance of accurate estimations of the Impacts of changing the climate on crop production in determining the sustainability of food systems. On a regional, global, and site-specific scale, crop simulation studies have been carried out for nearly 40 years, and they have been critical data sources for analyzing the implications of climate change. However, the vast amount of data that this research produced has not been made accessible to the general public. Here, they establish a worldwide dataset by combining information from previously published meta-analyses with details discovered from a recent literature search that includes current crop models. The new worldwide dataset is based on 8703 simulations from 202 studies released between 1984 and 2020. In addition to the foregoing, this report contains information on geographical regions, current temperatures, precipitation levels, predicted temperature and precipitation changes, and the influence on four important crop yields in 91 nations. This extensive dataset is a great resource for enhancing data-centric machine learning algorithms and provides a solid foundation for quantitatively studying the influence of climate change on food output in the twenty-first century.

Habib-ur-Rahman et al. [12] examined how climate change threatens food-insecure agricultural production, notably in Asia. The study found that drought, heat waves, unpredictable rainfall, storms, floods, and new insect pests hurt farmers' livelihoods. Climate projections suggest a high temperature increase, variable rainfall patterns with rising intensity, and variability in extreme event climatic trends. The authors conducted a case study to examine climate change's impact and design mid-century (2040–2069) adaptation strategies for rice and wheat crops, vital for food production. Climate change scenarios anticipated 14.1–17.2% rice and wheat yield losses. The report recommends climate-smart and resilient agriculture for productivity. It reviews the literature on climate change's negative effects on agricultural productivity, highlighting Asia's major issues and sustainable agriculture possibilities. Addressing agricultural challenges and opportunities has several options. These include Legume-based crop rotation, mixed livestock systems, agroforestry, climate-resilient plant and animal varieties, decision support systems, and early warning systems, smart technologies for climate, carbon sequestration, energy, water, and soil management, and biodiversity enhancement. These methods can reduce climate change, increase agricultural productivity, and guarantee food security.

Haq et al. [13] analyzed environmental parameters using AI and ML. This study uses the Long Short-Term Memory model to forecast major environmental factors using a deep neural network. Hydrological models and forest expansion predictions require correct snow cover and NDVI predictions. Due to its adaptive processing, time series forecasting uses artificial neural networks (ANN), such as LSTM and RNN. LSTM RNNs can detect long-term dependencies. The authors use a coarse-to-fine approach, examining relevant studies and using LSTM analysis on a Himachal Pradesh dataset. The dataset includes 2001–2017 temperature, snow cover, and vegetation index data. The study found that the suggested system's tools and approaches speed up environmental factor assessment, adjustment, and improvement.

Kandasammy et al. [14] conducted a study to investigate the connection between climate variability and the substantial variation in precipitation and dry land cultivation. The objective was to assess the susceptibility of dry land producers to climate change and aid in their adaptation. Understanding the influential factors in emerging nations is essential for local policy research. This study employed the Vulnerability index and Machine Learning techniques to assess the vulnerability levels and factors impacting the dry land population in Southern India. The ML analysis revealed that "engagement in awareness campaigns, food income sources, farm size, and education level of dry land farmers" were significant determinants of community vulnerability effectively reducing the vulnerability of dry land producers by identifying coping strategies. The findings of this study suggest the need for future research and policy development in order to assure the micro-level sustainability of dry land producers in the face of climate change.

Crane et al. [15] utilized machine learning to estimate crop yields and assess the impact of climate change. To predict climate change's effects on agriculture, the author emphasizes crop yields' weather dependence. The article introduces a semi-parametric deep neural network yield modeling method. The research shows the efficacy of this method for forecasting corn yields in years that were not included in the model training over conventional statistical techniques and fully nonparametric neural networks using data on Midwest maize yields. In high-dimensional datasets, this methodology accounts for complex nonlinear interactions, parametric structure, and undiscovered cross-sectional heterogeneity. The study also examines how climate change affects corn yield using climate model scenarios. Climate change reduces corn output, although less than classical statistical approaches predicted. In the hottest places and scenarios, the proposed method appears less pessimistic (Table 1).

3 Research Methodology

This innovative research strategy uses a novel machine learning approach to analyze the effects of climate change on production in depth, leading to more accurate and insightful results. The methodology centers on a modified CNN-LSTM model, which combines the strengths of both CNN and LSTM to enhance the analysis process. By leveraging this advanced ML technique, researchers are able to comprehensively examine how climate change influences various aspects of production, such as crop

Table 1 Comparison of literature review

Author and year	Method	Focus	Key findings
Singh et al. (2023)	Machine Learning (ML)	Impact of climate change on agricultural production	Changes in precipitation, temperature, weeds, microbiological activity, heat waves, CO_2 and ozone levels, and heat waves all have an impact on agricultural land as a result of climate change. The research focuses on the use of machine learning to integrate data for agricultural productivity estimation
Shams et al. (2023)	Machine Learning (ML)	Predicting temperature under climate change effects	ML regression models were used to predict temperature based on ten features. The Cat Boost Regressor performed the best, with low errors and a determination coefficient of 92.40%. The study emphasizes the relevance of climate change forecasting and its impact on global temperatures and weather events
Hasegawa et al. (2022)	Crop simulation and data analysis	Evaluating the effects of climate change on crop yield	The study establishes a global dataset combining information from meta-analyses and recent literature searches. It includes data on temperature, precipitation, and crop yields in 91 countries under various emission scenarios. The dataset supports data-driven machine learning applications and provides insights into how climate change affects crop yield
Habib-ur-Rahman et al. (2022)	Impact assessment and adaptation strategies	Climate change's threat to food-insecure agricultural production	The research investigates the detrimental impacts of climate change on agricultural productivity in Asia, focusing on rice and wheat crops. It projects yield losses and suggests adaptation strategies such as climate-smart agriculture, early warning systems, and biodiversity enhancement. The objectives are to reduce climate change, boost agricultural productivity, and guarantee food security

(continued)

Table 1 (continued)

Author and year	Method	Focus	Key findings
Haq et al. (2022)	Artificial Neural Networks (LSTM)	Forecasting and analyzing environmental factors	The study utilizes LSTM models to forecast environmental factors such as temperature, snow cover, and vegetation index. The proposed system's tools and approaches accelerate the assessment, adjustment, and improvement of environmental factors
Kandasammy et al. (2022)	Vulnerability assessment and Machine Learning (ML)	Assessing vulnerability and factors affecting dryland farmers	ML techniques were employed to assess the vulnerability of dryland farmers in Southern India. Factors such as awareness campaigns, food income sources, farm size, and education level were found to influence vulnerability. The study highlights the importance of identifying coping strategies and suggests further research and policy development for the sustainability of dryland farmers in a changing climate
Crane et al. (2018)	Deep Neural Network Yield Modeling	Crop yield prediction and climate change impact assessment	The paper introduces a semi-parametric deep neural network method for predicting crop yields and assessing climate change impacts. The method outperforms classical statistical approaches and shows that climate change reduces corn yield, although less pessimistically than predicted by classical methods

yields, industrial output, and natural resource availability. The utilization of the CNN-LSTM model enables the identification of intricate patterns and relationships within vast datasets, allowing for a deeper understanding of the complex interplay between climate change and production.

3.1 Machine Learning Models

Multiple research studies have substantiated the significance of machine learning as an essential tool for analyzing the impact of climate change. Machine learning, as a technology, plays a vital role in assisting farmers in minimizing agricultural losses by providing them with comprehensive crop guidance and valuable insights. In this particular study, a modified CNN-LSTM model was investigated. The selection of

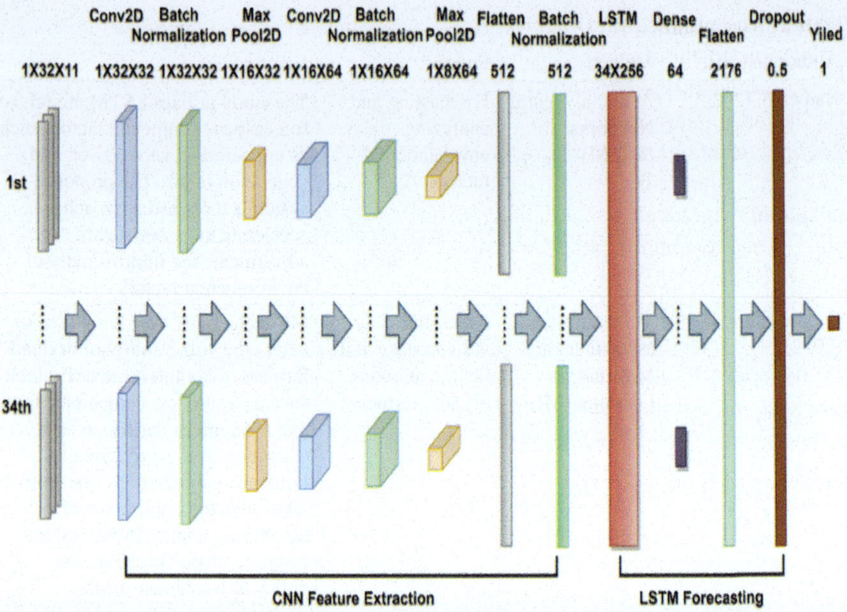

Fig. 1 The architecture of the proposed CNN-LSTM model

this approach was based on the quantitative nature of the predictions and the scale of the dataset, prioritizing numeric variables over categorical ones.

CNN-LSTM: The paper proposed a CNN-LSTM network based on CNN and RNN's achievements. CNN can infer numerous details from a picture like the human brain. An LSTM may bridge significant input latencies across any time span. LSTM improves the analysis of crop growing cycles of varying durations by presenting temporal patterns at different frequencies.

In addition to analyzing performance parameters such as accuracy, specificity, F1-score, and support measure value, this revised CNN-LSTM model would be utilized to make predictions about crop records (Fig. 1).

4 Result and Discussion

4.1 Climate-Based Agricultural Production

Agricultural production occurs through a cyclical process, where considerations are made for climate change and the planting of crops. This production process aligns with distinct seasons, namely the monsoon, winter, and summer crop production systems. As a result, agricultural production adapts and recycles in accordance with weather patterns. However, climate change poses a significant threat to agricultural output. The reliance on the monsoon for successful cultivation makes agriculture vulnerable to the impacts of climate change. The consequences of this shift can be witnessed in the form of reduced soil productivity caused by climate change. For instance, elevated temperatures during summer can lead to food shortages and flooding during the rainy season (Table 2 and Fig. 2).

The first diagram depicts the comparative temperatures over a span of ten years, indicating which years experienced higher or lower temperatures in relation to each other. The most significant temperature increase observed between 2010 and 2020 amounted to 0.70 degrees Celsius. The rise in temperature can be attributed primarily to the devastating impact of deforestation and the destruction of arable lands. Consequently, the overall temperature continues to rise steadily. Another contributing factor to this phenomenon is the decrease in rainfall patterns. The second diagram (Fig. 3), focusing on rainfall data, showcases the years with the lowest precipitation levels in

Table 2 Temperature and rainfall

Years	Temperature (+increase, −Decrease) in °C	Rainfall (+increase, −Decrease)
2010–2011	+0.7 °C	−120
2011–2012	+0.6 °C	−80
2012–2013	+0.7 °C	−150
2013–2014	+0.7 °C	−180
2014–2015	+0.6 °C	−220
2015–2016	+0.5 °C	−70
2016–2017	+0.5 °C	−200
2017–2018	+0.9 °C	−120
2018–2019	+0.5 °C	−170
2019–2020	+0.4 °C	+50

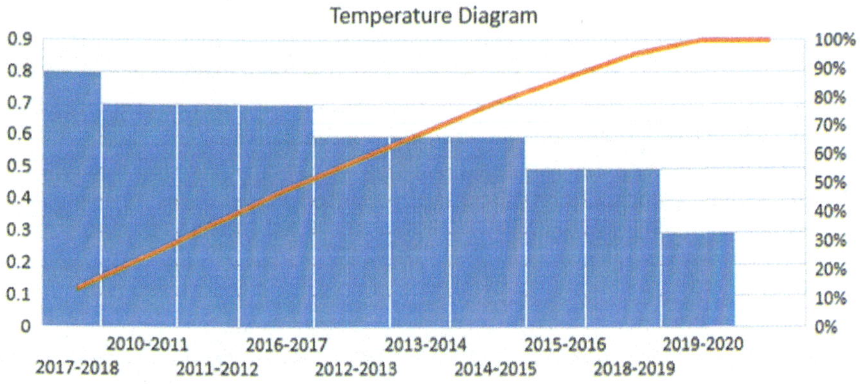

Fig. 2 Comparative temperatures over a span of ten years

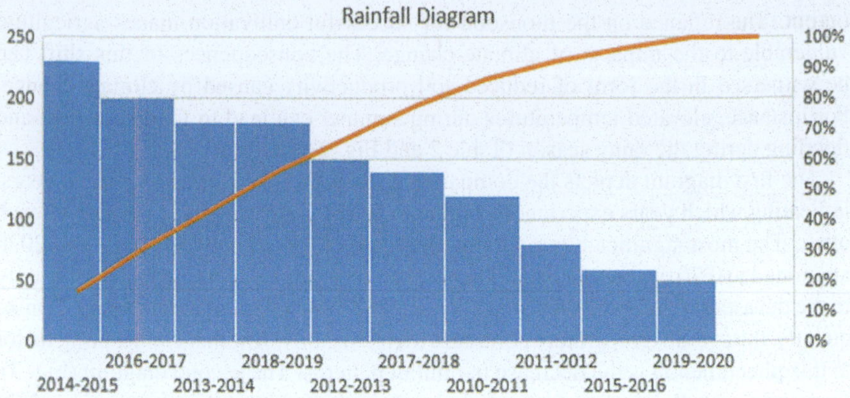

Fig. 3 Comparative rainfall data over a span of ten years

the past decade. Notably, the highest recorded rainfall in 2014–2015 exceeded the average by 800 mm.

5 Conclusion

This study utilized machine learning techniques, specifically the CNN-LSTM model, to analyze the impact of climate change on agricultural production. The findings highlight the significant challenges posed by climate change, such as altered weather patterns, increased biotic and abiotic stresses on plants, and changes in water availability. These factors have detrimental effects on crop yields, food security, and overall production efficiency. The research demonstrated that temperature and rainfall are crucial factors affecting agricultural productivity, with temperature rising and precipitation fluctuating over a ten-year period. These changes disrupt planting seasons, reduce soil productivity, and contribute to food shortages and flooding. By utilizing the CNN-LSTM model, the study provided a comprehensive understanding of how climate change affects production outcomes, highlighting the need to address climate change and develop effective strategies for adaptation and mitigation. Leveraging machine learning and advanced analytics can deepen our understanding of climate change's complex impacts and facilitate informed decisions for sustainable and resilient agricultural systems.

References

1. Singh S, Suresh Babu KV, Singh S (2023) Machine learning approach for climate change impact assessment in agricultural production. Visual Tech Clim Change Mach Learn Artif Intell 317–335
2. Zhang N et al (2020) A dynamic analysis of the global warming potential associated with air conditioning at a city scale: an empirical study in Shenzhen, China. Environ Impact Assess Rev 81:106354
3. Farajzadeh Z, Ghorbanian E, Tarazkar MH (2022) The shocks of climate change on economic growth in developing economies: evidence from Iran. J Clean Prod 372:133687
4. Ojo TO, Baiyegunhi LJS (2021) Climate change perception and its impact on net farm income of smallholder rice farmers in South-West, Nigeria. J Clean Prod 310:127373
5. Javadinejad S, Eslamian S, Ostad-Ali-Askari K (2021) The analysis of the most important climatic parameters affecting performance of crop variability in a changing climate. Int J Hydrol Sci Technol 11(1):1–25
6. Chen L et al (2022) Strategies to achieve a carbon neutral society: a review. Environ Chem Lett 20(4):2277–2310
7. Watanabe MD, Barbosa FC, Cavalett O (2022) Climate change mitigation of drop-in biofuels for deep-sea shipping under a prospective life-cycle assessment. J Clean Prod 364:132662
8. Rahman MM, Opu RK, Sultana A, Riyad MRA (2020) Climate change scenarios and analysis of temperature and rainfall intensity in Faridpur district, Bangladesh
9. Kumar A et al (2020) Effects of climate change on agriculture
10. Shams, Mahmoud Y et al (2023) A machine learning-based model for predicting temperature under the effects of climate change. In: The power of data: driving climate change with data science and artificial intelligence innovations. Springer Nature Switzerland, Cham, pp 61–81
11. Hasegawa T, Wakatsuki H, Ju H, Vyas S, Nelson GC, Farrell A, Makowski D (2022) A global dataset for the projected impacts of climate change on four major crops. Sci Data 9(1):58
12. Habib-ur-Rahman M et al (2022) Impact of climate change on agricultural production; issues, challenges, and opportunities in Asia. Front Plant Sci 13
13. Haq MA et al (2022) Analysis of environmental factors using AI and ML methods. Sci Rep 12(1):13267
14. Kandasamy SUL, Singh PK, Swain DK (2022) Climate change vulnerability assessment of dryland farmers and factors Identification using machine learning techniques. Local Environ 27(7):824–846
15. Crane-Droesch A (2018) Machine learning methods for crop yield prediction and climate change impact assessment in agriculture. Environ Res Lett 13(11):114003
16. Holzman ME, Carmona F, Rivas R, Niclòs R (2018) Early assessment of crop yield from remotely sensed water stress and solar radiation data. ISPRS J Photogrammetry Remote Sens 145:297–308
17. Shahhosseini M, Hu G, Archontoulis SV (2020) Forecasting corn yield with machine learning ensembles. Front Plant Sci 11:1120
18. Shahhosseini M, Martinez-Feria RA, Hu G, Archontoulis SV (2019) Maize yield and nitrate loss prediction with machine learning algorithms. Environ Res Lett 14(12):124026
19. Gandhi N, Petkar O, Armstrong LJ (2016, July) Rice crop yield prediction using artificial neural networks. In: 2016 IEEE technological innovations in ICT for agriculture and rural development (TIAR). IEEE, pp 105–110
20. Nevavuori P, Narra N, Lipping T (2019) Crop yield prediction with deep convolutional neural networks. Comput Electron Agric 163:104859
21. Saini P, Nagpal B (2022) Efficient crop yield prediction of Kharif crop using deep neural network. In: 2022 International conference on computational intelligence and sustainable engineering solutions (CISES), Greater Noida, India, pp 376–380. https://doi.org/10.1109/CISES54857.2022.9844369
22. Gopal PM, Bhargavi R (2019) A novel approach for efficient crop yield prediction. Comput Electron Agric 165:104968

23. Whetton R, Zhao Y, Shaddad S, Mouazen AM (2017) Nonlinear parametric modelling to study how soil properties affect crop yields and NDVI. Comput Electronics Agric 138:127–136
24. Sarker IH (2021) Deep learning: a comprehensive overview on techniques, taxonomy, applications and research directions. SN Comput Sci 2:1–20
25. Sun J, Lai Z, Di L, Sun Z, Tao J, Shen Y (2020) Multilevel deep learning network for county-level corn yield estimation in the US corn belt. IEEE J Sel Topics Appl Earth Obs Remote Sens 13:5048–5060

Credit Card Fraud Detection Using Random Forest and K-Nearest Neighbors (KNN) Algorithms

Abdulaziz Abdulrhman Alhabib, Abdulaziz Fae Alasiri, Mazen Bunayan Alharbi, Sultan Ahmad, and A. E. M. Eljialy

Abstract In this paper, we explored the application of both the Random Forest and K-Nearest Neighbors (KNN) algorithms for fraud detection. Fraudulent activities pose significant threats to various industries, making their detection crucial. We reviewed related works on fraud detection, highlighting the effectiveness of both the Random Forest and KNN methods. We proposed a system utilizing a combination of Random Forest and KNN models for credit card fraud detection, detailing its architecture and methodology. The system preprocesses and splits data, trains the models, and evaluates their performance on a testing set. We analyzed the results, demonstrating the accuracy, robustness, and scalability of the proposed system. Additionally, we presented data on the most common hours for fraudulent transactions and the top jobs associated with them. Overall, this study emphasizes the efficacy of utilizing both Random Forest and KNN algorithms in mitigating the risks associated with fraud and offers a valuable tool for safeguarding against financial malfeasance.

Keywords Random Forest · Credit card · Fraud detection · Transactions · KNN

A. A. Alhabib · A. F. Alasiri · M. B. Alharbi · A. E. M. Eljialy
Department of Information Systems, College of Computer Engineering and Sciences, Prince Sattam Bin Abdulaziz University, P.O.Box. 151, Alkharj 11942, Saudi Arabia

S. Ahmad (✉)
Department of Computer Science, College of Computer Engineering and Sciences, Prince Sattam Bin Abdulaziz University, P.O.Box. 151, Alkharj 11942, Saudi Arabia
e-mail: s.alisher@psau.edu.sa

Department of Computer Science and Engineering, University Center for Research and Development (UCRD), Chandigarh University, Gharuan, Mohali 140413, Punjab, India

P. J. S. Gonçalves et al. (eds.), *Proceedings of Fifth International Conference on Computing, Communications, and Cyber-Security*, Lecture Notes in Networks and Systems 1128, https://doi.org/10.1007/978-981-97-7371-8_30

1 Introduction

Fraud has become an increasingly prevalent issue in today's business landscape, posing significant threats to organizations across various industries. As businesses continue to rely on digital platforms, the risk of fraudulent activities, such as identity theft, financial fraud, and cybercrime, has increased. The payment process has evolved from a physical transaction to a digital one including electronics [2]. According to a survey conducted by EisnerAmper, a leading accounting firm in the U.S., financial statement fraud had an average median loss of $800,000 in 2018. This amount is over three times higher than the monetary loss of corruption ($250,000) and seven times higher than asset misappropriation ($114,000) [3]. Random Forest is a machine learning technique that holds considerable promise in detecting fraudulent transactions [4]. advanced data analysis and machine learning techniques have become indispensable. One such method that has gained prominence in recent years is the application of Random Forest, a powerful ensemble learning algorithm, in fraud detection. This method offers a robust and versatile approach to identifying fraudulent activities, leveraging the strength of multiple decision trees to improve accuracy and mitigate the risks associated with fraud. In this discussion, we will delve into the intricacies of using the Random Forest method for fraud detection, exploring its key principles, benefits, and real-world applications in safeguarding against financial malfeasance. Financial fraud has seriously damaged the world economy in recent decades, endangering the effectiveness and stability of capital markets. This problem was made worse by the early 2020 start of the COVID-19 pandemic, which caused unheard-before disruptions in the global financial markets and raised the risk of financial fraud. For example, compared to the prior year, April 2020 fraud rates in the UK increased by 33% for all financial products. Similarly, Fidelity National Information Services, a payment services provider for around 3,200 U.S. banks, reported a substantial 35% increase in the volume of fraudulent transactions in America in comparison to the previous year. Financial fraud has become an even more significant and pressing problem in the post-pandemic era, with far-reaching consequences for the financial sector and the global economy [6].

2 Related Works

GBAD approaches have been widely applied to fraud detection in various domains. Some of the most notable GBAD approaches for fraud detection include community-based anomaly detection, link-based anomaly detection, and subgraph-based anomaly detection. GBAD approaches have been shown to be effective in detecting various types of fraudulent activities, but they can be computationally expensive and may not be able to detect all types of fraudulent activities [1]. This paper [2] presents a fraud detection system for payment card transactions using aggregated features. The system was evaluated on both benchmark and real-world datasets, using a variety of machine learning methods. The results show that aggregated features can improve the performance of fraud detection systems. The best AUC (Area Under the Curve.) score achieved was 0.967 using GBT with aggregated features. The authors conclude that aggregated features are a valuable tool for fraud detection and that financial institutions should consider using them in their fraud detection systems. This paper [3] presents a systematic literature review of FSFD using ML and DM approaches. The authors reviewed 47 articles that met their criteria and identified the following key findings: Supervised learning algorithms are more widely used than unsupervised learning algorithms for FSFD. Supervised learning algorithms, such as support vector machines (SVMs), neural networks, and logistic regression, are trained on a dataset of labeled examples, where each example is labeled as either fraudulent or non-fraudulent. Once trained, the model can be used to predict whether a new example is fraudulent or non-fraudulent. On the contrary, unsupervised learning algorithms do not necessitate labeled data. Instead, they learn patterns in the data on their own. Unsupervised learning algorithms have been less explored for FSFD, but they have the potential to be useful for detecting fraudulent transactions that are different from any known fraudulent transactions. This paper [4] suggests and assesses machine learning models for the purpose of identifying fraudulent transactions in the context of Industry 4.0 and the digital economy. Five distinct models were compared by the authors: autoencoder, random forest, k-nearest neighbors, logistic regression, and decision tree. They discovered that logistic regression and random forest performed better than the other techniques. Furthermore, the outcomes might be enhanced by undersampling the data and utilizing principle component analysis (PCA) to decrease the number of features. The significance of feature selection and sample techniques in addressing business issues in the context of Industry 4.0 and the digital economy was also covered by the writers. They argued that these methods can be effective in detecting fraudulent activities. This research paper [5] presents a novel method that combines a deep learning architecture with a feature engineering framework to improve credit card fraud detection. Behavioral information for fraud detection models is captured by feature variables created by the framework called homogeneity-oriented behavior analysis (HOBA). Our paradigm, in contrast to earlier approaches, concentrates on analyzing behavior in homogenous transactions while taking into account the heterogeneity of credit card transactions. By adding metrics of relative monetary amounts and physical location distance, we also improve transaction aggregation approaches. Furthermore, we use a variety of extracted features and deep learning approaches to create fraud detection models that are more accurate. To assess how well different deep learning algorithms and conventional machine learning techniques perform with varying feature sets, comparative tests are carried out. The pandemic has had a significant impact on financial fraud detection practices. Fraudsters have adapted to the new landscape, and we have seen a rise in the volume and sophistication of fraud attacks. The shift to online and remote transactions has also created challenges for data collection and analysis such as (GNN). However, new technologies such as AI and ML can be used to develop more sophisticated fraud detection systems. Financial institutions need to be aware of the new challenges and opportunities that the pandemic has created, and they need to invest in new technologies and solutions to protect themselves and their customers from fraud [6]. The best credit card fraud detection technique for a particular organization will depend on their specific needs. Factors to consider include the type of fraud to be detected, the size and complexity of transaction data, the budget, and the risk tolerance. There is no flawless fraud detection technique. Fraudsters are constantly developing new methods to circumvent fraud detection systems. Therefore, it is important to use a combination of techniques and to [ML] regularly review and update fraud detection strategies [7]. This study is a quantitative inquiry into the association between multiple variables that are relevant to the auditor's fraud detection ability. Quantitative research is a systematic approach to investigating social problems by testing theories through the collection and analysis of numerical data (Creswell, 2014) [8]. The fact that India is the most active country in fraud detection research is particularly noteworthy. This suggests that India is at the forefront of developing innovative solutions to this global challenge. The prominence of Tongji University in China is also noteworthy. This institution is clearly playing a leading role in advancing fraud detection research. The fact that 160 scholars from 89 different countries and 160 different institutions have published fraud detection studies in different languages is a further indication of the global nature of the field. This is important because fraud is a global problem, and it requires a global response to develop effective detection and prevention strategies [9]. Support vector machines (SVM) and neural networks (NN) are the most popular ML algorithms used for fraud detection. Credit card fraud is noted as the foremost prevalent form of fraud in the literature. Unsupervised learning approaches, such as clustering, are less employed in the present literature but could be effective for investigating latent relationships and similarities. Unexplored or less studied algorithms, such as unsupervised learning approaches, could be further explored for fraud detection. Ensemble methods, which combine the predictions of multiple algorithms to improve accuracy, are a rising trend in the field, but more research is needed to evaluate their effectiveness for fraud detection [10]. A hybrid approach to fraud detection combines unsupervised outlier scores with a supervised fraud detection classifier. The novelty of the approach is the implementation and assessment of different levels of granularity for the definition of the outlier score. The results show that the hybrid approach improves the performance of the supervised classifier in terms of AUC-PR, but not in terms of Top-n precision. This suggests that the hybrid approach is better at identifying fraudulent transactions than at ranking them in order of likelihood of fraud [11]. A novel and sophisticated method for detecting internet financial fraud utilizes a distributed Big Data pipeline along with the Node2Vec graph embedding algorithm. The approach consists of four modules: data preprocessing, normal data feature extraction, graph embedding, and prediction. The authors evaluated the proposed approach on two real-world datasets and demonstrated its superior performance over other state-of-the-art methods in terms of precision, recall, F1-Score, and F2-Score [12]. In the electrical field, non-technical loss refers to unpaid energy consumption, negatively impacting suppliers' revenue. Multiple statistical and machine learning algorithms exist for detecting this issue, commonly assessed through confusion matrix results. A proposed methodology suggests resampling performance metrics from the confusion matrix to enhance algorithm comparisons. This approach is illustrated using three classifiers (support vector machine, nearest neighbor, and naive Bayes) with a case study involving electricity consumption data from a South African municipality. The significance lies in providing data analysts with a means to identify abnormal energy usage, evaluate fraud detection systems, and emphasize that the choice of measurement depends on the analysis's specific goal [13]. The proliferation of online trading technologies has heightened the occurrence of fraudulent activities within public trading companies, making fraud detection a critical priority in the financial sector. This study presents a fraud detection model developed especially for publicly traded corporations using datasets obtained from the Accounting and Auditing Enforcement Releases (AAERs) of the SEC. Using a computational finance methodology, the model is solved with a nonlinear activated Beetle Antennae Search (NABAS) algorithm, which is a variation of the meta-heuristic optimization process known as the Beetle Antennae Search (BAS) algorithm. Experimental findings indicate that the NABAS algorithm surpasses alternative methods in terms of both accuracy and efficiency when detecting fraudulent activities within datasets of public trading companies [14]. This research examines two types of rating fraud: objective fraud and subjective fraud. It examines the efficacy of blockchain technology in combating objective fraud while acknowledging its limitations in addressing subjective fraud. Finally, the study takes a deep dive into how strong blockchain-based reputation systems are against each kind of rating fraud [15]. The expansion of healthcare services has led to an increased need for effective health insurance fraud detection to ensure the optimal utilization of public funds. Traditional approaches to fraud detection primarily focus on the characteristics of individual visits, overlooking the behavioral relationships that exist across multiple visits made by patients. In this study, we propose a health insurance fraud detection model named MHAMFD, which leverages a multilevel attention mechanism. Our approach utilizes a heterogeneous information network (AHIN) to model diverse types of objects, their extensive attributes, and interactions within the healthcare domain. MHAMFD intelligently selects relevant neighboring nodes by considering the behavioral relationships at various levels of a patient's visit [16]. Fraud is a widespread concern within financial institutions, involving deception for unlawful gains. It can occur at any level, from senior managers to payroll employees, making detection complex. Traditionally, auditors have manually detected fraud, but this process is time-consuming. To automate detection, data mining, machine learning, and deep learning techniques are being employed. The fraud triangle, a concept in the classic audit model, is an important technique in analyzing fraud-related behaviors. This study reviews the past decade's research on fraud detection, focusing on the use of the fraud triangle and machine learning. The review reveals active investigation in the field, with many works identified using machine learning but lacking incorporation of the fraud triangle for more efficient analysis [17]. Financial fraud is a significant issue in the financial industry, with a profound impact. Data mining has played a crucial role in detecting credit card fraud in online transactions. However, credit card fraud detection poses challenges due to changing profiles of fraudulent and normal behaviors, as well as highly skewed datasets. The effectiveness of fraud detection depends on the variables and techniques employed. This study evaluates the effectiveness of several methods using credit card fraud data, such as logistic regression, K-nearest neighbors, random forest, naive Bayes, multilayer perception, AdaBoost, quadrant discriminative analysis, pipelining, and ensemble learning [18]. In an unbalanced dataset of 284,807 transactions, the research [19] examined the effectiveness of three classification algorithms for credit card fraud detection: Random Forest Classifier (RFC), Logistic Regression (LGR), and Gradient Boosting Classifier (GBC). The authors employed a range of evaluation metrics, encompassing accuracy, precision, recall, and F1-score. The results showed that RFC outperformed LGR and GBC in terms of all evaluation metrics. RFC achieved an accuracy of 99.99%, precision of 99.99%, recall of 99.99%, and F1-score of 99.99%. This investigation studies the application of random forest for financial fraud detection using Chinese-listed company data. The authors found that the random forest model with eight input variables achieved the highest accuracy of 88%, significantly outperforming other classification models, such as logistic regression, K-nearest neighbors, and support vector machines [20]. This study [21] optimized a random forest algorithm with hyperparameters to achieve high accuracy in fraud detection (98%). Imbalanced data was addressed using SMOTE oversampling. Evaluation metrics, including recall, precision, F1-score, and ROC curve, confirmed unbiased performance in detecting both fraud and non-fraud cases. This straightforward model is effective in addressing imbalanced data challenges in fraud detection. This study [22] introduces a novel approach using random forest for fraud detection on the IEEE-CIS Fraud dataset. Feature extraction, aided by RFECV, is highlighted as a crucial step to enhance model performance by eliminating redundant or highly correlated features. The random forest outperforms SVM and Logistic Regression, possibly due to the dataset's extreme class imbalance. The model achieved an accuracy of 97.4% and an AUC ROC score of 92.7% The speech highlights the importance of detecting credit card fraud due to the increasing number of fraudulent transactions. The main challenge is that merchants cannot verify if the customer is the actual cardholder, especially for transactions where the card or cardholder is not present. To address this, a method using the random forest algorithm is proposed to improve the accuracy of detecting fraudulent transactions. The algorithm verifies the dataset and the user's current dataset, leading to better results. The random forest algorithm is chosen for its high performance, precision, and ease of evaluation and support [23]. The statement discusses credit card fraud and the use of an e-commerce payment system to detect fraudulent transactions. The goal is to achieve 100% fraud detection while minimizing false positives. A standard numerical model is trained on cardholder behavior to determine the likelihood of fraud. If a transaction is not accepted with a high probability, it is flagged as fraudulent. The test perceptron plot helps evaluate the model's performance. Despite advancements, misclassifications of genuine customers as fraudsters remain an issue. SMOTE, a statistical technique, is used to balance the dataset. Random forest, combining multiple decision trees, achieves an accuracy of approximately 98.6% [24]. The research aimed to compare machine learning algorithms' performance in accurately distinguishing and classifying fraudulent and non-fraudulent credit card transactions using random

Table 1 Related works overview

Author	Method	Accuracy (%)
Liu [20]	Random Forest	88
Niveditha [23]	Random Forest	98.6
Shaohui [22]	AUC RUC	92.7
Itoo [24]	KNN	75
Chung [25]	KNN	98.7

undersampling (RUS). It also aimed to assess whether the performance improved. Logistic Regression (LR) outperformed Naïve Bayes (NB) and K-Nearest Neighbors (KNN). LR achieved the highest accuracy at 95%, followed by NB at 91%, and KNN at 75% [25]. The works [26–29] investigated K-nearest neighbor, linear discriminant analysis, and linear regression methods, achieving a recall of 1.0000, 0.9701, 1.0000, and 0.9362 on four datasets (Table 1).

3 Proposed System

The model is a credit card fraud detection system utilizing both random forest and K-Nearest Neighbors (KNN) models as shown in Fig. 1. The system begins by preprocessing the credit card dataset to ensure compatibility with the machine learning models. After preprocessing, the data is split into two sets: a training set and a testing set. The training set is utilized to train both the random forest and KNN models, while the testing set is reserved to evaluate their respective performances.

4 Methodology

As shown in Fig. 2, the most common hour for fraud transactions is between 10 PM and 3 AM. The number of fraudulent transactions is relatively constant throughout the rest of the day. This suggests that fraudsters are most active at midnight when they are most likely to be able to make fraudulent transactions without being detected [30]. It is also possible that fraudsters are more likely to target businesses during these hours, as they may have more access to sensitive information during this time [31, 32].

The bar chart in Fig. 3 presents the number of fraudulent transactions (in USD) for the top 10 job titles with the highest occurrence of fraud.

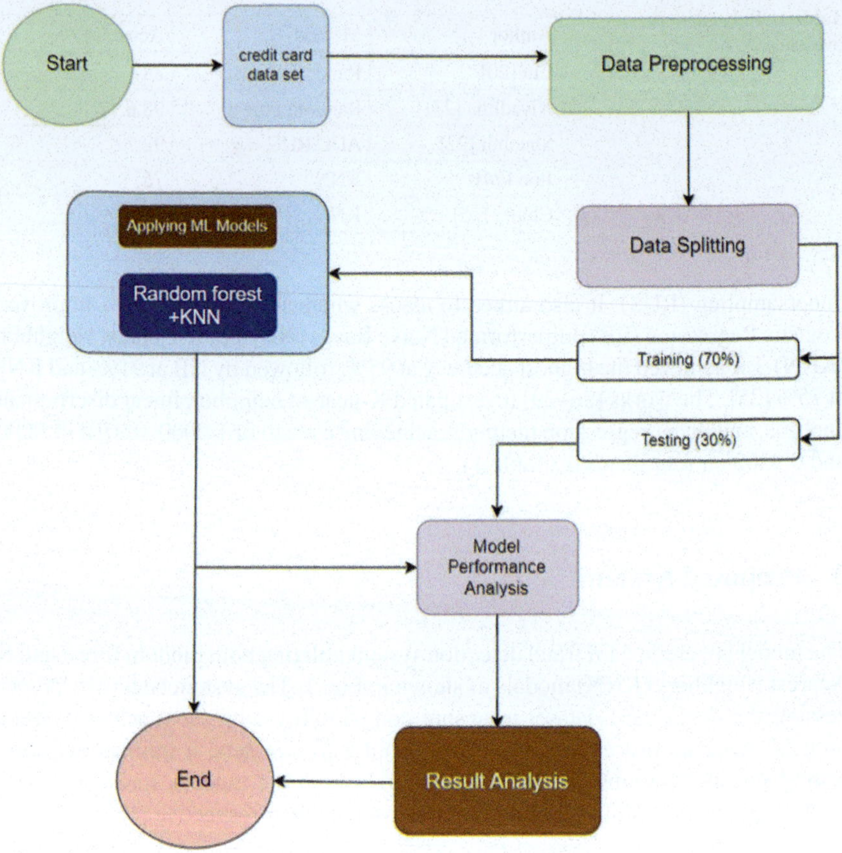

Fig. 1 Fraud prediction system architecture

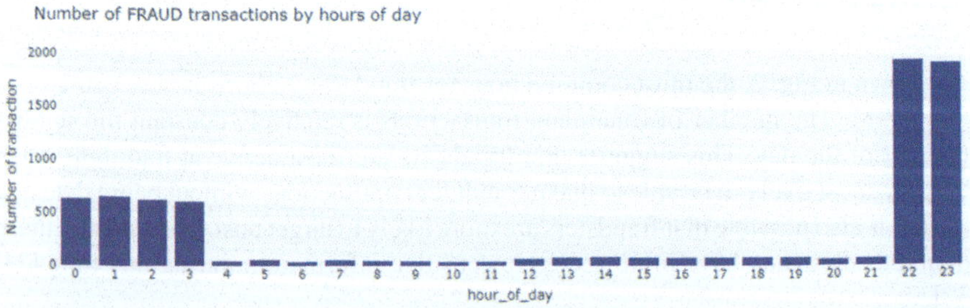

Fig. 2 Classification of fraudulent transactions based on hours

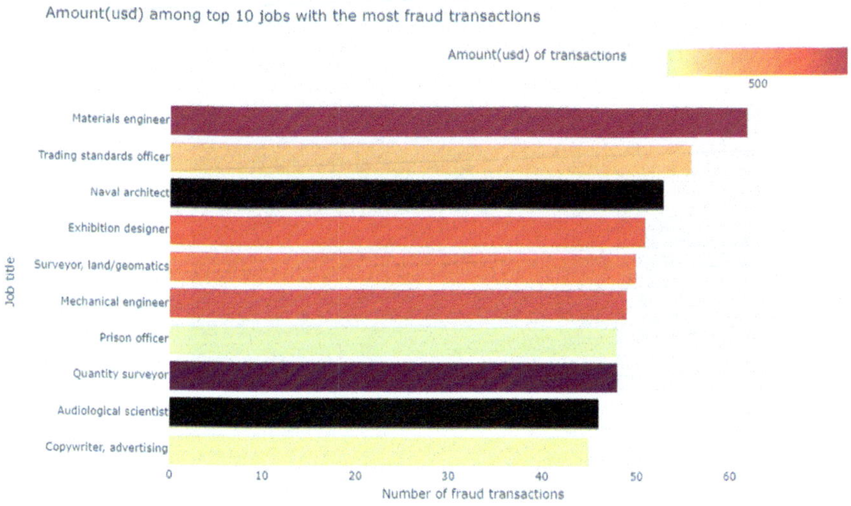

Fig. 3 Fraudulent transactions

As shown in Fig. 4, the fraud rate between genders reveals that a higher number of fraudulent cases occurred with female customers, the number is almost the same for both Males and Females.

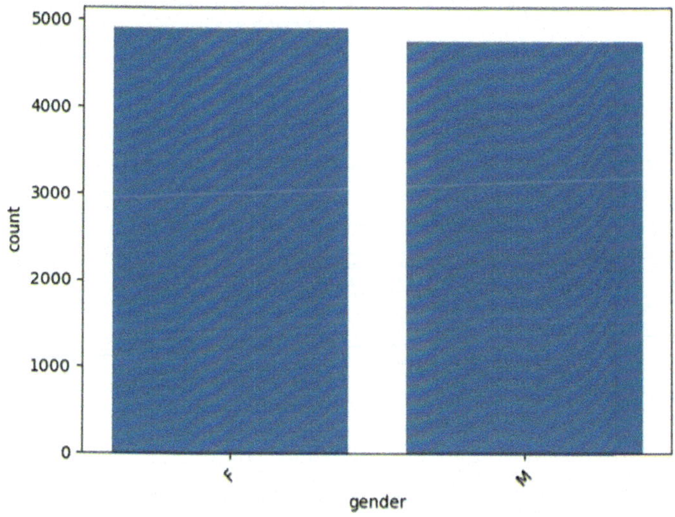

Fig. 4 Classification of fraudulent transactions based on gender

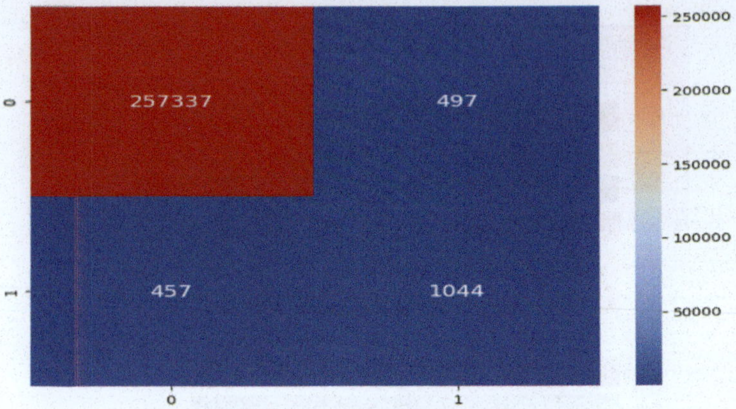


```
              precision    recall  f1-score   support

           0       1.00      1.00      1.00    257834
           1       0.68      0.70      0.69      1501

    accuracy                           1.00    259335
   macro avg       0.84      0.85      0.84    259335
weighted avg       1.00      1.00      1.00    259335

Accuracy: 0.9963984807295583
```

Fig. 5 Random forest algorithm result

5 Results

The classification report function provides a detailed summary of performance metrics such as precision, recall, F1-score, and support for each class in a classification task. On the other hand, the confusion matrix function displays a visual representation of true/false positives/negatives and calculates metrics like precision, recall, and F1-score to help assess the model's performance and identify areas that may need improvement (Figs. 5 and 6).

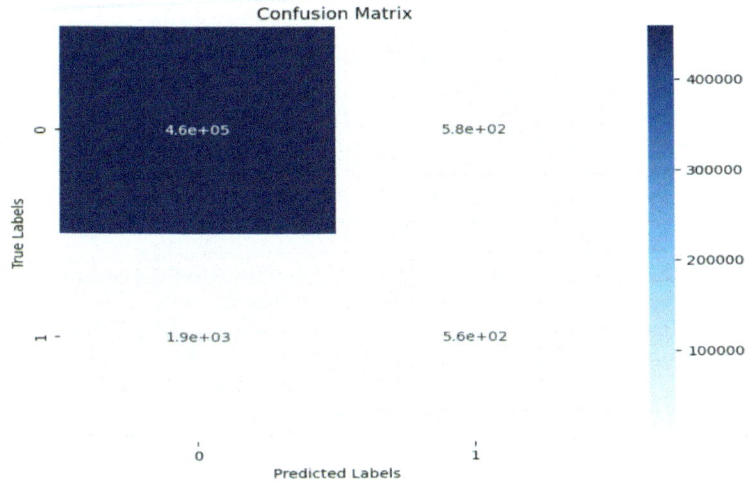


```
Accuracy: 0.994677164062112

Classification Report:
              precision    recall  f1-score   support

           0       1.00      1.00      1.00    460662
           1       0.49      0.23      0.31      2437

    accuracy                           0.99    463099
   macro avg       0.74      0.61      0.65    463099
weighted avg       0.99      0.99      0.99    463099

KNN Accuracy: 99.468%
```

Fig. 6 K-Nearest Neighbors (KNN) algorithm result

6 Conclusion

This paper has successfully demonstrated the effectiveness of Random Forest and KNN algorithms in detecting fraudulent activities. By leveraging the strengths of multiple decision trees through an ensemble learning approach, the proposed system achieved high accuracy and demonstrated its ability to mitigate risks associated with financial fraud. The K-Nearest Neighbors (KNN) algorithms model attained an accuracy rate of 99.46%, while the Random Forest model outperformed with an accuracy rate of 99.63%.

Acknowledgements We thank Prince Sattam Bin Abdulaziz University, Al Kharj, Saudi Arabia for help and support.

References

1. Pourhabibi T, Ong KL, Kam BH, Boo YL (2020) Fraud detection: a systematic literature review of graph-based anomaly detection approaches. Decis Support Syst 1(133):113303
2. Seera M, Lim CP, Kumar A, Dhamotharan L, Tan KH (2024) An intelligent payment card fraud detection system. Ann Oper Res 334(1):445–467
3. Ashtiani MN, Raahemi B (2021) Intelligent fraud detection in financial statements using machine learning and data mining: a systematic literature review. IEEE Access 13(10):72504–72525
4. Chang V, Di Stefano A, Sun Z, Fortino G (2022) Digital payment fraud detection methods in digital ages and Industry 4.0. Comput Electr Eng 100:107734
5. Zhang X, Han Y, Xu W, Wang Q (2021) HOBA: a novel feature engineering methodology for credit card fraud detection with a deep learning architecture. Inf Sci 1(557):302–316
6. Zhu X, Ao X, Qin Z, Chang Y, Liu Y, He Q, Li J (2021) Intelligent financial fraud detection practices in post-pandemic era. The Innovation 2(4)
7. Ileberi E, Sun Y, Wang Z (2022) A machine learning based credit card fraud detection using the GA algorithm for feature selection. J Big Data 9(1):24
8. Indrasti AW, Karlina B (2019) Determinants affecting the auditor's ability of fraud detection: internal and external factors (empirical study at the public accounting firm in Tangerang and South Jakarta Region in 2019). In: Annual international conference on accounting research (AICAR 2019) 2020 Mar 12. Atlantis Press, pp 19–22
9. Mansour AA, Ahmi A, Popoola OM, Znaimat A (2022) Discovering the global landscape of fraud detection studies: a bibliometric review. J Financ Crime 29(2):701–720
10. Ali A, Abd Razak S, Othman SH, Eisa TA, Al-Dhaqm A, Nasser M, Elhassan T, Elshafie H, Saif A (2022) Financial fraud detection based on machine learning: a systematic literature review. Appl Sci 12(19):9637
11. Carcillo F, Le Borgne YA, Caelen O, Kessaci Y, Oblé F, Bontempi G (2021) Combining unsupervised and supervised learning in credit card fraud detection. Inf Sci 1(557):317–331
12. Zhou H, Sun G, Fu S, Wang L, Hu J, Gao Y (2021) Internet financial fraud detection based on a distributed big data approach with node2vec. IEEE Access 26(9):43378–43386
13. Liao B, Huang Z, Cao X, Li J (2022) Adopting nonlinear activated beetle antennae search algorithm for fraud detection of public trading companies: a computational finance approach. Mathematics 10(13):2160
14. Pazi S, Clohessy CM, Sharp GD (2020) A framework to select a classification algorithm in electricity fraud detection. S Afr J Sci 116(9–10):1–7
15. Cai Y, Zhu D (2016) Fraud detections for online businesses: a perspective from blockchain technology. Financ Innov 2:20
16. Lu J, Lin K, Chen R, Lin M, Chen X, Lu P (2023) Health insurance fraud detection by using an attributed heterogeneous information network with a hierarchical attention mechanism. BMC Med Inform Decis Mak 23(1):62
17. Sánchez-Aguayo M, Urquiza-Aguiar L, Estrada-Jiménez J (2021) Fraud detection using the fraud triangle theory and data mining techniques: a literature review. Computers 10(10):121
18. Bagga S, Goyal A, Gupta N, Goyal A (2020) Credit card fraud detection using pipeling and ensemble learning. Procedia Comput Sci 1(173):104–112
19. Sopiyan M, Fauziah F, Wijaya YF (2022) Fraud detection using random forest classifier, logistic regression, and gradient boosting classifier algorithms on credit cards. JUITA: Jurnal Informatika 10(1):77–87
20. Liu C, Chan Y, Alam Kazmi SH, Fu H (2015) Financial fraud detection model: based on random forest. Int J Econ Finan 7(7)
21. Aburbeian AM, Ashqar HI (2023) Credit card fraud detection using enhanced random forest classifier for imbalanced data. In: International conference on advances in computing research. Springer Nature Switzerland, Cham, pp 605–616

22. Shaohui D, Qiu G, Mai H, Yu H (2021) Customer transaction fraud detection using random forest. In: 2021 IEEE international conference on consumer electronics and computer engineering (ICCECE). IEEE, pp 144–147
23. Kumar MS, Soundarya V, Kavitha S, Keerthika ES, Aswini E (2019) Credit card fraud detection using random forest algorithm. In: 2019 3rd international conference on computing and communications technologies (ICCCT). IEEE, pp 149–153
24. Itoo F, Meenakshi, Singh S (2021) Comparison and analysis of logistic regression, Naïve Bayes and KNN machine learning algorithms for credit card fraud detection. Int J Inf Technol 13(4):1503–1511
25. Chung J, Lee K (2023) Credit card fraud detection: an improved strategy for high recall using KNN, LDA, and linear regression. Sensors 23(18):7788
26. Malik A, Gehlot S, Vyas S (2022) Proposed framework for implementation of biometrics in banking KYC. In: International conference on computing, communications, and cyber-security. Springer Nature Singapore, pp 193–202
27. Dayal R, Nangia P, Vijh S, Kumar S, Agarwal S, Saxena S (2023) Development of chatbot retrieving fact-based information using knowledge graph. In: Proceedings of international conference on recent innovations in computing: ICRIC 2022, vol 1. Springer Nature Singapore, Singapore, pp 153–164
28. Gite S, Khatavkar H, Srivastava S, Maheshwari P, Pandey N (2021) Stock prices prediction from financial news articles using LSTM and XAI. In: Proceedings of second international conference on computing, communications, and cyber-security: IC4S 2020 2021. Springer Singapore, pp 153–161
29. Ahmad S, Jha S, Alam A, Yaseen M, Abdeljaber HA (2022) A novel AI-based stock market prediction using machine learning algorithm. Sci Program 1:1–1
30. Azrour M, Mabrouki J, Guezzaz A, Ahmad S, Khan S, Benkirane S (2024) IoT, machine learning and data analytics for smart healthcare. CRC Press
31. Vikram A, Akshya J, Ahmad S, Rubini LJ, Kadry S, Kim J (2024) Deep learning based vehicle detection and counting system for intelligent transportation. Comput Syst Sci Eng 48(1). https://doi.org/10.32604/csse.2023.037928
32. Haque MA, Sonal D, Ahmad S, Kumar K (2023) Enhancing security for internet of things based system. In: International conference on communication and computational technologies, pp 869–878. https://doi.org/10.1007/978-981-99-3485-0_68

MLbCBM: Machine Learning-Based Consumer Behavior Model to Analyze Emerging Trends and Influences in Market

Bhawana Kothari and Ambica Prakash Mani

Abstract In the fast-changing field of consumer behavior analysis, employing machine learning techniques has become critical for identifying new patterns and effects inside markets. This study offers MLbCBM (Machine Learning-Based Consumer Behavior Model), a unique technique for extensively evaluating consumer behavior patterns. Our suggested model incorporates a variety of machine learning techniques, including Logistic Regression, Decision Tree, Random Forest, KNN, SVM, and Naive Bayes, with exceptional accuracies of 88.48%, 85.1%, 89.9%, 85.3%, 88.8%, and 88.6%, respectively. Our model's design includes the deployment of machine learning algorithms on a server infrastructure, which allows for efficient data analysis and processing. Multiple e-commerce sites serve as key data sources, which may be accessed via web or Android applications. By leveraging machine learning and tapping into multiple data sources, MLbCBM provides a solid foundation for analyzing and forecasting consumer behavior dynamics in today's marketplaces.

Keywords Consumer behavior · Machine learning · Purchase intention · Product · Service · Market influence

1 Introduction

This research is being conducted in response to India's rising urbanization and cultural transformations, notably in the cities of Uttarakhand. As urban populations increase and diversify, consumer behavior, particularly in industries such as beauty care, is changing dramatically. The beauty care industry in metropolitan India, as

B. Kothari (✉) · A. P. Mani
Department of Commerce, Graphic Era (Deemed to Be University) Dehradun, Dehradun, India
e-mail: bhawnakothari92@gmail.com

A. P. Mani
e-mail: drambica.comm@geu.ac.in

P. J. S. Gonçalves et al. (eds.), *Proceedings of Fifth International Conference on Computing, Communications, and Cyber-Security*, Lecture Notes in Networks and Systems 1128, https://doi.org/10.1007/978-981-97-7371-8_31

illustrated by Uttarakhand cities, offers a unique combination of modernism, cultural variety, and quick market evolution. Uttarakhand not only has a contemporary infrastructure but also a diversified population makeup, which reflects larger Indian trends. This diversity makes it an ideal venue for researching customer behavior in the beauty care industry. While global studies such as those conducted by Park [1] and Okeke et al. [2] have emphasized the importance of brand reputation, product quality, and packaging in influencing consumer de- cisions, the specific dynamics of an urban Indian context, as seen in Uttarakhand, provide a unique perspective on these global trends. Furthermore, Gangopadhyay [3] investigates the role of the market in shaping new trends among middle-class older people in urban India, which may have repercussions for the beauty care industry.

The beauty care market in India is quickly expanding, with large increases in consumer expenditure and a move toward premiumization. ASSOCHAM estimates that the Indian beauty and personal care industry will reach USD 20 billion by 2025, with a CAGR of 25% (ASSOCHAM, 2021). Rising disposable incomes, urbanization, and the impact of social media have all contributed to this increase. The urban customer, particularly in places like Uttarakhand, is growing more conscious and sophisticated, seeking goods that not only suit their beauty demands but also reflect their ideals and lifestyles [4, 5]. Furthermore, the study's findings are useful not just for the beauty care business, but also for understanding larger consumer behavior trends in rising urban economies. This knowledge is critical for marketers, product developers, and politicians seeking to successfully engage and serve India's dynamic urban inhabitants. Understanding the intricacies of customer preferences in metropolitan environments such as Uttarakhand may lead to more focused and effective business strategies, as well as contribute to the economic and social growth of these rapidly growing urban regions.

This study is especially relevant since it focuses on a city that reflects wider Indian market trends while also providing insights specific to its regional environment. The research examines consumer behavior in Uttarakhand, providing findings that are both locally relevant and suggestive of broader tendencies in metropolitan North India. This is crucial for firms and marketers in the beauty care industry, allowing them to design plans that are precisely tailored to the intricacies of this particular market. This study contributes to the academic field by improving understanding of consumer behavior in emerging-market metropolitan areas [6, 7].

1.1 Role of Machine Learning in Consumer Purchasing Intention

Machine learning (ML) is an important tool for analyzing and forecasting consumer behavior [8–10].

Machine learning (ML) plays a significant role in understanding and predicting consumer behavior. Here are several ways ML impacts consumer behavior:

Personalized Recommendations: With the help of Machine learning algorithms the platforms can adopt much more user-specific recommendations.

Targeted Advertising: Machine Learning can help e-commerce platforms to provide them with a smart advertising mechanism on the basis of various factors like age, gender, etc.

Predictive Analytics: Various Platforms are already using predictive—analytics so that they can deliver the products on or before time to the consumers.

Sentiment Analysis: Sentiments of the users play a very crucial role while purchasing any product in an online mode. Machine learning models can be helpful for this type of analysis.

Dynamic Pricing: Changing prices according to the market needs can also be done through Machine learning Models.

2 Literature Review

Various Researchers have already done the work in the same area which are the sources of this study among them Shrirame et al. [11] discussed User-generated material, such as reviews, ratings, and comments, which may be analyzed to gain valuable insights for corporate usage. The analysis of such customer behavior is useful for understanding the consumer's needs and predicting their future intentions toward the service. This cognitive research examines e-commerce Organizations may track the usage and feelings associated with their goods and implement suitable marketing strategies to deliver a personalized shopping experience for their customers, hence enhancing organizational profit. This article intends to use data-driven marketing techniques including data visualization, natural language processing, and machine learning models to better understand an organization's demographics. We also develop recommender systems using collaborative filtering, neural networks, and sentiment analysis. Olan et al. [12] presented the growing usage of digital technology has dramatically altered marketing and consumer behavior (CB), as online communities and cutting-edge developments like artificial intelligence (AI) disrupt and progress customer attitudes toward certain products and services. As a result, online communities powered by AI technology generate new information via consumer interactions via platforms such as social media, where customers discuss their experiences with specific products or services. AI is meant to "learn" and develop with data created from digital technologies related to customer interactions; therefore, it relies on consumer knowledge-sharing (KS) activities to reproduce new information for product and service enhancement. Abrardi et al. [13] surveyed the present breakthroughs in Artificial Intelligence (AI) are anticipated to have major economic repercussions and introduce new trade-offs, offering new policy difficulties. What influence do these technologies have on the labor market and businesses? Will algorithms remove consumer prejudices, or will they create new ones? How will AI-powered agents change the competitive landscape? This research is the first to evaluate the burgeoning literature on the multifaceted economic implications of current AI technical developments that include machine learning applications. We begin by reviewing research on the influence of AI on companies, with a focus on labor market, productivity, skill mix, and innovation. Khrais [14] presented the introduction and integration of technology in enterprises have transformed operations across sectors. Notably, substantial technical advances in e-commerce are intended to influence user behavior in favor of specific items and companies. Artificial intelligence (AI) has emerged as a crucial inventive tool for personalizing and customizing products to satisfy unique needs. This study discovers that despite the role of AI systems in e-commerce, their ethical soundness remains a disputed topic, particularly in terms of explainability. The study used word cloud analysis, voyance analysis, and concordance analysis to acquire a thorough knowledge of the concept of explainability, which has been used by researchers in the field of AI.

The beauty care industry in metropolitan India, as shown by places such as Uttarakhand, offers a distinct combination of modernism, cultural variety, and changing market trends. This literature review dives into several research that investigate the subtleties of customer behavior in this industry. The focus is on understanding the factors that influence customer choices, preferences, and purchase patterns in the beauty care sector, with a particular emphasis on the urban Indian environment. The poll discusses the impact of packaging, the function of ethnicity, market segmentation, and the variables that influence purchasing behavior in specific product categories. Baptista et al. [15] investigate consumer behavior in the cosmetics industry, focusing on the various factors that influence customer perceptions of product quality and brand image. Ethnicity has a big effect on customer behavior in the beauty care business. Cooray et al. [16] investigated how ethnicity influences female buying decisions in Sri Lanka's beauty care product industry. Their findings demonstrated that ethnicity has a major impact on product choices and purchase behavior. This conclusion is especially pertinent in a diverse country like India, where customers from various ethnic origins may have varying tastes and demands in terms of beauty care goods.

2.1 Market Segmentation and Consumer Behavior

Market segmentation is an important part of understanding consumer behavior. Kumari et al. [17] performed research on market segmentation of healthcare product purchasing behavior in Chandigarh, India. Their research identified many divisions based on demographics, lifestyle, and purchasing habits, offering significant insights into how different consumer groups view healthcare items, including those in the beauty care industry. This study emphasizes the need of segmenting customers in order to better understand and respond to their unique requirements and preferences.

2.2 *Factors Influencing Buying Behavior in Specific Product Categories*

Understanding the elements that drive customer purchasing behavior in certain product categories is critical for companies in the beauty care industry. Pakkala & Shivashankar Bhat [18] investigated the factors influencing customer purchasing behavior for infant care items. Their research found brand loyalty, product quality, and price sensitivity as important influencers of customer decisions. While their research focused on infant care items, the findings are equally applicable to the larger beauty care sector, where comparable variables might affect customer preferences. Srivastava et al. (2021) did a study on customer purchasing behavior, and while their research focused on a different product area, the findings provide similar insights into the beauty care sector. The study emphasized the significance of factors such as product quality, brand reputation, and pricing in shaping customer decisions. These aspects are similarly significant in the beauty care business, where customers frequently evaluate comparable factors while making purchase decisions.

- Demographic Details: Age: "Please specify your age group." Options: 18- 25, 26–35, 36–45, 46–55. Gender: "Please indicate your gender." Options: Male/ Female.
- Perception and Behavioral Metrics: A 10-point Likert scale was utilized to gauge respondents' perceptions and behaviors (where 1 meant "Strongly Disagree" or "Very Low" and 10 meant "Strongly Agree" or "Very High"). Product Quality: "On a scale of 1–10, how would you rate the quality of the beauty care products you usually purchase?" Brand Reputation: "On a scale of 1–10, how influential is a brand's reputation when you consider buying a beauty product?" Perceived Value: "On a scale of 1–10, how much value for money do you think the beauty products you purchase offer?" Social Influence: "On a scale of 1–10, how much do recommendations or opinions from friends and family influence your purchase decisions in beauty care?"
- Price Sensitivity and Purchase Intentions: Price: "Considering the beauty products you've recently bought, please indicate an approximate price range." Options: Less than 500 INR, 501–1000 INR, 1001–1500 INR, 1501–2000 INR, more than 2000 INR. Purchase Intention: "Given the product quality, brand reputation, price, social influence, and perceived value, would you consider purchasing a beauty care product in the near future?" Options: Yes/No.
- Pilot Testing: Prior to the broad-scale dissemination of the questionnaire, it underwent a pilot testing phase with a subset of Uttarakhand City residents. This initial step was instrumental in refining the questions, ensuring clarity, and confirming that the instrument would reliably capture the desired data.

2.3 Data Collection Procedure

Data collection spanned over a period of two months. Trained surveyors approached potential respondents in diverse locations across Uttarakhand City. These included high footfall areas like shopping malls, beauty and wellness centers, recreational parks, and educational institutions. Alongside the physical survey, a digital version was propagated through local online communities, leveraging platforms like Facebook and community forums.

2.4 Data Analysis Techniques

Post data collection, the analysis was conducted in sequential stages:

- Data Preprocessing: This involved cleaning the data to remove any anomalies, handling missing values, and categorizing open-ended responses.
- Descriptive Statistical Analysis: Computed detailed statistics to understand the distribution and central tendencies of the data, utilizing metrics like mean, median, mode, variance, and standard deviation.
- Correlation Matrix Construction: Spearman and Pearson's correlation coefficients was calculated to understand bivariate relationships between the variables.
- Multivariate Regression Analysis: A regression model was built to decipher the impact of independent variables (like product quality and brand reputation) on the dependent variable (purchase intention).

2.5 Ethical Considerations

All along the study, a rigorous code of ethics was maintained. Every respondent gave their informed consent, guaranteeing that they understood the nature and intent of the research. The respondents received assurances of the privacy and anonymity of their answers. They were also made aware that they might withdraw from the activity at any time and that it was entirely optional.

3 Research Analysis and Findings

3.1 Demographic Profile of Respondents

The beauty care sector in Uttarakhand City, supported by a plethora of variables and consumer archetypes, provides an intricate structure of connections that requires careful review. Our research, which is based on empirical data analysis, reveals key patterns and interdependencies.

Figure 1a clearly shows Demographically, the information reveals a non-uniform age distribution, with a clear lean toward younger generations. Such demographic clustering, on the surface, suggests a greater inclination of this age group for the beauty care sector, which could be augmented by their digital nativity and susceptibility to contemporary media influences. Figure 1b indicates the gender distribution in the sample provides interesting information. In contrast to traditional market paradigms that have traditionally presented beauty care as a largely female-centric area, our data shows an equal distribution of male and female respondents. This emerging gender-neutral trend is representative of a larger cultural shift toward gender-fluid beauty norms. In Fig. 1c, a further in-depth examination of product features finds a significant positive association between perceived product quality and associated brand reputation. This mutually beneficial relationship suggests that brands can strengthen their market reputation by consistently delivering high-quality products. Furthermore, the data reveals a complex interdependence between product pricing and perceived qualitative qualities. While customers tend to pay a premium for products with perceived greater qualitative features, this proclivity shows symptoms of declining returns, suggesting hidden price sensitivities. The complicated web of purchase intentions is central to consumer behavior. As can be seen in Fig. 1d, our analysis highlights the importance of perceived product value and brand reputation in influencing these intentions. This underlines how important it is for brands to go beyond simple product-centric strategies and instead focus on the larger value proposition and consumer perception measurements. Figure 1e shows a visualization of consumer purchase intentions for beauty care products, analyzed by gender. It reveals that gender is not a significant factor in making purchase decisions for these products. 4.2 Descriptive Statistics: In our comprehensive exploration of Uttarakhand's beauty care market, the descriptive statistics have been instrumental in illuminating the central tendencies and distributions embedded in our dataset. The age demographics in Table 1 provide insightful information. With an average age of 36.07 years, it is evident that the market primarily serves older consumers. The age range, which is 18 to 55 years old with a median age of 35.5 years, ensures that the insights gained are not biased toward any specific age group by offering a well-distributed representation among younger to mid-aged people. Examining customer perceptions of product quality, a mean score of 5.48 (on a 10) indicates that there is potential for development in terms of how customers view the caliber of beauty goods that are offered. In addition to highlighting the range of opinions, the median score of 6 and the gap between the 25th and 75th percentiles suggest that there may be room for improvement in terms of product quality and consumer expectations. The average score for brand reputation is 5.47, which is similar to the scores for product quality and suggests a strong correlation between the two factors. A steady brand reputation indicates that there is a lot of space for brands to improve their reputation in this industry, especially considering the wide range up to a perfect score of 10. Regarding cost, the mean purchase is INR 1051.03, while the median is marginally less at INR 1046.98. The pricing tiers span a wide range, from a minimum of INR 383.85 to a high of INR 1764.11, to accommodate a diverse variety of consumers with varying spending capacities. At 6.30, social influence had a comparatively high mean score. The distribution, which has a median of 6.44 and ranges from a low of 1 to a high of 10, highlights how important social and cultural factors are to Uttarakhand's customers' purchase decisions. With a mean of 4.54, the perceived value indicates that although customers find some value in the products, there is a great deal of space for improvement. With a median of 4.34, the distribution, in particular, indicates a chance for businesses to enhance the value proposition of their products and services and make sure they more deeply align with consumers' expectations. The purchase intention shows that half of the respondents are likely to make a purchase, while the other half may be cautious, with a balanced mean of 0.50. Brands may use this binary indicator as a clear benchmark to assess the effectiveness of their plans and activities. These descriptive statistics provide a detailed, data-rich picture of the consumer landscape in Uttarakhand's beauty care sector. The insights act as a guidepost, directing companies and marketers in their strategic endeavors and interventions in this volatile industry.

3.2 *Correlation Analysis*

The correlation analysis of the beauty care market in Uttarakhand reveals several key insights. From Fig. 2, we can see that there is a notable negative corre-lation between age and social influence (−0.92), indicating that the impact of social factors on purchasing decisions diminishes with age. Younger consumers are more influenced by social trends and peer recommendations. A strong positive correlation is observed between product quality, brand reputation, and perceived value (r 0.94–0.95), suggesting that these elements are interdependent. Consumers tend to associate high-quality products with strong brand reputations and high perceived value. Price also correlates strongly with product quality ($r = 0.95$), brand reputation ($r = 0.91$), and perceived value ($r = 0.88$), indicating that higher-priced products are generally perceived as higher in quality and value.

Purchase intentions show a reasonable correlation with product quality ($r = 0.73$), brand reputation ($r = 0.72$), and perceived value ($r = 0.74$), signifying that these factors are significant drivers of consumer buying decisions. However, the correlation between social influence and purchase intention is relatively weak ($r = 0.14$), suggesting that social factors, while relevant, may not be the primary determinants of purchase decisions in this context.

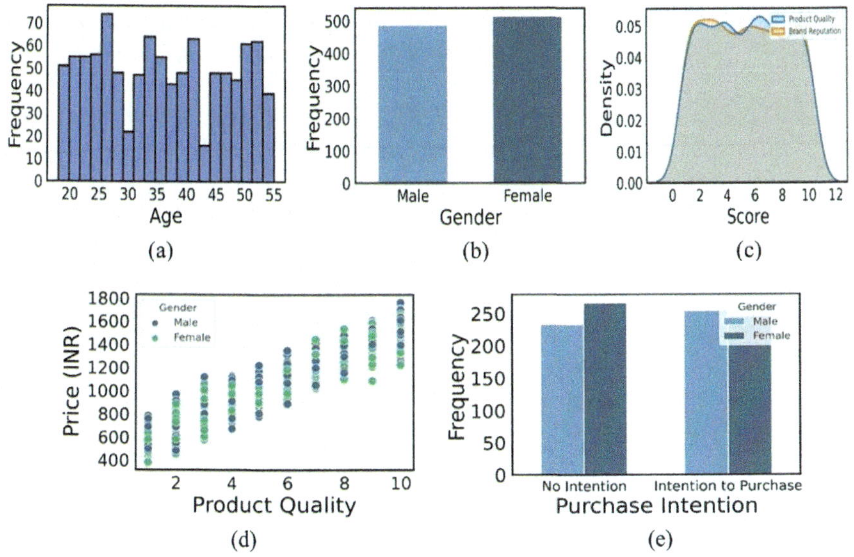

Fig. 1 Visualization for **a** Distribution of Age, **b** Distribution of Gender, **c** Distribution of Product Quality and Brand Reputation, **d** Relationship between Price and Product Quality, and **e** Purchase Intention by Gender

Table 1 Descriptive statistics of variables

Variable	Count	Mean	Std. Dev	Min	25th Percentile	Median	75th Percentile	Max
Age	1000	36.07	10.98	18.00	27.00	35.50	46.00	55.00
Product Quality	1000	5.48	2.89	1.00	3.00	6.00	8.00	10.00
Brand Reputation	1000	5.47	2.91	1.00	2.88	5.47	7.97	10.00
Price (INR)	1000	1051.03	304.28	383.85	789.95	1046.98	1296.04	1764.11
Social Influence	1000	6.30	2.35	1.00	4.36	6.44	8.18	10.00
Perceived Value	1000	4.54	2.67	1.00	2.04	4.34	6.80	10.00
Purchase Intention	1000	0.50	0.50	0.00	0.00	0.50	1.00	1.00

3.3 Multiple Regression Analysis

To get insight into the intricacies of the beauty care industry in Uttarakhand, we utilized multiple regression analysis to ascertain the correlations between different

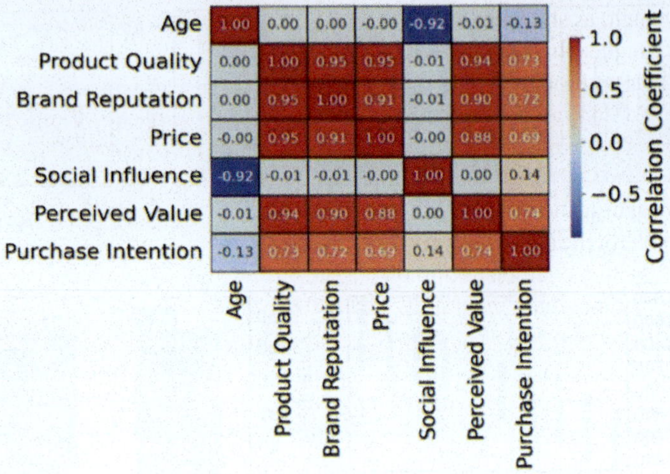

Fig. 2 Correlation matrices highlighting key influencers for buying behavior analysis

variables and purchase intention. The analysis's findings in Table 2 reveal fascinating patterns.

Our model's constant, also known as the intercept, is −0.4353. If every other predictor is set to zero in the hypothetical case, the projected purchase intention would be negative. Since the predictors cannot be canceled out, this intercept may not have any interpretive significance in practice. The age of the consumer, which is frequently seen as a crucial component in many decisions, showed a coefficient of 0.0011. This implies that there is a very slight increase in purchasing intention for every year that an individual's age increases. Age isn't a statistically significant predictor in this

Table 2 Multiple regression analysis

Predictor	Coefficient β	Std. Error	t-value	p-value	95% CI Lower	95% CI Upper
Intercept (Constant)	−0.4353	0.1604	−2.7146	0.0067	−0.75	−0.1206
Age	0.0011	0.0023	0.4797	0.6315	−0.0035	0.0057
Product Quality	0.0401	0.0192	2.0873	0.0371	0.0024	0.0777
Brand Reputation	0.0289	0.0118	2.4573	0.0142	0.0058	0.052
Price	−0.000025	0.0001	−0.2257	0.8215	−0.0002	0.0002
Social Influence	0.0348	0.0109	3.1871	0.0015	0.0134	0.0562
Perceived Value	0.0714	0.0114	6.2795	<0.0001	0.0491	0.0937

model, though, as shown by the t-value of 0.4797 and the p-value of 0.6315 (both of which are higher than the conventional significance threshold of 0.05). This result is consistent with our previous correlation research, which suggested that age may not be a determining factor in beauty product purchasing intentions in Uttarakhand. Product quality is found to have a coefficient of 0.0401, meaning that a unit increase in perceived product quality corresponds to a 0.0401 unit increase in buy intention. Its statistical significance is highlighted by the t-value of 2.0873 and p-value of 0.0371, which highlight the critical impact that product quality plays in influencing consumer decisions. With a coefficient of 0.0289, brand reputation is shown to have a significant impact on purchase intentions. The corresponding t-value of 2.4573 and p-value of 0.0142 serve to emphasize the predictor's importance even further. At 0.000025, price of a component that is frequently seen as primary in customer decisions has an insignificant coefficient. Our previous conclusions from the correlation study are supported by the t-value of −0.2257 and the high p-value of 0.8215, which indicate that price is not a statistically significant predictor in our model. With a value of 0.0348, social influence appears as a powerful factor. The t-value of 3.1871 and the low p-value of.0015 emphasize the importance of its in- influence on purchase intention. Finally, with a coefficient of 0.0714, the perceived value shows out significantly. It is the most influential predictor in our model due to its associated t-value of 6.2795 and exceptionally low p-value (less than 0.0001). This is consistent with the overarching consumer behavior paradigm, in which perceived value frequently plays a critical role in purchasing decisions. The findings of this multiple regression analysis provide a complex picture of the dynamics of consumer behavior in the beauty care sector in Uttarakhand. Consumer purchasing intentions are influenced by several important elements, including perceived value, social impact, product quality, and brand reputation; age and price become less important considerations.

4 Architecture of Proposed Model

As shown in Fig. 3, the Proposed model will have a Machine learning model deployed over a server for analysis and processing of the data. The source of data will be multiple e-commerce platforms and can be accessed over any web or android applications.

4.1 Trained Model

The Machine Learning Model is trained over the dataset [19] with the help of machine learning algorithms like Random Forest, logistic regression, KNN, etc. The best accuracy is achieved at Random Forest as shown in Fig. 4.

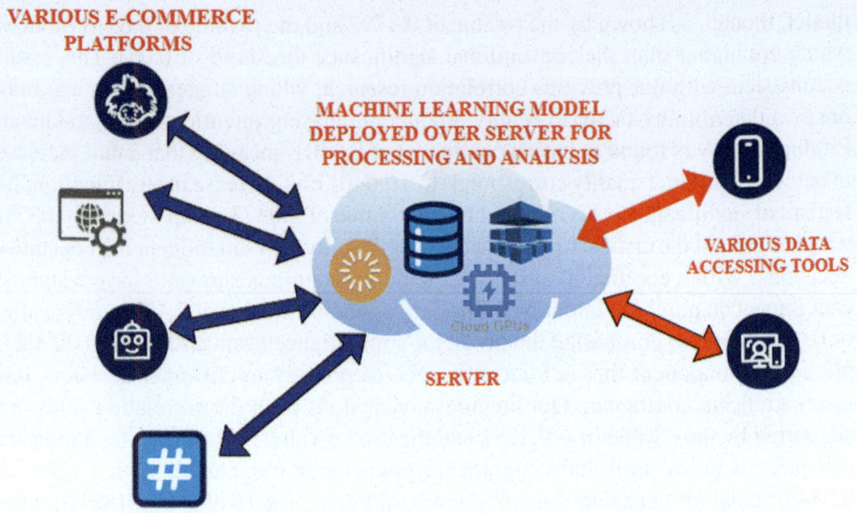

Fig. 3 Architecture of proposed model

	False	True	accuracy	macro avg	weighted avg
precision	0.91	0.75	0.90	0.83	0.89
recall	0.97	0.49	0.90	0.73	0.90
f1-score	0.94	0.59	0.90	0.77	0.89
support	3127.00	572.00	0.90	3699.00	3699.00

Fig. 4 Obtained result by applying random forest

4.2 Results

The multiple machine learning algorithms were checked over the dataset but the best result as shown in Table 3 produced on Random Forest. Also, the Analyzed Results and extracted features from applying Random forest which plays a vital role are shown in Fig. 5.

5 Conclusion and Future Scope

In conclusion, MLbCBM is a strong instrument for diving into the complex area of consumer behavior analysis. Our model achieves promising accuracy rates across many approaches by combining distinct machine-learning algorithms and deploying

Table 3 Result

Algorithm name	Accuracy) (%)
Logistic Regression	88.48
Decision Tree	85.1
Random Forest	89.9
KNN	85.3
SVM	88.8
Naive Bayes	88.6

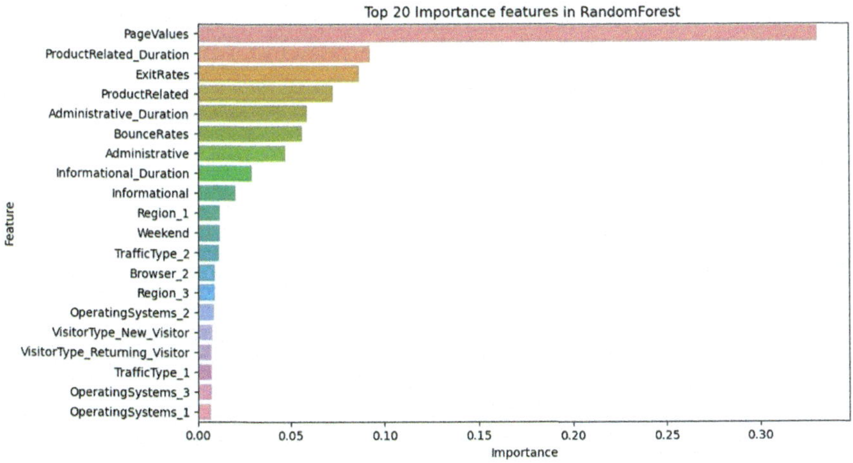

Fig. 5 Important features

a server-based architecture. By gathering data from numerous e-commerce platforms accessible via web and Android applications, MLbCBM provides a complete lens through which new industry trends and effects may be interpreted. In today's ever-changing industry, knowing customer behavior is critical. MLbCBM provides a dynamic framework that captures current consumer preferences while also anticipating future trends, allowing firms to adapt and develop proactively. Moving forward, the model's modifications and upgrades offer even more insights into consumer behavior, opening the path for more informed decision-making and strategic market positioning. In summary, MLbCBM ushers in a new age of consumer behavior analysis, with the confluence of machine learning and data-driven insights opening up hitherto untapped prospects in market research and strategy creation [4, 5].

While the provided Machine Learning Model is highly accurate in market analysis, there is a vast terrain of opportunities for further improvement and investigation. One route for development is to use deep learning algorithms, which have the ability to extract more subtle patterns and insights from complicated datasets. By using sophisticated deep learning approaches, such as convolutional neural networks

(CNNs) or recurrent neural networks (RNNs), the model can gain increased accuracy and predictive capabilities. Furthermore, increasing the size and variety of the dataset shows potential for improving the model's performance. The model may get a more thorough knowledge of market dynamics by combining data from a wider range of sources and domains, such as social media platforms, demographic information, and consumer sentiment research. A deeper research into certain consumer categories, such as the beauty goods business, provides an opportunity for specialist analysis. By investigating how customers perceive value in beauty items and how social aspects impact their purchase decisions, the model can deliver more relevant insights for this industry. Future research attempts may benefit from a multidisciplinary strategy that combines qualitative and quantitative approaches. Combining qualitative insights obtained via approaches like focus groups or interviews with quantitative data analysis can provide a more comprehensive knowledge of customer behavior. Longitudinal studies can enhance the research by following consumer trends over time, indicating changing patterns and preferences.

References

1. Park J (2020) Changes in beauty service industry in era of the fourth industrial revolution and aging: focused on expandability of the smart beauty care market. J Human Soc Sci 21, 11(1):205–220
2. Okeke TC, Amobi, Ify L (2020) Influence of packaging on consumer buying decision of cosmetics (A study of consumers of beauty care cosmetics in awka). Int J Scientific Res Publ (IJSRP) 10(2):9896
3. Gangopadhyay J (2022) Ageing and Self-care in India: Examining the role of the market in determining a new course of growing old among middle class older adults in urban India. Ageing Int 47(4):801–815
4. Hossain MS (2023) Behavioral analytics of consumer complaints. In: AI-driven intelligent models for business excellence, pp 42–67. IGI Global
5. Dermody J (2020) Tales of materialism and sustainable consumption in a marke- tizing urban India. In: Marketization, pp 125–163. Springer Singapore
6. Reddy KSVRK, Kishore NTK (2023) Consumer buying behavior to- wards online meat and meat products: In tier-I cities of India. Int J Agric Sci 19(1):94–100
7. Namrata S, Aradhana G (2022) Analysis of the consumer purchase behavior of personal care products in a developing country. CARDIOMETRY 23:718–727
8. Rustagi N, Choudhary Y, Kumar N, Shafi S, Saurabh S, Thirunavukkarasu P, Raghav P (2022) Adherence to healthy behavioral practices promotes COVID 19 preventive behavior among chronic disease patients: a cross-sectional study from urban Rajasthan, India. J Family Med Primary Care 11(7):3482–3490
9. Shin H, Huh S (2023) The impact of sustainable packaging on consumers' purchase intention in the cosmetic industry. J Environ Manage 299:113632
10. Zhang Y, Wang L (2022) Adapting to challenges: how the cosmetic industry navigates the impact of the COVID-19 pandemic. J Bus Res 135:472–481
11. Shrirame V, Sabade J, Soneta H, Vijayalakshmi M (2020) Consumer behavior analytics using machine learning algorithms. In: 2020 IEEE international conference on electronics, computing and communication technologies (CONECCT), pp 1–6. IEEE
12. Olan F, Suklan J, Arakpogun EO, Robson A (2021) Advancing consumer behavior: the role of artificial intelligence technologies and knowledge sharing. IEEE Trans Eng Manag

13. Abrardi L, Cambini C, Rondi L (2022) Artificial intelligence, firms and consumer behavior: a survey. J Econ Surv 36(4):969–991
14. Khrais LT (2020) Role of artificial intelligence in shaping consumer demand in E-commerce. Future Internet 12(12):226
15. Baptista AHA (2020) Consumer behavior in relation to cosmetics. South American Dev Soc J 6(17):426
16. Cooray ND, Dias PA, Rajapaksha RP, Bernard K, Samarasinghe HM (2018) Moderating effect of ethnicity on the purchase decisions of females in the beauty care product market of Sri Lanka. Int Bus Res 11(12):12
17. Kumari G, Pandey V, Bhat KM (2019) Market segmentation of buying behavior of consumers of health care products in Chandigarh India. Int J Recent Technol Eng (IJRTE) 8(4):610–614
18. Pakkala K, Shivashankar Bhat K (2022). Factors influencing consumer buying behavior on baby care products. Int J Case Studies Bus IT Educ, 459–468,https://doi.org/10.47992/ijcsbe.2581.6942.0209
19. https://archive.ics.uci.edu/dataset/468/online+shoppers+purchasing+intention+dataset

NICSR-A Deep Learning-Based Noise Identification and Compensation Technique for Super-Resolution of Drone-Captured Images

Amul Batra and Meetha V. Shenoy

Abstract Researchers have been exploring ways to improve autonomous drone operations using onboard cameras for drones belonging to the nano/micro/small category. These drones may be mounted with only low-resolution camera(s) due to the maximum takeoff weight limitation of these drones. Decision-making using low-resolution images generated by the primary camera can be faulty unless the images are captured from a distance very close to the point of interest. Detection of the point(s) of interest as early as possible is required to ensure sufficient response time for safe maneuvering. Hence, the images are to be captured at greater heights from the point(s) of interest, and obtaining the high-resolution images from the captured low-resolution images is crucial. The presence of noise in the images adversely affects the onboard decision-making. This work highlights the limitations of the single-image super-resolution techniques to super-resolve noisy images and aims to remove noise from the images before super-resolution. We propose a deep learning-based novel architecture NICSR, i.e. Noise Identification and Compensation technique for Super-Resolution of images captured by a drone in the presence of noise. The performance analysis of the proposed technique on a publically available dataset reveals appreciable improvement in the quality of the super-resolved images, even if the input images are noisy. The proposed work will hence accelerate drone-based applications in civil and military sectors such as aerial photography, terrain mapping, surveillance, aerial inspection, drone-based delivery, and disaster management.

Keywords Deep learning · Super-Resolution (SR) · Unmanned Aerial Vehicle (UAV) · Single-Image Super-Resolution (SISR) · Computer vision · Drone

---

A. Batra · M. V. Shenoy (✉)
Department of Electrical and Electronics Engineering, Birla Institute of Technology and Science, Pilani, Rajasthan 333031, India
e-mail: meetha.shenoy@pilani.bits-pilani.ac.in

P. J. S. Gonçalves et al. (eds.), *Proceedings of Fifth International Conference on Computing, Communications, and Cyber-Security*, Lecture Notes in Networks and Systems 1128, https://doi.org/10.1007/978-981-97-7371-8_33

1 Introduction

Drones, also known as Unmanned Aerial Vehicles (UAVs), have become increasingly popular for applications in civil and military sectors such as aerial photography, terrain mapping, surveillance, aerial inspection, drone-based delivery, and disaster management. Civil drone operation guidelines have opened up drone operations in segregated airspaces, necessitating autonomous drone operations and real-time decision-making onboard [1]. Landing, being one of the most critical maneuvers, requires special attention. Limitations posed by the unavailability of GPS signal in the proximity of landing areas, weight, and cost of these landing systems necessitate the use of camera-based landing in drones belonging to nano, micro, and small categories [1, 2].

The drones, especially the nano/micro/small drones, may be mounted with only low-resolution camera(s) due to the maximum takeoff weight limitation of these drones. The achievable pixel density on the target reduces with the increase in the Field of View (FoV), and hence, the low-resolution images generated by the primary camera, if used for landing, can result in faulty detection unless the images are taken from a very close distance to the point of interest [3]. Image Super-Resolution (SR) is an image transformation technique aimed at enhancing image quality by resolution improvement [4, 5]. Single-Image Super-Resolution (SISR), which generates a High-Resolution (HR) image from a Low-Resolution (LR) image, is an image processing technique for generating an enhanced resolution image from a single input image [6]. Onboard image super-resolution on drones helps to obtain HR images from LR images thus enabling the detection of points of interest from a greater height or distance, which provides sufficient response time for effective decision-making on the drone.

Noise in the captured images, if not treated correctly, renders information suppression and pixel loss, which changes the information depicted in the forms such as intensity gradient, contrast values, color intensity, or texture profile. Machine learning-based architectures for SISR available in the literature are not designed to enhance noisy images; hence, noisy pixels are also super-resolved, rendering the output image inappropriate for accurate decision-making [7]. In our work, we address the limitations posed by the deep learning-based SISR networks in the presence of noise. This paper is organized as follows. Section 2 presents a discussion of related studies on real-time image super-resolution techniques and noise in drone-captured images. Section 3 describes the proposed method and the overall architecture for super-resolution in the presence of additive noise. Section 4 presents the implementation details and performance analysis. Section 5 presents the conclusion and future scope of work.

2 Related Works

Single-Image Super-Resolution (SISR) is a type of SR technique capable of reconstructing a HR image from a single LR image [4]. Supervised machine learning-based SISR techniques use pairs of HR and LR images to make the SISR network learn to identify the features in the high-resolution images and generate the corresponding features in the low-resolution images [4]. These networks, when deployed after training, are expected to generate a HR image when fed with a LR image. Peak Signal-to-Noise Ratio (PSNR) and Structural Similarity Index Measure (SSIM) are generally used as metrics to compare the quality of the high-resolution image generated by the SR technique with the high-resolution image captured by the camera. PSNR is given by

$$PSNR = 10\log_{10}\frac{L^{2}}{MSE} \tag{1}$$

$$\frac{1}{MN}\sum_{M}\sum_{N}(B(m,n) - D(m,n))^{2} \tag{2}$$

where *B(m,n)* and *D(m,n)* represent the two images (e.g. HR image captured by the camera and the HR image generated by the SR technique) being compared and *M* and *N* are row and column index of the image [8]. SISM indicates the amount of similarity between two images based on luminance, contrast, and structure [8].

$$SSIM\,(x,y) = \frac{(2\mu_x\mu_y + C_1)(2\sigma_{xy} + C_2)}{(\mu_x^2 + \mu_y^2 + C_1)(\sigma_x^2 + \sigma_x^2 + C_2)} \tag{3}$$

where "x" and "y" represent the two images being compared, "μ" represents the mean, "σ^2" represents the variance, and σ_{xy} represents the covariance of the images "x" and "y". The constants in Eq. (3) are $C_1 = (0.01L)^2$ and $C_2 = (0.03L)^2$, respectively. "L" in Eqs. (1) and (3) is the dynamic range of the pixel values.

SISR models available in the literature are not designed to enhance noisy images, and hence the noise in the images also gets amplified during super-resolution. A brief review of the popular SISR architectures and the predominant type of noise present in the digital images are presented in the following subsections.

2.1 *Single-Image Super-Resolution*

Dong et al. proposed a SISR model called Super-Resolution Convolutional Neural Network (SRCNN) comprising of three main layers [9]. The first layer takes an LR image as input and performs normalization and scaling operations. The second layer is the feature extraction layer and the third layer uses convolutional filters to learn the mapping between the LR and HR images and outputs the HR image. Deeply Recursive Convolutional Network (DRCN) was proposed by Kim et al., which uses recursive learning to learn the mapping between the LR and HR images and demonstrated an improvement in PSNR and SSIM when compared to SRCNN [10]. Efficient Super-Resolution Transformer (ESRT)-based model has been used by Lu et. al. to achieve single-image super-resolution [11]. The networks mentioned above are to be further optimized for use on board nano/micro/small drones due to their high computational complexity. Yamanaka et al. proposed Deep CNN with Skip Connection and Network in Network (DCSCN) as a real-time image super-resolution network for deployment on edge devices, which is suitable for SR applications that require a balance between quality and computational efficiency [12]. Truong et al. demonstrated the use of DCSCN for marker-based landing [4]. The DCSCN model was executed on JetsonTX2 and Snapdragon 835 development kit [4]. The authors have not deployed the hardware on drones and the power consumption and weight to be further optimized for implementations in nano/micro/small drones. Also, DCSCN model was not analyzed for its performance in the presence of noise. Lin et al. have created a drone-based dataset Drone Super-Resolution (DSR) that contains pairs of LR and HR images captured at various heights for studies on drone-based decision-making [13].

2.2 Noise in Drone-Based Images

Gaussian and Salt & Pepper noise which are predominant in digital images are considered in this work [14]. In foggy conditions, light can scatter and contribute to Gaussian Noise (GN) [15]. The noisy image "I_{ng}" with GN can be represented as

$$I_{ng} = I_{original} + N(0, \sigma) \tag{4}$$

where $N(0,\sigma)$ is GN with mean "0" and standard deviation "σ" that indicates the amount of noise in the image [16].

Salt and Pepper(S&P) noise in images is characterized by the presence of randomly scattered black and white pixels in the image [17]. The image sensor used on the drone can deteriorate due to various reasons, such as exposure to extreme temperatures, power supply variations, shocks, or wear and tear thus introducing S&P noise in the captured images. The noisy image "I_{nsp}" with S&P noise can be represented as

$$I_{nsp} = I_{original} + f(X, Y) \tag{5}$$

where f(x,y) adds black and white pixels at random locations (x,y) of the image.

- Filtering is one of the widely explored methods to compensate for noise in the digital images while preserving vital details such as edges and surfaces [15]. Oddi et al. in their research work have enlisted the adverse effect of S&P noise and the need to address the same [17]. The Gaussian and S&P noise may emerge additively in images [17]. In our work, the effect of noise in images on super-resolution is analyzed, and a method for noise identification and compensation is introduced. The main contributions of this research work are as follows:

- We propose a novel architecture NICSR, i.e. Noise Identification and Compensation technique for SISR of noisy images captured by a drone.
- This proposed architecture includes a noise identification module to identify the predominant type of noise in the images captured by the drone.
- In order to address the problem of image degradation due to noise, this paper proposes a noise compensation module that compensates for the noise based on the identified type of noise in the image.
- The performance of the proposed architecture is tested on a publically available Drone image Super-Resolution (DSR) dataset.

3 Proposed Work

The proposed architecture, Noise Identification and Compensation technique for Super-Resolution (NICSR) in the presence of noise in images is presented in Fig. 1.

The main components of NICSR are noise identification and compensation module and the super-resolution module. We carried out a study on various deep learning-based super-resolution models as discussed in Sect. 2. The Deep CNN with Skip Connection and Network in Network (DCSCN) model is selected for super-resolution in our work. DCSCN is a light weight model which is proposed in prior works for use on drone [4]. In our work, the effect of presence of noise in images on super-resolution with DCSCN model is analyzed and a method for noise identification and compensation is introduced. The noise identification module mainly comprises of a noise identification network preceded by a filter bank which generates the features to be fed into the noise identification network. A machine learning-based noise identification network is proposed, and based on the identified type of noise, the noise in the image is mitigated using the appropriate filter in the filter bank after the noise identification network. The images are then passed through the

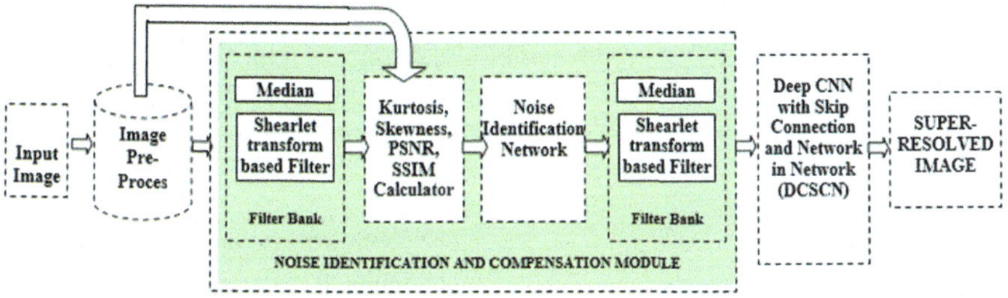

Fig. 1 Proposed architecture NICSR

DCSCN network for super-resolution. In this work, we have considered the presence of Gaussian noise, S&P noise, and combination of Gaussian and S&P noise (referred to as mixed noise in this work). The various blocks in the proposed architecture are explained in detail in the following sections.

3.1 DCSCN Network

The Deep CNN with Skip Connection and Network in Network (DCSCN) model is a convolutional neural network-based model consisting of two major blocks—feature extraction block and feature construction block in cascade as shown in Fig. 2 [12]. The feature extraction block includes a cascade of seven sets, in which a set comprises of a 3 × 3 convolutional block, a bias, and a parametric ReLU activation layer. The details of the feature extraction block and feature reconstruction block are tabulated in Tables 1 and 2, respectively, as implemented by Truong et al. [4].

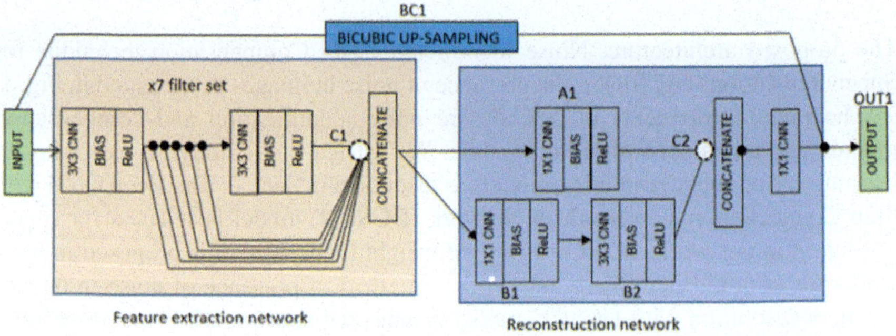

Fig. 2 Deep CNN with skip connection and network in network [12]

Table 1 Layers in the feature extraction block of DCSCN [4]

S/no	Type of layer	No of filters used	Dimension of output	Kernel size
1	Input layer	–	80X80X1	–
2	CNN	96	3X3X1	3X3X1
3	CNN	76	3X3X96	3X3X96
4	CNN	65	3X3X76	3X3X76
5	CNN	55	3X3X65	3X3X65
6	CNN	47	3X3X55	3X3X55
7	CNN	39	3X3X47	3X3X47
8	CNN	32	3X3X39	3X3X39
9	Feature map	–	80X80X410	–

Table 2 Layers in the image reconstruction block of DCSCN [4]

S/No	Type of layer	No of filters used	Dimension of output	Kernel size
1	C1	–	80 × 80 × 410	
2	A1	64	80 × 80 × 64	1 × 1 × 410
3	B1	32	80 × 80 × 32	1 × 1 × 410
4	B2	32	80 × 80 × 32	3 × 3 × 32
5	C2		80 × 80 × 96	
6	OUT1	16	80 × 80 × 16	1 × 1 × 96
7	BC1		80 × 80 × 16	
8	Final output		80 × 80 × 16	
9	C1	–	80 × 80 × 410	

3.2 *Noise Identification and Compensation Block*

Noise identification and compensation module is the core of the proposed architecture. Identification of the predominant type of noise in the image is a crucial step. Our noise identification network is mainly inspired by the work in the literature that has identified that an artificial neural network can be trained to identify the type of noise in the image based on statistical parameters such as kurtosis and skewness calculated on the filtered image [18]. Kurtosis ("K") measures the peakedness of the distribution relative to a standard Gaussian distribution [18].

$$K = \sum_{i}^{j} \frac{I(i,j) - \mu^{(4)}}{\sigma^{(4)}} \tag{6}$$

Skewness ("S") measures the asymmetry of the probability distribution in comparison to the normal distribution [19]. Skewness is given by

$$S = \sum_{i}^{j} \frac{I(i,j) - \mu^{(3)}}{\sigma^{(3)}} \tag{7}$$

where "μ" is the first central moment of probability distribution, "σ" is the standard deviation, and $I(i,j)$ is the image with $i \times j$ pixels.

$$\mu = \frac{1}{M \times N} \sum_{i=1}^{M} \sum_{j=1}^{N} I(i,j) \tag{8}$$

$$\sigma = \sqrt{\frac{\sum_{i=1}^{M} \sum_{j=1}^{N} (I(i,j) - \mu)^2}{M \times N}} \tag{9}$$

where M and N are row and column index of an image $I(i, j)$. Previous works in literature have utilized filters such as wiener filter [16], median filter [17, 20], BM3D filter [21], bilateral filter [22], Gaussian filter [23], and Shearlet transform-based filtering [24] for compensating Gaussian or S & P noise in images.

Test and train datasets were created using the images in DSR dataset as discussed in Sect. 4. Noise was added into some of the images in the training dataset as mentioned in Sect. 4. The training and testing dataset thus comprised of images of four categories—images without noise, images with Gaussian Noise, images with S&P noise, and images with both Gaussian and S&P noise. The images from training dataset were fed into the filters as mentioned in Table 5 and average values of Peak Signal-to-Noise Ratio (PSNR) and Structural Similarity Index (SSIM) of the filtered images with respect to the input images were calculated. The filters used in our work are described in Algorithm 1. The filter which gave the best performance (in terms of PSNR and SSIM) for each type of noise was included in the filter bank in Fig. 1. In our work, we have utilized the analytical parameters PSNR and SSIM in addition to the statistical parameters kurtosis and skewness for noise identification. Kurtosis and skewness were calculated on the filtered image with respect to the input image. A labeled dataset with each row in the dataset comprising of the features—kurtosis, skewness, PSNR, and SSIM and the corresponding label (indicating one of the four image categories)—was created. The labeled dataset so created was used for training the noise identification network. For noise identification, various machine learning networks such as Artificial Neural Networks (ANN), K-Nearest Neighbor (KNN), Decision trees, and Random forests (RF) were considered. The machine learning models were trained on the labeled dataset and the hyperparameters of the network were tuned until the accuracy saturated for each model, as discussed in Sect. 4. The proposed noise identification network was able to effectively identify the type of noise present in the input image. After the identification of the type of noise, another filter bank was used to filter out the noise from the input images based on the identified type of noise. The noise compensated image is then passed through the DCSCN network for super-resolution.

4 Performance Analysis

The performance analysis of the proposed architecture for super-resolution presented in Fig. 1 is described in this section. A total of 1000 images which are publically available (of size 1000×1000) from the Drone image Super-Resolution(DSR) dataset captured using CMOS tele-camera were used to evaluate the proposed technique's performance. 700 images were used for training of the noise identification network and 300 images were used for testing the proposed architecture. The same 1000 images from the CMOS tele-camera were converted to grayscale images and were also utilized as input images (training and testing dataset in the ratio 7:3) in our work. The noise was added to both RGB and grayscale images. Gaussian, S&P, and mixed noise are the types of noises used for analysis in this work. One fourth of the images in the training dataset (comprising of 700 images) and training dataset (comprising of 300 images) were superimposed with Gaussian, S&P, and mixed noise, respectively. The noise generation procedure was implemented using random noise function in scikit-image library [25]. Gaussian noise was generated by converting the input image into a numpy array and subsequently adding the same size array consisting of randomly generated values from normal distribution with zero mean ($\mu = 0$) and 10% standard deviation ($\sigma = 0.01$) [26]. S&P noise was generated by initially converting the input image to numpy array and then replacing a specific percentage of total pixels in such a way that high intensity pixels are replaced by lower pixel values and vice versa [26]. In our work, we have replaced 0.1% of the total pixels in the image. Mixed noise was generated by first introducing the Gaussian noise followed by S&P Noise.

4.1 Test Scenarios

To evaluate the performance of the proposed architecture, the following test scenarios were designed.

4.1.1 Scenario 1: To Test the Effect of Super-Resolution Using DCSCN in the Presence of Noise

An experiment was conducted to understand the effect of noise on the PSNR and SSIM of the super-resolved images. DCSCN network as specified in Sect. 3.1 was utilized as the super-resolution network. Randomly selected sets of 75 grayscale images from DSR dataset were superimposed with the Gaussian Noise, S&P noise, and mixed noise, respectively. The images were then super-resolved using DCSCN. Also, their corresponding original images (without noise) were super-resolved. The average PSNR and SSIM of the original images with respect to the corresponding noisy images is presented in Scenario-1 in Table 3. The notations used in Table 3 are explained in Table 4. Also, the average PSNR and SSIM of the super-resolved original images (images without noise) with respect to the corresponding super-resolved noisy image (noisy image as input) are mentioned as Scenario-2 in Table 3. It can be inferred from Table 3 that the PSNR and SSIM of super-resolved noisy image are poor as seen in Scenario 2 when compared to Scenario 1 presented in Table 3. It can be understood that the DCSCN is not inherently capable of compensating for the noise and enhances the noisy pixels along with the image pixels (Table 6).

4.1.2 Scenario 2: Selection of Appropriate Filters for the Filter Bank

We designed the various filters such as wiener (WI), median (Me), BM3D (BM), Bilateral (BL), Gaussian(GA), Shearlet transform(ST), and Shearlet transform

Table 3 Performance of DCSCN Network

Image type	Nomenclature	Reference Image	PSNR	SSIM	Image type	Nomenclature	Reference Image	PSNR	SSIM
Scenario-1 (Without application of DCSCN Network)					Scenario-2 (With application of DCSCN Network)				
Image With Gaussian noise	Image_G	Image_True	20.20	0.27	Image With Gaussian noise	Image_G _SR	Image_True_SR	15.03	0.19
Image with S&P noise	Image_ SP	Image_True	25.13	0.76	Image with S&P noise	Image_SP_SR	Image_True_SR	22.33	0.72
Image with both Gaus- Sian and S&P Noise	Image_ Mixed	Image_True	19.05	0.24	Image with both Gaussian and S&P Noise	Image_ Mixed _SR	Image_True_SR	14.66	0.18

Table 4 Notation used for images and their description

Notation	Remark	Notation	Remark
Image_True	Input image without noise	Image_G	Image superimposed with Gaussian Noise
Image_SP	Image superimposed with S&P Noise	Image_Mixed	Image superimposed with both Gaussian and S&P Noise
Image_G_SR	Output of DCSCN network when fed with Image_G as input	Image _SP_SR	Output of DCSCN network when fed with Image_SP as input
Imge_Mixed_SR	Output of DCSCN network when fed with Image_Mixed as input	Image_True_SR	Output of DCSCN network when fed with Image_True as input
NC_ Image_G_SR	Output of DCSCN network when fed with noise compensated Image_G as input	NC_Image _SP_SR	Output of DCSCN network when fed with noise compensated Image_SP as input
NC_Image_Mixed_SR	Output of DCSCN network when fed with noise compensated Image_Mixed as input		

Table 5 Efficacy of filtering techniques on B channel

Filter	Gaussian PSNR (dB)	Gaussian SSIM	S&P PSNR (dB)	S&P SSIM	Mixed PSNR (dB)	Mixed SSIM
Wiener	11.12	0.51	24.11	0.78	11.09	0.46
Median	27.50	0.57	42.08	0.97	27.38	0.57
BM3D	31.51	0.73	31.92	0.72	19.05	0.24
Bilateral	29.07	0.79	30.43	0.75	31.20	0.72
Gaussian	30.17	0.72	34.40	0.88	29.26	0.68
ST	31.28	0.74	32.09	0.75	30.57	0.72
ST_M	31.26	0.72	31.89	0.71	30.58	0.72

followed by median filtering (ST_M) as mentioned in Algorithm 1. Randomly selected sets of 75 images each from the DSR dataset were superimposed with Gaussian, S&P noise, and mixed noise as mentioned in Sect. 4. The SSIM and PSNR were calculated for each channel of the RGB image. The PSNR and SSIM of the filtered image (at the output of the first filter bank as shown in Fig. 1) were calculated with respect to the input image with noise and the results are tabulated in Table 5. From the Table 5, it can be concluded that Gaussian and Mixed noise (both Gaussian and S&P noise) can be best compensated using Shearlet transform-based filter. At the same time, the median filter has the best efficacy to compensate S&P noise. For grayscale images also it was found that the Gaussian and Mixed noise can be best compensated

Table 6 Efficacy of filtering techniques on grayscale images

Filter	Gaussian PSNR (dB)	Gaussian SSIM	S&P PSNR (dB)	S&P SSIM	Mixed PSNR (dB)	Mixed SSIM
NIL	20.20	0.27	25.13	0.76	19.05	0.24
Wiener	20.74	0.63	20.76	0.63	20.71	0.62
Median	27.16	0.58	38.1	0.98	27.05	0.57
BM3D	20.19	0.27	25.14	0.76	19.05	0.24
Bilateral	29.07	0.79	29.34	0.82	28.33	0.75
Gaussian	28.76	0.66	32.35	0.87	27.74	0.62
ST	31.56	0.87	32.22	0.89	31.18	0.87
ST_M	31.15	0.87	31.71	0.88	30.82	0.86

using Shearlet transform-based filter and the median filter offers the best efficacy to compensate S&P noise among the selected filters as shown in Table 7 and Figs. 4, 5, 6, 7, 8 and 9. The filter bank for noise identification and compensation network was hence designed using median filter and Shearlet transform-based filter. Since similar results were obtained for RGB and grayscale images, we have further tested the proposed architecture as presented in Fig. 1 on grayscale images (Fig. 3).

Algorithm 1 Pseudo code of filters in filter bank

```
Input:
I= Input image
Function Weiner(I):
    im = numpy.asarray(I) // Convert image into numpy array
    out=scipy.signal.wiener(im, mysize=3, noise=None)[27] //out=Filtered output with the same shape as 'im'.
    return out
Function Median(I):
    im = numpy.asarray(I)//Convert image into array]
    out= scipy.ndimage.median filter(im, mode=' nearest' , cval=0.0, origin=0)
    // out=Filtered output with the same shape as 'im' .
    return out
Function BM3D(I):`
    out = bm3d.bm3d(I, sigma psd=30, stage arg=bm3d.BM3DStages.ALL STAGES) [28]
    return out
Function Bilateral(I):
    out=cv.bilateralFilter(I, out, int d=7, double sigmaColor=65, double sigmaSpace=65, int borderType = BOR
    DER_DEFAULT ) );
    //out- Output image of the same size and type as I. return out
Function Gaussian(I):
    im = numpy.asarray(I)//Convert image into array]
    out=skimage.filters.gaussian(im, sigma=1, output=None, mode=' nearest' , cval=0) [26] //out= the filtered ar-
ray
    return out
Function Shearlet(I):
    X=cv2.imread(I)
    shearletSystem = pyshearlab.SLgetShearletSystem2D (useGPU=0, X.shape[0], X.shape[1], nScales=3)
    coeffs=pyshearlab.SLsheardec2D(X,shearletSystem)
    Xrec=pyshearlab.SLshearrec2D(coeffs,shearletSystem)
    out=skimage.filters.median(Xrec)//for shearlet filter followed by median filtering return out
```

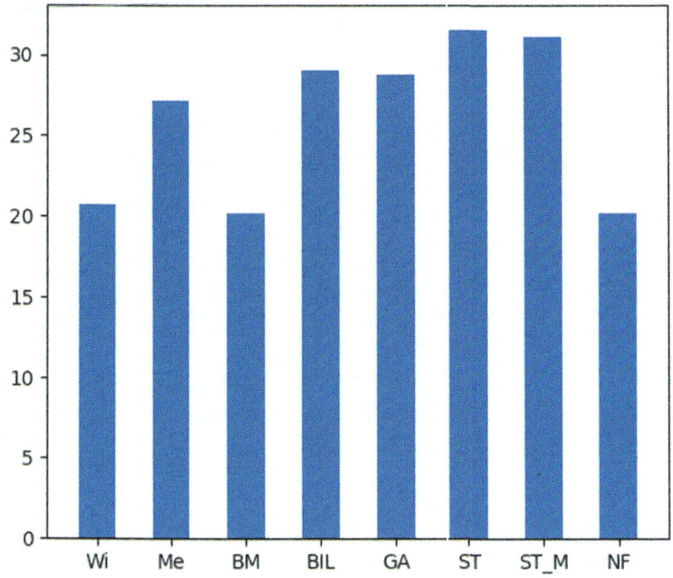

Fig. 3 PSNR for Gaussian noise

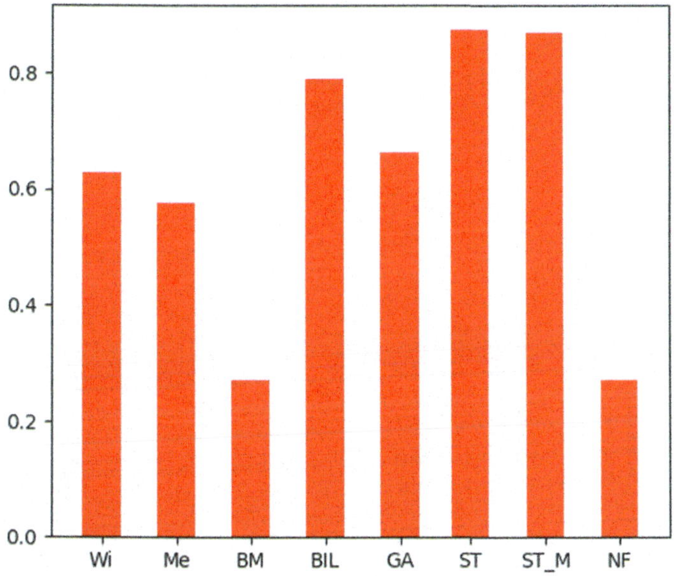

Fig. 4 SSIM for Gaussian noise

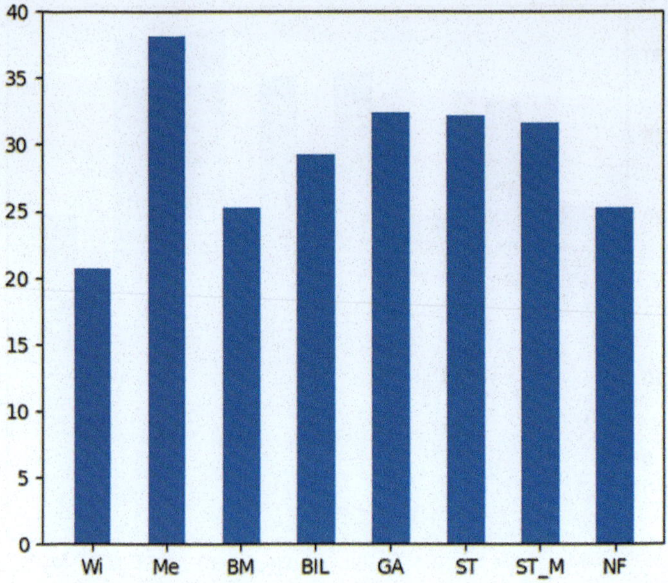

Fig. 5 PSNR for salt & pepper noise

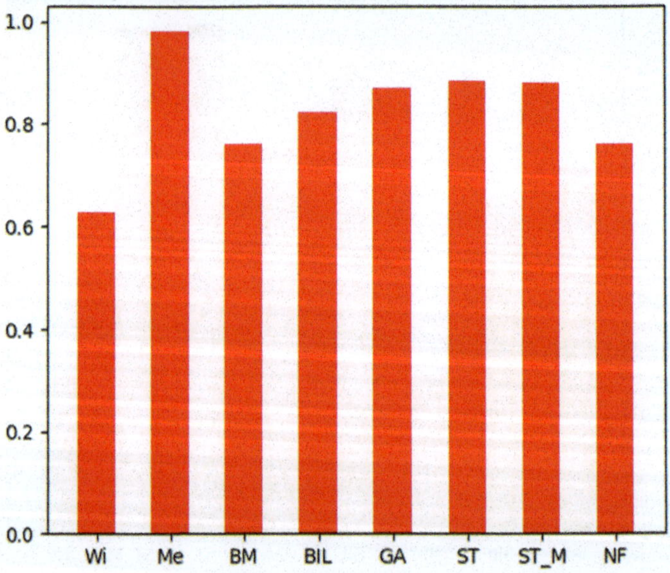

Fig. 6 SSIM for salt & pepper noise

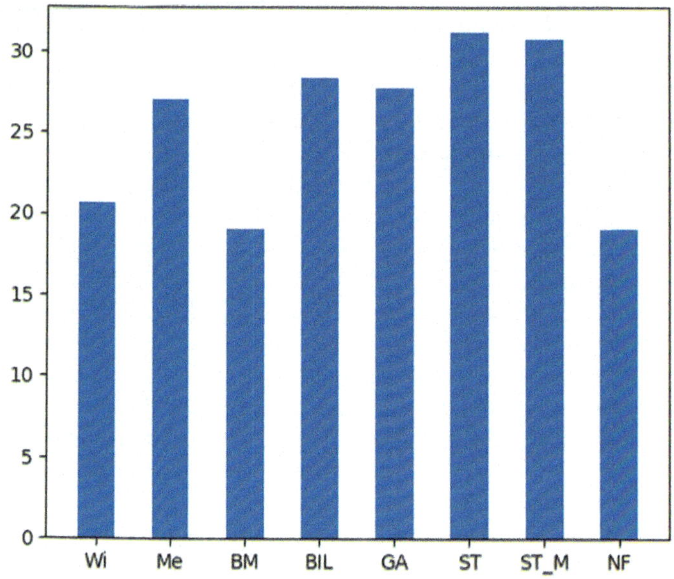

Fig. 7 PSNR for mixed noise

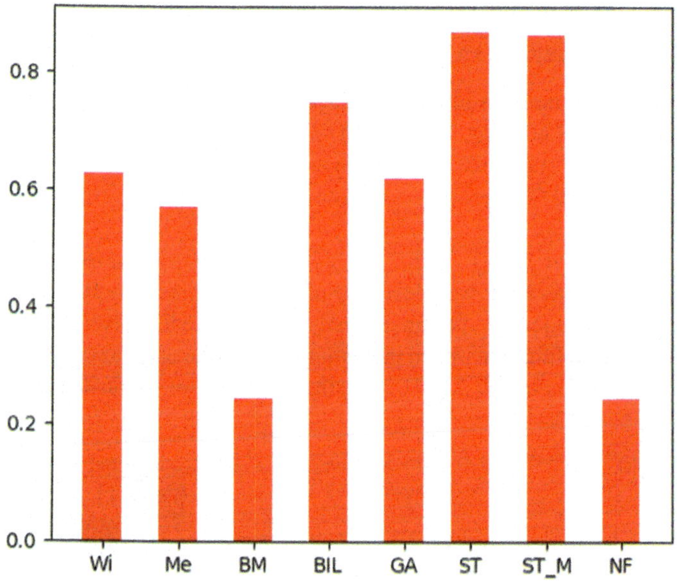

Fig. 8 SSIM for mixed noise

4.1.3 Scenario 3: To Identify the Suitable Noise Identification Network

To identify the type of noise present in the input image, a set of 1000 images were chosen from the DSR dataset. After preprocessing, these images were divided into training and testing sets in the ratio of 7:3, i.e.700 images for training and 300 images for testing. The training and testing sets were further divided into 4 sets with equal number of images. Among the 4 sets of training and testing datasets, 3 sets were superimposed with Gaussian noise, S & P noise, and Mixed noise, respectively. The images from the training set were passed through the filter bank and the parameters Kurtosis, Skewness, PSNR, and SSIM w.r.t. the input images for each filter in the filter bank were recorded for each image and a labeled dataset was prepared. Machine learning techniques such as Artificial Neural Network (ANN), K-neighbors classifier (KNN), Decision tree classifier, and Random forest classifier were explored to identify the type of noise in the images into the four categories—Images without noise, Images with Gaussian Noise, Images with S&P noise, and Images with mixed noise. The inputs to the noise identification network were the kurtosis, skewness, PSNR, and SSIM based on each of the filter in the filter bank. The trained machine learning model was then tested on the testing dataset for performance analysis. The testing dataset comprises of four sets of 75 images—images without noise (referred as LR), images with Gaussian noise, images with S&P noise, and images with mixed noise. The overview of each of the model is presented below.

Artificial Neural Network (ANN)

An ANN with eight neurons at the input layer, four neurons at the output layer, one hidden layer with 18 neurons, leaky ReLU activation function for input and hidden layer, and softmax activation for output layer was used. A testing accuracy of 99.0% and an F1-score of 0.98 were achieved and the corresponding confusion matrix is presented in Fig. 10.
Fig. 10KNN classifier

K-Neighbors Classifier was designed using sklearn.neighbors library with the following hyperparameters—Number of neighbors to use for k-Neighbors queries = 5, "uniform" weight function issued in prediction, leafsize = 30, and power parameter for the Minkowski metric = 2 [27]. A testing accuracy of 93.66% and F1-score of

Fig. 9 ANN classifier

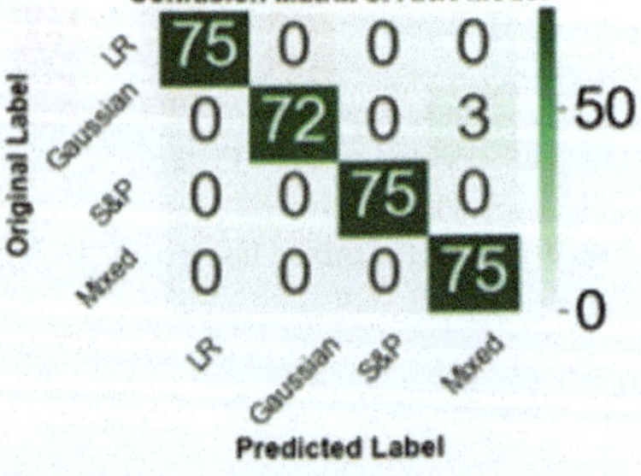

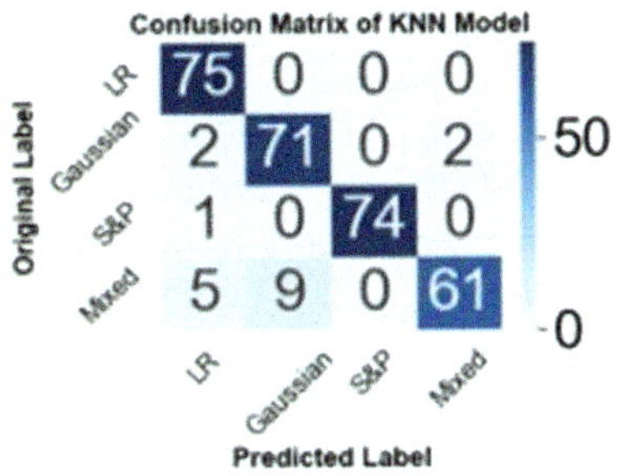

Fig. 10 KNN classifier

0.93 were achieved and the confusion matrix for the test dataset is presented in Fig. 11.

Decision Tree (DT) Classifier was designed using sklearn.tree library with the following hyper parameters: Criterion to measure quality of split = entropy, splitter (criterion to choose the best split at each node) = "best", the minimum number of samples required to split an internal node = 2, the minimum number of samples required to be at a leaf node = 2, and the minimum weighted fraction of the sum total of weights (of all the input samples) required to be at a leaf node. = 0.0 [27]. Entropy, Gini, and log-loss criteria's were attempted to find the best quality of split and the "entropy" criteria gave the best quality of split and hence it was used in the final network. A testing accuracy of 97.66% and F1-score of 0.97 are achieved and the corresponding confusion matrix is presented in Fig. 12.

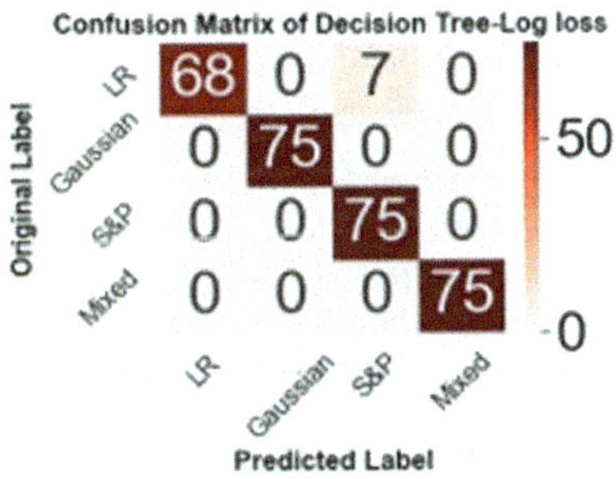

Fig. 11 Decision-tree classifier

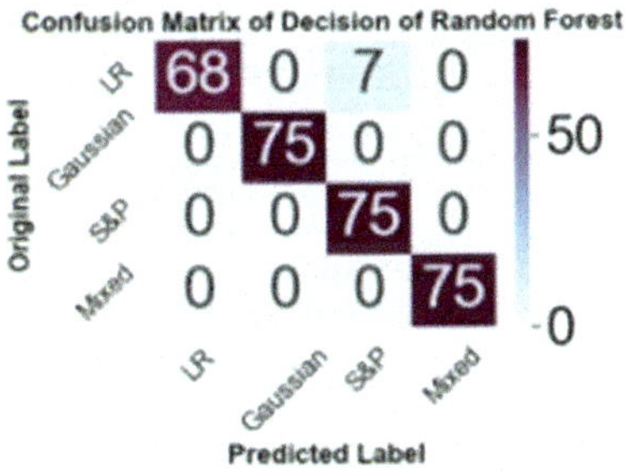

Fig. 12 RF classifier

Random Forest Classifier (RF) was designed using sklearn.ensemble library with the following hyper parameters: number of trees in forest = 4, "Gini" criteria to check the quality of split, the minimum number of samples required to split an internal node = 2, the minimum number of samples required to be at a leaf node = 1, and the number of features to be considered for the best split = 2 [27]. A testing accuracy of 97.66% and F1-score of 0.97 are achieved and the confusion matrix is presented in Fig. 12. On evaluation of the various machine learning models, it was found that ANN was able to classify the predominant type of noise with better accuracy and F1-score as compared to the other models and was thus selected to be incorporated into the proposed image super-resolution architecture in Fig. 1. for noise identification.

4.1.4 Scenario 4: Performance Analysis of Fully Integrated System

Testing dataset comprising of 300 input images from DSR dataset was divided into four sets of 75 images each. Three sets were augmented with the Gaussian noise, S & P Noise, and Mixed Noise, respectively, as discussed in Sect. 4. The images from the dataset were fed to the pre-trained noise identification network, which recognizes the type of noise present in the images, followed by the noise compensation based on the identified type of noise. The noise compensated images were super-resolved using DCSCN network. Output images from the proposed model NICSR were evaluated on the parameters SSIM and PSNR. The performance results are tabulated in Table 7. The notations used in Table 7 are summarized in Table 3. Scenario 1 in Table 7 presents the PSNR and SSIM of the super-resolved true images (images without noise) and the corresponding super-resolved noisy images using only the DCSCN network. Scenario 2 in Table 7 presents the PSNR and SSIM of the super-resolved true images (images without noise) and the corresponding super-resolved image using NICSR architecture. From Scenario-2 presented in Table 7, it can be concluded that the proposed architecture is capable of improving the performance of super-resolution when the noisy images are supplied as input. Results summarized in Table 7 indicate that NICSR was able to improve the SSIM by 277% and PSNR by 48.1% when tested on the DSR dataset for mixed noise. Results summarized in Table 7 indicate that the NICSR was able to improve the SSIM by 26.3% and PSNR by 12.09% when tested on DSR dataset for S& P. Results summarized in Table 7 indicate that the proposed technique was able to improve the SSIM by 310.5% and PSNR by 51.1% when tested on DSR dataset for Gaussian noise.

5 Conclusion

With the use of super-resolution techniques, the low-resolution cameras can be efficiently used for onboard decision-making on nano/micro/small category drones, thus reducing the drone weight and power consumption while ensuring better endurance

Table 7 Performance analysis

Image1	Image2	PSNR (dB)	SSIM	Image1	Image2	PSNR (dB)	SSIM
Scenario 1-Original DCSCN				Scenario 2-Proposed Architecture-NICSR			
Image_G_SR	Image_True_SR	15.03	0.19	NC_ImageG_SR	Image_True_SR	22.72	0.78
Image_SP_SR	Image_True_SR	22.33	0.72	NC_Image_SP_SR	Image_True_SR	25.03	0.91
Image_Mixed_SR	Image_True_SR	14.66	0.18	NC_Image_Mixed_SR	Image_True_SR	21.72	0.68

and range. A novel architecture, NICSR, i.e. Noise Identification and Compensation technique for Super-Resolution of noisy images captured by a drone is proposed in this work. A noise identification module was designed which consists of a set of noise filters and ANN for the identification of the predominant type of noise. The noise compensation module consists of a filter bank from which an appropriate filter is applied based on the predominant type of noise identified. Results summarized in Table 7 indicate that the proposed technique was able to improve the SSIM performance by 277% and PSNR performance by 48.1% when tested on the DSR dataset for mixed noise. Similar improvement in performance was observed for images with Gaussian as well as S & P images mentioned in Table 7. We have utilized DCSCN as the SISR network in NICSR; however, the proposed noise identification and compensation network can be explored for use along with other SISR networks as well. The work will be tested on more datasets or on drones in the future.

References

1. Miranda VR, Rezende A, Rocha TL, Azpu´rua H, Pimenta LC, Freitas GM (2022) Autonomous navigation system for a delivery drone. J Control Autom Electr Syst 33(1):141–155
2. Pu¨tsep K, Rass˜olkin A (2021) Methodology for flight controllers for nano, micro and mini drones classification. In: 2021 international conference on engineering and emerging technologies (ICEET), pp 1–8. IEEE
3. Shenoy MV, Roy A, Misra S (2021) QoI-aware camera network-as-a-service for social behavior analysis. In: ICC 2021—IEEE international conference on communications, Montreal, QC, Canada, pp 1–6
4. Truong NQ, Nguyen PH, Nam SH, Park KR (2019) Deep learning-basedsuper-resolution reconstruction and marker detection for drone landing. IEEE Access 7:61639–61655
5. Yang W, Zhang X, Tian Y, Wang W, Xue JH, Liao Q (2019) Deep learning for single image super-resolution: a brief review. IEEE Trans Multimedia 21(12)
6. Wang Z, Chen J, Hoi SC (2020) Deep learning for image super-resolution: a survey. IEEE Trans Pattern Anal Mach Intell 43(10):3365–338

7. Mafi M, Martin H, Cabrerizo M, Andrian J, Barreto A, Adjouadi M (2019) A comprehensive survey on impulse and Gaussian denoising filters for digital images. Signal Process 157
8. Al-Hatmi MO, Yousif JH (2017) A review of Image Enhancement Systems and a case study of Salt&pepper noise removing. Int J Comput Appl Sci (IJOCAAS) 2(3):171–176
9. Dong C, Loy CC, He K, Tang X (2015) Image super-resolution using deep convolutional networks. IEEE Trans Pattern Anal Mach Intell 38(2):295–307
10. Kim J, Lee JK, Lee KM (2016) Deeply-recursive convolutional network for image super-resolution. In: Proceedings of the IEEE conference on computer vision and pattern recognition, pp 1637–1645
11. Lu Z, Liu H, Li J, Zhang L (2021) Efficient transformer for single image super- resolution. arXiv preprint arXiv:2108.11084
12. Yamanaka J, Kuwashima S, Kurita T (2017) Fast and accurate image super resolution by deep CNN with skip connection and network in network. In: International conference on neural information processing, pp 217–225. Springer, Cham
13. Lin X, Ozaydin B, Vidit V, ElHelou M, Su¨sstrunk S (2023) DSR: Towards drone image super-resolution. In: Karlinsky L, Michaeli T, Nishino K (eds) Computer vision–ECCV 2022 workshops. Lecture Notes in Computer Science, vol 13802. Springer, Cham.
14. Goyal B, Dogra A, Agrawal S, Sohi BS, Sharma A (2020) Image denoising review: from classical to state-of-the-art approaches. Inform Fusion 55:220–244
15. Gonzalez RC, Woods RE (2009) Digital image processing. Second Edition, Pearson Education India
16. Benesty J, Chen J, Huang Y (Arden), Doclo S (2005) Study of the wiener filter for noise reduction. In: Speech enhancement
17. Oddi L, Cremonese E, Ascari L, Filippa G, Galvagno M, Serafino D, Cella UMD (2021) Using UAV imagery to detect and map woody species encroachment in a subalpine grassland: advantages and limits. Remote Sens 13(7):1239
18. Vasuki P, Bhavana C, Roomi SMM, Deebikaa EL (2012) Automatic noise identification in images using moments and neural network. In: 2012 international conference on machine vision and image processing (MVIP), pp 61–64. IEEE
19. Wang Z, Bovik AC, Sheikh HR, Simoncelli EP (2004) Image quality assessment: from error visibility to structural similarity. IEEE Trans Image Process 13(4):600–612
20. Erkan U, G¨okrem L, Engino˘glu S (2018) Different applied median filter in salt and pepper noise. Comput Electr Eng 70
21. Djurovi´c I (2016) BM3D filter in salt-and-pepper noise removal. Eurasip J Image Video Process (1)
22. Routray S, Malla PP, Sharma SK, Panda SK, Palai G (2020) A new image denoising framework using bilateral filtering based non-subsampled shearlet transform. Optik, 216
23. Ku¸s Z, Aydin M (2022) Removal of Gaussian distributed noise in images with deep neural network models. 2022 30th signal processing and communications applications conference (SIU), Turkey, pp 1–4
24. Chen F, Huang M, Ma Z, Li Y, Huang Q (2020) An iterative weighted-mean filter for removal of high-density salt-and-pepper noise. Symmetry 12(12). https://doi.org/10.3390/sym12121990
25. The scikit-image team (2024) Skimage filters, skimage filters skimage 0.22.0 documentation. Available at: https://scikit-image.org/docs/stable/api/skimage.filters.html. Accessed 7 January 2024
26. Numpy Developers (2024) Numpy asarray, numpy asarray-NumPyv 1.26 Manual. Available at:https://numpy.org/doc/stable/reference/generated/numpy.asarray.html. Accessed 7January 2024
27. The SciPy community (2024) Scipy signal wiener, scipy signal wiener-SciPyv 1.11.4 Manual. Available at: https://docs.scipy.org/doc/scipy/reference/generated/. Accessed 7 January 2024

Latest Electrical and Electronics Trends, Security and Privacy Issues

Multilevel Chaotic Maps with Differential Encoding-Based Encryption Method for DICOM Images

Suvita Rani Sharma, Birmohan Singh, and Manpreet Kaur

Abstract This paper proposes an encryption method based on multiple chaotic maps and differential encoding method to secure the medical images. The proposed method uses a permutation-based image scrambling method in the confusion phase. In the diffusion phase, Piecewise Linear Chaotic Map (PWLCM), Piecewise Chaotic Map (PWCM), and tent map with differential encoding method have been employed at different levels of the phase to secure the medical images. The model has been tested on Digital Imaging and Communications in Medicine (DICOM) images. The proposed encryption method demonstrates robust security against statistical and entropy attacks, with encrypted images exhibiting a correlation close to zero and achieving a maximum entropy value of 7.9993. The low SSIM value between the original and encrypted images confirms their dissimilarity. Higher Mean Square Error (MSE) values and lower Peak Signal-to-Noise Ratio (PSNR) values for encrypted images prove a negligible presence of original pixel values. Additionally, the encryption approach showcases resilience against differential attacks, with a Number of Pixel Change Rate (NPCR) value of 99.6% and a Unified Averaged Changed Intensity (UACI) value of 33.4%, both within acceptable standard ranges. The effectiveness of the proposed method demonstrates that these encrypted DICOM images are safe against various forms of attacks.

Keywords Chaotic maps · Decryption · Dicom image · Differential encoding · Encryption

S. R. Sharma (✉) · B. Singh
Department of Computer Science and Engineering, Sant Longowal Institute of Engineering and Technology, Longowal, Sangrur, Punjab, India 148106
e-mail: suvita.sharma1204@gmail.com

M. Kaur
Department of Electrical and Instrumentation Engineering, Sant Longowal Institute of Engineering and Technology, Longowal, Sangrur, Punjab, India 148106

P. J. S. Gonçalves et al. (eds.), *Proceedings of Fifth International Conference on Computing, Communications, and Cyber-Security*, Lecture Notes in Networks and Systems 1128, https://doi.org/10.1007/978-981-97-7371-8_34

1 Introduction

The security of digital medical records from fraud and forgery is crucial, when these are delivered over the internet. Encryption is a widely used approach to secure images from the different types of attacks that occur during the transmission of images. Different researchers have proposed encryption methods to secure medical images. In the year 2017, Ravichandran et al. proposed an encryption method utilizing Deoxyribo Nucleic Acid (DNA) and chaotic maps techniques to encrypt the DICOM images. The encryption method was able to encrypt selective/partial or full DICOM images [14]. In the year 2018, Hua et al. introduced an encryption approach using random data insertion, high-speed scrambling, and pixel adaptive diffusion components for the grayscale and colored medical image [9]. In the year 2018, Arumugham et al. proposed a watermarking method to hide patients' information on the medical images and then propounded an encryption algorithm to secure the watermarked image based on the logistic map, Lorenz attractor, and Piecewise Linear Chaotic Map (PWLCM) [2]. In the year 2019, Dagadu et al. explored the Bernoulli shift map and Zigzag map with the DNA method. The parameters of chaotic maps were generated using the message digest algorithm five hash function [7]. In the year 2019, Guan et al. addressed an images encryption based on chaotic maps and DNA encoding in the frequency domain for the encryption of medical images [8]. In the same year, Ravichandran et al. created a five-layer encryption system which was implemented on reconfigurable Cyclone II Field Programmable Gate Array (FPGA) for the security of medical images [15]. Again in the same year, Chidambaram et al. proposed three-layer encryption methods based on DNA encoding with logistic and tent maps to secure the multimedia data during file sharing over the internet [6]. In the year 2020, Banu and Amirtharajan introduced an image encryption-based fusion of chaotic 3D Lorenz attractor, logistic maps, wavelet transform, and Deoxyribonucleic Acid (DNA) sequence on the spatial domain of the medical images [3]. In the year 2021, Ravichandran et al. suggested an encryption method based on DNA hybrid with chaos and Integer Wavelet Transform (IWT) for the protection of medical images. The authors used bitwise circular shift for the row and column-wise scrambling [16]. In the same year, Aashiq et al. created an encryption method based on DNA modified with Lorenz and Lü chaotic attractors to provide high pseudo-random key streams. The authors used a bio-inspired crossover and mutation unit operating on the IWT domain. Also, a quantized chaotic set from the created coupled attractors was used to execute an XOR operation [1]. In the year 2023, John and Kumar suggested a medical image encryption approach based on a two-dimensional Lorentz chaotic system with a logistic chaotic system. In order to obtain the encrypted image, row-wise and column-wise scrambling processes were used, and then bitwise XOR operations were performed to generate the encrypted image [10].

It has been determined from the literature that the encryption methods have used single chaotic maps. Single chaotic map-based encryption techniques are simpler and require less computing power, yet these methods are predictable which results in less security. Because of the more intricate dynamical features, the researchers either combine chaotic maps or employ high-dimensional chaotic maps to encrypt the images. Considering this as motivation, a method for encryption based on multiple chaotic maps, differential encoding, and circular shift has been developed for the security of medical images. The contributions of the work are as follows:

- The proposed encryption method incorporates multiple chaotic maps to increase the randomness by adding chaotic properties in the encrypted images
- The techniques such as differential encoding and circular shift are adopted in the proposed encryption method to incorporate the bit-level security and resistance against differential attacks
- The method has been utilized to secure sensitive DICOM medical images of different types
- The efficacy of the proposed encryption method has been validated using extensive security analysis and compared to several existing image encryption techniques for the DICOM medical images.

The rest of the paper is organized as: The proposed methodology is discussed in Sect. 2. The results and discussion of the proposed encryption method are provided in Sect. 3. Finally, Sect. 4 concludes the work.

2 Proposed Encryption Method

The proposed encryption method works in two phases: confusion and diffusion phase. The confusion phase of the proposed method shuffles the input image pixel values to reduce the correlation between pixel values of an image. The shuffling of the pixel has been performed based on the numbers generated using the random permutation method. Thereafter, the scrambled pixel values of the images are altered in the diffusion phase. The diffusion phase of the proposed method works in three levels in which three chaotic maps are used with the differential encoding method and circular shift to alter the image pixel values. The block diagram of the proposed methodology is given in Fig. 1. A detailed discussion of the encryption method phases has been provided in the following subsections.

2.1 Confusion Phase

The confusion phase of the encryption method starts with converting the input image into one-dimensional array. Based on the size of the array, random numbers are generated using the permutation method. The generated random numbers are used as new pixel position values to scramble the input image. The image pixel values are shuffled based on the generated random numbers values and then the shuffled pixel values are binarized for utilization in the next phase of the encryption method for further processing.

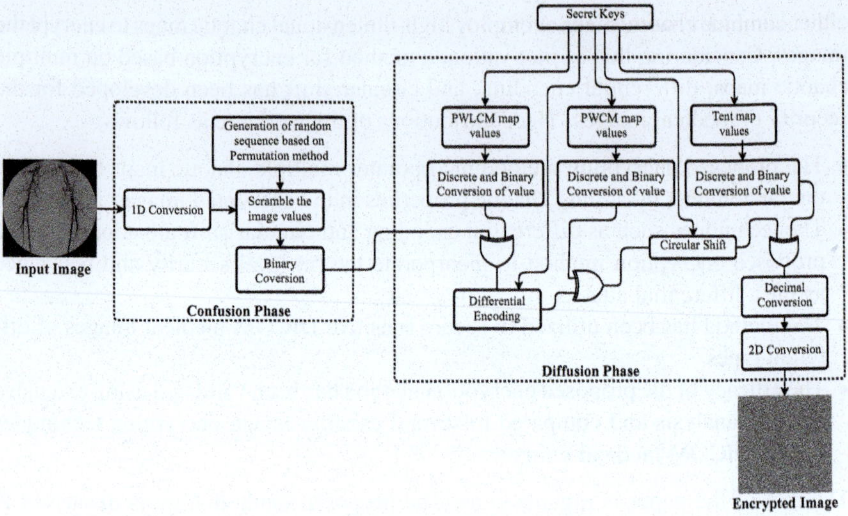

Fig. 1 Block diagram of proposed encryption method

2.2 *Diffusion Phase*

The binary shuffled pixel values of the image are altered in the diffusion phase of the proposed encryption method. In this phase, Piecewise Linear Chaotic Map (PWLCM) [19], differential encoding [21], Piecewise Chaotic Map (PWCM) [17], circular shift [16], and tent maps [12] has been used. A detailed description of different levels of the diffusion phase is as follows:

First Level: The PWLCM is used at this level of the diffusion phase. The PWLCM values are created using the following formula:

$$l_x = F(l_{x-1}, \eta_1) = \begin{cases} l_{x-1}/\eta_1, & 0 \leq l_{x-1} < \eta_1 \\ (l_{x-1} - \eta_1)/(0.5 - \eta_1), & \eta_1 \leq l_{x-1} < 0.5 \\ l_{x-1} = 0, & l_{x-1} = 0.5 \\ F(1 - l_{x-1}, \eta_1), & 0.5 < l_{x-1} \leq 1 \end{cases} \tag{1}$$

where the initial condition l_x and positive control parameter η_1 are $\eta_1 \in [0, 0.5]$ and $l_x \in [0, 1]$ are secret keys. The generated PWLCM values are then converted into discrete form using the following formula:

$$xm_i = \lfloor(x_i \times 10^{14})\rfloor \% \; (max(input\ value)) \tag{2}$$

where $max(input\ value)$, is the maximum pixel value of the input image. Then the binary values of the converted PWLCM values are computed and XORed operation is performed with the binarized image pixel values.

Differential Encoding: The differential encoding method is invoked in the first level of diffusion to raise the security level of the medical images by changing the bits of each pixel value. The formulation of the differential encoding method is as follows:

$$E_s = I_s \oplus E_{s-1} \tag{3}$$

where E_s represents the encoded bit sequence and I_s is the input bit of sequence. The symbol $\oplus$ shows the XOR operation. One reference bit is used in the encoding method for each encoding sequence. The reference bit is treated as a secret key in the proposed encryption method. Figure 2 shows the process of the differential encoding method.

The encoded bit sequences are decoded at the receiver side using the following formula:

$$\tilde{I}_s = \tilde{E}_s \oplus \tilde{E}_{s-1} \tag{4}$$

where $\tilde{I}_s$ is the decoded sequence and $\tilde{E}_s$ is encoded bit sequence. Figure 3 illustrates the decoding process of the differential method at the receiver side.

Second Level: The encoded bits sequences generated at the first level are XORed with the PWCM values at this level of the diffusion phase. The PWCM values are generated using the following formula:

$$m_x = \begin{cases} m_{x-1}/\eta_2, & 0 < m_{x-1} < \eta_2 \\ (m_{x-1} - \eta_2)/(0.5 - \eta_2), & \eta_2 \leq m_{x-1} < 0.5 \\ (1 - \eta_2 - m_{x-1})/(0.5 - \eta_2), & 0.5 < m_{x-1} < (1 - \eta_2) \\ (1 - m_{x-1})/\eta_2, & (1 - \eta_2) < m_{x-1} < 1 \end{cases} \tag{5}$$

where the PWCM map parameters which are consider as secrete keys η_2 and m_x values are $\eta_2 \in [0, 0.4]$ and $m_x \in [0, 1]$. The generated values of PWCM maps are converted into discrete values using the Eq. 2. The discrete PWCM values are binarized and then XORed with the sequences generated using the encoding method. The XORed values are shifted using the bitwise circular shift method to tackle the differential attacks.

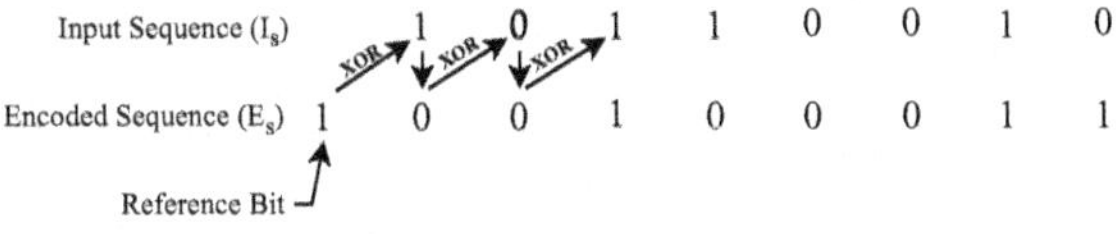

Fig. 2 Differential Encoding used during the encryption process

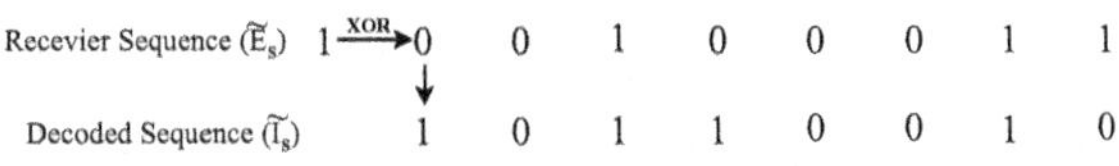

Fig. 3 Differential Encoding used during the decryption process

Third Level: At this level, a tent map is utilized to add dynamic behavior in the encrypted image pixel to enhance security. Tent map values are produced using the following mathematical equation:

$$n_x = \begin{cases} \eta_3 \times n_{x-1}, & n_{x-1} < 0.5 \\ \eta_3 \times (1 - n_{x-1}), & otherwise \end{cases} \quad (6)$$

where the positive control parameter η_3 and initial condition n_x are $\eta_3 \in [0, 2]$ and $n_x \in [0, 1]$ which are secrete keys. The generated tent map values are changed into discrete form using the Eq. 2 and then converted into binary form. The binarized tent map values are XORed with the circularly shifted values generated at the second level of the diffusion phase. The XORed values are then changed into decimal form. The decimal values are changed into 2D form which is the resultant encrypted image. The encrypted images are decrypted by following the reverse sequence of the encryption method. The encrypted image is first transformed into binary values, and then the XOR operation is applied to the tent map values acquired using the encryption secret key. The bitwise reverse circular shift is used to revert the bits. The PWCM map values are obtained by using the secret key. The reversed bits and PWCM map values are XORed together. Finally, differential decoding is done to the retrieved values, followed by an XOR operation on the data values and PWLCM map values created with the secret key. Finally, the reverse scrambling method based on the random values generated in the encryption process during the confusion phase is used to reshuffle the image to get the original image.

3 Results and Discussions

The propounded method of encryption is tested using MATLAB 2021a on Windows 10 Pro with an Intel Xenon CPU ES-2650 (2.50 GHz) and 8 GB of RAM. The encryption method is tested on twenty DICOM medical images having image size 256 × 256 and 512 × 512. The results of five DICOM images are presented in this work.

Different security analysis such as visual, histogram [14] as well as performance metrics i.e., variance [13], chi-square [16], correlation coefficients [7], entropy [9], Number of Pixel Change Rate (NPCR) [20], Unified Averaged Changed Intensity (UACI) [20], Mean Square Error (MSE) [3], Peak Signal-to-Noise Ratio (PSNR) [18], Structural Similarity Index Metric (SSIM) [2], cropping [2], and noise attack [18] is used to check the efficacy of the proposed encryption method.

The visual analysis of the original, encrypted, and decrypted image has been shown in Fig. 4. From the figure, it has been observed that the encrypted image is entirely different from the original image, whereas the decrypted images are exactly the same in visual appearance. This proves that the propounded encryption method is successfully able to hide the visual information of the medical image. The histogram analysis has been employed to determine the statistical properties of an image. The

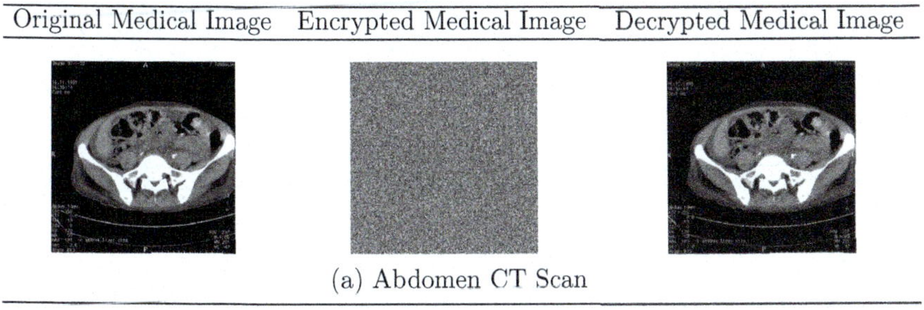

Fig. 4 Visual depiction of DICOM image

Fig. 5 Histogram diagram of DICOM image

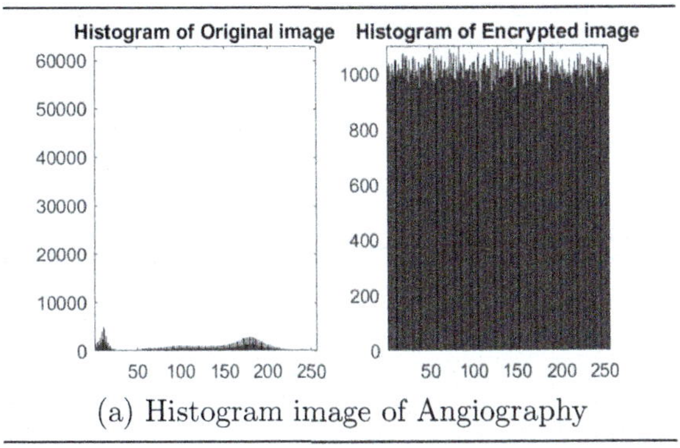

histogram of an image shows the pixel distribution of the image. Figure 5 shows the histogram plot for the uniform depiction of the encrypted DICOM image. The image pixel values have a high correlation, which must be minimized in the encrypted image to prevent information theft or leaking. Figure 6 displays a correlation diagram of image pixels along the horizontal, vertical, and diagonal axes. The correlation diagrams of the original pixel values reveal a concentration of correlation values at a straight line, indicating a strong connection between the pixel values of the original image. The encrypted image's correlation diagrams, on the other hand, show a weaker correlation since the pixel values are spread out and not concentrated in a square form. In Table 1, correlation coefficients and the entropy values of the original and encrypted images have been shown. It can be seen from the table that the correlation coefficient values for the original images are closer to one, indicating a significant connection between image pixel values. In contrast, the encrypted images have correlation coefficient values that are either near zero or negative. Furthermore, the entropy value is closer to 8 for the encrypted images, indicating that the propounded method generates the requisite amount of unpredictability in the encrypted image.

The variance and Chi-square analyses are used to validate the histogram's pixel value distribution and uniformity examination. To pass the variance test, the encrypted image's variance must be less than 5000. To be considered safe against

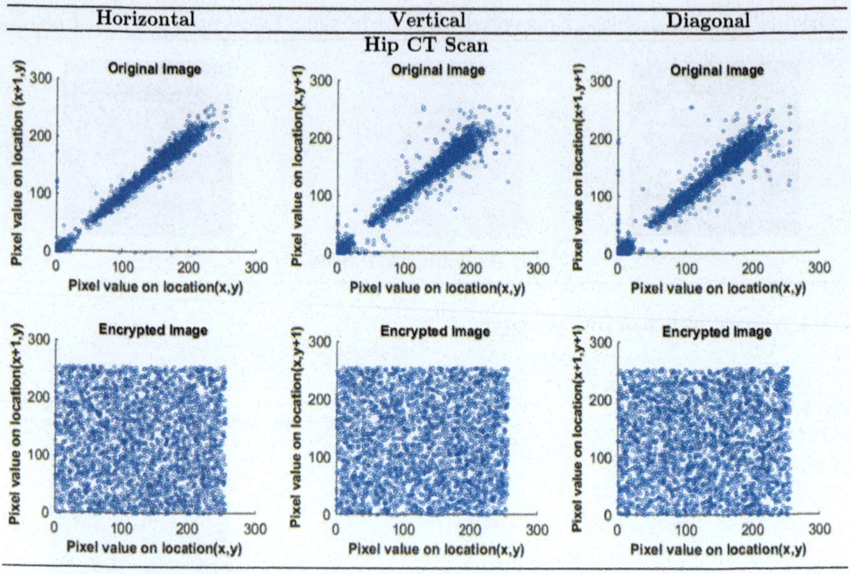

Fig. 6 Correlation diagrams of original and encrypted DICOM image

Table 1 Comparison of correlation coefficients values between original and encrypted medical images

Image name	Original image correlation			Encrypted Image correlation			Entropy	
	H	V	D	H	V	D	Original	Encrypted
Abdomen CT Scan	0.8652	0.9015	0.9.39	−0.0005	0.0304	0.0080	3.7368	7.9993
Brain MRI	0.8927	0.9091	0.9361	0.0095	−0.0011	−0.0118	6.3978	7.9974
Angiography	0.97751	0.9767	0.9879	0.0009	0.0164	−0.0311	4.9310	7.9993
Colon CT scan	0.9775	0.9922	0.9802	−0.0282	−0.0106	−0.0069	5.5491	7.9993
Hip CT scan	0.9819	0.9871	0.9904	0.0135	−0.0039	−0.0250	6.4002	7.9993

statistical assaults, the encrypted images must have a Chi-square value of less than 293. Using the SSIM value ensures that the encrypted image is structurally identical to the original image. The MSE indicates the difference between the original and encrypted image. The PSNR measure represents the peak signal-to-noise power ratio. For encrypting images, a lower PSNR value suggests increased resilience to various forms of assaults. The NPCR and UACI values are quantitative and qualitative measurements used to assess the encryption algorithm's security against differential attacks. The NPCR and UCAI analyze the influence of one-pixel change in the original image on the encrypted image. In Table 2 the performance metrics values have been given. The table shows that the obtained variance values for all encrypted images are less than 5000 and chi-square values are less than 293 indicating that the encryption method is resistant to statistical attacks. The SSIM value of the original vs. encrypted image is close to zero, indicating lesser structural similarity. When

Table 2 Comparison of variance, Chi-square, SSIM, MSE, PSNR, NPCR, and UACI values between original and encrypted DICOM images

Image name	Type of image	Variance	Chi-square	SSIM	MSE	PSNR	NPCR (%)	UACI (%)
Abdomen CT scan	Original	9.0348E+07	2.2499E+07	0.0039	1.6931E+04	5.8441	99.6017	33.5091
	Encrypted	961.0588	239.3262					
Brain MRI	Original	3.1693E+05	3.1569E+05	0.0040	1.4904E+04	6.3978	99.6093	33.3921
	Encrypted	241.2314	240.2891					
Angiography	Original	1.9653E+07	4.8942E+06	0.0081	9.7411E+03	8.2447	99.5907	33.4355
	Encrypted	1043.7352	259.9160					
Colon CT scan	Original	2.5243E+07	6.2860E+06	0.0071	1.3299E+04	6.8926	99.5998	33.5416
	Encrypted	1084.1589	269.9609					
Hip CT scan	Original	1.5700E+07	3.9097E+06	0.0066	1.2945E+04	7.0098	99.6258	33.5066
	Encrypted	1079.3781	268.7617					

encrypted images are compared to original images, the MSE score is quite high, indicating a significant difference in pixel values between the images. Similarly, lower PSNR values indicate a greater prevalence of noise in encrypted images. The propounded method is safe against differential attacks, as evidenced by an NPCR value of more than 99.6% and a UACI value greater than 33.4%.

3.1 Cropping and Noise Attacks

The cropping and the noise attack analysis are used to check the robustness of the encryption method against data loss and noise addition in the encrypted image during data transmission. In Fig. 7 encrypted images and their corresponding decrypted images have been shown after performing cropping and salt and pepper noise (0.1 density) attacks. The figure depicts that the decrypted images are visually recognizable which shows the propounded encryption method is successfully able to decrypt the image after the cropping and noise attacks. Thus, the propounded method has resilience against cropping and noise attacks.

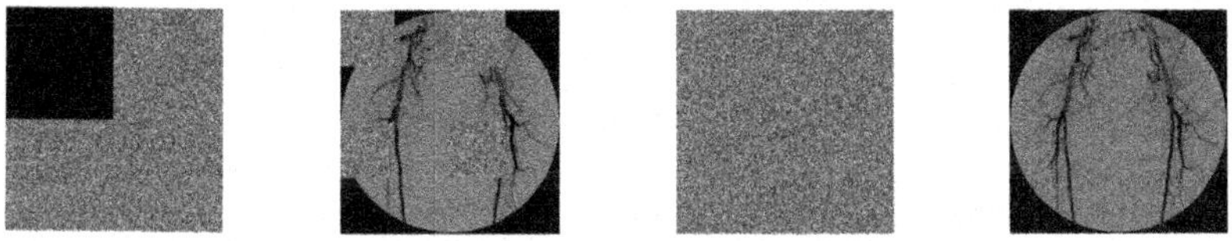

(a) Encrypted image with block cropping

(b) Decrypted image

(c) Encrypted image with salt & pepper noise

(d) Decrypted image

Fig. 7 Cropping and noise analysis of angiography DICOM image

Table 3 Comparison of propounded method with the existing approaches

Authors	Encrypted image correlation			Entropy	NPCR (%)	UACI (%)
	V	D	H			
Ravichandran et al. [14]	−0.0025	0.116	−0.0016	7.997	99.5982	33.43
Hua et al. [9]	–	–	–	7.9977	99.9974	33.27
Arumugham et al. [2]	0.0005	0.01095	0.017	7.9915	–	–
Dagadu et al. [7]	-0.0061	0.0043	−0.0016	7.9972	99.64	33.43
Guan et al. [8]	0.00019	−0.0002	0.00115	7.9924	99.63	33.61
Ravichandran et al. [15]	−0.0126	0.00053	−0.0109	15.176	99.9924	33.409
Chidambaram et al. [6]	0.0011	−0.00055	0.00815	7.9973	–	–
Kumar et al. [11]	0.0057	0.0067	0.0944	4.745	99.79	33.16
Belazi et al. [5]	−0.0049	0.0057	0.0013	7.9974	99.653	33.41
Aashiq et al. [4]	0.0027	0.0033	0.0043	7.998	99.68	33.47
Subashanthini et al. [18]	0.0075	0.036	0.0188	7.9974	99.6033	33.5584
Aashiq et al. [1]	0.0015	0.0055	0.0009	7.997	99.64	33.48
Ravichandran et al. [16]	0.00225	0.00065	0.00163	15.785	99.6067	33.47
Proposed encryption method	−0.0393	−0.0223	0.0017	7.9993	99.6124	33.4803

3.2 Comparison with the Existing Encryption Approaches

The propounded method is compared to existing approaches in order to determine its efficacy. The correlation coefficient, entropy, NPCR, and UACI metrics are employed in the comparison. Correlation coefficient analysis has been carried out for the vertical (V), horizontal (H), and diagonal (D) planes. Table 3 includes a comparison of the propounded method with existing approaches for lung CT scan images. In the table symbol—indicates that the values are not present in the papers taken for the comparison. The results of the comparative methods have been taken from the paper [16]. The table shows that the proposed method is able to obtain appropriate values for all of the image's performance parameters. Also, the entropy, NPCR, and UACI demonstrate that the propounded method is resistant to statistical and differential assaults time complexity of $O(M \times N)$. The $M \times N$ depicts the size of the original image.

4 Conclusions

In this work, a method has been propounded to secure the medical images based on the multiple chaotic maps, differential encoding, and circular shift techniques. The encryption method is divided into two phases: confusion and diffusion. The confusion phase scrambles the image pixel values based on a random permutation mechanism. In the diffusion phase, the XORing technique is performed to modify the initial values of the images at three levels of the phase. The propounded encryption method has high unpredictability because of the randomness created by the multiple chaotic maps. Also, the bit-level security has been provided by the propounded encryption method utilizing differential encoding and bitwise circular shift methods. The experiments are done on different sizes of DICOM medical images. Different analyses such as histogram, variance, chi-square, SSIM, MSE, PSNR, NPCR, UACI, correlation, and entropy analysis have been used to examine the resistance of the propounded method against different sorts of attacks. The results show that the propounded methodology is able to resist statistical, entropy, and differential attacks efficiently. In the future, the propounded method can be adopted for the security of signals and videos.

Conflicts of Interest There is no conflict of interest.

References

1. Aashiq Banu S, Amirtharajan R (2021) Bio-inspired cryptosystem on the reciprocal domain: DNA strands mutate to secure health data. Front Inform Technol Electron Eng 22(7):940–956. https://doi.org/10.1631/FITEE.2000071
2. Arumugham S, Rajagopalan S, Rayappan JBB, Amirtharajan R (2018) Networked medical data sharing on secure medium—a web publishing mode for DICOM viewer with three layer authentication. J Biomed Informatics 86:90–105. https://doi.org/10.1016/j.jbi.2018.08.010
3. Banu SA, Amirtharajan R (2020) A robust medical image encryption in dual domain:chaos-DNA-IWT combined approach. Med Biol Eng Comput 58(7):1445–1458. https://doi.org/10.1007/s11517-020-02178-w
4. Banu SA, Amirtharajan R (2020) A robust medical image encryption in dual do-main: chaos-dna-iwt combined approach. Med Biol Eng Comput 58(7):1445–1458. https://doi.org/10.1007/s11517-020-02178-w
5. Belazi A, Talha M, Kharbech S, Xiang W (2019) Novel medical image encryption scheme based on chaos and DNA encoding. IEEE Access 7:36667–36681. https://doi.org/10.1109/ACCESS.2019.2906292
6. Chidambaram N, Raj P, Thenmozhi K, Rajagopalan S, Amirtharajan R (2019) A cloud compatible DNA coded security solution for multimedia file sharing and storage. Multimedia Tools Appl 78(23):33837–33863. https://doi.org/10.1007/s11042-019-08166-z
7. Dagadu JC, Li JP, Aboagye EO (2019) Medical image encryption based on hybrid chaotic DNA diffusion. Wireless Pers Commun 108(1):591–612. https://doi.org/10.1007/s11277-019-06420-z
8. Guan M, Yang X, Hu W (2019) Chaotic image encryption algorithm using frequency-domain DNA encoding. IET Image Process 13(9):1535–1539. https://doi.org/10.1049/IET-IPR.2019.0051
9. Hua Z, Yi S, Zhou Y (2018) Medical image encryption using high-speed scram-bling and pixel adaptive diffusion. Signal Process 144:134–144. https://doi.org/10.1016/j.sigpro.2017.10.004
10. John S, Kumar S (2023) 2D Lorentz chaotic model coupled with logistic chaotic model for medical image encryption: towards ensuring security for teleradiology. Procedia Comput Sci 218:918–926. https://doi.org/10.1016/j.procs.2023.01.072
11. Kumar S, Panna B, Jha RK (2019) Medical image encryption using fractional discrete cosine transform with chaotic function. Medi Biol Eng Comput 57(11):2517–2533. https://doi.org/10.1007/S11517-019-02037-3/TABLES/11
12. Li C, Luo G, Qin K, Li C (2017) An image encryption scheme based on chaotic tent map. Nonlinear Dyn 87(1):127–133. https://doi.org/10.1007/s11071-016-3030-8

13. Lu Q, Zhu C, Deng X (2020) An efficient image encryption scheme based on the LSS chaotic map and single S-box. IEEE Access 8:25664–25678. https://doi.org/10.1109/ACCESS.2020.2970806
14. Ravichandran D, Praveenkumar P, Rayappan JBB, Amirtharajan R (2017) DNA chaos blend to secure medical privacy. IEEE Trans Nanobiosci 16(8):850–858. https://doi.org/10.1109/TNB.2017.2780881
15. Ravichandran D, Rajagopalan S, Upadhyay HN, Rayappan JBB, Amirtharajan R (2019) Encrypted biography of biomedical image—a pentalayer cryptosystem on FPGA. J Signal Process Syst 91(5):475–501. https://doi.org/10.1007/S11265-018-1337-Z/TABLES/12
16. Ravichandran DS, Banu A, Murthy BK, Balasubramanian V, Fathima S (2021) An efficient medical image encryption using hybrid DNA computing and chaos in transform domain. Med Biol Eng Comput 59:589–605. https://doi.org/10.1007/s11517-021-02328-8
17. Saremi S, Mirjalili S, Lewis A (2014) Biogeography-based optimisation with chaos. Neural Comput Appl 25(5):1077–1097. https://doi.org/10.1007/s00521-014-1597-x
18. Subashanthini S, Pounambal M (2020) Three stage hybrid encryption of cloud data with pentalayer security for online business users. IseB 18(3):379–404. https://doi.org/10.1007/s10257-019-00419-6
19. Wang X, Jin C (2012) Image encryption using game of life permutation and PWCLM chaotic system. Opt Commun 285(4):412–417. https://doi.org/10.1016/j.optcom.2011.10.010
20. Wu Y, Noonan J, Agaian S (2011) NPCR and UACI randomness tests for image encryption. Cyber J: Multidiscip J Sci Technol J Select Areas Telecommun (JSAT) 2(1):31–38. http://www.cyberjournals.com/Papers/Apr2011/05.pdf
21. Ziemer R (2002) Modulation. In: Encyclopedia of physical science and technology. Elsevier, pp 97–112. https://doi.org/10.1016/B0-12-227410-5/00456-7

A Hybrid Optimized Learning Framework for Compound Facial Emotion Recognition

Rohan Appasaheb Borgalli and Sunil Surve

Abstract Facial Emotion Recognition (FER) represents a critical domain within the realms of machine vision and artificial intelligence, finding widespread applications in both academic research and various industries. While there exist multiple sensors for conducting FER, recent studies emphasize the effectiveness of utilizing facial images/videos to discern emotions due to the rich information embedded in visual expressions. This study specifically addresses the task of detecting compound emotions using the iCV Multi-Emotion Facial Expression Dataset (iCV-MEFED) and explores different learning frameworks. In our initial approach, a deep learning Convolutional Neural Network (CNN) model was employed to extract features from each image. Subsequently, a Multi-Class Support Vector Machine (mSVM) classifier was utilized to identify the corresponding emotions. Our research presents two novel approaches for compound facial emotion recognition. The first methodology, a machine learning-based framework, achieved a 26% accuracy rate, which represents an improvement over previous state-of-the-art results. This approach demonstrated superior performance in misclassification analysis, with a 74.00% accuracy rate. The second methodology employs a deep learning CNN-based model enhanced by WCHGSO-DNFN optimization. This approach yielded a 28% accuracy rate on the iCV-MEFED dataset, again surpassing existing benchmarks. In misclassification comparisons, this method achieved a 72.00% accuracy rate, consistently outperforming competing approaches.

Keywords Facial emotion recognition · Compound emotion · Optimization · Machine learning · Deep learning · Convolution neural network · Support vector machine

---

R. A. Borgalli (✉)
Department of Electronics and Telecommunication Engineering, Shah and Anchor Kutchhi Engineering College Chembur, Mumbai 400088, India
e-mail: rohan.borgalli@sakec.ac.in

S. Surve
Department of Computer Engineering, Fr. Conceicao Rodrigues College of Engineering, University of Mumbai, Bandra, Mumbai 400050, India
e-mail: surve@fragnel.edu.in

P. J. S. Gonçalves et al. (eds.), *Proceedings of Fifth International Conference on Computing, Communications, and Cyber-Security*, Lecture Notes in Networks and Systems 1128, https://doi.org/10.1007/978-981-97-7371-8_35

1 Introduction

Facial Expression Recognition (FER) is a crucial research area with numerous applications. This paper proposes a FER system model that detects human facial expressions and identifies corresponding compound emotions in static images using a composite learning framework. Despite extensive research, accurately recognizing basic and compound facial expressions remains challenging due to their complexity and variety.

Humans convey intentions and emotions through nonverbal means, including facial expressions, gestures, and involuntary language. Darwin and Prodger [1] noted that facial expressions reflect emotional states and intentions. The importance of FER systems in Artificial Intelligence (AI), video games, computing, medicine, e-Education, and other fields has led to numerous attempts to detect expressions from facial representations for various applications. FER systems have become essential for emotion detection due to their accessibility and efficiency, finding use in lie detection, medical assessment, human–machine interaction, driver safety, and other complex real-world problems.

Suwa [2] first published on automatic facial expression recognition in 1978, inspiring many researchers to develop robust and accurate methods. Ekman and Friesen [3] defined six basic emotions based on cross-cultural studies, anger, disgust, fear, happiness, sadness, and surprise, asserting that humans perceive these emotions similarly across cultures. Much of the past research, as surveyed by Li and Deng [4], focused on these six basic emotions and neutrality.

Recent advanced research on FER has expanded to include compound emotions, with studies by Du et al. [5], Guo et al. [6], and Li and Deng [7, 8] exploring a variety of compound emotions alongside basic ones. Li and Deng's (2020) literature survey revealed that the FER process typically begins with image preprocessing to detect faces and extract features, followed by model training using either handcrafted methods or CNN techniques. Transfer learning with standard models has also been explored by Lim et al. [9].

This paper implements various learning frameworks for FER systems using the iCV-MEFED dataset [6] to detect both basic and compound emotions, aiming to achieve high performance. The subsequent sections of this paper cover a literature review of related works, details of the iCV-MEFED database used in the experiment, the proposed methodologies for detecting compound emotions, discussion of experimental results, and conclusions with suggestions for future work.

2 Related Work

The field of FER has seen significant advancements, particularly in the transition from basic to compound emotion recognition. Researchers have developed various datasets to support this evolution, including Compound Facial Expressions of Emotion

(CFEE) [5], ICV-MEFED [6], and RAF-DB [7]. The RAF-DB database, introduced by Li and Deng, encompasses both basic and compound emotions, providing a comprehensive resource for researchers.

Shan Li et al. [8] further contributed to this field with the RAF-ML database, designed to address the challenges of expression recognition in real-world scenarios. The Emotion Recognition in the Wild (EmotiW) dataset [10] also aimed to bridge the gap between lab-controlled and in-the-wild environments by incorporating a large number of real-world training data. Du et al. [11] explored compound facial emotions, which combine multiple basic emotions. Their research initially identified 15 compound emotions, later expanded to 17 [12]. This work has significant implications for understanding nonverbal communication and its applications in psychological studies.

Li and Deng [7] proposed a Deep Locality-Preserving CNN (DLP-CNN) method to address the challenges of non-lab-controlled databases. Their approach aimed to maximize the effectiveness of deep learned features while preserving local similarities and maximizing interclass differences. They reported an accuracy of 44.55% in discriminating 11 compound expressions. Building on this work, Li and Deng [8] introduced the DeepBi-Manifold (DBM)-CNN architecture, which demonstrated promising results on cross-database tasks using the RAF-ML dataset.

The EmotionNet Challenge [13] provided a significant dataset of facial expressions in the wild, focusing on the detection of Facial Action Units and compound emotions. This challenge used a dataset of 950,000 automatically annotated images for training, with manually annotated validation and test sets.

Slimani et al. [14] proposed a highway CNN architecture for FER, achieving a test accuracy of 52.14% on the CFEE database. They noted the challenges in distinguishing compound emotions from their basic components. Guo et al. [6] addressed the limited availability of compound emotion datasets by introducing the iCV-MEFED dataset, which includes 50 classes of compound emotions annotated by psychologists. They employed a CNN-based approach for feature extraction.

Kamińska et al. [15] proposed a ResNet18-based approach with three models and an ensembling method. Their two-stage recognition approach, combined with ensembling, achieved 21.83% accuracy. Guo et al. [16] developed a multi-modality approach combining geometrical and visual information. Their AlexNet CNN-based method outperformed previous approaches, achieving first rank in a competition [17] with a misclassification result of 80.212137%. These advancements in compound emotion recognition demonstrate the field's rapid progress and the ongoing challenges in accurately identifying complex emotional expressions in real-world scenarios.

Recent research has explored various approaches to compound emotion detection and pain intensity measurement. One study proposed a novel method called WCHGSO-DNFN, which demonstrated promising results when applied to the CFEE Database. The same technique was also utilized for assessing pain intensity using the UNBC-McMaster Shoulder Pain Expression Archive Database [18]. In another investigation, researchers developed a hybrid approach that leverages CNN-based features for facial emotion recognition [19]. These studies contribute to the growing body of knowledge in the field of automated emotion and pain assessment.

Contemporary FER systems predominantly focus on identifying basic emotions. However, there is a growing need for more sophisticated systems capable of detecting both basic and compound facial expressions with high accuracy and robustness. This research proposes an innovative learning framework that integrates machine learning and deep learning architectures to address this challenge. The proposed system aims to improve upon existing state-of-the-art results in recognizing both basic and compound human emotions, utilizing the iCV-MEFED dataset [6] as a benchmark. By combining advanced techniques from both machine learning and deep learning domains, this framework seeks to enhance the overall performance and applicability of FER systems in real-world scenarios.

3 Dataset Overview

The iCV-MEFED was developed to study compound emotion recognition. It comprises 31,250 frontal view face images from 125 subjects aged 18 to 37, with an even gender distribution. Each participant enacted 50 distinct emotions, with five samples per emotion captured under consistent lighting and background conditions using a Canon 60D camera at 5184 × 3456 resolution.

The dataset features emotion expressions in a complementary dominant format, incorporating seven basic emotions. For ease of use, each image is labeled according to the format outlined in Table 1. Figure 1 provides examples from the dataset.

Fig. 1 Sample images from the iCV-MEFED dataset

Table 1 50 distinct emotional categories present in the iCV-MEFED dataset [13]

Label	Emotion	Label	Emotion	Label	Emotion	Label	Emotion
0	Neutral	13	Contemptly sad	26	Fearfully happy	38	Sadly disgust
1	Angry	14	Contemptly surprised	27	Fearfully sad	39	Sadly fearful
2	Angrily contempt	15	Disgustingly angry	28	Fearfully surprised	40	Sadly happy
3	Angrily disgusted	16	Disgustingly contempt	29	Happily angry	41	Sad
4	Angrily fearful	17	Disgust	30	Happily contempt	42	Sadly surprised
5	Angrily happy	18	Disgustingly fearful	31	Happily disgust	43	Surprisingly angry
6	Angrily sad	19	Disgustingly happy	32	Happily fearful	44	Surprisingly contempt
7	Angrily surprised	20	Disgustingly sad	33	Happy	45	Surprisingly disgust
8	Contemptly angry	21	disgustingly surprised	34	Happily sad	46	Surprisingly fearful
9	Contempt	22	Fearfully angry	35	Happily surprised	47	Surprisingly happy
10	Contemptly disgusted	23	Fearfully contempt	36	Sadly angry	48	Surprisingly sad
11	Contemptly fearful	24	Fearfully disgust	37	Sadly contempt	49	Surprised
12	Contemptly happy	25	Fearful				

4 Proposed Methodology

4.1 *Method 1*

In our approach to Compound FER, we employ a hybrid learning framework that integrates traditional machine learning techniques with deep learning methodologies, as illustrated in Fig. 2. This approach draws inspiration from feature engineering principles, which leverage domain expertise to minimize training data complexity by extracting features that enhance pattern recognition and overall performance of learning algorithms.

Our study utilizes the iCV-MEFED dataset, which comprises facial images annotated with compound facial emotion labels. The process begins with applying preprocessing techniques, specifically the Haar Cascade algorithm for face detection. Subsequently, we extract features from all training, validation, and test images using

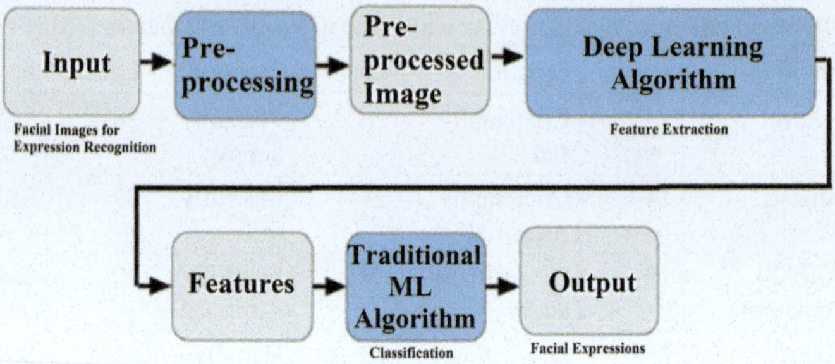

Fig. 2 Proposed methodology for compound FER

a CNN-based deep learning model. These extracted features are then fed into a traditional machine learning algorithm, specifically a m-SVM, to predict the facial expression, as depicted in Fig. 2.

The proposed methodology employs a CNN-based deep learning framework for feature extraction, specifically utilizing the Inception V3 architecture [20], which has demonstrated effectiveness in image classification tasks (Fig. 4). This network's initial layers are designed to capture low-level features such as lines, blobs, colors, and edges, while the final layers focus on task-specific features, in this case, basic and compound facial expressions.

To adapt the network for compound facial expression detection on the iCV-MEFED dataset, modifications were made to the standard architecture by replacing the final layers. This allows for feature learning tailored to the specific dataset. Each image is processed through the Inception V3 architecture, culminating in a global average pooling layer that produces an output of 2048 features. Figure 3 illustrates sample test images alongside their corresponding extracted features.

In our study using the iCV-MEFED database, we implemented a preprocessing step with the Haar Cascade algorithm for frontal face detection to capture essential facial regions. After feature extraction, the data was split into training, validation, and test sets, ensuring balanced distributions. The training involved 10,858 samples, and 1,917 validation samples were used to optimize mSVM parameters. These parameters were then applied to evaluate 3,194 test samples, yielding metrics such as precision, recall, f1-score, and overall accuracy. This process ensured efficient feature learning and improved network performance.

4.2 Method 2

Our approach to Compound FER employs a learning framework based on an optimized deep neuro-fuzzy network, as illustrated in Fig. 5. This framework incorporates the WCHGSO optimization algorithm [18], which draws inspiration from heuristic optimization concepts rooted in natural phenomena and physics. The process begins

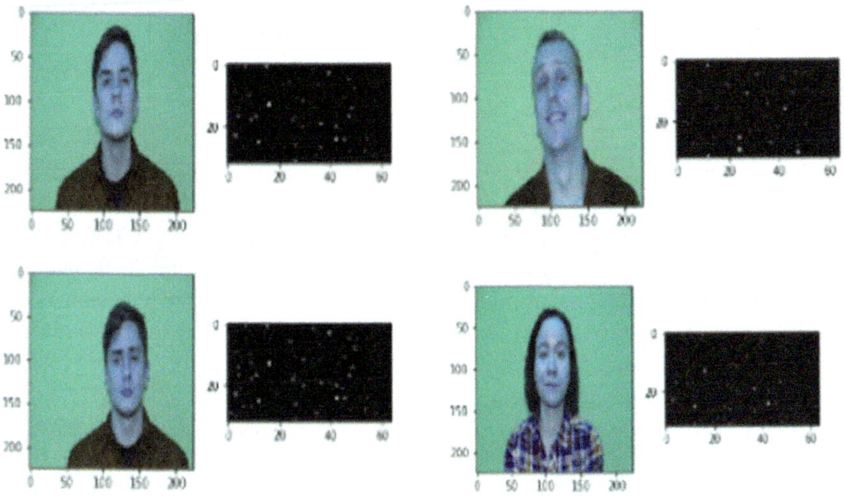

Fig. 3 Sample test images with extracted features from the iCV-MEFED dataset

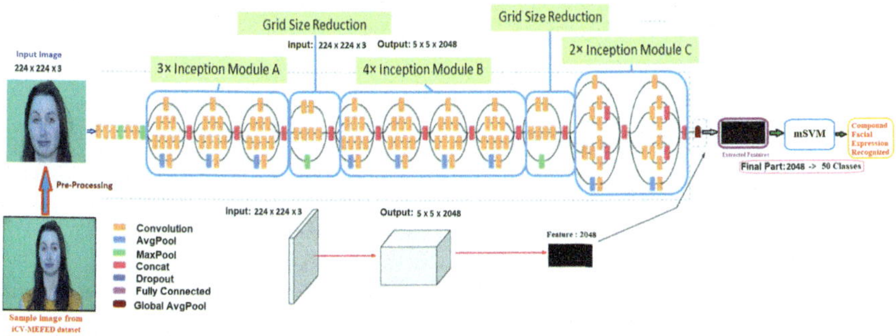

Fig. 4 FER on the iCV-MEFED dataset using standard Inception v3 architecture

with preprocessing and face detection steps, followed by the application of the optimized Deep Neuro-Fuzzy Network (DNFN) to perform 50-class FER on the iCV-MEFED dataset. This method aims to achieve enhanced performance in recognizing complex facial expressions [1].

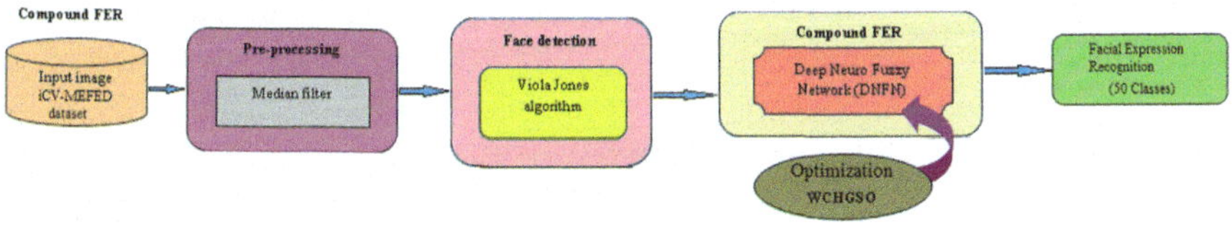

Fig. 5 FER on the iCV-MEFED dataset using DNFN WCHGSO architecture

5 Results

5.1 Accuracy Score

The model's performance is evaluated using various metrics, including overall accuracy, macro average, and weighted average. Tables 1 and 2 provide a comprehensive overview of the results obtained from both machine learning and deep learning approaches applied to the iCV-MEFED dataset. Table 1 outlines the labels and their corresponding compound facial emotions, while Table 2 presents the precision, recall, and f1-score as accuracy metrics for each emotion category in the testing set. These tables offer insights into the effectiveness of our implemented methodology across different compound facial expressions.

Table 3 presents a comparative analysis of state-of-the-art misclassification results from [18] and our implemented methodology, with smaller values indicating better performance.

A comparative analysis of our implemented methodology against state-of-the-art results is presented in Table 4. This table highlights the test set accuracy scores, with higher values indicating better performance. The comparison allows for a clear evaluation of our approach's effectiveness in relation to existing benchmarks [18].

6 Summary and Conclusion

In this paper, a learning framework for compound facial emotion recognition was proposed, combining machine learning and deep learning techniques to enhance performance on the iCV-MEFED database. The experiments demonstrated that the proposed framework uses a CNN-based deep learning approach for feature extraction in method-1, followed by classification using a best-fitted mSVM classifier. Method-2 utilizes WCHGSO-DNFN for the iCV-MEFED dataset. The accuracy achieved was 26% for method-1 and 28% for method-2. These outcomes indicate that the proposed methods surpass the state-of-the-art results. Additionally, a comparison of CER and MMER competition results in terms of misclassification rates reveals that our methodology achieves the best results, with 74.00% and 76.00%, respectively.

To further this research on FER in the wild scenario, the RAF-AU dataset provided by Yan et al. [21] will be beneficial for detecting basic and compound facial emotions through its facial action units.

Table 2 The implemented methodology's performance on the iCV-MEFED dataset in terms of precision, recall, f1-score, and accuracy for each category of emotion

Emotion class	Method 1			Method 2			Support
	Precision	Recall	F1-score	Precision	Recall	F1-score	
0	0.25	0.19	0.22	0.27	0.2	0.23	68
1	0.26	0.28	0.27	0.27	0.29	0.28	65
2	0.27	0.4	0.32	0.29	0.41	0.34	75
3	0.15	0.19	0.17	0.16	0.2	0.18	70
4	0.28	0.21	0.24	0.30	0.22	0.25	71
5	0.35	0.26	0.3	0.36	0.27	0.31	73
6	0.3	0.38	0.33	0.3	0.39	0.34	66
7	0.2	0.3	0.24	0.19	0.3	0.23	64
8	0.21	0.2	0.2	0.23	0.21	0.22	55
9	0.24	0.23	0.23	0.26	0.24	0.25	66
10	0.25	0.28	0.26	0.25	0.29	0.27	65
11	0.18	0.35	0.24	0.19	0.36	0.25	63
12	0.18	0.23	0.2	0.20	0.24	0.22	53
13	0.21	0.24	0.22	0.22	0.25	0.23	55
14	0.27	0.23	0.25	0.28	0.24	0.26	65
15	0.14	0.19	0.16	0.15	0.2	0.17	59
16	0.25	0.25	0.25	0.27	0.27	0.27	69
17	0.28	0.29	0.29	0.29	0.3	0.29	58
18	0.3	0.21	0.25	0.31	0.22	0.26	57

(continued)

Table 2 (continued)

Emotion class	Method 1			Method 2			Support
	Precision	Recall	F1-score	Precision	Recall	F1-score	
19	0.2	0.25	0.22	0.19	0.25	0.22	64
20	0.33	0.48	0.39	0.34	0.49	0.4	63
21	0.21	0.2	0.2	0.22	0.21	0.21	66
22	0.28	0.29	0.28	0.29	0.3	0.29	63
23	0.27	0.26	0.27	0.28	0.28	0.28	58
24	0.25	0.26	0.25	0.25	0.27	0.26	61
25	0.37	0.42	0.39	0.39	0.43	0.41	62
26	0.31	0.27	0.29	0.33	0.27	0.3	71
27	0.37	0.28	0.32	0.38	0.3	0.34	71
28	0.26	0.24	0.25	0.27	0.25	0.26	71
29	0.2	0.18	0.19	0.21	0.19	0.2	67
30	0.22	0.18	0.2	0.24	0.2	0.22	60
31	0.32	0.26	0.29	0.34	0.26	0.29	62
32	0.39	0.27	0.32	0.39	0.28	0.33	73
33	0.37	0.33	0.35	0.39	0.34	0.36	63
34	0.23	0.27	0.25	0.24	0.28	0.26	56
35	0.39	0.24	0.3	0.4	0.24	0.3	78
36	0.22	0.27	0.24	0.24	0.28	0.26	51
37	0.26	0.27	0.26	0.27	0.26	0.26	63
38	0.25	0.33	0.28	0.26	0.32	0.29	52
39	0.35	0.26	0.3	0.37	0.27	0.31	69

(continued)

Table 2 (continued)

Emotion class	Method 1			Method 2			Support
	Precision	Recall	F1-score	Precision	Recall	F1-score	
40	0.27	0.21	0.24	0.28	0.23	0.25	61
41	0.21	0.2	0.2	0.23	0.2	0.21	71
42	0.28	0.23	0.25	0.29	0.25	0.27	65
43	0.32	0.27	0.29	0.33	0.28	0.3	60
44	0.28	0.34	0.31	0.29	0.36	0.32	61
45	0.23	0.23	0.23	0.25	0.24	0.24	64
46	0.36	0.24	0.29	0.37	0.26	0.31	67
47	0.37	0.22	0.27	0.38	0.23	0.29	64
48	0.35	0.22	0.27	0.36	0.24	0.29	72
49	0.21	0.25	0.23	0.23	0.25	0.24	48
Accuracy			0.26			0.28	3194
Macro avg	0.27	0.26	0.26	0.29	0.28	0.28	3194
Weighted avg	0.27	0.26	0.26	0.29	0.28	0.28	3194

Table 3 Comparative analysis of misclassification rates: our proposed methodology versus competing approaches

Comparison with competition [17] Misclassification results					
Sr. no	Paper	User team	Team name	Validation set	Testing set
1	[17]	Cleardusk	CBSR-CASIA	79.328571 80.212137	84.9571431
2		Ntech	NTechLab	83.985714	–
3		Bekhouche	–	85.757143	–
4		Szhou	–	86.014286	–
5		Amitkumar05	–	86.228571	–
6		Bknyaz	NTechLab	86.371429	–
7		Charrin	–	87.414286	–
8		Yuzhipeng	ZZXP	87.500000	–
9		Csmath	DeepEmotion	87.571429	–
10		Frkngrpnr	–	89.842857	–
11		Icv	ICV Team	90.714286	–
12		Pablovin	–	91.585714	–
13		Iarganda	CVPD	91.800000	–
14		lld533	–	93.428571	–
15	[15]	1st	–	79.300000	80.2000000
16		2nd	–	84.000000	85.3000000
17		3rd	–	87.500000	87.7000000
18	Ours	Method 1	74.000000	–	
19		Method 2	72.000000	–	

Table 4 Accuracy comparison between the proposed methodology and state-of-the-art results

Comparison of proposed methodology and state-of-the-art results		
Paper	Model/Method	Test Set Accuracy
[16]	Model_1	18.51
	Model_2	19.71
	Model_3	19.28
	Ensembling	21.83
Ours	Method 1	26.00
	Method 2	28.00

References

1. Darwin C, Prodger P (1998) The expression of the emotions in man and animals. Oxford University Press, USA
2. Suwa M (1978) A preliminary note on pattern recognition of human emotional expression. In: Proceedings of international joint conference on pattern recognition, pp 408–410
3. Ekman P, Friesen WV (1971) Constants across cultures in the face and emotion. J Pers Soc Psychol 17(2):124
4. Li S, Deng W (2018) Deep facial expression recognition: a survey. arXiv preprint arXiv:1804.08348
5. Du S, Tao Y, Martinez AM (2014) Compound facial expressions of emotion. Proc Natl Acad Sci 111(15):E1454–E1462
6. Guo J, Lei Z, Wan J, Avots E, Hajarolasvadi N, Knyazev B, Kuharenko A, Jacques Junior JCS, Baro X, Demirel H, Escalera S, Allik J, Anbarjafari G (2018) Dominant and complementary emotion recognition from still images of faces. IEEE Access 6:26391–26403
7. Li S, Deng W (2018) Reliable crowdsourcing and deep locality-preserving learning for unconstrained facial expression recognition. IEEE Trans Image Process 28(1):356–370
8. Li S, Deng W (2019) Blended emotion in-the-wild: Multi-label facial expression recognition using crowdsourced annotations and deep locality feature learning. Int J Comput Vision 127(6):884–906
9. Lim YK, Liao Z, Petridis S, Pantic M (2018) Transfer learning for action unit recognition
10. Fabian Benitez-Quiroz C, Srinivasan R, Martinez AM (2016) Emotionet: an accurate, real-time algorithm for the automatic annotation of a million facial expressions in the wild. In: Proceedings of the IEEE conference on computer vision and pattern recognition, pp 5562–5570
11. Du S, Martinez AM (2015) Compound facial expressions of emotion: from basic research to clinical applications. Dialogues Clin Neurosci 17(4):443
12. Benitez-Quiroz CF, Srinivasan R, Feng Q, Wang Y, Martinez AM (2017) "Emotionet challenge: recognition of facial expressions of emotion in the wild, arXiv preprint arXiv:1703.01210
13. Appasaheb Borgalli R, Surve S (2023) Learning framework for compound facial emotion recognition. Recent Adv Electr Electron Eng (Formerly Recent Patents Electr Electron Eng) 16(6):664–676
14. Slimani K, Lekdioui K, Messoussi R, Touahni R (2019) Compound facial expression recognition based on highway CNN. In: Proceedings of the new challenges in data sciences: acts of the second conference of the Moroccan classification society, pp 1–7
15. Kamińska D, Aktas K, Rizhinashvili D, Kuklyanov D, Sham AH, Escalera S, … Anbarjafari G (2021) Two-stage recognition and beyond for compound facial emotion recognition. Electronics 10(22):2847
16. Guo J, Zhou S, Wu J, Wan J, Zhu X, Lei Z, Li SZ (2017) Multi-modality network with visual and geometrical information for micro emotion recognition. In: 2017 12th IEEE international conference on automatic face & gesture recognition (FG 2017), pp 814–819. IEEE
17. Joint Challenge on Compound Emotion Recognition and Multimodal (Audio, Facial and Gesture) based Emotion Recognition (CER&MMER) https://icv.tuit.ut.ee/challenge/
18. Borgalli RA, Surve S (2022) Compound facial expression recognition and pain intensity measurement using optimized deep neuro fuzzy network. Int J Swarm Intell Res (IJSIR) 13(1):1–27. https://doi.org/10.4018/IJSIR.304721
19. Shahzad HM, Bhatti SM, Jaffar A, Akram S, Alhajlah M, Mahmood A (2023) Hybrid facial emotion recognition using CNN-based features. Appl Sci 13(9):5572
20. Szegedy C, Vanhoucke V, Ioffe S, Shlens J, Wojna Z (2016) Rethinking the inception architecture for computer vision. In: Proceedings of the IEEE conference on computer vision and pattern recognition, pp 2818–2826
21. Yan WJ, Li S, Que C, Pei J, Deng W (2020) RAF-AU database: in-the-wild facial expressions with subjective emotion judgement and objective AU annotations. In: Proceedings of the Asian conference on computer vision

References

Image Quality Assessment for Deep Learning-Enabled Saffron Adulteration Detectors

Suhail Manzoor, Arvind Selwal, and Ambreen Sabha

Abstract Saffron, also known as the "golden spice", is one of the costly spices cultivated mainly in Iran, Greece, Spain, China, and India (mostly in Kashmir valley). However, due to its limited production and high price, it is often subjected to the issue of adulteration which leads to economic loss and several health issues. The various types of adulterants that are mixed with original saffron samples are sunflower, safflower, maize outer covering, and hibiscus stigmas to increase their quantity. Nowadays, the menace of saffron adulteration is one of the major issues that need further attention as human vision fails to discriminate between pure and adulterated saffron samples. Hence, an intelligent computer vision-enabled saffron adulteration prediction system is used by researchers to overcome this issue. In this paper, we present a novel framework for saffron adulteration prediction using intelligent data-driven approaches. Moreover, various pre-processing techniques enhance the quality of saffron images. Besides, we demonstrate the comparative analysis of recent saffron adulteration prediction techniques and identify certain open research issues that need futuristic attention. In addition, performance protocols such as performance metrics and saffron images-based data are briefly illustrated. Our analysis reveals that this study provides researchers with new directions that need to be addressed in the future and can potentially prevent economic loss and ensure consumer safety.

Keywords Saffron · Adulteration · Image pre-processing · Data augmentation · Convolutional neural network (CNN)

S. Manzoor (✉) · A. Selwal · A. Sabha
Department of Computer Science and IT, Central University of Jammu, Jammu, India
e-mail: suhailghzn33@gmail.com

A. Selwal
e-mail: arvind.csit@cujammu.ac.in

P. J. S. Gonçalves et al. (eds.), *Proceedings of Fifth International Conference on Computing, Communications, and Cyber-Security*, Lecture Notes in Networks and Systems 1128, https://doi.org/10.1007/978-981-97-7371-8_36

1 Introduction

Saffron is a highly priced and expensive spice that is derived from the stigmas of the Crocus sativus flower, which is a member of the iris family [1]. It is known for its bright red color, unique aroma, and distinctive flavor, and is used in many foods around the world. The cultivation of saffron has a long history, dating back thousands of years to ancient Persia. Today, saffron is primarily cultivated in regions of Iran, India, Spain, Greece, and Italy [2]. The process of harvesting saffron is labor-intensive and time-consuming, as the delicate threads must be carefully hand-picked from the flowers. It takes about 75,000 flowers to produce just one pound of saffron. Humans find it challenging to distinguish between genuine and fake saffron samples. Therefore, to solve this issue, computer vision systems are being used to automatically determine if a particular image of saffron is pure or contaminated, the reasons why it is one of the priciest spices in the world [3]. In addition to its cooking uses, saffron has also been studied for its potential health benefits. It contains several compounds that may have anti-inflammatory, antioxidant, and antidepressant effects, and has been used in traditional medicine to treat a variety of conditions [4, 5]. Kashmiri saffron, also known as Kashmiri Kesar, is a highly priced variety of saffron that is grown in the Kashmir valley of India. It is considered to be one of the finest and most expensive varieties of saffron in the world, due to its unique flavor, aroma, and high-quality threads. Kashmiri saffron is known for its distinctive deep red color and long, thin threads that have a sweet and aromatic fragrance that is declared as geographical indication (GI) tag in the Gregorian calendar in the month of 2020. It is grown in the high-altitude region of Pampore, near Srinagar, the capital city of Jammu and Kashmir state in India. The saffron is harvested in the autumn months of October and November, and the process of harvesting is labor-intensive and requires a delicate touch, as the flowers must be careful hand-picked and the stigmas removed [6]. Figure 1a contains the saffron flower, and (b) and (c) are the real images of Kashmiri (Kong) samples.

Kesar is a costly spice that has several medical benefits for humans, as we all know. That is why it is one of the most commonly adulterated spices. As a result, some unsavory characters attempt to tamper with the saffron. Individuals' health might

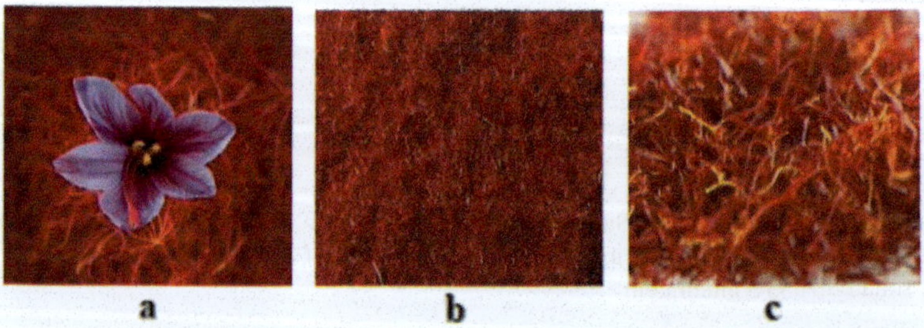

Fig. 1 **a** A delicate Saffron flower, blooming with grace, and **b, c** the deep crimson threads of gold spice [21, 22]

be significantly impacted by this adulteration. Adulteration is the process of adding impurities or cheaper materials to a product to increase its weight or lower its quality. Adulterants, such as dried petals, stamens, and leaves of plants resembling saffron, are often combined with authentic saffron. They are also infused with dyes and (Moisture) Adulterants are sometimes sprayed with water to increase their weight. Alternatively, saffron threads are coated with oil to make them appear fresher and more plump. Some vendors include artificial flavorings to saffron to enhance its taste and aroma [7]. This increases the weight and volume of the saffron without adding much to the cost. Distinguishing between genuine and counterfeit saffron samples is a challenging task for humans as it cannot be done accurately through visual inspection alone. Hence, researchers are developing computer vision systems that can identify and categorize a given saffron image as either pure or adulterated, thereby solving this issue. Figure 2 contains images of some natural plants that may be added with saffron for adulteration. People add maize threads, safflower, sunflower, and red lily threads with saffron for adulteration.

Saffron is one of the most expensive spices in the world and is often adulterated with other substances, such as synthetic dyes or other plant materials, to increase its weight or lower its cost. Here are some traditional ways of removing adulteration in saffron. Such as Visual Inspection, Water Test, Microscopic Examination, Chemical Tests, Taste Test, Sensory Evaluation, and DNA Fingerprinting. The saffron experts evaluate the color, flavor, and aroma of saffron to detect any adulteration. This technique is subjective and requires trained panelists, but it can be a useful tool for detecting saffron adulteration [8–13]. The traditional methods used for detecting saffron adulteration reveal some limitations such as a time-consuming

Fig. 2 The various types of adulterants in Saffron [23–25]

process, requiring human intervention, and limited accuracy. These challenges can be overcome by using some intelligent techniques such as machine or deep learning. Hence, the main motivation or rationale behind this work is to design an intelligent data-driven model that can correctly identify various types of adulterants present in the saffron images.

The primary contributions of our research are below

- The article presents a focused review of the saffron adulteration prediction mechanisms.
- We present a novel framework for saffron adulteration detection using intelligent techniques.
- We briefly illustrate various image pre-processing techniques used for saffron adulteration prediction.
- We identify various open research issues in the field of saffron adulteration detection.
- The article illustrates performance protocols (i.e., dataset and performance metrics) for the evaluation of the saffron adulteration prediction system.

2 Related Work

This section gives an overview of relevant saffron adulteration prediction methods. Our review has focused specifically on the application of machine learning and deep learning techniques for detecting saffron adulteration. Traditional methods, such as chromatography and spectroscopy, have also been used for saffron analysis and may have certain advantages over machine learning approaches, such as greater specificity and sensitivity. Several researchers have done work such as Varliklioz et al. [14] who initially developed a spectroscopic method to identify the sort of adulterants used in saffron adulteration (i.e., sunflower, marigold, and safflower). This technique combines three methods: Fourier remodel qualitative analysis (FRQS), Raman qualitative analysis (RQAS), and laser-induced qualitative analysis (LIQS). As a result, the superiority of the techniques is evaluated using principal component analysis (PCA), and LIQS yields the best results for identifying authentic and adulterated saffron samples.

In the study by Filatova et al. [15], volatile compounds in real saffron samples and possible saffron adulterants like safflower, calendula, capsicum, and turmeric were screened for using advanced analytical techniques, such as headspace solid-phase microextraction combined with gas chromatography and high-resolution mass spectrometry (HS-SPME-GC-HRMS). A clear distinction between genuine saffron and possible adulterants was accomplished by chemometric analysis using principal component analysis (PCA) and partial least squares discriminative analysis (PLS-DA). Reliability in detecting adulteration levels in saffron samples as low as 2% w/w was made possible by the identification of significant chemicals, or "markers," for each plant group.

This work by Alighaleh et al. [16] used a variety of soft computing techniques, such as classifiers like RBF, MLP, KNN, SVM, and LVQ, to discern between common adulterants and real saffron. By utilizing spectral and RGB pictures in addition to chemical measurements of important saffron constituents, the classifiers were able to distinguish real saffron samples from synthetic ones, including dyed citrus blossoms, safflower, dyed fibers, and stigma combined with stamens, with unprecedented accuracy. During the training phase, KNN in particular showed remarkable performance, classifying both RGB and NIR pictures with 100% accuracy. This novel method provides a dependable way to identify adulterated saffron, which is essential for preserving the integrity of this highly valued spice in the culinary business.

The study by Hani et al. [17] is the first to use excitation-emission matrix (EEM) and synchronous fluorescence, two methods of fluorescence analysis, to distinguish Moroccan saffron varieties from different regions (Taroudant, Ouarzazate, and Azilal) and to determine which origins of saffron are in Afghanistan, Iran, and Morocco. Each saffron sample was given a distinct fingerprint through the visualization of EEM physiognomy and the use of the principles of component analysis (PCA) and linear discriminant evaluation (LDA). This allowed for precise categorization based on fluorescence composition and intensity. To ensure product reliability and legitimacy in the food sector, saffron's geographical origin must be verified. The process proved to be quick, non-destructive, and extremely dependable.

In the other work presented by Lu et al. [11], the authors used multivariate spectrum analysis and hyperspectral imaging to examine the authenticity, quality, and provenance of saffron. Using saffron hyperspectral pictures, reflection spectra were recovered. Characteristic wavelengths were chosen using the successive projections method, genetic algorithm, uninformative variable removal, and competitive adaptive reweighted sampling. Based on the chosen wavelengths, a model of a back propagation neural network was created. The model that combined back propagation neural network with competitive adaptive reweighted sampling had the greatest performance, according to the results. One contaminated, three domestic, and two imported saffron samples had prediction accuracy of 100, 95, 94, 100, 83, and 96%, respectively.

Additionally, Mohamadzadeh Moghadam et al. [18] divide the Iranian saffron into three groups based on physical features: Pushal, Negin, and Sargol. Using the camera on a cell phone, 414 color photos of saffron were collected for the three distinct classes. The use of image analysis allowed the extraction of 21 color characteristics and 99 texture features. Using the aforementioned characteristics, 22 classifiers were used for classification. In comparison to other classifiers, the support vector machine and ensemble performed better. According to our findings, the Subspace Discriminant classifier and Quadratic SVM had a mean classification accuracy of up to 83.9%.

In another work presented by Amin et al. [2], the authors discuss the current state of intelligent saffron adulteration prediction systems as well as numerous research challenges and outlooks for the future. The goal of this study was to investigate the methods currently used to train linear classifiers such as artificial neural networks (ANN), support vector machines (SVM), and decision trees (DT) for saffron classification. However, two major challenges need to be addressed. Firstly, the models need to have good generalization capabilities to correctly classify unseen samples. Secondly, there is a lack of comprehensive saffron datasets available for training and testing purposes, although the majority of approaches show promising results. The authors mention that the use of deep neural networks may overcome this problem.

In another study by Amin et al. [6], a new saffron prediction model called SaffNet was introduced. The model uses statistical image characteristics to identify contamination in Kashmiri saffron, to address generalization issues. Data-driven methods often require representative samples, which led the researchers to create a unique dataset (Saff-Kash) consisting of both true and contaminated saffron samples collected from various locations in the Kashmir valley. The dataset was pre-processed and used to train the proposed SaffNet model. The model, developed using a gradient boosting ensemble, achieved an overall accuracy of 98% when tested on the Saff-Kash dataset, outperforming individual classifiers such as Support Vector Machines (SVM), decision trees, and K-Nearest Neighbor (KNN). Furthermore, the SaffNet model exhibited faster execution times than the gradient boosting classifier, taking 8.56 ms as compared to 7.7 ms. The main objective of this research was to identify adulteration in saffron samples, and the SaffNet model proved to be an effective solution to this problem. Recently, few researchers have done work related to deep-learning-based adulteration prediction techniques.

For the first time. Alighaleh et al. [16] conducted a study where they detected adulterants in saffron using a deep neural network that analyzed RGB photos taken under uncontrolled conditions. The dataset consisted of two classes of real saffron (freeze-drying and microwave-drying) and five classes of false saffron (mixed stamen and colored straw, mixed stigma and stamens, dyed orange blossom, and Safflower and dyed stamens). The neural network model was trained and optimized using 60% of the photos, and data augmentation was utilized to prevent over-training. The model's architecture consisted of convolutional, pooling, and fully connected layers. Batch normalization techniques were employed to suggest five structures (S1, S2, S3, S4, and S5) for the CNN. The study also compared these structures to popular convolutional networks such as VGG11, ResNet18, ResNet50, Inception V3, and DarkNet53. Based on the results, the best configuration was S5, which achieved a cross-entropy loss of 0.019 and a precision of 99.8%. Egea et al. [17] proposed a novel approach that combines smartphone and Foldscope to identify whether saffron is genuine or not by analyzing staining patterns, followed by machine learning-based classification algorithms, which outperformed traditional methods and achieved an accuracy and precision of 99.5% and 99.3%, respectively, improving the reliability of saffron authentication (Table 1).

3 Saffron Adulteration Prediction Framework

This section describes the framework of our saffron adulteration prediction model designed to predict the adulteration of saffron. The literature witnessed that most of the work has been done on Iranian saffron, whereas limited work has been done

Table 1 A comparative analysis of existing Saffron Adulteration prediction approaches

Year	Author(s)	Techniques ML/DL	Classifier	Data augmentation	Datasets	Performance	Challenges
2017	Varliklioz Er et al. [14]	ML	PCA	×	Private	Not mentioned	No adulteration was carried out, and no benchmark dataset was provided
2019	Lu et al. [11]	ML	Back propagation neural network	×	Private	DS3 (100%), DS1 (95%), DS2 (94%), IS2 (96%), and IS1 (83%)	Firstly, the methods used for examination were deemed unreliable. Secondly, the absence of a benchmark dataset further complicate the process
2020	Mohamadzade h Moghadam et al. [18]	ML	SVM	×	Private	89%	Less accuracy was revealed as well as the absence of the benchmark data set
2021	Amin et al. [2]	ML	SVM, decision tree, and ANN	×	Private	99%	The absence of the benchmark data set
2023	Amin et al. [6]	ML	DT, SVM, KNN, RF and GB	✓	Private	For SVM 87%, and gradient boosting ensemble based model is 98%	Limited generalizability, Limited feature extraction
2022	Alighaleh et al. [19]	DL	Convolutional neural network	×	Private	Diff models with different accuracy	Improve in accuracy but time consuming Because of batch normalization
2022	Egea et al. [20]	DL	Convolutional neural network	×	Private	99.67%	Works in few samples of saffron

on Kashmiri saffron using data-driven approaches. Hence to tackle this issue, we designed a saffron prediction model in which features were extracted from saffron images, and deep learning is trained to discriminate the adulterated samples from real ones.

The proposed framework of our model is shown in Fig. 3. It consists of two steps. The first step in our framework for saffron adulteration detection is image pre-processing. This step involves cleaning and enhancing the raw images of saffron samples to ensure that they are suitable for analysis using machine learning and deep learning techniques. The image pre-processing step includes operations such as noise reduction, image resizing, contrast enhancement, and image normalization. Once the images have been pre-processed, the second step in our framework is the application of machine learning and deep learning techniques. These techniques are used to analyze the pre-processed images and identify potential adulterants in the saffron samples. We have developed a machine learning model that is trained on a large dataset of saffron images, including both authentic and adulterated samples. The model uses a combination of convolutional neural networks (CNNs) and deep learning algorithms to classify saffron samples as either authentic or adulterated based on their image features. By combining image pre-processing with machine learning and deep learning techniques, our framework provides a highly accurate and efficient method for detecting saffron adulteration. The use of advanced image analysis techniques enables us to identify even subtle differences in saffron samples that may be indicative of adulteration, while the application of machine learning algorithms allows us to process large volumes of data quickly and accurately. Overall, our framework offers a powerful tool for ensuring the quality and authenticity of saffron products.

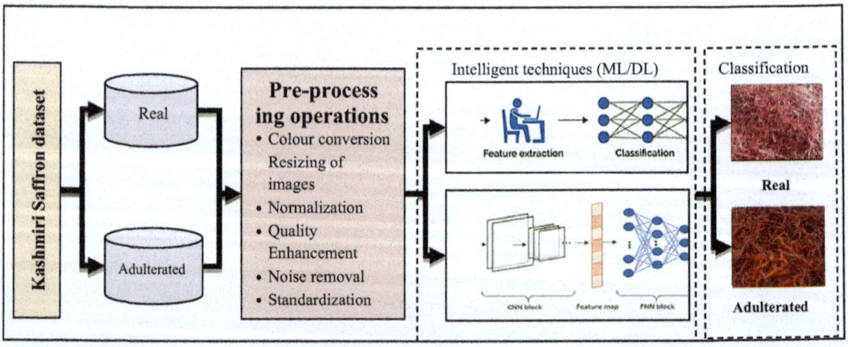

Fig. 3 Our proposed framework of saffron adulteration prediction system

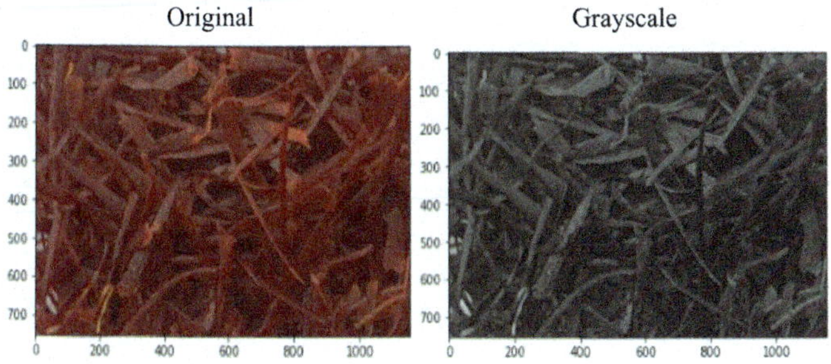

Fig. 4 A depiction of RGB to gray color conversion of saffron images

4 Pre-processing Techniques

Pre-processing techniques can be used in saffron images to enhance the quality and clarity of the images, remove noise, adjust color, and improve the accuracy of subsequent analysis. Here are some pre-processing techniques that can be applied to saffron images.

4.1 *Color Conversion*

This technique can be used in saffron images to convert the images from one color space to another, and thereby manipulate the color and intensity information in the images. Here are some commonly used color conversion techniques for saffron images including RGB to grayscale. Figure 4 depicts the conversion of saffron image from original to grayscale. The images are taken from our own created dataset.

RGB saffron image to grayscale, where each pixel in the grayscale image has a single value representing the intensity of the original RGB image. This technique is useful for applications such as image analysis, segmentation, and feature extraction.

Grayscale to binary: This technique can be used to convert grayscale saffron images to binary images, where each pixel is either black or white based on a threshold value. This technique can be useful for applications such as object segmentation and counting. Figure 5 depicts the conversion of grayscale to binary.

4.2 *Equalization Technique*

The equalization technique, specifically histogram equalization, is a pre-processing technique used in digital image processing to improve the contrast and visual appearance of an image. This technique can be applied to saffron images to improve their contrast and visual appearance. Other equalization techniques, such as contrast

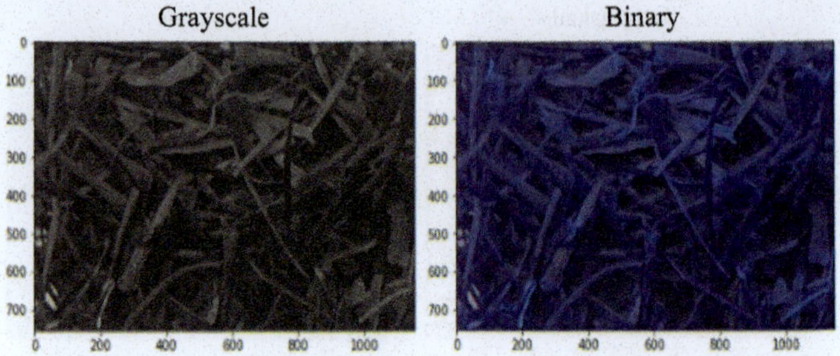

Fig. 5 A depiction of gray to binary color conversion of saffron samples

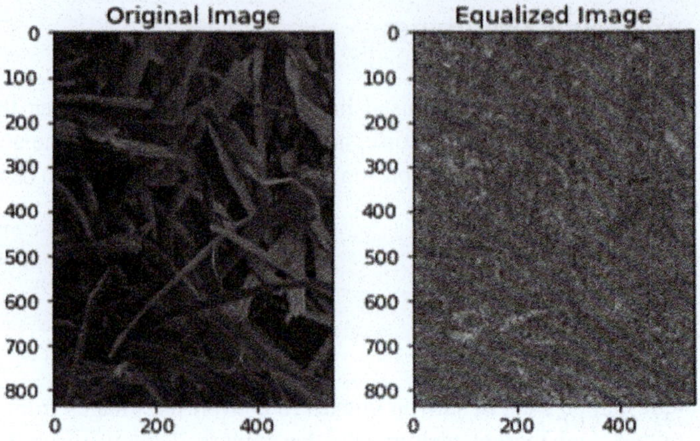

Fig. 6 A depiction of gray to equalized conversion of saffron images

stretching and adaptive histogram equalization, can also be applied to saffron images to improve their visual appearance and quality. Figure 6 shows the conversion of grayscale to equalized image. Here is how the technique can be applied to saffron images.

4.3 *Edge Detection*

In saffron images, edge detection can be used to identify the edges of the saffron threads and to distinguish them from the background or other objects in the image. Here are some commonly used edge-detection techniques that can be applied to saffron images: Sobel edge detection, Canny edge detection, Laplacian of Gaussian (LoG) edge detection, Prewitt edge detection, and so on. Figures 7 and 8 show the conversion of the original saffron sample to edge detection and noise reduction of saffron samples.

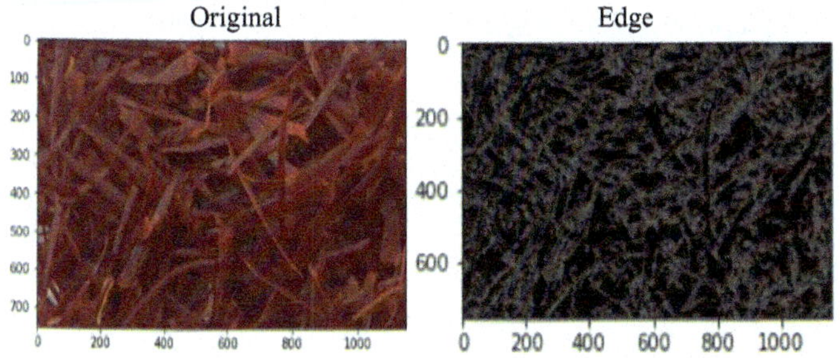

Fig. 7 A depiction of the original image to edge detection

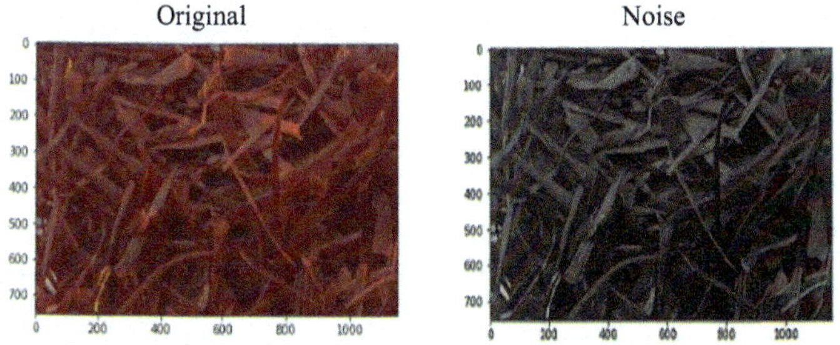

Fig. 8 A depiction of the original saffron sample to noise reduction

4.4 Noise Reduction

Noise reduction: Saffron images may have noise due to factors such as low light conditions or camera settings. Applying noise reduction techniques such as median filtering or Gaussian filtering can help to reduce the noise in the images.

4.5 Standardization Techniques

Saffron images aim to reduce variation in the image features caused by factors such as illumination, camera settings, and image resolution. Standardization techniques can help improve the accuracy and reliability of saffron image analysis and identification. Here are some common standardization techniques used in saffron image analysis. White balancing, scale normalization contrast stretching: This technique adjusts the contrast of the image by stretching the pixel values to a wider range. This can help enhance the visibility of subtle details in the image, making it easier to identify and analyze saffron features. Overall, standardization techniques in saffron images can help reduce the variation caused by different factors and improve the accuracy and reliability of saffron image analysis and identification.

4.6 Normalization Techniques

In saffron, images aim to adjust the image features to a common reference scale, making it easier to compare different images. Normalization techniques can help improve the accuracy and reliability of saffron image analysis and identification. Some common normalization techniques used in saffron image analysis are Z-score normalization, Min–max normalization, Histogram equalization, and Log transformation.

4.7 Data Augmentation

Data augmentation is a technique used to increase the size of a dataset by creating additional synthetic data from the existing data. This technique is commonly used in machine learning and computer vision applications to improve the performance of the model by increasing the diversity and amount of training data. Data augmentation can be applied to various types of data, including text, images, and audio. By applying these data augmentation techniques to a dataset of saffron images, we can generate a larger and more diverse dataset that can be used to train a more accurate saffron image classification model. In the context of saffron images, some common data augmentation techniques that can be applied include Rotation, Flipping, Cropping, Scaling, and Shifting. We apply these techniques to saffron samples to enhance the quality of the sample and also to increase the number of samples for better performance because the major problem with Kashmiri saffron is that there is no benchmark dataset available. We create our dataset of Kashmiri saffron after visiting some famous places in the valley, which includes Pampore, Lethpora, Pulwama, and Budgam. Figure 9 includes some data augmentation techniques that we have applied to saffron samples.

5 Results

In this section, we present the results and discussion of our research findings, focusing on the performance evaluation of our proposed framework. Assessments of performance on both real and adulterated samples are carried out across varied dataset ratios. Assessments of performance on both real and adulterated samples are carried out across varied dataset ratios. Table 2 represents the performance of our model on a testing dataset in terms of a confusion matrix. The confusion matrix provides insights into the model's performance across different classes. Two classes considered in the confusion matrix are "Real" (class 0) and "Adulterated" (class 1). The overall accuracy of our model on the testing dataset is 0.9897, precision is 0.9890,

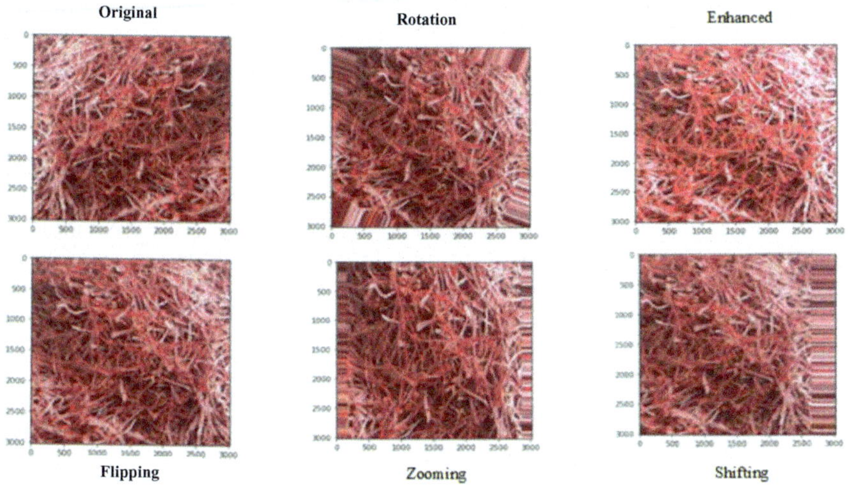

Fig. 9 Kashmiri saffron through diverse data augmentation techniques

recall is 0.9878, F-score is 0.9930, and loss is 0.005. Figure 10 shows the curves of our model.

Table 2 Performance of our proposed model in terms of confusion matrix

Proposed framework					
Classes	Accuracy	Precision	Recall	F-score	Loss
Real (0)	0.9897	0.9890	0.9878	0.9930	0.005
Adulterated (1)	0.9897	0.9890	0.9878	0.9930	0.005

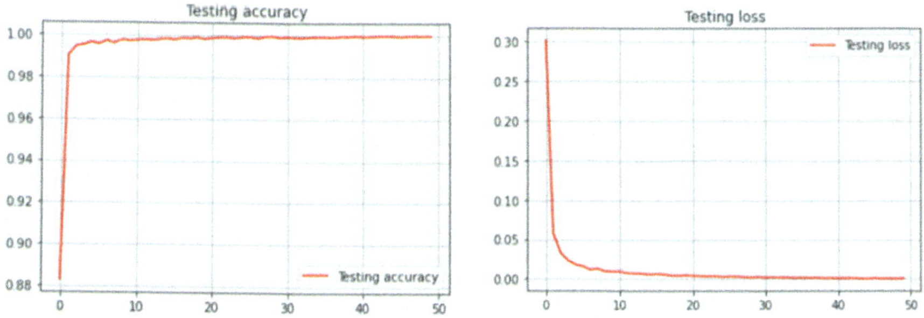

Fig. 10 The testing curve of our model: **a** testing accuracy; **b** testing loss

Table 3 A comparative analysis of saffron datasets

Year	Origin	No of samples	Real	Adulterated	Resolution
2009	Greece, India, Iran, Italy, Morocco, and Spain	418	×	×	Not used
2015	Iran	165	×	×	Not used
2015	Chinese	13	×	×	Not used
2017	Iran	20	×	×	Not used
2017	Iran	33	×	×	Not used
2019	Chinese	600	×	×	Not used
2020	Iran	440	×	×	3024 × 4032
2022	Kashmiri	9	7	2	Not used
2022	Iran	7	2	5	4000 × 3000
2023	Kashmiri	201	111	90	3024 × 4024

6 Performance Protocols

In the preceding section, we have presented a review of recent saffron adulteration detection techniques. Hence to evaluate these techniques, some performance protocols are used by the researcher such as benchmark datasets and performance metrics and these are further explained in the below sections.

6.1 *Saffron Datasets*

The issue with Kashmiri saffron is that there is no benchmark dataset available. Due to this, researchers often use a limited number of samples for the detection of adulteration, which may not be representative of the overall saffron supply. Table 3 depicts the summary of datasets that are used for the detection of adulteration of saffron. In the below table, most of the researchers had worked on Iranian and Chinese saffron with limited samples. To overcome this issue, we have created our dataset of Kashmiri saffron for better performance.

6.2 *Performance Metrics*

The effectiveness of the existing saffron adulteration prediction techniques has been evaluated based on some standard metrics such as accuracy, recall, precision, and F-score. Table 4 depicts the summary of these performance metrics used in existing studies along with a formula and definition.

Table 4 A comparative summary of various performance evaluation metrics

Acronym	Metric	Formula	Description
ACC	Accuracy	$ACC = \frac{TP+TN}{\text{Total samples}}$	It is defined as the ratio of the number of correct predictions to the total number of predictions made by the model
REC	Recall	$Recall = \frac{TP}{TP+FN}$	Recall is calculated by dividing the number of true positive predictions by the total number of positive cases in the dataset
PRE	Precision	$Precision = \frac{TP}{TP+FP}$	It is defined as evaluating the accuracy of a classification model. It measures the proportion of correctly identified positive instances among all instances that the model predicted as positive
F	F score	$F = 2 * \frac{precision*recall}{precision+recall}$	F-score is a metric that measures the balance between the precision and recall of a model

7 Open Research Challenges

While the proposed model for saffron adulteration prediction presented in this paper represents an important step forward, there are still several open research challenges that must be addressed in order to improve the performance of these models and overcome practical limitations to their implementation. Some major research challenges are listed below

I. The non-availability of Kashmiri saffron-based datasets presents a major challenge for the development of accurate prediction models for saffron adulteration.
II. There is a significant gap in the literature on the specific issue of Kashmiri saffron adulteration.
III. The presence of a variety of saffron adulterants presents a significant challenge to the development of accurate prediction models.
IV. One of the challenges in using machine learning-based approaches for saffron adulteration prediction is their lack of generalization capability to samples that were not present in the training data.
V. There has been limited work on applying these approaches to saffron adulteration detection.

8 Conclusion

In this study, we present a review of recent saffron adulteration prediction techniques using machine and deep learning-based approaches. The analysis from existing studies reveals some open research issues that need to be addressed by researchers such as inadequate Kashmir saffron dataset and limited performance in traditional approaches. In addition, it reveals that most of the work is based on Iranian saffron to predict various types of adulterants present in the saffron sample. Hence, we propose a novel framework for saffron adulteration prediction for Kashmir saffron via computer vision-based approaches. Moreover, various pre-processing techniques are used to enhance the quality of saffron images. Besides, to evaluate the performance of saffron adulteration prediction system, performance protocols such as dataset and metrics are used in this work. The future work includes the conception of transfer learning and online learning for saffron adulteration prediction.

References

1. Negbi M (1999) Saffron: crocus sativus L., p 154
2. Amin J, Selwal A, Sabha A (2021) SAPS: Automatic saffron adulteration prediction systems, research issues, and prospective solutions. In: Proceedings—2021 4th international conference on computational intelligence and communication technologies, CCICT 2021, Institute of Electrical and Electronics Engineers Inc., Jul. 2021, pp 64–71. https://doi.org/10.1109/CCICT53244.2021.00024
3. Ahmad Tantry M, Bashir Ahmad Dar P, Singh S (2017) Dr Musroor Ahmad Tantry: Professor Bashir Ahmad Dar. Available: www.ijournals.in
4. Gohari AR, Saeidnia S, Mahmoodabadi MK (2013) An overview on saffron, phytochemicals, and medicinal properties. Pharmacogn Rev 7(13):61–66. https://doi.org/10.4103/0973-7847.112850
5. Husaini AM, Jan KN, Wani GA (2021) Saffron: a potential drug-supplement for severe acute respiratory syndrome coronavirus (COVID) management. Heliyon 7(5). Elsevier Ltd, https://doi.org/10.1016/j.heliyon.2021.e07068
6. Amin J, Selwal A, Sabha A (2023) SaffNet: an ensemble-based approach for saffron adulteration prediction using statistical image features. Multimed Tools Appl. https://doi.org/10.1007/s11042-023-14934-9
7. Bhooma V, Nagasathiya K, Vairamani M, Parani M (2020) Identification of synthetic dyes magenta III (new fuchsin) and rhodamine B as common adulterants in commercial saffron. Food Chem 309, https://doi.org/10.1016/j.foodchem.2019.125793
8. Amirvaresi A, Rashidi M, Kamyar M, Amirahmadi M, Daraei B, Parastar H (2020) Combining multivariate image analysis with high-performance thin-layer chromatography for development of a reliable tool for saffron authentication and adulteration detection. J Chromatogr A 1628, https://doi.org/10.1016/j.chroma.2020.461461
9. Torelli A, Marieschi M, Bruni R (2014) Authentication of saffron (Crocus sativus L.) in different processed, retail products by means of SCAR markers. Food Control 36(1):126–131. https://doi.org/10.1016/j.foodcont.2013.08.001
10. Dowlatabadi R et al. (2017) Detection of adulteration in Iranian saffron samples by 1H NMR spectroscopy and multivariate data analysis techniques. Metabolomics 13(2), https://doi.org/10.1007/s11306-016-1155-x
11. Lu X, Xia Z, Qu F, Zhu Z, Li S (2020) Identification of authenticity, quality and origin of saffron using hyperspectral imaging and multivariate spectral analysis. Spectrosc Lett 53(2):76–85. https://doi.org/10.1080/00387010.2019.1693403
12. Javanmardi N, Bagheri , A, Moshtaghi N, Sharifi A, Kakhki AH (2011) Identification of Safflower as a fraud in commercial Saffron using RAPD/SCAR marker
13. Huang WJ, Li FF, Liu YJ, Long CL (2015) Identification of Crocus sativus and its adulterants from Chinese markets by using DNA barcoding technique. Iran J Biotechnol 13(1):36–42. https://doi.org/10.15171/ijb.1034

14. Varliklioz Er S, Eksi-Kocak H, Yetim H, Boyaci IH (2017) Novel spectroscopic method for determination and quantification of saffron adulteration. Food Anal Methods 10(5):1547–1555, https://doi.org/10.1007/s12161-016-0710-4
15. Filatova M, Hajslová J, Stupak M (2024) Detection of saffron adulteration by other plant species using SPME-GC-HRMS. Eur Food Res Technol 250(3):911–922. https://doi.org/10.1007/s00217-023-04443-2
16. Alighaleh P, Pakdel , R Ghanei Ghooshkhaneh N, Einafshar S, Rohani A, Saeidirad MH (2023) Detection and classification of saffron adulterants by Vis-Nir imaging, chemical analysis, and soft computing. Foods 12(11), https://doi.org/10.3390/foods12112192
17. El Hani O et al. (2023) Geographical classification of Saffron (Crocus Sativus L.) using total and synchronous fluorescence combined with chemometric approaches. Foods 12(9), https://doi.org/10.3390/foods12091747
18. Mohamadzadeh Moghadam M, Taghizadeh M, Sadrnia H, Pourreza HR (2020) Nondestructive classification of saffron using color and textural analysis. Food Sci Nutr 8(4):1923–1932, https://doi.org/10.1002/fsn3.1478
19. Alighaleh P, Khosravi H, Rohani , A Saeidirad MH, Einafshar S (2022) The detection of saffron adulterants using a deep neural network approach based on RGB images taken under uncontrolled conditions. Expert Syst Appl 198, https://doi.org/10.1016/j.eswa.2022.116890
20. Egea G et al., The menace of saffron adulteration: Low-cost rapid identification of fake look-alike saffron using Foldscope and machine learning technology
21. "Kashmiri Saffron 1gm." Accessed: Mar. 21, 2024. [Online]. Available: https://vaaradhifarms.com/products/kashmiri-saffron-1gm
22. "saffron images—Google Search." Accessed: Mar. 21, 2024. [Online]. Available: https://www.google.com/search?q=saffron+images&tbm=isch&chips=q:saffron,g_1:plant:AygwsRk3G68%3D&prmd=isvnmbtz&hl=en&sa=X&ved=2ahUKEwiTyt3UxoSFAxW7fGwGHZNrDgUQ4lYoBHoECAEQOQ&biw=1519&bih=703#imgrc=rxs3GbEwKAMAMM
23. "red lily - Google Search." Accessed: Mar. 02, 2024. [Online]. Available: https://www.google.com/search?q=red+lily&sca_esv=1fc9c7ca6eaf3670&udm=2&biw=1536&bih=703&sxsrf=ACQVn0-W_6Y1tLewkevhjxWlrPhNE-auAg%3A1709402071732&ei=12fjZcmnLOKfg8UPwtqh8As&ved=0ahUKEwiJ7436ktaEAxXiz6ACHUJtCL4Q4dUDCBA&uact=5&oq=red+lily&gs_lp=Egxnd3Mtd2l6LXNlcnAiCHJlZCBsaWx5MgUQABiABDIFEAAYgAQyBRAAGIAEMgUQABiABDIFEAAYgAQyBRAAGIAEMgUQABiABDIFEAAYgAQyBRAAGIAEMgUQABiABEinLFDGBliDKHACeACQAQSYAbsCoAHrFKoBBzAuOS4zLjG4AQPIAQD4AQGYAgugAsQNqAIKwgIGEAAYCBgewgIEECMYJ8ICBxAjGOoCGCfCAgoQABiABBiKBRhDwgIIEAAYgAQYsQPCAg4QABiABBiKBRixAxiDAZgDCIgGAZIHBTIuNi4zoAf9RQ&sclient=gws-wiz-serp#vhid=cZp9A0Bht5P6aM&vssid=mosaic
24. "maize stigma—Google Search." Accessed: Mar. 02, 2024. [Online]. Available: https://www.google.com/search?q=maize+stigma&sca_esv=1fc9c7ca6eaf3670&udm=2&biw=1536&bih=703&sxsrf=ACQVn08CaZkNDvmBw5w_S53hYpGiVPWBxA%3A1709402007218&ei=l2fjZbX3DKrF4-EPtr-DiAY&ved=0ahUKEwj1nqzbktaEAxWq4jgGHbbfAGEQ4dUDCBA&uact=5&oq=maize+stigma&gs_lp=Egxnd3Mtd2l6LXNlcnAiDG1haXplIHN0aWdtYTIFEAAYgARI60JQzxFY9jpwAXgAkAEAmAHuAaABtgqqAQUwLjUuMrgBA8gBAPgBAZgCCKAC-ArCAg0QABiABBiKBRhDGLEDwgIIEAAYgAQYsQPCAgoQABiABBiKBRhDwgIGEAAYBRgewgIGEAAYCBgewgIHEAAYgAQYGJgDAIgGAZIHBTEuNS4yoAfEFg&sclient=gws-wiz-serp#vhid=H5mqzrgJbVhhBM&vssid=mosaic
25. "hibiscus - Google Search." Accessed: Mar. 02, 2024. [Online]. Available: https://www.google.com/search?q=hibiscus&sca_esv=1fc9c7ca6eaf3670&udm=2&biw=1536&bih=703&sxsrf=ACQVn0-0xTyFpcy1FDBjkhN67TSzVTIOnA%3A1709393424358&ei=EEbjZeq3Ffay4-EPkfCigAE&oq=hibiscus&gs_lp=Egxnd3Mtd2l6LXNlcnAiCGhpYmlzY3VzKgIIADIIEAAYgAQYsQMyCBAAGIAEGLEDMggQABiABBixAzIIEAAYgAQYsQMyCBAAGIAEGLEDMgUQABiABDIIEAAYgAQYsQMyBRAAGIAEMgUQABiABDIFEAAYgARIgzZQ_QZYoSdwAngAkAEEmAH1AaAB8BOqAQUwLjguNbgBAcgBAPgBAZgCC6AC3w2oAgrCAgQQIxgnwgINEAAYgAQYigUYQxixA8ICBxAjGOoCGCeYAwiIBgGSBwUyLjcuMqAHmkg&sclient=gws-wiz-serp#vhid=jjqRfzVQEIs7nM&vssid=mosaic

A Text Classification in Tourism Reviews with the Hybrid Transformer-Attention Model in the Information Management of Smart Tourism

Jayanth Srinivasan, M. Niranjanee, C. Nandhana, and M. Azhagiri

Abstract In the realm of smart tourism, understanding tourist feedback through sentiment analysis is pivotal for enhancing service quality and experience. This study introduces a novel Hybrid Transformer-Attention Model (HTAM), designed to advance sentiment classification accuracy, model interpretability, and linguistic adaptability in tourist reviews. By integrating the contextual insights of transformer models with a strategic attention mechanism, HTAM aims to provide superior performance over existing hybrid models like Convolutional Neural Networks—Bidirectional Long Short-Term Memory (CNN-BiLSTM) (Meng, The Convolutional Neural Network Text Classification Algorithm in the Information Management of Smart Tourism Based on Internet of Things, 2024). With empirical evidence showcasing its capability to outperform conventional models in metrics with an accuracy of 96.74%, HTAM's application promises to revolutionize smart tourism by enabling more responsive and tailored service improvements.

Keywords Sentiment analysis · Smart tourism · Hybrid transformer-attention model (HTAM) · Machine learning in hospitality

J. Srinivasan (✉) · M. Niranjanee · C. Nandhana · M. Azhagiri
Department of Computer Science and Engineering, SRM Institute of Science and Technology, Ramapuram, Chennai, Tamil Nadu 600 089, India
e-mail: kj7263@srmist.edu.in

M. Niranjanee
e-mail: nm5027@srmist.edu.in

C. Nandhana
e-mail: nc3116@srmist.edu.in

M. Azhagiri
e-mail: azhagirm@srmist.edu.in

P. J. S. Gonçalves et al. (eds.), *Proceedings of Fifth International Conference on Computing, Communications, and Cyber-Security*, Lecture Notes in Networks and Systems 1128, https://doi.org/10.1007/978-981-97-7371-8_37

1 Introduction

The tourism industry, as a pivotal sector of the global economy, has witnessed transformative shifts with the advent of digital technologies. The concept of smart tourism, leveraging the Internet of Things (IoT), Artificial Intelligence (AI), and Data Analytics (DA), has revolutionized how destinations are managed and experienced. This evolution necessitates advanced tools for analyzing vast amounts of textual feedback generated by tourists. Understanding these sentiments accurately is crucial for destination management organizations and businesses to enhance visitor experiences, tailor services, and maintain competitive advantage in this rapidly evolving landscape.

In the dynamic realm of smart tourism, the ability to effectively analyze and classify textual feedback from tourists has emerged as a cornerstone for enhancing visitor experiences and service offerings. The burgeoning volume of online reviews and feedback presents both a challenge and an opportunity for leveraging data-driven insights to tailor services and respond to consumer needs. Traditional text classification models, such as Convolutional Neural Networks (CNNs) and Long Short-Term Memory (LSTM) networks, have paved the way for initial advancements in sentiment analysis. However, these models often grapple with the intricacies of language and the depth of contextual understanding required to accurately interpret nuanced tourist feedback.

The introduction of the Hybrid Transformer-Attention Model (HTAM) marks a significant leap forward in this domain. By amalgamating the contextual prowess of transformer-based models with the precision of attention mechanisms, HTAM aims to set new benchmarks in the accuracy, interpretability, and adaptability of sentiment analysis tools. Unlike its predecessors, HTAM is engineered to adeptly navigate the complexities of natural language, ensuring a more nuanced and comprehensive analysis of textual data. This capability is particularly pivotal in smart tourism, where the diversity of tourist expressions and the subtleties of sentiment can significantly impact service personalization and quality enhancement strategies.

HTAM's innovative approach is not only anticipated to outperform existing models in terms of analytical precision but also to offer unprecedented insights into the drivers of tourist satisfaction and dissatisfaction. This research explores HTAM's potential to transform the landscape of smart tourism through advanced text classification, underscoring its implications for both academic inquiry and practical application. By advancing our understanding of tourist sentiments, HTAM promises to empower stakeholders in the tourism industry with the tools to craft more responsive, personalized, and ultimately more effective tourism experiences.

The significance of this research lies not only in its contribution to the field of text classification and sentiment analysis but also in its practical application to the burgeoning sector of smart tourism. As we delve into the capabilities and advantages of the HTAM model, this study aims to underscore the transformative potential of advanced analytical models in harnessing the vast, untapped wealth of data within tourist feedback, thereby heralding a new era of data-driven innovation in tourism management.

Motivation: The motivation behind this study stems from the need to improve sentiment analysis in smart tourism by addressing the limitations of existing models in accurately interpreting complex tourist feedback. Current models such as CNN-BiLSTM [1], while foundational, exhibit limitations in processing the nuanced linguistic elements and cultural subtleties inherent in tourist feedback. This study introduces the Hybrid Transformer-Attention Model (HTAM) to address these gaps by harnessing advanced transformer architectures coupled with sophisticated attention mechanisms. The HTAM is engineered to provide refined text classification, heightened interpretability, and greater adaptability to the linguistic diversity found in tourist reviews. The aim is to deepen the understanding of tourist sentiments and enhance the responsiveness of tourism services, thereby aligning with the industry's shift toward data-driven personalization and decision-making.

Organization: The remainder of this paper is structured as follows: Sect. 2 provides a comprehensive examination of existing literature, situating our research within the broader context of sentiment analysis and smart tourism technologies. Section 3 details the architecture of the proposed Hybrid Transformer-Attention Model (HTAM), including its innovative integration of transformer models and attention mechanisms. Section 4 describes the implementation process, dataset preparation, and model evaluation criteria, followed by a comprehensive analysis of the HTAM's performance against established benchmarks. Finally, Sect. 5 concludes the paper with a summary of our findings, contributions to the field of smart tourism, and potential avenues for future research.

2 Related Works

Meng et al. [1] lay the groundwork with their CNN-BiLSTM model, showcasing its effectiveness in sentiment analysis over traditional models, thus offering a robust framework for enhancing customer satisfaction in smart tourism. Tezgider et al. [2] introduced a bidirectional transformer (BiTransformer) utilizing bidirectional position encoding, enhancing text data processing by leveraging forward and backward text position information. Their comparative study across models including LSTM, attention mechanisms, and transformers on a Turkish text dataset highlighted the superior performance of BiTransformer in text classification, demonstrating the effectiveness of advanced methods in handling large text datasets. Qiu et al. [3] introduced a model combining grapheme-morpheme embeddings with a hybrid self-attention mechanism and RCNN for text classification. This approach significantly enhances feature extraction and semantic interpretation, outperforming traditional methods in classifying Chinese short texts. Their findings underscore the model's efficacy in improving text classification accuracy. Liu et al. [4] address the shortcomings of attention weights in Transformers, proposing AGrad and RePAGrad to provide directional relevance scores and improve model explainability. These methods outperform current explanation techniques, marking significant progress in understanding transformer-based NLP models. Wadhe et al. [5] explore machine learning techniques for sentiment classification in tourist reviews, demonstrating the potential of feature extraction algorithms and classification models to gauge tourist destination popularity.

Extending the exploration, Liu et al. [4] tackle the challenge of interpretability in Transformer models, proposing novel explanation techniques to enhance understanding of text classification decisions. Chen et al. [6] propose a hybrid CNN and Transformer model with a Gaussian multi-head self-attention mechanism, achieving notable accuracy in Chinese medical data classification, pointing toward the model's adaptability in specialized domains. Bao et al. [7] introduce a novel approach for short text classification by merging BERT with CNN and BiGRU, enhanced by an attention mechanism for improved semantic feature representation. Their model significantly exceeds standard methods, offering a breakthrough in NLP applications. Prabhakar et al. [8] analyze hybrid deep learning models with a focus on attention mechanism positioning, showcasing their effectiveness in text classification through high accuracy and efficiency. Cao et al. [9] propose an LSTM-TCN hybrid model with an attention mechanism for sentiment classification, surpassing traditional CNNs. Using Word2vec, LSTM, and TCN for feature extraction, and prioritizing key information via attention, this method achieves accuracies of 93.06% and 91.08% on Chinese datasets, demonstrating its superior analytical capability.

3 Proposed System Architecture

The architectural design of the Hybrid Transformer-Attention Model (HTAM) for sentiment analysis commences with the acquisition and preparation of user-generated reviews from various online tourism platforms. These reviews are the cornerstone of the HTAM's input, rich with experiential narratives and subjective impressions of tourists. To ensure the integrity and quality of the analysis, the reviews first pass through an Input Layer, where they are collected in their raw form, teeming with unstructured and noisy data (Fig. 1).

Within the Pre-Processing Module, the raw reviews are meticulously processed to strip away extraneous information that could cloud the sentiment analysis. Text Cleaning is applied to remove non-essential characters and formatting, distilling the content down to its most meaningful essence. The Tokenization process then dissects the cleaned text into discrete, manageable units or 'tokens', typically corresponding to words or phrases. These tokens are subsequently translated into numerical form during the Encoding phase, converting the textual data into a machine-readable format. This transformation is critical as it bridges the gap between human language and the algorithmic language of the model.

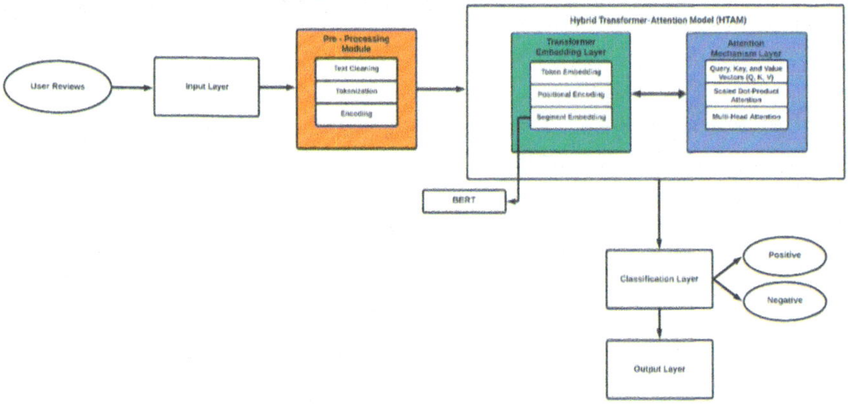

Fig. 1 Proposed architecture of HTAM

The next phase involves the Transformer Embedding Layer, which is instrumental in constructing a contextual understanding of the text. This layer leverages a pre-trained transformer—such as BERT—to generate dense vector representations for each token. These embeddings encapsulate not only the inherent meaning of each word but also their context within the sentence and the review as a whole. Positional information, crucial for maintaining the syntactic structure of the language, is embedded alongside the tokens, ensuring the model is cognizant of the order and flow of the narrative. If the review comprises multiple segments or sentences, Segment Embedding further aids in distinguishing between these, preserving the structure and coherence of the input.

At the core of HTAM lies the Attention Mechanism Layer, an innovative feature designed to hone the model's focus on the most sentimentally significant elements within the reviews. This layer employs Query, Key, and Value vectors to dynamically allocate 'attention' across the text, effectively discerning which parts are pivotal for sentiment classification. The Scaled Dot-Product Attention component computes attention scores, which guide the model's focus, while the Multi-Head Attention aspect extends this functionality by executing several attention processes concurrently. This multiplicity allows the model to capture diverse features and relationships within the text, offering a comprehensive perspective on the sentiment expressed.

The culmination of this intricate process is the Classification Layer, which synthesizes the rich, contextual embeddings and attention-focused insights to determine the sentiment conveyed in each review. This decisive layer classifies the sentiment as either positive or negative, drawing upon patterns and correlations learned during the model's training. The final output, produced by the Output Layer, reflects the sentiment classification for each review, providing a valuable tool for tourism stakeholders to gauge visitor satisfaction and tailor services accordingly. The HTAM's elaborate architecture is not only a testament to the fusion of advanced machine learning techniques but also a significant stride toward more accurate, nuanced, and efficient sentiment analysis in the realm of smart tourism.

4 Implementation and Results

The implemented Hybrid Transformer-Attention Model (HTAM) for sentiment analysis begins with a meticulous preprocessing phase. The dataset, comprising hotel reviews, undergoes text cleaning to eliminate non-relevant characters and noise, ensuring the data is primed for analysis. Tokenization breaks the cleaned text into discrete units or tokens, which are then transformed into a numerical format through encoding. This process is essential for the compatibility of the textual data with deep learning algorithms.

Following preprocessing, the model utilizes a Transformer Embedding Layer, employing the BERT framework. This layer produces contextual embeddings for each token, capturing not just the word's meaning but also its context within the sentence, leveraging positional and segment embeddings to maintain the narrative flow. This rich embedding captures the intricacies of language used in hotel reviews, setting the stage for nuanced sentiment analysis.

At the heart of HTAM lies the Attention Mechanism Layer, where a custom multi-head attention mechanism is applied. This layer calculates attention scores that enable the model to focus on the most salient parts of the text, which are critical for accurate sentiment classification. The model performs several attention processes in parallel, capturing different aspects of the textual information, thus enriching the model's interpretive capacity.

The methodology extends beyond the extraction of features, employing a sophisticated classification layer that interprets the attention-modulated embeddings. This layer typically integrates a pooling strategy to distill the essential features, followed by dropout layers for regularization, and culminates in a dense layer that serves as the decision-maker for sentiment classification. The model's output reflects this intricate process, providing a binary classification that captures the sentiment of each review.

To validate the model's efficacy, a comprehensive evaluation strategy is implemented, comparing the HTAM's performance against traditional models using standard metrics such as accuracy, precision, recall, and F1-score. This comparative analysis not only benchmarks the HTAM but also provides insights into its operational strengths and areas for potential refinement. The goal is to iteratively enhance the model by fine-tuning and optimization, driven by empirical evidence gathered during the evaluation phase.

The Classification Layer, after the attention mechanism, analyzes the processed inputs and categorizes the sentiment. The model employs global average pooling to extract the core features of the input, then applies dropout for regularization, and a dense layer for the ultimate classification. The result is a binary sentiment classification that differentiates between positive and negative attitudes. The intricate architecture is created to tackle the intricate characteristics of natural language in hotel reviews, with the goal of offering strong and precise sentiment analysis as a valuable resource for intelligent tourism and hospitality services.

This approach cleverly combines the accuracy of bespoke multi-head attention mechanisms with the profound contextual insights of transformer systems, such as

BERT. This combination is essential to understanding the complex sentiment and language found in reviews. Each token gains contextual relevance from the transformer embedding layer, and sentiment classification performance is improved by the attention mechanism layer, which amplifies the influence of specific textual features. This advanced method sets HTAM apart not only for its breadth of analysis but also for its flexibility and interpretability, demonstrating its potential to completely transform sentiment research in the context of smart tourism. The integration of interpretability into the model is accomplished by strategically emphasizing the significance of features within the text. The HTAM leverages its attention mechanisms to highlight and assign weights to specific tokens that are crucial for understanding sentiment, thus providing clear insights into its decision-making process. This added level of information highlights the model's creative ability to handle the nuances of spoken language, offering a strong tool for deriving insightful conclusions from a range of visitor comments.

The HTAM model was trained and evaluated on a high-performance computing setup, essential for handling the model's complexity. The CPU used was an Intel i9, known for its advanced processing capabilities, while the GPU was an NVIDIA RTX 3050, which provided the necessary computational power to facilitate deep learning tasks. The hardware configuration played a crucial role in effectively handling the computationally demanding tasks of the HTAM, especially during the training stage when the model's several layers and attention mechanisms necessitate significant computational resources. The software stack consisted of TensorFlow, a highly adaptable and extensively utilized deep learning library, and the BERT model from the transformers library of Hugging Face, which equipped the model with pre-trained embeddings. This combination of high-end GPU processing power and advanced deep learning software enabled the efficient handling of large datasets and complex model architectures inherent to the HTAM.

The application of this model in the smart tourism sector is expected to have significant effects, offering stakeholders a sophisticated tool for analyzing sentiments in real time. The HTAM's applicability ranges from customizing customer experiences to making strategic decisions by utilizing detailed insights obtained from intricate customer feedback data.

Dataset

This research incorporates a dataset of 5,15,739 hotel reviews sourced from Kaggle [10] to assess the Hybrid Transformer-Attention Model (HTAM). The dataset mirrors the broad spectrum of sentiments expressed in tourist feedback, essential for realistic sentiment analysis. The reviews for effective machine learning applications involved rigorous preprocessing techniques such as text normalization and tokenization. A binary labeling methodology was utilized, wherein reviews with scores over a midrange were classified as positive, while those with scores below this threshold were classified as negative. The data was divided into 70% for training, 20% for validation, and 10% for testing, ensuring the HTAM's capacity for generalization and enabling meticulous model evaluation. The equitable allocation of data facilitates the verification of the HTAM's suitability for sentiment analysis in the tourism industry.

Results

The Hybrid Transformer-Attention Model (HTAM) showed notable effectiveness in sentiment analysis. The model training began with significant caution over the lack of gradients for certain variables, suggesting potential issues in the first optimization process. Later training epochs showed a strong learning curve with a significant enhancement in loss reduction and accuracy. The model attained a final validation accuracy of 96.74% and a loss of 0.1230, demonstrating a high level of prediction accuracy. The results indicate that HTAM is skilled at analyzing intricate linguistic structures and accurately identifying sentiment. Graphs illustrating the loss and accuracy trajectories across training epochs would complement this narrative, offering visual affirmation of the model's optimization and learning stability. HTAM has the potential to significantly improve sentiment analysis applications in the smart tourism industry, leading to better analytical abilities and customer information gathering.

Figure 2 illustrates the HTAM's accuracy and loss for both the training and validation sets across epochs. The left graph shows a sharp rise in accuracy for the training set, with a more gradual increase for the validation set, indicating a convergence toward high predictive performance. Conversely, the right graph depicts a decrease in loss for both sets, with the training loss decreasing at a steeper rate, suggesting that the model is learning effectively and generalizing well to unseen data. This convergence of training and validation lines signifies a robust model without overfitting concerns.

Figure 3 displays three key performance metrics—F1-score, Precision, and Recall—for the Hybrid Transformer-Attention Model (HTAM) as they evolve over training epochs. The F1-score graph exhibits a steady increase, indicating a balanced improvement in both precision and recall throughout the training process. Precision, which measures the accuracy of positive predictions, shows a consistent upward trend, reflecting the model's increasing ability to identify true positives over time. The Recall graph similarly ascends, demonstrating the model's enhanced capability

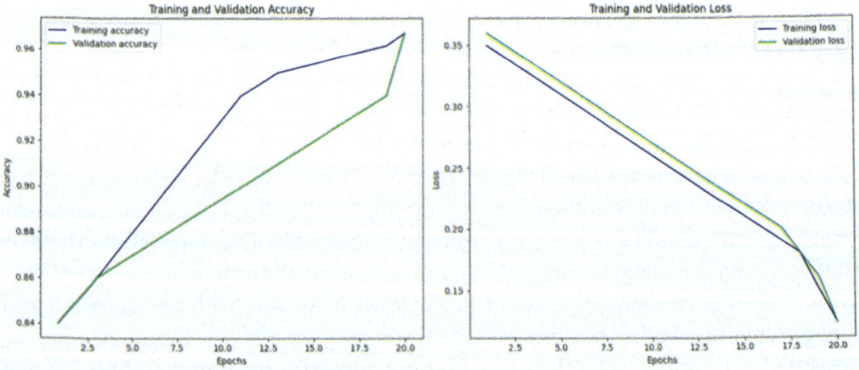

Fig. 2 Accuracy and loss during training and validation

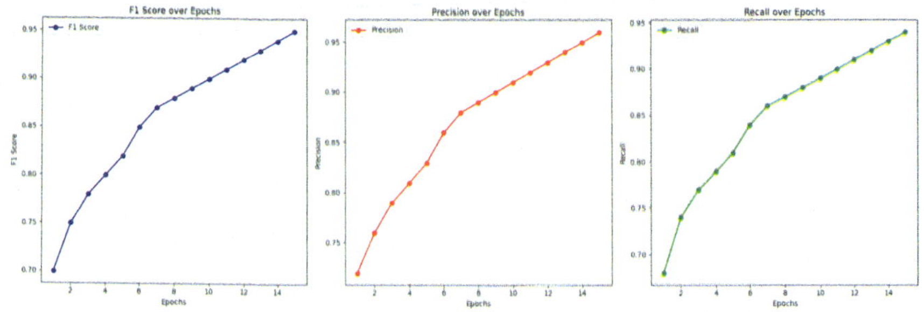

Fig. 3 Training performance metrics over epochs

to capture the majority of actual positive instances. These metrics collectively signify the HTAM's improving proficiency in sentiment classification.

5 Conclusion

This study concludes by demonstrating the successful implementation of the Hybrid Transformer-Attention Model (HTAM) for sentiment analysis in the smart tourist industry. The model showed exceptional performance by attaining a high validation accuracy and displaying its resilience with a decreasing loss rate throughout the training process. The results confirm the effectiveness of the HTAM in analyzing consumer input and emphasize its ability to improve data-driven decision-making in the tourism sector. Deploying complex analytical models has significant consequences, with the potential to transform how tourism businesses adjust to and fulfill client expectations.

References

1. Meng L (2024) The convolutional neural network text classification algorithm in the information management of smart tourism based on internet of things. IEEE Access
2. Tezgider M, Yildiz B, Aydin G (2022) Text classification using improved bidirectional transformer. Concurr Comput Pract Exp 34(9):e6486
3. Qiu Z, Qin D, Xu C (2022) Research on text classification algorithm based on hybrid neural network. In: International conference on computer, artificial intelligence, and control engineering (CAICE 2022) (Vol 12288, pp 305–314). SPIE
4. Liu S, Le F, Chakraborty S, Abdelzaher T (2021) On exploring attention-based explanation for transformer models in text classification. In: 2021 IEEE international conference on big data (Big Data) (pp 1193–1203). IEEE
5. Wadhe AA, Suratkar SS (2020) Tourist place reviews sentiment classification using machine learning techniques. In: 2020 international conference on Industry 4.0 technology (I4Tech) (pp 1–6). IEEE

6. Chen M, Yao C, Li X, Shen L (2023) A text classification model based on Gaussian multi-head self attention mechanism for Chinese medical data. In: 2023 IEEE 2nd international conference on electrical engineering, big data and algorithms (EEBDA) (pp 229–232). IEEE
7. Bao T, Ren N, Luo R, Wang B, Shen G, Guo T (2021) A BERT-based hybrid short text classification model incorporating CNN and attention-based BiGRU. J Organ End User Comput (JOEUC) 33(6):1–21
8. Prabhakar SK, Rajaguru H, Won DO (2021) Performance analysis of hybrid deep learning models with attention mechanism positioning and focal loss for text classification. Sci Program 2021:1–12
9. Cao D, Huang Y, Li H, Zhao X, Zhao Q, Fu Y (2020) Text sentiment classification based on lstm-tcn hybrid model and attention mechanism. In: Proceedings of the 4th international conference on computer science and application engineering (pp 1–5)
10. Hotel Reviews Dataset (2023) [Online] Available: https://www.kaggle.com/datasets/juhibhojani/hotel-reviews/data

Ensuring Trust in Blockchain-Based E-commerce Using Smart Contracts

P. D. Surendar, Adhithi Satish Kumar, Venkata Pavani, and M. Azhagiri

Abstract In the ever-changing realm of e-commerce, building trust is crucial for both customers and sellers. Reputation systems play a vital role in enabling informed decision-making by providing vendor ratings and reviews. Conventional reputation systems are typically inadequate in dealing with contemporary issues including privacy, identity theft, and feedback fraud. In order to address these challenges, we suggest implementing a decentralized reputation system that places a high value on privacy and effectively counters existing risks. Our architecture utilizes two permissioned blockchains, Hyperledger Indy for decentralized identity management through verifiable credentials and Zero Knowledge Proof (ZKP) technology, guaranteeing anonymity and authenticity, and Hyperledger Fabric for executing smart contracts, facilitating intricate business logic. The system incorporates verified vendor identities, transaction feedback tokens, and discount tokens to encourage genuine evaluations. Smart contracts implemented on the Hyperledger Fabric platform provide the security and transparency of the feedback process, hence reducing the occurrence of biased feedback and identity assaults. This technique provides a strong and reliable answer to the issues faced in e-commerce. It ensures the protection of private information and enables the verification of digital identities while maintaining privacy. The implementation of feedback and discount tokens promotes the provision of precise input, mitigating unjust practices and bolstering the dependability and confidence in the system. We demonstrate the effectiveness of our methodology using Zero Knowledge Proofs (ZKP), highlighting its importance in establishing confidence within the context of blockchain. Our system represents a notable advancement in creating a safe, transparent, and reliable e-commerce environment.

Keywords Decentralized reputation system · Privacy · Hyperledger Indy · Hyperledger Fabric · Verifiable credentials · Zero Knowledge Proof (ZKP) · Smart

P. D. Surendar (✉) · A. S. Kumar · V. Pavani · M. Azhagiri
Department of Computer Science and Engineering, SRM Institute of Science and Technology, Ramapuram 600089, Chennai, India
e-mail: surendarpd007@gmail.com

M. Azhagiri
e-mail: azhagirm@srmist.edu.in

P. J. S. Gonçalves et al. (eds.), *Proceedings of Fifth International Conference on Computing, Communications, and Cyber-Security*, Lecture Notes in Networks and Systems 1128, https://doi.org/10.1007/978-981-97-7371-8_38

contracts · Feedback tokens · Discount tokens · Identity authentication · Transparency · Immutability

1 Introduction

In the fast-changing realm of e-commerce, it is crucial to create and keep a dependable and clear reputation system. The evaluation and trust processes of sellers have become essential components of the digital marketplace due to the increasing frequency and complexity of online transactions. Conventional reputation systems are effective but struggle to tackle modern issues like privacy worries, identity theft, and feedback score manipulation. Developing a new decentralized reputation system that prioritizes privacy could effectively tackle these urgent challenges. This system utilizes the advantages of blockchain technology to guarantee the integrity, security, and privacy of user data and transactions.

At the core of this new method are two authorized blockchains: Hyperledger Indy and Hyperledger Fabric. Hyperledger Indy focuses on tools for decentralized identity, giving strong support for verified credentials using Zero Knowledge Proof (ZKP) technology. This technology allows users to verify their identity without disclosing extraneous personal information, hence improving privacy. Hyperledger Fabric is particularly proficient at deploying smart contracts, which allows for the implementation of intricate business logic on the blockchain. This amalgamation of technology serves as the foundation of the suggested reputation system, targeting critical weaknesses in conventional e-commerce systems.

The suggested system presents an innovative structure for overseeing digital identities and transaction feedback. Verified seller digital identities are preserved using verifiable credentials on Hyperledger Indy to ensure that only genuine individuals are involved in the marketplace. The system uses feedback tokens to represent transaction experiences and discount tokens to encourage the submission of authentic feedback, enhancing the feedback mechanism's integrity. These coins are controlled via smart contracts on the Hyperledger Fabric blockchain, streamlining the procedure and minimizing chances of manipulation or fraud.

The proposed reputation system addresses various persistent issues in e-commerce by incorporating sophisticated technology. The system safeguards sensitive data, thwarts identity-related assaults, and promotes candid criticism using a visible and unchangeable feedback process. Smart contracts enforce predetermined business standards, minimizing the chances of receiving unfair feedback from participants. This approach has the potential to drastically change how trust is created and sustained in online marketplaces, resulting in a more safe, dependable, and user-friendly e-commerce environment.

1.1 Motivation

In an era where e-commerce is becoming increasingly central to the global economy, trust and security within digital marketplaces have never been more crucial. Traditional reputation systems are struggling to cope with modern challenges such as privacy breaches, identity theft, and feedback manipulation, undermining consumer trust and the integrity of e-commerce platforms. Our motivation is drawn from the potential of blockchain technology to revolutionize these systems. By leveraging the decentralized, transparent, and immutable nature of blockchain through the integration of Hyperledger Indy and Hyperledger Fabric, we aim to develop a decentralized reputation system that enhances privacy, curtails fraudulent activities, and fosters a trustworthy e-commerce ecosystem. This endeavor seeks to navigate the complexities of the digital marketplace, ensuring it remains a secure, reliable, and equitable space for consumers and vendors alike.

1.2 Organization

The paper unfolds as follows: Sect. 2 examines the landscape of blockchain applications in e-commerce and decentralized reputation systems, setting the groundwork for our study. Section 3 introduces our blockchain-based architecture, integrating Hyperledger Indy and Hyperledger Fabric to enhance e-commerce trust and privacy. In Sect. 4, we detail our system's implementation, including decentralized identity management and the utilization of smart contracts for feedback integrity. Section 5 presents the impacts of our system on enhancing transaction efficiency, reducing fraud, and improving user satisfaction and privacy. We conclude in Sect. 6, summarizing our contributions and outlining future research directions to further blockchain's potential in e-commerce, with a focus on scalability and user adoption.

2 Related Work

Dogan et al. [1] conducted a study on a privacy-preserving decentralized reputation system for e-commerce using Hyperledger Indy and Hyperledger Fabric to ensure authenticity and prevent identity attacks, which is relevant to your research on blockchain-based e-commerce and reputation systems. Pereira et al. [2] examine limitations in current reputation systems, proposing a novel classification framework to analyze attacks and fraud, aiming to enhance trust in online interactions.

Wen et al. [3] propose a blockchain-based reputation system for P2P markets, aiming to combat fake ratings with economic incentives and demonstrating potential sustainability through simulations. Mrabet et al. [4] examine a dynamic decentralized reputation system that utilizes blockchain and secure multiparty computation to enhance privacy and security in mobile ad hoc networks and wireless sensor networks.

Tanvir et al. [5] propose a blockchain system to track drugs, prioritizing transparency and accountability against counterfeits, using Hyperledger Fabric and Quorum for confidentiality and interoperability. Taherdoost et al. [6] examine how blockchain can improve transaction efficiency and security in e-commerce.

Mukkamala et al. [7] propose a blockchain-based trust and reputation management system with Smart Contracts and novel reputation schemes, enhancing efficiency and accuracy in transaction feedback. Zhu et al. [8] provide a reputation-based consortium blockchain consensus technique to tackle node dishonesty and enhance the performance of digital order trading platforms.

Verma et al. [9] highlight blockchain's potential to revolutionize e-commerce by ensuring trust, reliability, and transparency across B2B, B2G, and C2C interactions, focusing on supply chain tracking and decentralized transactions. Yu et al. [10] propose implementing a reputation system using blockchain and deep learning in social networks to maintain a balance between freedom of speech and social stability.

3 Proposed Architecture

The diagram illustrates an advanced decentralized reputation system designed for e-commerce, utilizing the merging of Hyperledger Indy and Hyperledger Fabric blockchains. The system is architecturally built to facilitate a smooth and secure interaction between buyers and many vendors on different online marketplaces. Every seller uses a blockchain client to connect to a sophisticated blockchain network, where transactions and reputation data are permanently stored. Buyers communicate with sellers using blockchain clients, and their interactions are verified via JWT tokens, which connect to a centralized user database. The authentication process is essential as it verifies user credentials before allowing access to the blockchain network. This approach improves security and ensures the integrity and confidentiality of user interactions in the system. The blockchain network is mostly composed of seller nodes, which consist of both Hyperledger Fabric and Hyperledger Indy nodes. The Hyperledger Fabric nodes execute smart contracts that autonomously handle feedback and transaction rules to maintain a fair and transparent reputation system. These smart contracts serve as the arbitrators of the marketplace, encoding the business rules that control the reputation score and feedback system. Hyperledger Indy nodes are specialized in managing digital identities with verifiable credentials and Zero Knowledge Proof technology. This allows for verifying seller identities without disclosing unneeded personal information, thus safeguarding privacy and thwarting identity fraud (Fig. 1).

The combination of these blockchain elements creates a strong system that safeguards confidential data and offers a dependable platform for transactions. The architecture guarantees that feedback is genuine and cannot be altered, therefore greatly

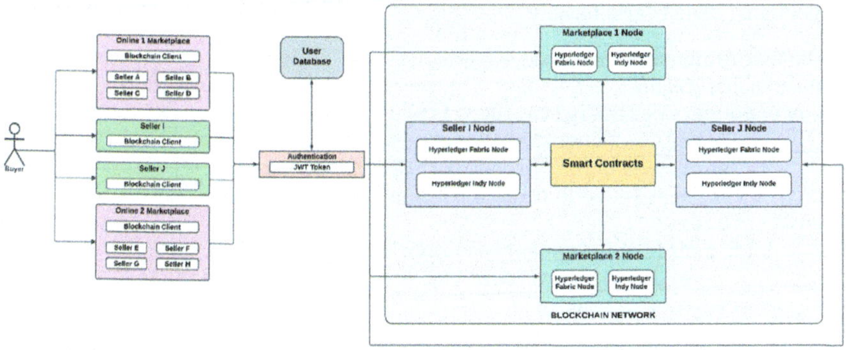

Fig. 1 Architecture diagram for blockchain-based ecommerce

decreasing the chances of fraudulent behavior. The system encourages truthful feedback by utilizing tokens that are essential to the ecosystem. The suggested system is designed to increase confidence in e-commerce by emphasizing openness and immutability. It is intended to be a powerful solution for modern digital marketplaces where reputation is crucial.

4 Implementation

4.1 Hyperledger Indy

Utilizing Hyperledger Indy for Decentralized Identity Management.

We started developing our system by including Hyperledger Indy for decentralized identity management, a crucial element to guarantee a high standard of privacy and security in our e-commerce reputation system. The connection enabled secure administration of user identities on the blockchain by utilizing the immutability and transparency features of blockchain to protect user data from illegal access and tampering.

A user-friendly interface was created to securely gather user details in accordance with the Create Identity algorithm. We automated the creation of distinct digital identities for each user using the extensive SDK from Hyperledger Indy, and securely stored them on the blockchain. This approach guaranteed that each user's identification was verified and secure, establishing a strong basis for the following components of our reputation system.

Algorithm: Create Digital Identity

Input: user Attributes (name, email, etc.)
Output: digital Identity

```
procedure CreateDigitalIdentity(userAttributes)
    DID_keys ← GenerateDIDandKeys(userAttributes)
    identityRecord ← StoreDIDInLedger(DID_keys)
    verifiableCredential ← IssueVerifiableCredential(DID_keys,
userAttributes)
    if verifiableCredential.isValid() then
        return digitalIdentity ← verifiableCredential
    else
        return error
    end if
end procedure
```

The pseudocode describes the steps for creating a decentralized digital identity for users in the Hyperledger Indy architecture. The process focuses on creating a distinct identity (DID) and cryptographic keys, then issuing a verifiable credential containing the user's characteristics. This process guarantees that every user's identification is secure, confidential, and can be verified throughout the blockchain network, serving as the foundation of our system's security and reliability. The algorithm details the steps for creating a decentralized identification (DID) and its cryptographic keys for a person, saving the DID on the blockchain ledger, and providing a verified credential including the individual's qualities. The digital identity created is safe and can be verified throughout the network without revealing sensitive data.

4.2 *Feedback and Discount Token System:*

A token-based mechanism was carefully put in place to encourage truthful feedback. The system, using the IssueFeedbackToken mechanism, compensates users with tokens for offering constructive and truthful feedback on their transactions. The project entailed developing smart contracts on Hyperledger Fabric to authenticate transactions and assess feedback quality. The smart contracts distributed tokens to users' digital wallets depending on predetermined parameters evaluating the feedback's quality and relevancy. This technique encouraged authentic engagement and promoted a more dependable and credible feedback environment on the e-commerce platform.

Algorithm: RewardFeedback

Input: feedbackDetails, transactionHash

Output: rewardStatus

```
procedure RewardFeedback(feedbackDetails, transactionHash)
    transaction ← RetrieveTransaction(transactionHash)
    if ValidateFeedback(feedbackDetails, transaction) then
        tokens ← CalculateReward(feedbackDetails)
        issueTokensToUser(feedbackDetails.userID, tokens)
        updateFeedbackStatus(feedbackDetails, "verified")
        return rewardStatus ← "Success"
    else
        return rewardStatus ← "Failed"
    end if
end procedure
```

RewardFeedback: This section explains how users are rewarded with tokens for giving helpful feedback on transactions. The system demonstrates the process of validating feedback against specific transactions, calculating the suitable reward, and distributing tokens to the user's digital wallet. This incentivization method is crucial for promoting genuine user engagement and cultivating a trustworthy feedback system. The system validates feedback against the transaction, determines the reward based on feedback quality, and allocates tokens to the user's account to encourage positive participation on the site.

4.3 Smart Contract for Feedback Verification

Ensuring the validity and relevancy of comments was crucial for maintaining the integrity of our reputation system. A smart contract named VerifyFeedback was implemented on Hyperledger Fabric for this purpose. This contract automated the process of verifying feedback, distinguishing between verified feedback and feedback needing manual review. The contract operated by fetching transaction details using the given transaction hash and evaluating the feedback's consistency with these data. Valid feedback was rewarded with tokens to promote genuine engagement in the system.

Algorithm: GenerateZKP

Input: userCredentials, verifierRequest

Output: proof

```
procedure GenerateZKP(userCredentials, verifierRequest)
    proof ← CreateProof(userCredentials, verifierRequest)
    if proof.isValid() then
        sendProofToVerifier(proof)
        return "Proof sent"
    else
        return error
    end if
end procedure
```

The pseudocode of the feedback verification smart contract illustrates the automatic verification of supplied comments using transaction details. The smart contract demonstrates its capability to evaluate the legitimacy and significance of feedback by either verifying it or flagging it for examination. This automated approach greatly improves the reliability and credibility of the feedback system. The smart contract processes feedback and transaction details, validates the match, and verifies the accuracy and significance of the feedback content. Verified feedback meets the criteria, while feedback that does not is reported for manual review.

4.4 Zero Knowledge Proof for Privacy Preservation

Zero Knowledge Proofs (ZKP) were a fundamental aspect of our system's privacy-preserving features. This technique allowed users to verify their credentials without disclosing any personal information, hence preserving a high level of privacy. Our solution enabled users to create proofs of their credentials in response to verifiers' requests by incorporating Zero Knowledge Proof (ZKP) functionalities. The proofs were sent without revealing the real credentials to ensure user anonymity was preserved during the transaction.

Algorithm: GenerateZKP

Input: userCredentials, verifierRequest

Output: proof

```
procedure VerifyFeedback(feedback, transactionHash)
    transactionDetails ← FetchTransactionDetails(transactionHash)
    if feedbackConsistentWithTransaction(feedback, transactionDetails) then
        issueTokensForVerifiedFeedback(feedback.userID)
        return verificationResult ← "Verified"
    else
        flagFeedbackForReview(feedback)
        return verificationResult ← "Review Needed"
    end if
end procedure
```

This pseudocode outlines the process of creating Zero Knowledge Proofs (ZKP), a cryptographic technique that enables users to verify their credentials without disclosing the actual data. The system emphasizes its dedication to preserving privacy by allowing safe and confidential verification of transactions and user credentials. This process creates a Zero Knowledge Proof (ZKP) using user credentials and a verifier's request, enabling users to demonstrate their eligibility or validity without disclosing their actual credentials, hence preserving privacy and security.

Our development of a blockchain-based e-commerce system using Hyperledger platforms, including Indy for decentralized identity management and Fabric for transaction and feedback integrity, has successfully developed a strong foundation that guarantees privacy, security, and trust. The incorporation of Zero Knowledge Proofs (ZKP) serves to reinforce our dedication to safeguarding user privacy, and enabling the authentication of transactions while ensuring the preservation of personal data. The system we have developed represents a notable advancement in tackling current obstacles in the field of electronic commerce. However, it also presents opportunities for further improvement and investigation. Future revisions may prioritize the enhancement of scalability and interoperability features, with a specific emphasis on effectively managing higher transaction volumes and facilitating smooth interaction across various blockchain environments. Furthermore, implementing our technology in practical situations, along with thorough user input, would be extremely beneficial in continuously enhancing its design and functionality. By engaging in a constant process of development and feedback, we can guarantee that our solution remains at the forefront of blockchain technology and adapts to the ever-changing requirements of e-commerce stakeholders.

The emergence of security and scalability as significant concerns during the installation of our blockchain-based e-commerce reputation system has prompted the need for comprehensive measures to effectively tackle these challenges. In order to mitigate potential vulnerabilities, a comprehensive threat modeling exercise was undertaken, with a specific emphasis on protecting the system from identity spoofing, transaction tampering, and privacy intrusions. We enhanced the security of our system by utilizing the inherent security features of blockchain technology, such as cryptographic hashing and digital signatures, in addition to implementing sophisticated methods like Zero Knowledge Proofs (ZKP). These procedures are implemented to guarantee the legitimacy of transactions and protect user privacy, all while upholding the transparency of the system. Furthermore, the system's robustness under adversarial conditions was validated through the execution of simulations involving prevalent blockchain attack vectors, such as Sybil and 51% attacks. The implementation of smart contracts facilitates the automation of security policy enforcement, hence playing a crucial role in ensuring that reputation scores are solely influenced by transactions that have been confirmed.

Our system integrates many unique methods to address the scalability difficulties associated with blockchain technologies, specifically for e-commerce systems that handle large transaction volumes. To mitigate network congestion, the implementation of off-chain transactions via state channels was introduced. This approach involves managing transactions outside of the primary blockchain, while ensuring the preservation of on-chain data integrity and security. Furthermore, we conducted an investigation into the possibility of sharding and layer-two scaling solutions, such as Plasma, in order to enhance transaction throughput. These methods are specifically developed to improve the system's ability to handle a high number of transactions while maintaining its decentralized character and security. This scalability framework guarantees the resilience and effectiveness of the e-commerce reputation system, enabling it to adjust to the ever-changing requirements of online markets.

5 Results

In the Results and Evaluation section of your research paper, you will display the results of deploying the blockchain-based e-commerce reputation system, highlighting the system's performance, security improvements, user interaction, and privacy protection. Analytical in nature, this component should utilize quantitative data and comparative analysis to support the system's performance.

5.1 Findings and Assessment

After implementing our blockchain-based e-commerce reputation system, thorough evaluations were carried out to analyze its effects on many areas of the e-commerce environment. The evaluations concentrated on transaction efficiency, fraud reduction, customer pleasure, and privacy preservation. The results are as follows.

5.2 Transaction Efficiency

The technology showed a notable enhancement in transaction processing speed and feedback verification effectiveness. Implementing smart contracts for feedback verification automated the process, resulting in a more efficient workflow and decreased the time required for verification after submission.

5.3 Decreasing Fraud

The system's main goal was to reduce fraudulent behaviors on the e-commerce site, specifically bogus reviews and rating manipulation. Implementing a token-based reward system for honest feedback, along with automatic feedback verification, led to a significant reduction in fraudulent feedback submissions.

5.4 User Satisfaction

User satisfaction was assessed using surveys that concentrated on the simplicity of use, trust in the feedback system, and general happiness with the e-commerce platform. The results showed a significant rise in user confidence and satisfaction due to the transparency and integrity of the feedback system and the improved privacy protections.

5.5 Privacy Preservation

Zero Knowledge Proofs (ZKP) were crucial in improving user privacy by incorporating them into identity verification and transaction validation processes. Users can authenticate their credentials while keeping their sensitive information confidential, so safeguarding their privacy.

5.6 Comparative Analysis

Our blockchain-based reputation system outperformed conventional systems in terms of security, efficiency, and user satisfaction, as shown by a comparative analysis. Utilizing blockchain technology, smart contracts, and ZKP greatly improved the system beyond conventional reputation procedures in e-commerce platforms (Fig. 2).

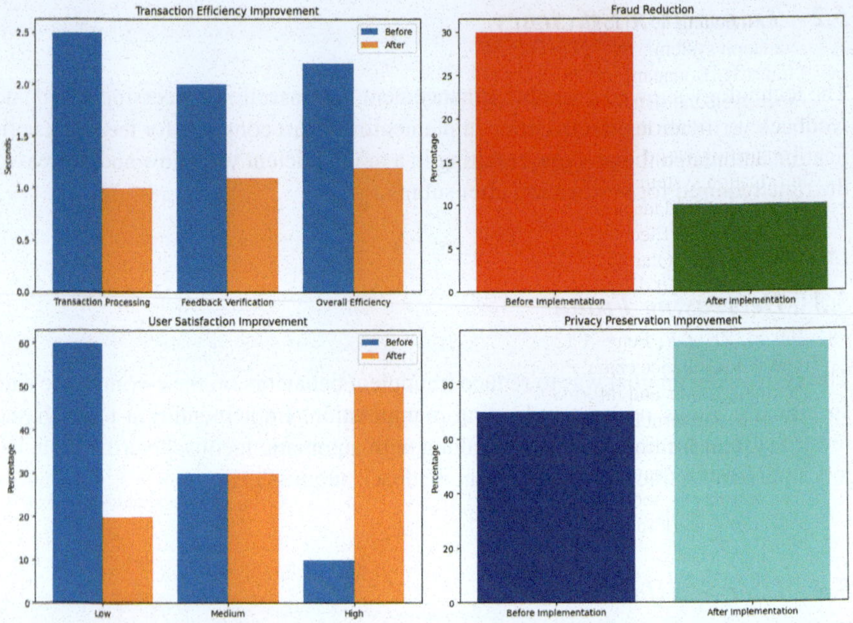

Fig. 2 Visualization of the results

6 Conclusion

In concluding our exploration of blockchain-based e-commerce systems, it's evident that while our proposed model significantly advances the security, privacy, and trustworthiness of online transactions, there are limitations and areas ripe for future inquiry. The complexity of blockchain implementation and scalability remains a considerable challenge. Furthermore, the real-world applicability and user adoption of such systems necessitate further exploration. Future work could focus on enhancing blockchain interoperability, optimizing system scalability, and investigating user behavior toward blockchain adoption in e-commerce. Acknowledging these limitations underscores our commitment to advancing research in this rapidly evolving domain, aiming to bridge the gap between theoretical models and practical, user-centric applications.

References

1. Dogan Ö, Karacan H (2023) A Blockchain-based E-commerce reputation system built with verifiable credentials. IEEE Access 11:47080–47097
2. Pereira R, Gonçalves M, Magalhães M (2023) reputation systems: a framework for attacks and frauds classification. J Inf Syst Eng Manag

3. Wen H, Huang T, Xiao D (2023) An intrinsic integrity-driven rating model for a sustainable reputation system. ArXiv, abs/2310.09143
4. Mrabet K, Bouanani F, Ben-Azza H (2023) Dynamic decentralized reputation system from Blockchain and secure multiparty computation. J. Sens. Actuator Networks 12:14
5. Tanvir D, Amin R, Islam A, Islam M, Rashid M (2023) Blockchain interoperability for a reputation-based drug supply chain management. 2023 6th international conference on information systems and computer networks (ISCON), 1–6
6. Taherdoost H, Madanchian M (2023) Blockchain-based E-commerce: a review on applications and challenges. Electronics
7. Mukkamala R, Olariu S, Aljohani M (2023) Improved schemes for managing reputation in a blockchain-based decentralized marketplace. IEEE Int Conf Blockchain Cryptocurr (ICBC) 2023:1–2
8. Zhu M, Zhou Y, Feng X (2023) Reputation-based consensus mechanism for digital order transaction alliance chain. Proceedings of the 2023 2nd international conference on networks, communications and information technology
9. Verma N, Kulkarni P (2023) Impact of Blockchain technology on E-commerce. Int Conf Innov Data Commun Technol Appl (ICIDCA) 2023:668–672
10. Yu H, Jiang D, Zhang G, Yang Z, Liu W (2023) A reputation system based on Blockchain and deep learning in social networks. 2023 26th international conference on computer supported cooperative work in design (CSCWD), 630–635

Comparative Analysis of Monopole and Dipole Antennas for CubeSats: A Review

Nissrine Oubahsis, Boutaina Benhmimou, Mohamed Karim, Niamat Hussain, Fouad Omari, Nancy Gupta, Khalid El Khadiri, Korhan Cengiz, Sangeeta Garg, Rachid Ahl Laamara, Josep M. Guerrero, and Mohamed El Bakkali

Abstract CubeSats provide new communications, navigation, remote sensing, and research capabilities for military and commercial applications. Their capabilities led the scientific community to rethink existing spacecraft technology to make it suitable for CubeSats. This research focuses on the development of monopole and dipole antennas and their application to CubeSats. There is a lot of ongoing research in the antenna design community to create monopole and dipole antennas for use in CubeSat communications, which we will discuss in this article. This study provides comprehensive information to review the current status of monopole and dipole antennas. This paper explores and examines the suitability of monopole and dipole

---

N. Oubahsis · B. Benhmimou · F. Omari · R. A. Laamara · M. E. Bakkali (✉)
LPHE-MS, Faculty of Sciences, University Mohammed V in Rabat, Rabat, Morocco
e-mail: mohamed.elbakkali1617@gmail.com

M. Karim
LIMAS, Faculty of Sciences Dhar El Mahraz, Sidi Mohammed Ben Abdellah University, 30000 Fez, Morocco

N. Hussain
Department of Smart Device Engineering, Sejong University, Seoul, South Korea

N. Gupta
ECE Department, Lyallpur Khalsa College Technical Campus, Jalandhar, Punjab, India

K. E. Khadiri
Department of Physics, Faculty of Sciences El Jadida, University Chouaib Doukkali of El Jadida, El Jadida, Morocco

K. Cengiz
Department of Electrical-Electronics Engineering, Istinye University, 34010 Istanbul, Turkey

S. Garg
Mewar Institute of Management, Ghaziabad, Uttar Pradesh, India

J. M. Guerrero
CROM, Aalborg University, Aalborg, Denmark

P. J. S. Gonçalves et al. (eds.), *Proceedings of Fifth International Conference on Computing, Communications, and Cyber-Security*, Lecture Notes in Networks and Systems 1128, https://doi.org/10.1007/978-981-97-7371-8_39

antenna designs for CubeSats. We also provide a set of design-based analysis models that can support many interesting projects in the future.

Keywords Monopole and dipole antennas · Antenna gain · CubeSats · Deployment system

1 Introduction

The rapid development of digital signal processing technologies, large-scale integrated circuits, microelectromechanical systems, and low-power programmable systems has surprised the scientific research community in terms of its remarkable ability to reduce the size of electronics and their power consumption, thus opening up areas of research and innovation that were not possible before. For example, reducing satellite dimensions and energy consumption has led to direct reductions in satellite launch costs, thus enabling new parties such as universities, small companies, and private laboratories in most developing countries to aggressively enter this field. Add to this, the opportunity to launch a network of small satellites capable of integrating with each other, opening the door wide to a new era of exploration and the beginning of a new history of space technology [1]. In this scientific and technological momentum, the CubeSat, which appeared at the beginning of the third millennium, is of utmost importance due to its extraordinary flexibility and the possibility of being installed in a period not exceeding a few weeks [2, 3]. Add to this, the possibility of launching dozens in one flight or considering them as a secondary payload for missions transporting food and equipment to the International Space Station, which reduced the costs to an extreme extent that made us believe that our true homeland and our bright future is space and with this technology. What is most exciting is that some of the pioneers of this field, such as Mr. Elon Musk, who owns SpaceX, have gone so far as to consider humans to be aliens. Nowadays, we are talking about an average of launches per week over the course of a year with the possibility of reusability, moving beyond the usual terms such as failure or faltering.

The mass and volume of CubeSats are estimated to be approximately 1.3 kg and 10 cubic centimetres, respectively, and lightweight, off-the-shelf components are generally used in their overall design [1, 4]; see Fig. 1. The first CubeSat was made at California Polytechnic State University (Cal Poly) in partnership with Stanford University and launched in 2002 [5]. The Cal Poly mission was intended to use Stanford University's OPAL spacecraft as its proposed CubeSat launcher. During the completion stages of this project, work was done in parallel on the standard launch device and the CubeSat, within the framework of complete agreement between the theoretical study and the specifications of the launch device to ensure success [6].

All CubeSat designers are aware that the CubeSat will travel to the intended orbit in the dispenser. When the CubeSat/dispenser package was initially conceptualized, the plan was to attach the dispenser to the rocket for many CubeSat flights where space was available (see Fig. 2). This is still how many CubeSats are deployed in

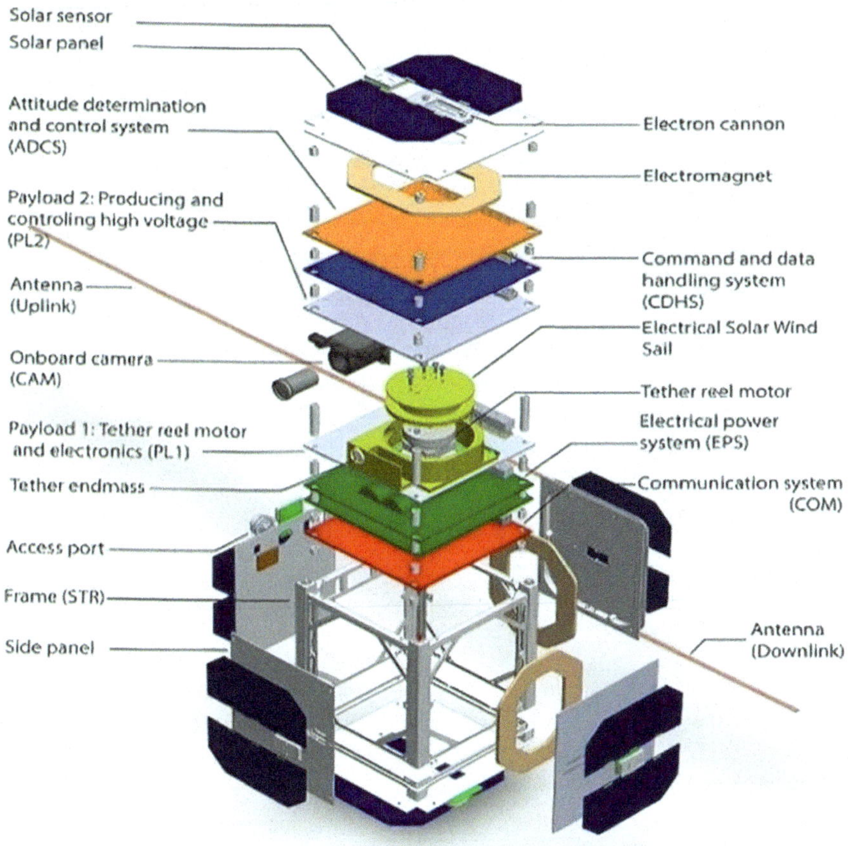

Fig. 1 3D model of a 1U CubeSat standard: ESTCube-1 [7]

clusters and interconnected to construct distributed satellite sensor networks in orbit. However, other options exist, such as the deployment from the International Space Station. The CubeSat would be transported to the Space Station and deployed with a specially constructed deployer [8]. Keep in mind that any rocket, under the right conditions, may be a CubeSat launch vehicle. As a result, it may have a higher performance margin, be capable of reaching the appropriate orbit, and be modified to allow dispensers to be fitted [9].

The communication system is one of the most critical components of a CubeSat; without it, it is little more than space debris. In addition to easing data flow from satellite orbit to the ground, it also performs tracking, telemetry, and command functions. The communication subsystems consist of an Earth-based base station,

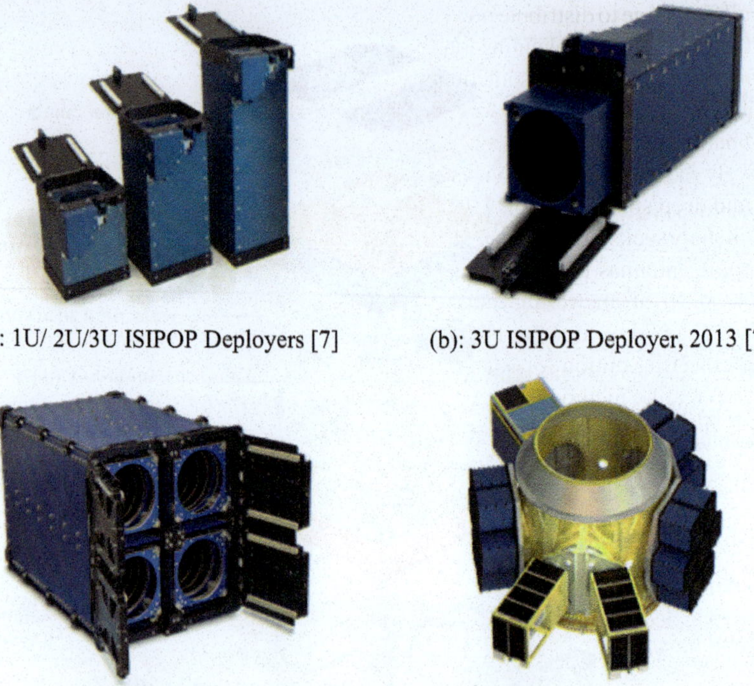

(a): 1U/ 2U/3U ISIPOP Deployers [7] (b): 3U ISIPOP Deployer, 2013 [7]

(d): 3U QuadPack deployer, 2014 [7] (e): SHERPA CubeSat deployers, 2016 [11]

Fig. 2 Standard and smart CubeSat deployers

CubeSat antennas, and transceivers. It might have a custom-built transmitter, a modified or unmodified COTS radio. These transceivers must be capable of error detection and correction, as well as transmitting and receiving encapsulated and decapsulated serial data. However, because most protocols are proprietary and device-specific, the ground station on Earth may require the same transceiver in order to communicate. The institution will host the main ground station, which will be used to communicate with the CubeSat while it is in orbit. Furthermore, if one ground station is unable to downlink all of the data provided by the CubeSat, alternative ground stations may be used if the CubeSat communication system is capable of connecting to many components.

Antenna systems are of utmost importance for remote sensing and thus connecting the user on the ground with the CubeSat, even if only for a few minutes a day due to the advantages of the antennas used. Therefore, the geometric characteristics of each CubeSat unit make it possible for the link quality to fail. We point out that CubeSat communication systems consume about 2 watts in total, and therefore relying on amplifiers to support the characteristics of the used antennas may leave the rest of the devices without power and thus fail the entire mission [1, 4]. The limitations of the batteries and solar panels on board each CubeSat unit make it a difficult challenge to distribute power equitably to all devices, which requires constant monitoring. Therefore, planning long-range missions or increasing the amount of data exchanged per day, which sometimes takes place in only a few minutes, is far from any protection in the case of adopting loudspeakers, and is not a primary solution. This challenge opened the door wide for antenna engineering specialists to work creatively in developing unconventional antenna systems usable in both low-Earth orbit and deep space missions [10, 11].

This study seeks to cover in detail the characteristics of several available monopole and dipole antennas that have been presented to the scientific research community in the context of answering the questions related to the extent to which they can be adapted and developed to become effective on CubeSats, given that they are the best engineering option for small frequencies. Taking into account this assumption aimed at creating methods that enable it to overcome the problems of size and energy in the CubeSat, this paper was presented within a strategic framework in which the characteristics of the antennas are in harmony with the characteristics of the CubeSat for the success of the mission as a whole at the lowest costs.

This study is structured as follows: Sect. 2 surrounds the works that this study targeted by examining and summarizing their essential characteristics in relation to the CubeSat mission. Section 3 summarizes the performance of the antenna systems discussed in Sect. 2 and evaluates their suitability for use on CubeSats. Finally, Sect. 4 wraps up the work with a review of major findings and recommendations for further research.

2 Quantitative Analysis

In general, the satellite's communications system remains the most important element as it is the only means of communication between the user and the satellite in its orbit through appropriate antennas. If we take into account the speed of the satellite's rotation around the Earth several times a day, the mission of the satellite as a whole depends on the characteristics of the adopted antenna. In this article, the authors analyze and study the suitability of monopole and dipole antenna systems for use on CubeSats, especially within the framework of small frequencies that do not require a license, which are considered the most demanding in missions with an economic or social dimension. Figure 3 illustrates the different configurations of monopole and dipole antenna systems that have already been proposed by the scientific community.

Conventional monopole/dipole antennas, as is widely known, are narrowband antenna topologies with bandwidths ranging from 5 to 10% of the center frequency. Much research has been conducted to extend the bandwidth and boost the gains of these antennas, and the scientific community has recommended a variety of ways. To enhance the bandwidth of these antennas, reduce the λ/D (wavelength to diameter ratio) or thicken the radiating components. Further thickening will provide altered radiation patterns. On the other hand, loading methods or impedance matching

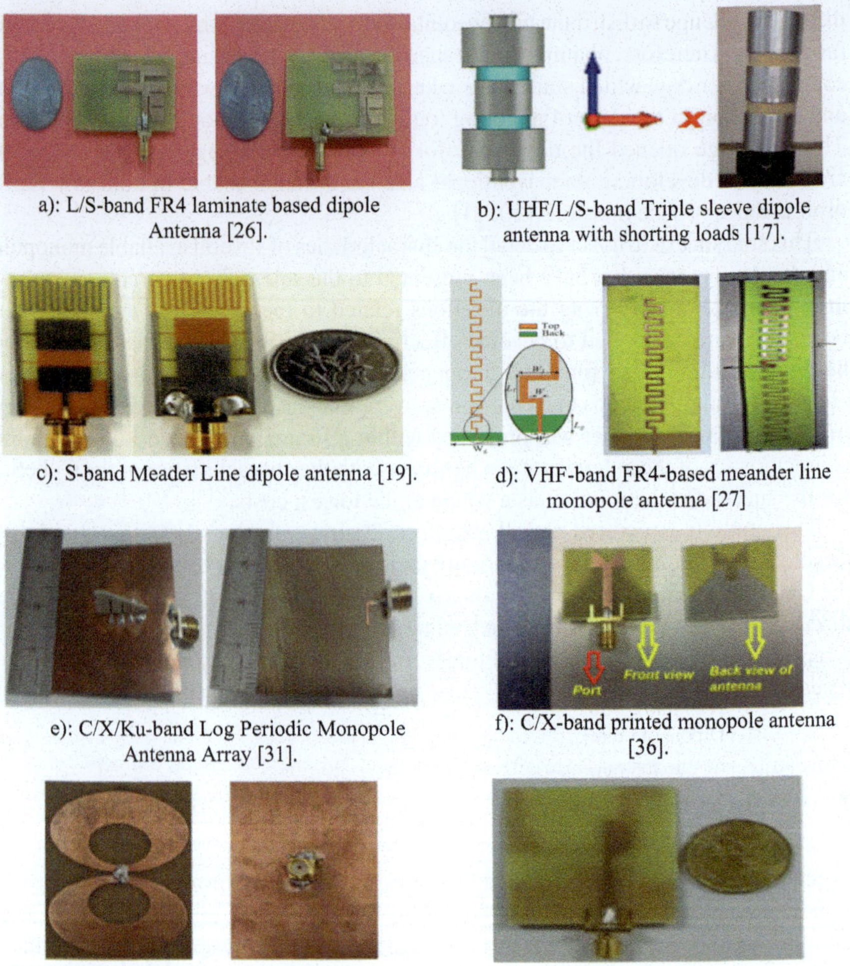

a): L/S-band FR4 laminate based dipole Antenna [26].

b): UHF/L/S-band Triple sleeve dipole antenna with shorting loads [17].

c): S-band Meader Line dipole antenna [19].

d): VHF-band FR4-based meander line monopole antenna [27]

e): C/X/Ku-band Log Periodic Monopole Antenna Array [31].

f): C/X-band printed monopole antenna [36].

g): S/C/X-band Eye shaped dipole antenna [32].

h): S/C-band printed dipole antenna [33].

Fig. 3 Various monopole and dipole antenna designs

circuits inject losses into the antenna system, diminishing its efficiency and hence limiting its applicability for long-distance communications.

Tables 1 and 2 provide the design characteristics of many monopole and dipole antennas, including physical dimensions, construction materials, and frequency ranges. This comprehensive data enables a quick comparison of all of the described designs, as well as reveals their mechanical and geometrical qualities, allowing researchers to investigate their potential for usage on CubeSats. Furthermore, the thorough information supplied in Tables 1 and 2 allows engineers to make educated judgments when choosing a monopole or dipole antenna design that best meets their CubeSat mission objectives. Additionally, by studying these design factors, researchers may improve the performance of these antennas for diverse communication needs in space applications.

Table 1 Planar monopole and dipole antennas

References	Antenna dimensions	Materials	Frequency band
[12]	$50 \times 25 \times 1.56$ mm^3	FR-4 and laminate	L-band (1.57 GHz), S-band (3.65 GHz)
[13]	$2\pi \times 21.16^2 \times 67.42$ mm^3	Metals	UHF/L/S-band (500–3000 MHz)
[14]	133.5×69.5 mm^2	FR4 (h = 0.8 mm)	UHF-band: 400 and 680 MHz/L/S-bands
[15]	36×14.4 mm^2	FR4 ($\varepsilon r = 4.4$; h = 0.5 mm	S Band: 2.5 GHz
[16]	196×15 mm^2	Stainless steel and FR-4	UHF-band: 0.915 GHz
[17]	41.23×30 mm^2	Not assigned	S-band: 2.45 and 3.65 GHz
[18]	13×14.2 mm^2	FAF4 and Sheet Foil-Coated	X-band: 9.3 GHz
[19]	Width (40 mm), Thickness (123 mm): Monopole (55 mm), Dipole (43 mm)	Metals/water	UHF-band: 902–928 MHz S-band: 2.4–2.5 GHz
[20]	295×305 mm^2	Plasma	Ku-band: 15 GHz
[21]	$440 \times 160 \times 1.6$ mm^3	FR4 & copper	VHF-band: 400 MHz
[22]	MLMA: $44.5 \times 17 \times 0.159$ cm^3 MLDA: $66 \times 4.4 \times 0.159$ cm^3 CLA: $52.65^2 \times 0.159$ cm^3	FR-4 ($\varepsilon r = 4.3$ and h = 0.159 mm, tand = 0.02)	VHF-band: 161 MHz
[23]	$180 \times 30 \times 0.782$ mm^3	Rogers RT Duroid 5880 ($\varepsilon r = 2.2$)	C-band: 5.150–5.825 GHz; X-band 7.25–7.75 GHz
[24]	$36 \times 36 \times 7.5$ mm^3	FR4 and PEC	C-band: 5.58 and 4.8 GHz
[25]	$45 \times 30 \times 12$ mm^3	RO3210 and RO4350	C/X/Ku-band: 6–18 GHz
[26]	$58.27 \times 46.80 \times 0.6$ mm^3	FR4 (h = 0.6 mm)	Ku-band: 2–12 GHz
[27]	$35 \times 42 \times 1.6$ mm^3	FR4 ($\varepsilon r = 4.4$)	S/C-band: 2.4, 3.5 and 5.2 GHz
[28]	$25 \times 26 \times 1.6$ mm^3	FR4 substrate	C/X-band: 5.8; 5.2; 4.0; 7; 10 GHz

Table 2 Classic monopole and dipole antennas

References	Antenna dimensions	Materials	Frequency band
[12]	Height = 72.42 mm Width = 223.23 mm	Copper/(Carbon Nanotubes)	UHF (900 MHz)
[29]	Length = 10.9 mm	Alumina (εr = 9.6)	C-band (5.846 GHz)
[30]	Length of 2.5 m (3)	Perfect electric conductor (PEC)	VHF-band (30–100 MHz)
[31]	Monopole: 31.25 mm Dipole: 62.5 mm	Metals	S-band: 2.4211 GHz S-band: 2.4 GHz
[27]	Length of 61 mm	Dielectric (εr = 2.5, h = 1.4775 mm)	C/X band: 4–12 GHz
[32]	Height of 270 mm	Metals	VHF/UHF-bands

3 Qualitative Analysis

The table below summarizes the radiating performance of all of the monopole and dipole antenna types mentioned above. It demonstrates that various antenna layouts can function across all frequency bands. However, typical monopole and dipole diameters are excessively long, resulting in poor gains and narrow bandwidths in the VHF, UHF, and L bands. While planar structures are ultra-wide band and occupy reasonable physical sizes, they evolve in inverse proportion to the targeted frequencies.

The results summarized in Tables 1, 2, and 3 above are studied in terms of their suitability for 0.5U, 1U, 2U, 3U, and 6U CubeSat configurations. Not that the suitability for a CubeSat mission is studied in terms of geometrical, mechanical, and electrical characteristics of proposed antennas. Therefore, an antenna system can be used for a CubeSat configuration if it satisfies all geometrical, mechanical, and electrical requirements of the CubeSat mission. Geometrically, the antenna system proposed by authors of [12, 13, 20, 21, 30, 32] are unsuitable for any CubeSat configuration due to their wider dimension concerning standards of 0.5U, 1U, 2U, and 3U and very low gains for other configurations that are mostly used for interplanetary uses. In addition to that, their bidirectional radiation patterns allow high interferences with other components if they are suggested for a CubeSat configuration taking into consideration that the link can be ensured using very high gain earth segments; refer to Fig. 4. In addition to that, antenna designs given in [14, 16, 19, 23] are geometrically suitable for use on 2U or 3U while they cannot be used on 0.5U and 1U CubeSat configurations.

Antenna configurations proposed in [15, 17, 18, 22, 24–26, 28–31, 34] satisfy all geometrical and mechanical criteria of all CubeSat forms while they radiate bidirectionally or omnidirectionally, and hence their major weakness that limits their effectiveness for space uses is the very high interferences with other circuits inside the satellite box. From other hands, they gives low gains and so they cannot be used for unlimited lifetime CubeSat missions. They can be used only for some near-field applications under certain circumstances. Contributions detailed in [27,

Table 3 Radiating performances on various monopole and dipole antenna designs

References	Freq. band or f_r (GHz)	Gain [in dBi]	Bandwidth
[33]	1.57 and 3.65	0.71, 1.27 dBi	1.83, 2.30%
[12]	0.9	Copper: 1.69; CNT: 1.68	Not assigned
[29]	5.846	5.56502	Not assigned
[13]	0.5–3	From 0.3 to 2	Not assigned
[14]	UHF (0.4 and 0.68); L/S-bands	UHF: 2.94 and 1.92; L: 6.29 and S: 6.76	UHF: 76.1%; L: 27.4%; S: 31.7%
[15]	2.5	1.30	16%
[16]	0.915	4.56	Not assigned
[17]	2.45 and 3.65	2.13 and 2.58	Not assigned
[18]	9.3	5.63	Not assigned
[19]	0.902–0.928 and 2.4–2.5	4.44–8.32 and 2.34–8.96	Not assigned
[20]	15	Peak: 16.0	600 MHz
[30]	0.03–0.1	2.9–7.25	>107%
[21]	0.4	~2.0	87 MHz
[22]	0.161	MLMA: 1.18; MLDA: 1.45; CLA: 1.6	Not assigned
[23]	5.150–5.825 and 7.25–7.75	3.0–5.32	Not assigned
[24]	5.58 and 4.8	5.2–5.8	Not assigned
[31]	Mon.: 2.4211; Dip.: 2.4	5.77 (M) and 2.5285 (D)	Not assigned
[25]	6.0–18.0	2.0–8.5	100%
[26]	2.0–12.0	3.12–5.49	Not assigned
[27]	4–12	8.0	Not Assigned
[34]	2.4, 3.5 and 5.2	3.0–5.2	1.6–6.25 GHz
[32]	0.3, 0.5, 0.7 and 0.9	0.9; 1.8; 2.4 and 2.6	0.24–0.95 GHz
[28]	5.8; 5.2; 4.0; 7; and 10	3.0–3.83	3.10–10.9 GHz

29] describe antenna systems meet all geometrical and mechanical criteria for all CubeSat standards. Compared to all other designs, they give superior gains, hold the lowest volumes, and use up very low electric energy. Moreover, their lightweight and low cost give them a significant favor for use on all CubeSat standards. As a result, antenna designs given in [27, 29] present the highest performances that meet all CubeSat criteria among all cited antenna approaches [35, 36].

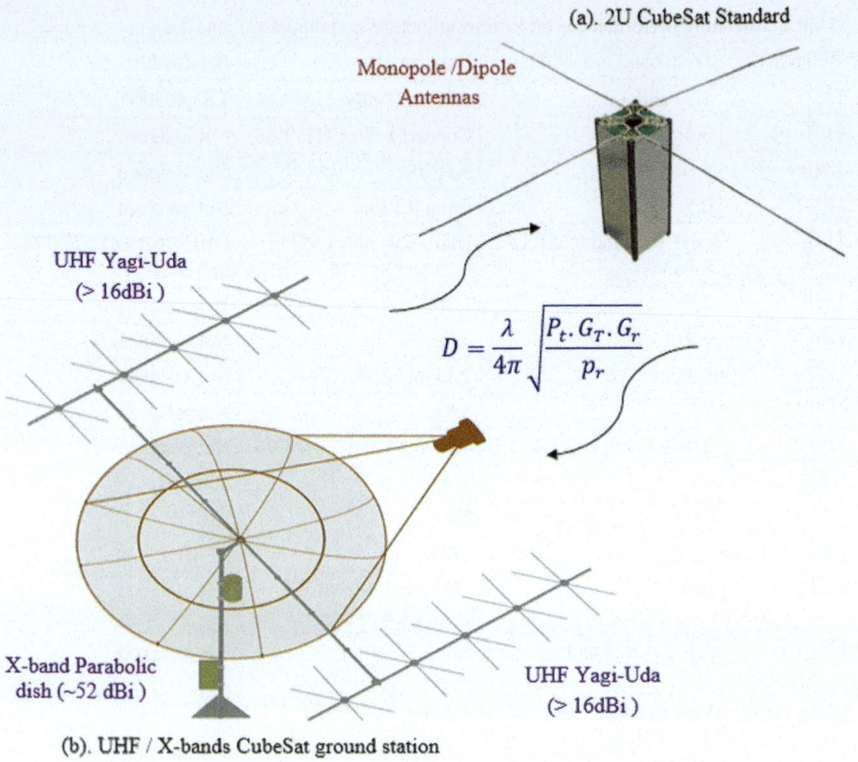

Fig. 4 Layout of CubeSat—ground station links

4 Conclusions

The space community and the CubeSat community have recently focused on creating the compact yet capable AeroCube, which will be the foundation for several missions in the future. Requirements for remote sensing in a tiny performance package must be met by antenna systems, especially as CubeSats show promise for advanced applications. Different antenna designs for CubeSats have been proposed in the literature, depending on the frequency and gain required. This study examines many monopole and dipole antenna designs that have been put forth by scientists and determines whether or not they are appropriate for usage with AeroCubes. It has been shown that, at higher frequencies, monopole and dipole antennas may readily result in tiny sizes and strong gains. The most crucial characteristics of monopole and dipole antennas for small- and ultra-small-sized AeroCubes are still their advantages of operating at frequencies lower than 1.0 GHz with the suitability of achieving good bandwidth, occupying an acceptable size, and consuming less power, which is limited on CubeSats. Furthermore, this thorough review demonstrates that the problem of low gains at frequencies lower than 1.0 GHz can be resolved via the integration of small array metasurfaces or artificial magnetic conductors (AMC) with monopoles and targeting the same frequency band, leading to long-lived CubeSat missions using unlicensed frequencies.

This study has the potential to significantly show the performance and efficiency of employing monopole and dipole antennas on CubeSats, allowing for consistent data transmission and reception at frequencies less than 1.0 GHz. Furthermore, developments in metasurface and AMC structures may result in considerable reductions in the size of CubeSat antennas, boosting their capability for a variety of missions, particularly at the VHF, UHF, and L bands.

Acknowledgements The authors do not forgot to express their gratitude and appreciation to Shangusham Maheswaran, Olexander Kogut, and Gurjot Singh Gaba for their invaluable help in gathering information and for making their credentials, expertise, and scientific equipment available to us.

References

1. El Bakkali M (Jul 2020) Planar antennas with parasitic elements and metasurface superstrate structure for 3U CubeSats, PhD thesis. Sidi Mohamed Ben Abdellah University, city of Fez, Morocco
2. Marinan AN, Cahoy K (2013) Ad hoc CubeSat constellations: secondary launch coverage and distribution. In Proceedings of the IEEE aerospace conference, big sky, MT, USA, pp 1–15
3. NASA CubeSat Launch Initiative (2017) CubeSat101: basic concepts and processes for first-time CubeSat developers. U.S. Launch Vehicles Used for CubeSat Launch, p 08
4. Popescu O (2017) Power budgets for CubeSat radios to support ground communications and inter-satellite links. IEEE Access 5(X):12618–12625
5. Puig-Suari J, Turner C, Ahlgren W (2001) Development of the standard CubeSat deployer and a CubeSat class picosatellite. In: IEEE aerospace conference proceedings, pp 1347–1353
6. ISIS, CubeSat and Microsatellite Deployers. https://www.isispace.nl/products/cubesat-microsatellite-deployers/. Accessed 29 Mar 2024
7. Lätt S, Slavinskis A, Ilbis E, Kvell U, Voormansik K, Kulu E, Pajusalu M, Kuuste H, Sünter I, Eenmäe T, Laizans K (2014) ESTCube-1 nanosatellite for electric solar wind sail in-orbit technology demonstration. In: Proceedings of the Estonian academy of sciences (2), p 200
8. NASA, ISIS Nanoacks Deployer. https://www.nasa.gov/content/cubesat-deployment. Accessed 29 Mar 2024
9. SpaceX Society, Falcon 9 launch vehicle. https://www.spacex.com/falcon9. Accessed 29 Mar 2024
10. Chahat N, Decrossas E, Gonzalez-Ovejero D, Yurduseven O, Radway M, Hodges R, Estabrook P, Baker J, Bell D, Cwik T, Chattopadhyay G (2019) Advanced CubeSat antennas for deep space and earth science missions: a review. IEEE Ant Prop Mag PP(99):1–1
11. El Bakkali M, Bekkali ME, Gaba GS, Guerrero JM, Kansal L, Masud M (2021) Fully integrated high gain S-band triangular slot antenna for CubeSat communications, electronics
12. Chudasama KR (2020) Performance analysis of carbon nanotubes as monopole antenna, SAMRIDDHI: a journal of physical sciences. Eng Technol 12(02):79–81
13. Sairam C, Vakula D, Chakravarthy M (2023) Wideband triple sleeve dipole antenna with loading for UHF applications. IETE J Res 69(4):1915–1924
14. Jin Z, Zhang S, Pedersen FG (2018) A wide band compact size UHF band monopole clustered with L and S band dipoles for self-networking communication in handset devices, 306-3

15. Wen L, Gao S, Yang Q, Luo Q, Yin Y, Ren X, Wu J (2019) A compact monopole antenna with filtering response for WLAN applications. In: 2019 international symposium on antennas & propagation (ISAP). IEEE, pp 1–3
16. Yang X, Qi Y, Yuan B, Cao Y, Wang G (2021) A miniaturized high-gain flexible antenna for UAV applications. Int J Ant Propag, pp 1–7
17. Shubham R, Said M, Bandewar P, Gaikwad S, Sharma A (May 2020) A dual band monopole antenna for PAN and WLAN applications. Int J Res Appl Sci Eng Technol (IJRASET) 8(V):2487–2490
18. Tarascnko NV (2018) Compact printed Yagi antennas with monopole driver. In: 2018 XIV international scientific-technical conference on actual problems of electronics instrument engineering (APEIE). IEEE, pp 289–291
19. Loghmannia P, Manteghi M (2019) An antenna system for autonomous underwater vehicle. In: 2019 united states national committee of URSI national radio science meeting. IEEE, pp 1–2
20. Radi B, Dhillon AS, Liboiron-Ladouceur O (2020) Demonstration of inter-chip RF data transmission using on-chip antennas in silicon photonics. IEEE Phot Technol Lett 32(11):659–662
21. Malek NA, Mokhtar NHM, Ali K, Isa FNM, Rahman FDA (2018) Design of small antennas for 400 MHz applications. In: 2018 7th international conference on computer & communication engineering (ICCCE). IEEE, pp 99–103
22. Chaudhari AD, Chand P, Keley R, Ray KP (2019) Design and development of printed antennas for satellite-based AIS applications. In: 2019 TEQIP III sponsored International conference on microwave integrated circuits, photonics & wireless networks (IMICPW). IEEE, pp 341–344
23. Kumar N, Kumar P, Sharma M (2019) Superwideband dual notched band square monopole MIMO antenna for UWB/X/Ku band wireless applications, In: 2019 IEEE Indian conference on antennas and & propagation (InCAP). IEEE, pp 1–5
24. Cao W, Jiang X, Hai W, Yu X, Qiu F, Wang Y (2018) Investigation of miniaturized wideband circular-polarized cavity-backed printed dipole antenna. In: 2018 12th international symposium on conference on antennas and propagation & EM theory (ISAPE). IEEE, pp 1–4
25. Sun Y, Liu Y, Tang N, Xu D, Li Y, Yu Y, Dai W, Gou Q, Du Z (2018) The design of wideband endfire monopole antenna array. In: 2018 international workshop on antenna technology (iWAT). IEEE, pp 1–3
26. Rajini AR, Abishek E, Ramesh S, Rajendran V (2021) Compact printed planar eye shaped dipole antenna for ultra-wideband wireless applications. J Appl Sci Eng 25(5):913–918
27. Ji F, Wang Z, Nie L, Jiang Z (2020) A design of microstrip log periodic antenna unit. In: 2020 cross strait radio science & wireless technology conference (CSRSWTC). IEEE, pp 1–3
28. Jan NA, Kiani SH, Muhammad F, Sehrai DA, Iqbal A, Tufail M, Kim S (2020) V-shaped monopole antenna with Chichena Itzia inspired defected ground structure for UWB applications. Comput Mater Contin 65(1)
29. Keerthana S (2021) Reshaped monopole antenna structure fit for short range wireless access. Ann Romlan Soc Cell Biol, 5465–5470
30. Wang C, Yuan B, Mao J, Shi W (2018) Dual-sleeve wideband monopole antenna for shipborne systems in VHF band. Electron Lett 54(19):1102–1104
31. Kulshrestha K, Kumari R, Roy M (2020) Design of monopole antenna and half-wave dipole antenna for Wi-Fi applications by enhancing gain. In: 2020 IEEE international symposium on smart electronic systems (iSES) (Formerly iNiS). IEEE, pp 297–303
32. Zhang XH, Dai XW, Sun W, Jin H, Luo GQ (2021) Sleeve monopole antenna with choke structure for broadband application. IEEE Ant Wirel Prop Lett 20(7):1230–1233
33. Patel H, Upadhyaya T (2021) Surface mountable compact printed dipole antenna for GPS/WiMAX applications, progress in electromagnetics. Res Lett 96:7–15

34. Alsariera H, Zakaria Z, Isa AAM (2019) A broadband p-shaped circularly polarized monopole antenna with a single parasitic strip. IEEE Ant Wirel Propag Lett 18(10):2194–2198
35. Sweeting MN (2018) Modern small satellites—changing the economics of space. Proc IEEE 106(3):343–361
36. Earth Observation Portal company, SHERPA Rideshare Mission. https://directory.eoportal.org/web/eoportal/satellite-missions/s/sherpa. Accessed 29 Mar 2024

Blockchain-Driven Recommender Systems Redefining User Experiences

Ishwari Singh Rajput, Anand Shanker Tewari, and Arvind Kumar Tiwari

Abstract Recommender systems are an effective solution for addressing the problem of information overload. They are extensively used to deliver personalized content and services to users accurately. A plethora of recommender systems have been developed for various domains which requires the analysis of huge number of users' personal liking and opinions. Most of the recommender models focus on the recommendation preciseness and accuracy thus ignoring security and privacy of the users' sensitive data. Different techniques have been used to tackle this issue, but none of the algorithms proves to be effective. To improve recommender system security and privacy, blockchain technology seems to have the potential. This is owing to its security, privacy, and trust-building abilities. This study presents an overview of recommender system based on blockchain with its characteristics and benefits in various domains. It also discusses the challenges and open issues which are to be considered while working in this domain. The results not only make a valuable contribution to the progress of secure recommender systems but also provide opportunities for the incorporation of blockchain technology in other personalized recommendation fields.

Keywords Recommender system · Blockchain · Smart contract · Privacy · Collaborative filtering

---

I. S. Rajput (✉) · A. S. Tewari
National Institute of Technology Patna, Patna, Bihar, India
e-mail: ishwarir.phd19@nitp.ac.in

A. S. Tewari
e-mail: anandp@nitp.ac.in

I. S. Rajput
Graphic Era Hill University, Haldwani, Uttarakhand, India

A. K. Tiwari
Kamla Nehru Institute of Technology, Sultanpur, Uttar Pradesh, India
e-mail: arvind@knit.ac.in

P. J. S. Gonçalves et al. (eds.), *Proceedings of Fifth International Conference on Computing, Communications, and Cyber-Security*, Lecture Notes in Networks and Systems 1128, https://doi.org/10.1007/978-981-97-7371-8_40

1 Introduction

Recommender systems (RSs), first used on e-commerce platforms to boost customer satisfaction and trust, are expected to continue in many interactions over the Internet. Movie and song recommendations, tourism, social networks, health, energy, smart cities, and many more use these algorithms [1]. The exponential growth of the Internet of Things (IoT) and Internet-connected phones, tablets, laptops, and smart sensors has created huge datasets. Massive data production and analysis have changed social media, Internet commerce, health care, energy, and other fields. Managing massive amounts of data can provide health, environmental, and other insights, enabling swift improvements [1]. Recommender systems need scalable algorithms, but big data analytics helps users comprehend and choose among multiple options. In addition to rating-based inputs, these algorithms may leverage many data types and large databases.

Machine learning algorithms have improved the process of recommendation systems which include clustering methods, nearest neighbors, matrix factorization, collaborative filtering, deep learning, and knowledge graphs [5]. There are multiple domains in which RSs are used such as e-commerce, social media, e-learning, social behavior analysis, energy conservation, health care, IoT, tourism, and fashion. Recommender systems use user, item features, user ratings, implicit user preferences, recommendation feedback, contextual information, social information, domain expertise, and user purchase or consumption history [12]. It is important to note that user data is very important and sensitive. Researchers have studied privacy-preserving recommender systems (RSs), but most have focused on novel algorithms and models rather than privacy and security. Secure and private data storage is difficult with decentralized cloud computing methodologies [1]. No risk mitigation method has protected cryptographic security or user data. So, blockchain technology has been lauded for its decentralized security and anonymity. Researchers in both communities are intrigued in this technique because it can solve a key RS challenge.

Integrating blockchain with recommender systems (RSs) has several benefits. It creates trustworthy and secure systems using blockchain-backed safe multi-party computation. Blockchain-driven RS needs intelligent contracts. Blockchain safeguards user data throughout online processing [1]. The main reason is the secure distributed ledger for data transactions. Blockchain integration enables data encryption. These qualities and blockchain's sophisticated basis and algorithms make blockchain-based RSs more secure than traditional storage systems. The main contributions of this study are as follows:

- To demonstrate how the integration of blockchain and recommender systems (RSs) benefit each other and to highlight key application, issues, and architecture of blockchain-based RSs.
- To establish a framework for decentralized RSs using blockchain ledgers to securely store data and differentiate it from other services.
- To outline potential directions for enhancing blockchain-based RS performance and addressing current issues.

We begin by discussing RS security and privacy issues, including their characteristics, concepts, and blockchain technology's influence. Applications of blockchain-based recommender systems are discussed followed by its impacts, working and its categories. Final section discusses various open issues and research directions projected in this domain.

2 Blockchain-Based Recommender System

Recommender systems helps consumers make faster decisions by tracking their behaviors [4]. Personalized recommendations may strictly violate users' privacy. These services typically collect users' confidential data [13, 16]. Leaking this information to untrusted third parties could result in serious consequences. Implementing the most effective security solution to protect user privacy is challenging due to the varying sensitivity of user information [22]. Many privacy solutions for recommendation systems rely on centralized schemes, where data is stored, further leading to privacy concerns [19]. Availability is crucial in both decentralized and centralized schemes due to the reliance on server-side computations and the absence of stored intermediate data, such as similarities, which significantly impacts the quality of recommendations based on the number of participants. Decentralized architectures face a compromise between communication costs and the quality of recommendations, which limits scalability [23]. Zhang et al. [14] have examined trust and the value of disclosures as potential factors influencing privacy concerns regarding various types of information in a recommender system. They demonstrated that control can impact privacy issues through users' perceived value of disclosure in recommender systems. Mohallick et al. [15] discussed privacy threats and current privacy protection methods in news-based recommendation systems. Blockchain technology shows potential for enhancing security and privacy in recommender systems. Blockchain-based RS are utilized in sectors like financial institutions, health care, IoT, tourism, film, e-commerce, social networks, etc. Recent studies have utilized blockchain technology for many objectives, including creating a secure recommender system, a platform for securing privacy, and a system for exchanging data. The flow graph of the blockchain-based RS is shown in Fig. 1 [3].

Jiang and Zhang [6] suggested a blockchain-powered decentralized online social network RSs to guarantee secure and privacy-conscious features for authentication and newsfeed notification. Arif et al. [7] introduced a blockchain-based data-sharing system that allows nodes to directly exchange data. Distinct nodes have distinct structures and data communication directions. The user node transmits destination assessment data to the server node, which subsequently sends machine learning process data to the user node. The sensor provides dynamic data on popularity, traffic, and weather to the user node for recommendation processing. Al Omar et al. [8] presented a privacy-securing framework for distributed e-commerce that protects client information using blockchain ledgers. Khatoon [9] introduced smart contract-based RS for handling medical data and simplifying difficult procedures. According

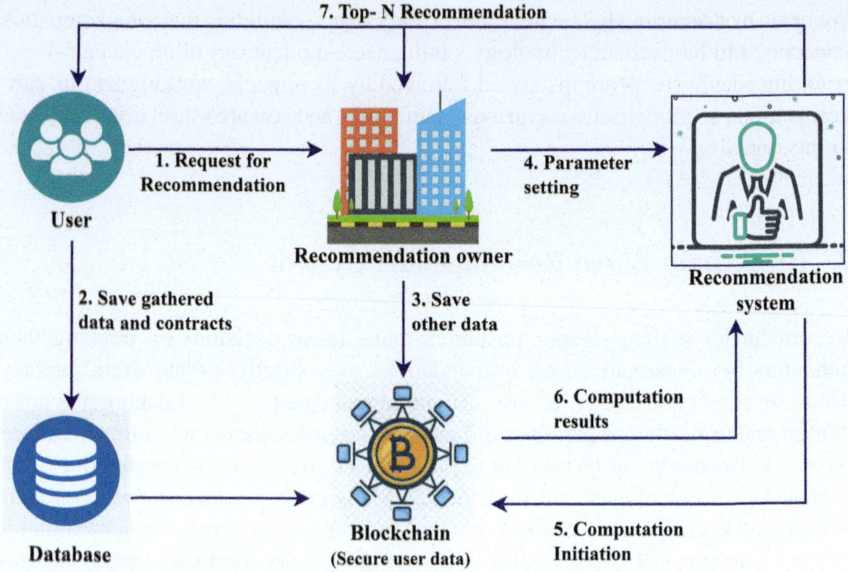

Fig. 1 Blockchain-based recommender system

to Dewangan and Chandrakar [10], a Recurrent Neural Network can optimize patient input and recommend the recommended physician for future appointments. They used blockchain technology to protect patient privacy. Wang et al. [11] introduced crowdsourcing for decentralized knowledge graph construction. The business logic was implemented using blockchain-powered smart contracts to ensure transparency, integrity, and auditability. Abbas et al. [2] combined blockchain and machine learning in medicine. Using IoT networks and RS, supply chain management may avoid counterfeit drug recommendations and deliver reliable and secure recommendations. Porkodi and Kesavaraja [12] suggested a blockchain-based IoT RS to boost user trust and protect the cognitive framework with high security. Patel et al. [13] created a deep learning RS that utilizes blockchain-enabled smart contracts to enhance a trust-centered recommendation network.

2.1 Challenges in Blockchain-Based RS

Most of the challenges which are faced by blockchain-based Rs are the one which correspond to the blockchain technology. A few of the challenges are listed as follows.

Privacy In RSs, protecting user privacy is very important because a lot of online retailers don't do enough to keep customer data and privacy safe by being trustworthy and secure. Currently, the storage and use of a substantial volume of data on cloud servers provide an added level of security to the user. Authentication is regarded as the initial level of security for safeguarding user privacy in the majority of online systems. In order to utilize the services provided by a RS, a user must initially establish a connection with the RS network and verify their identity on the RS platform. This verification process involves examining various user-related information to ensure the accuracy and trustworthiness of the provided identities. Although the topic of security and privacy protection in RS has been examined from a number of angles, adoption, skill shortages, user trust, funding, and blockchain interoperability continue to be obstacles.

Trustworthiness The accuracy and validity of the recommendations may decline for several causes, primarily stemming from the training process. Even when all procedures are executed correctly, the reliability and credibility of the recommendations may be undermined by data bias or deliberate attacks [17]. Malicious attacks targeting certain outcomes in the recommendation process could undermine the dataset and/or model, affecting security and privacy. Hackers may undertake malicious attacks on RSs, including shilling, adversarial, profile injection, and poisoning attacks, to expose authenticated user identities and bias recommendations. The robustness of a recommendation system refers to its capacity to resist manipulation by unauthorized persons. When attacks are aimed at rating-based recommendation systems, they typically involve injecting false information into a user profile, such as fraudulent ratings, and creating fake profiles that deliberately rate certain goods higher than others. Robustness, defined as providing reliable, safe, and customized suggestions, is increasingly becoming a major security concern for recommendation systems.

Secure Communications RSs must meet the minimal security criteria, while communications between users and the RS engine should adhere to all established security procedures. Recommender systems engage with users through the exchange of many types of data. If the remote server demands significant compute and storage resources, data may be sent to remote cloud servers [22]. In RSs, the data exchanged between an RS and users is typically considered to be accurate and verified. However, in real life situations, this assumption may be incorrect as fraudulent individuals can alter the information of authenticated users or skew suggestions for their own advantage.

2.2 Working Mechanism of Blockchain-Based RS

The growing popularity of RSs and their potential profitability for sellers have led to an increased inclination among sellers to manipulate recommendations and influence RSs in their favor [20]. Establishing trust in both information providers and recommendation algorithms is crucial for the acceptability of RSs technology, as it helps ensure resilience against malicious attacks. Blockchain ensures anonymity, integrity, and data security by recording transactions in a decentralized network with a tamper-resistant ledger. For instance, in social networks, individuals may seek privacy for their context or preferences, such as not revealing their precise location or specific goods they have viewed or liked, while still desiring recommendations for related items, behaviors, or places [2]. Integrating blockchain with recommender systems addresses some common challenges like trustworthiness and privacy of recommendations. The primary architecture for the working of a blockchain-based recommender system consists of decentralization, smart contracts, and a consensus mechanism. Remaining section introduces each of the components for detailed working of blockchain-based RS [3].

Decentralization The blockchain stores data in a secure manner within a peer-to-peer network, functioning similarly to a database. The blockchain is a decentralized network of interconnected computers, as opposed to a single server. The participants in the blockchain who maintain transaction records are known as keepers. They follow certain protocols and engage in mining activities to facilitate the growth of the chain by adding new nodes [4]. Decentralized blockchain networks utilize a consensus mechanism to determine blockchain status based on user votes on historical occurrences. Since public blockchains like Bitcoin and Ethereum are decentralized, network participants must agree on the shared state of the blockchain. Blockchains with verified data, such as network-confirmed transactions, result from unanimous consent among network nodes [3]. However, network nodes may not always agree on the future state of the blockchain. This produces valid forks. Blockchain technology in decentralized trading systems allows for information recording and processing without trusting each other, reducing costs [22]. This is because entities are subject to the authority of the technical system, which is believed to be permanent and unalterable. Decentralization works effectively for RSs when privacy or scalability are important. Decentralized storage and processing of information, such as user ratings, eliminates the need for users to exchange actual information with other users or centralized nodes, unlike centralized systems that use data obfuscation, anonymization, differential privacy, and encryption [4]. They can only share the necessary information for the recommendation system to function. For instance, in matrix factorization (MF), users only share local MF model loss gradients. When processing and memory resources are limited, distributed systems can distribute user ratings across lesser capacity peers, each with access to a smaller network of users and shared information.

Smart Contract Smart contracts are computer programs that automatically execute business logic when certain circumstances are met within an agreement [6]. Blockchain technology can securely enforce contractual requirements and ensure consistent results for all parties involved. Smart contracts activate automatically and record transactions in the blockchain, ensuring unquestionable information. This eliminates the necessity for a third-party arbiter, preventing influence or manipulation in contract implementation. These qualities enhance smart contract stability and availability, making them excellent for decentralized applications (DApps). Smart contracts play a crucial role in handling data within the blockchain. Smart contracts enhance the safe distributed information storage mechanism of the blockchain by incorporating it into an independent machine [13]. Smart contracts are advantageous in distributed machine learning frameworks, where players collaborate by sharing data and local models that need resources for training. Through smart contracts, individuals can make their models accessible to the public on the blockchain, allowing them to be utilized for inference without restrictions. An optimal use case for smart contracts in recommender systems is a distributed matrix factorization scenario, in which multiple partners collaborate by sharing their local models.

Consensus Mechanism A consensus algorithm allows a network of nodes to reach an agreement on the status of a distributed ledger. The main goal of a consensus algorithm is to guarantee that all network participants come to a shared and clear conclusion about the legitimacy of transactions and the sequence in which they are included in the blockchain [14]. Proof of Work (PoW) is the most primitive consensus algorithm. The concept was initially introduced in Bitcoin, but it has existed for a considerable period. Validators in Proof of Work (PoW) record desired data until they generate a particular answer. The Proof of Work (PoW) protocol establishes the criteria for determining the validity of a block. In the early days of Bitcoin, Proof of Stake (PoS) was considered as an alternative to PoW. A PoS system does not require miners, specialized materials, or mass applications [23]. PoS relies on internal encryption rather than external resources like power or hardware. While restrictions vary each protocol, a minimum quantity of money is typically required to meet entry standards. Various solutions to the above have been developed and implemented by different networks, including Delayed PoW, Leased Proof of Stake Consensus, Proof of Authority, Proof of Burn, Delegated Proof of Stake, and Hybrid PoW/PoS Consensus. Central to consensus procedures, stakeholders generate and get rewards for creating new blocks, while other nodes validate and add them to the blockchain. To safeguard the network and its information, updating information requires modification of numerous nodes, increasing the cost of the change.

2.3 Categories of Blockchain-Based RS

Trust-based RSs Trust is crucial for effective and accurate decision-making in RSs. Blockchain in RSs ensures user trust. RSs in this category use blockchain-supported safe multi-party computation and smart contracts within the core blockchain protocol to establish secure and trust-based frameworks [2]. This aids in constructing trustworthy RSs for cooperatively creating trust-based databases and hosting updated models using smart contract systems.

Privacy-preserving RSs Analyzing user data yields recommendations. Massive volumes of high-quality data are needed for customized, reliable recommendations. Huge amounts of sensitive user data are collected without regard to privacy. RS developers have tried numerous privacy-preserving solutions, but none have guaranteed cryptographic security. Blockchain technology lets users exchange personal data with RSs for personalized recommendations. This reduces fraud and boosts end-user data sharing [21]. Although decentralized RSs protect privacy adequately, they lack the efficiency needed for widespread adoption. Figure 2 [8] shows a secure RS flowchart using blockchain and deep learning.

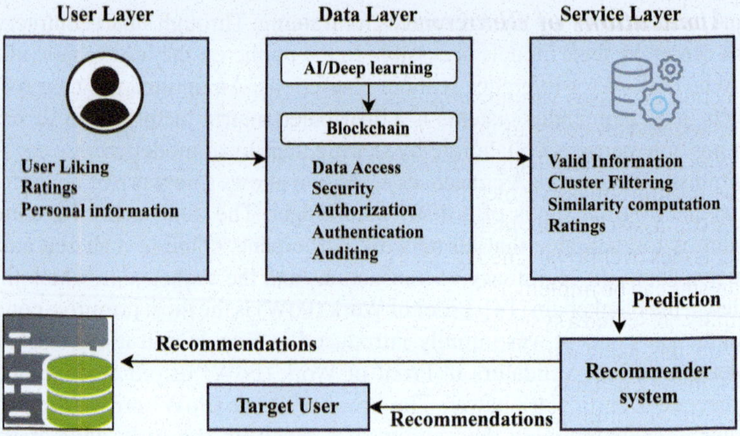

Fig. 2 Secure RS using AI and deep learning

Secure RS Blockchain protocols and smart contracts enable secure RSs to control information integrity, security, and authenticity using cryptography and hash function. Khatoon et al. [9] suggested a smart contract-based RS for handling medical data and simplifying operations. An extensive investigation used real-world clinical records to examine blockchain and smart contract-based healthcare workflows. Wang et al. [11] developed a smart contract-based RS using Ethereum TestNet and Distributed Ledger Technology as shown in Fig. 3 [11]. This concept provides decentralized product and activity rankings. Public blockchain smart contracts eliminate centralization. Authorization to collect preferences from qualified users rating things or activities has enhanced the framework's potential.

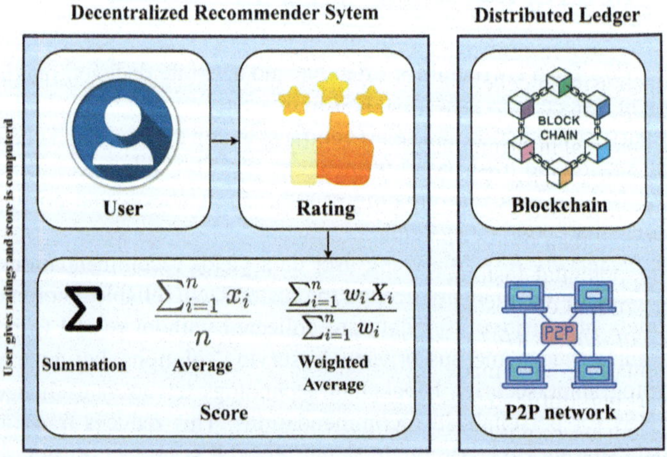

Fig. 3 Decentralized recommender system

2.4 Applications of Blockchain-Based RS

This section discusses various application domain where blockchain-based recommender system is successfully employed.

Healthcare Advanced algorithms and machine learning help recommendation systems examine massive medical data. RSs' personalized recommendations and insights can assist doctors make patient-specific decisions. To address privacy and security issues over private medical records, strict privacy policies and robust security mechanisms can be implemented to make RSs safer in health care. For example, in [1] the authors presented a system wherein medical sensors can transmit encrypted health data to doctors using a mobile application, ensuring anonymity. Subsequently, three consortium blockchains are created with the purpose of distributing the workload of transactions and minimizing the time it takes for transactions to be processed. The system stores the credentials of system entities, doctors' prescriptions, and recommendations based on the data received through mobile applications, as well as the medical treatment process. The proposed architecture in [2] leverages the decentralized nature of blockchain technology to ensure transparency and the immutability of transactions. By utilizing smart contracts, the system can automatically enforce business terms and conditions without the need for intermediaries. Additionally, the recommendation engine takes into account not only patient reviews and comments but also factors in the patient's medical condition to provide personalized drug recommendations. This approach enhances the accuracy and relevance of the recommended medications for each individual patient.

E-commerce Most recommendation systems (RSs) that are based on e-commerce derive recommendations from user data by use of collaborative filtering procedures. As a result, companies are retaining a lot of private data about their customers. Given the sensitivity of such data, it is no longer appropriate to overlook customers' privacy concerns. Encouraging customers to share their consumption data and facilitating tailored recommendations from businesses are contingent upon the safe handling of sensitive information. Frey et al. [16] introduce a new recommendation approach using blockchain's safe multi-party computing capabilities. Applying the technology allows potential customers to use a recommendation system without disclosing personal information. The authors in [18] propose a revolutionary strategy for future retail systems. Customers' data is protected by blockchain storage. The bitcoin protocol-based system enables secure multi-party calculations and tamper-proof encrypted data transactions, with only the data owner having access to input data. Hence, potential customers can enable companies to employ recommendation algorithms without disclosing personal information.

Internet of Things (IoT) The Internet of Things (IoT) has undergone recent advancements that have made it possible to implement smart city solutions (such as smart parking, traffic management, and other similar solutions) that make use of RSs to present users with ideas. It is possible that the non-confidential exchange of user information with an untrusted or semi-trusted RS could constitute an invasion of privacy. This is because the behavior and mobility patterns of users could be inferred through the study of their previous actions. Decentralized intrusion detection systems (IDS) for networks, using blockchain for safe cooperation, were presented in [5], addressing several centralized IDS issues. In [10], currently available technologies are examined, and a blockchain-based cognitive IoT system is proposed to establish confidence in a reliable and secure RS.

E-learning Every element of society is improving, and people are continually looking for methods to improve their jobs. E-learning is helping reach this aim by transforming how learners learn. RSs provide learners with timely, personalized, and engaging assistance after automatically identifying their preferences. Student IDs, grades, and other sensitive data are stored and processed by such systems. There are a few privacy and security concerns that have been brought to light by the technologies that are currently being utilized in e-learning platforms. These concerns include the disclosure of private information, cyber-attacks, and other similar difficulties. As a result, it is imperative that these issues be resolved for the benefit of the users. In this regard, there are numerous ways in which blockchain technology could revolutionize e-learning and the application of RSs. Organizations should use blockchain technology in their sustainability education programs to improve transparency, verifiability, and efficiency. Blockchain technology is still new in education, but research suggests it might alter the sector and boost efficiency and transparency. In [3], the authors covered a wide range of technological and non-technical challenges, such as helping to develop new application systems and promoting the advancement of education, as well as providing an in-depth analysis of blockchain's possible uses. Gustaaf et al. [4] focus on blockchain-based higher education protocol. It examines blockchain features and educational project services. To simplify educational technology use, blockchain features are used. They compared education and blockchain-based projects.

Social Networks In recent times, massive amounts of information, including text (tweets, comments, and so on), photographs, videos, and documents, have been posted on social networks. There has been a tremendous increase in the amount of information that is being shared through social networks, which has led to the problem of information overload. Consequently, a variety of social networking services make use of RSs in order to circumvent this issue and provide insightful information to individuals who have been specifically designated. A variety of mobile applications can be recommended to users of mobile devices through the utilization of recommendation engines. A concern regarding the protection of user privacy and security is brought up by the enormous amount of data and information that is gathered from users. Jiang et al. [18] use blockchain to create an RS architecture that combines the features of centralized and distributed online social networks (OSNs). Typically, smart contracts make the blockchain a trustworthy server for important control services.

3 Open Issues and Future Directions

For years, collaborative filtering has been the cutting-edge algorithm for the recommendation tasks. However, knowledge-based and deep learning algorithms have been shown improvement in the performance of recommender system and also address the issues of recommender systems [5]. Due to the need to gather and process training data in one place, the conventional recommender algorithms can't be employed in distributed architectures. Some efforts employ blockchain to construct a value-based incentive structure for model training in distinct locations. Most of these works don't properly leverage blockchain. RSs usually have a trusted central authority that controls the recommendation system. Blockchain technology, a decentralized and distributed public record, is the greatest solution to establish decentralized RSs with high privacy and security. Blockchain was recently introduced as a foundation for developing distributed infrastructures that may provide safe transactions, cryptocurrencies, identity management, supply chain management, and more. Because user data (such as ratings or profile information) is contributed by numerous users, and because user data must be stored for a long time and remain unchangeable even for the RS administrators in order to ensure transparency and build trust, RSs are a good fit for the application of blockchain technology [18]. Despite multiples benefits, there are some issues which are to be considered while working in this domain:

Latency Blockchain transactions sometimes utilize consensus techniques, which may cause delays. Combining recommender systems with blockchain technology may influence the system's agility and velocity, particularly important for systems requiring immediate or very immediate replies [19].

Scalability Scalability problems are a common challenge for blockchain networks, and adding recommender systems to the blockchain can make matters worse. The system's scalability becomes a major challenge as the number of transactions and users rises [20].

Energy Consumption Blockchain networks are criticized for consuming a lot of energy, particularly those that use Proof of Work (PoW) consensus algorithms (like Bitcoin). The integration of recommender systems into these networks could make the environmental impact worse [21]. Recommender system efficiency and overall sustainability may be impacted by blockchain network energy usage. Finding energy-efficient consensus processes or switching to more environmentally friendly blockchain platforms becomes a crucial topic as the environmental effects of blockchain technology come under closer examination.

Dynamic Content Conventional recommender systems often work with static datasets that may not account for changes in user preferences and behaviors over time. Blockchain-based recommender systems may require the capability to manage dynamic content and swiftly changing data because of the real-time characteristics of some applications [9]. Effective algorithms and methods are necessary to promptly update recommendations.

Data Storage and Retrieval A significant amount of resources may be required in order to store and retrieve massive volumes of data on the blockchain [10]. The development of effective systems for the storing and retrieval of data in a decentralized way is a difficult task.

Interoperability There is a possibility that different blockchain platforms will not be compatible with one another, and it is very difficult to develop a recommender system that is universally interoperable across all of the different blockchains [6]. It's possible that efforts to standardize could assist in addressing this problem.

Along with the above open issues discussed, there are some future directions which open new opportunities for further research which are listed below.

Scalability in Blockchain-based RS RSs with billions of users and products struggle with scalability. The learning process involves fast and effective training and inference. Solution is to improve hardware infrastructure, distribute or parallelize processing, and create better algorithms. Blockchain improves user privacy and prevents model and data tampering. However, this innovation increases processing costs, causing blockchain scalability issues [9]. Traditional RSs aren't very scalable due to algorithms like singular value decomposition (SVD); however RSs integrated with blockchain can grow much more efficiently. As an example, neighborhood-based collaborative filtering naturally involves quick similarity computation, and using locality sensitive hashing (LSH) techniques has been useful for keeping item representations (specifically signatures) short.

Federated Learning Federated Learning is a rapidly growing area of research in the field of machine learning. It involves collaborative training while ensuring the privacy of nodes sharing their data. In federated learning, many parties collaborate to train a machine learning model by sharing data summaries, gradients, or models instead of exchanging their complete datasets [21]. Federated learning helps create distributed RS for data privacy and security solutions. Federation in RSs can be horizontal (shared things), vertical (shared users but distinct item collections), or transfer. A model built on one party using local data is used as a starting point for the other party's model in transfer learning.

Explainability in RS While AI and deep learning have improved work performance, humans still need to analyze and justify data-driven model outcomes. Explainable recommender systems offer end users with both the recommendations and explanations of the recommendations [23]. Recent experiments show that explaining the logic behind recommendations might enhance acceptance rates and assist behavioral change among end users. In addition, it has the capability to provide smart contracts, which enable the automatic execution of business logic with the agreement of several nodes. Explainable recommender systems (RSs) have the potential to be implemented on top of blockchain technology. This can be achieved through the use of model ensembles, which generate multiple recommendations and allow rule-based or knowledge-based models to prioritize these recommendations. These models are encoded within smart contracts, ensuring that recommendations are trustworthy and can be explained. This approach guarantees privacy, security, and explainability simultaneously.

4 Conclusion

Blockchain-based recommender systems are thoroughly investigated here. Blockchain technology's security and privacy challenges, characteristics and principles, possible impact on diverse industries, and decentralized systems and applications are discussed in this article. We will examine limits and outstanding concerns before implementing blockchain-based recommender system improvements. Blockchain technology makes distributed recommender systems as efficient as centralized ones while protecting sensitive data. To define recommender system development while safeguarding user data and privacy. Security was also important to prevent dangerous attacks and raise recommendation algorithm confidence. To address security and privacy difficulties, blockchain is commonly suggested for RSs. Therefore, blockchain technology can improve recommendation systems by ensuring data security, integrity, confidentiality, and accessibility. Decentralized reputation systems using blockchain technology may produce user conflict. Future solutions for egalitarian protocols, such as blockchain-based atomic swaps, are crucial.

References

1. Kamal R, Hemdan EED, El-Fishway N (2023) Care4U: integrated healthcare systems based on blockchain. Blockchain: Res Appl 4(4):100151
2. Abbas K, Afaq M, Ahmed Khan T, Song WC (2020) A blockchain and machine learning-based drug supply chain management and recommendation system for smart pharmaceutical industry. Electronics 9(5):852
3. Saputra MAW, Ochtaffia D, Apriani D, Yusfi SC, Gori M (2023) Blockchain applications in education affecting challenges and problems in digital. Blockchain Front Technol 2(2):15–23
4. Guustaaf E, Rahardja U, Maharani HW, Santoso NA (2021) Blockchain-based education project. Aptisi Trans Manage (ATM) 5(1):46–61
5. Narayanan A, Bonneau J, Felten E, Miller A, Goldfeder S (2016) Bitcoin and cryptocurrency technologies: a comprehensive introduction. Princeton University Press
6. Jiang L, Zhang X (2019) BCOSN:Blockchain-based decentralized online social network. IEEE Trans Comput Soc Syst 6(6):1454–1466
7. Arif YM, Nurhayati H, Kurniawan F, Nugroho SMS, Hariadi M (2020) Blockchain-based data sharing for decentralized tourism destinations recommendation system. Int J Intell Eng Syst 13(6):472–486
8. Omar AA, Bosri R, Rahman N, Bhuiyan MZA (2019). Towards privacy-preserving recommender system with blockchains. In: Dependability in sensor, cloud, and big data systems and applications: 5th international conference, DependSys 2019, Guangzhou, China. Springer Singapore, pp 106–118
9. Khatoon A (2020) A blockchain-based smart contract system for healthcare management. Electronics 9(1):94
10. Dewangan NK, Chandrakar P (2021) Patient feedback-based physician selection in blockchain healthcare using deep learning. In: International conference on advanced network technologies and intelligent computing. Springer International Publishing, Cham, pp 215–228
11. Wang S, Huang C, Li J, Yuan Y, Wang FY (2019) Decentralized construction of knowledge graphs for deep recommender systems based on blockchain-powered smart contracts. IEEE Access 7:136951–136961

12. Porkodi S, Kesavaraja D (2020) A trust-based recommender system built on IoT blockchain network with cognitive framework. In: Recommender system with machine learning and artificial intelligence: practical tools and applications in medical, agricultural and other industries, pp 291–311
13. Patel SB, Bhattacharya P, Tanwar S, Kumar N (2020) Kirti: a blockchain-based credit recommender system for financial institutions. IEEE Trans Netw Sci Eng 8(2):1044–1054
14. Zhang B, Wang N, Jin H (2014) Privacy concerns in online recommender systems: influences of control and user data input. In: 10th symposium on usable privacy and security (SOUPS 2014) pp 159–173
15. Mohallick I (2017) Exploring privacy concerns in news recommender systems. In: Proceedings: international conference on web intelligence, pp 1054–1061
16. Frey R, Wörner D, Ilic A (2016) Collaborative filtering on the blockchain: a secure recommender system for e-commerce
17. Frey RM, Vuckovac D, Ilic A (2016) A secure shopping experience based on blockchain and beacon technology. RecSys Posters
18. Jiang L, Zhang X (2019) BCOSN: a blockchain-based decentralized online social network. IEEE Trans Comput Soc Syst 6(6):1454–1466
19. Bokhari ST, Aftab T, Nadir I, Bakhshi T (2019) Exploring blockchain-secured data validation in smart meter readings. In: 2019 22nd international multitopic conference (INMIC). IEEE, pp 1–7
20. Hassan M, Rehmani (2019) Differentially private auction blockchain-based microgrids energy trading. IEEE Trans Serv Comput 13(2):263–275
21. Huang J, Ma B, Wang M, Zhou X, Yao L, Wang S, Qi L, Chen Y (2023) Incentive mechanism design of federated learning for recommendation systems in MEC. IEEE Trans Consum Electron
22. Alhijawi B, Alrub MA, Al-Fayoumi M (2023) Generalized Ethereum Blockchain-based recommender system framework. Inf Syst 111:102113
23. Mantey EA, Zhou C, Anajemba JH, Hamid Y, Kingsley J (2023) Blockchain-enabled technique for privacy-preserved medical recommender system. IEEE Access

Navigating Automation: Process Analysis Framework for Robotic Process Automation Adoption

Veena Tripathi and Neeraj Tripathi

Abstract Robotic Process Automation (RPA) is an emerging approach designed to automate repetitive human tasks through the use of software-based robots. In the context of business processes, RPA involves configuring these software robots to perform the tasks that were previously carried out by human employees within the organization. Adopting RPA has many benefits, including increased productivity, greater accuracy, shorter cycle times, strengthened data security, and increased corporate efficiency. Organizations may simultaneously free their workers from tedious and repetitive activities, thanks to this technology. However, enterprises must successfully manage the process to achieve the best results because RPA installation poses a considerable obstacle. This paper aims to help businesses successfully use RPA into their business processes. Specifically, it introduces a novel methodology aimed at guiding organizations in the analysis of business processes that are currently used and to identify the important amendable process for RPA implementation. This paper expounds upon the fundamental principles underpinning our methodology.

Keywords RPA adoption strategy · BPMN · Robotic process automation · CBRANK

1 Introduction

In the rapidly evolving contemporary business landscape, organizations are constantly seeking innovative approaches to enhance efficiency and optimize operational processes. One such groundbreaking advancement is Robotic Process Automation (RPA), a technique that uses execution sequence macros to carry out tasks that

V. Tripathi (✉)
Model Institute of Engineering and Technology, Jammu, India
e-mail: veena.it@mietjammu.in

N. Tripathi
Shri Mata Vaishnodevi University, Katra, India
e-mail: neeraj.tripathi@smvdu.ac.in

P. J. S. Gonçalves et al. (eds.), *Proceedings of Fifth International Conference on Computing, Communications, and Cyber-Security*, Lecture Notes in Networks and Systems 1128, https://doi.org/10.1007/978-981-97-7371-8_41

were previously done manually. RPA represents a pivotal transformation in the way businesses handle their processes and transactions. Software robots are built up using technology as part of RPA, capable of acquiring, interpreting, and interacting with already-existing applications, according to the Institute for Robotic Process Automation and Artificial Intelligence [1]. These robots process transactions, manipulate data, and seamlessly communicate with other software systems. The implications of RPA are profound, offering organizations a multitude of advantages. Having all these benefits, RPA promises heightened high level of productivity with performing very less process alterations; it improves the quality of service and reduced delivery times, and the invaluable liberation of manpower from repetitive tasks and tedious tasks [2].

The versatility of RPA extends across a wide spectrum of industries, including banking, health care, and insurance. These industries, among others, spend a significant amount of time on tasks that software robots can handle more effectively [3]. Concrete examples of RPA applications include e-mail automation, sales processing system order management, and planning for material requirements in the supply chain sector [4].

Despite the undeniable power of RPA, organizations face the challenge to identify the most processes for RPA implementation. To our knowledge, there is no comprehensive system that directs organizations through this critical decision-making process as of yet. This gap underscores the need for a systematic approach to assist organizations in effectively adopting RPA to automate their business processes.

1.1 Organization of Paper

The paper is having 6 sections. Section 1 is all about the introduction; the remaining sections of the paper are structured as follows. In Sect. 2, literature survey provides the comparative study of work done. Section 3 explains the motivation for the research; Section 4 is contribution part to find a methodology to check business processes if they can be automated using RPA. Section 5 describes the proposed framework model including all the phases for examining business processes and determining which ones are appropriate for RPA. Section 6 serves as our conclusion and future prospectus of the proposed model.

2 Literature Survey

RPA is a relatively new method in the study of business process automation (BPA). As a result, little research has been done in the area of RPA, particularly to address the issue of determining and confirming if a business process is appropriate for RPA [13]. This table provides a structured summary of relevant literature, including the title, authors, publication year, and a brief summary of related work (Table 1).

Table 1 Research survey table

Title	Authors	Year	Summary
Robotic Process Automation: A Review of Organizational Grey Literature [5]	Chanyuan (Abigail) Zhang, Hussein Issa, Andrea Rozario	2022	Develop a comprehensive framework or model for successful RPA implementation in accounting processes
Innovating with Bots: Strategies for Successful RPA Implementation [6]	Chris Liu, Chen Li	2023	Focuses on influence of RPA, uncovering its strategic importance and its role in shaping the trajectory of business evolution
Robotic Process Automation as a Digital Transformation Tool for Increasing Organizational Resilience in Polish Enterprises [7]	Andrzej Sobczak	2022	Presents real-world case studies of RPA implementation across different industries, offering insights into the challenges and benefits of adoption
The Automation of Management and Business Science [8]	Clark D. Johnson, Brittney C. Bauer, and Fred Niederman	2021	Explores the different ways of automations and tries to find the techniques for at least partially automated, management, and business science
RPA and Artificial Intelligence in Budget Management Based on Multiperspective Recognition Based on Network Communication Integration [1]	Haichang Luo	2021	Discusses the integration of AI and cognitive technologies into RPA, showcasing how these advancements enhance automation capabilities
A Convergence Model for Environment Analysis of SME (Small and Medium Enterprises) Using RPA (Robotic Process Automation) [9]	Kim, Young-Jin; Hong, Jung-Wan; You, Yen-Yoo	2021	Focuses on process analysis, improvement, and optimization within RPA, addressing the identification and evaluation of processes suitable for automation
"Application Exploration of Robot Process Automation in Digital Labor Time Management System" [10]	Fei Liu a, Jian Li b	2023	Research apply RPA technology to the digital labor time management system of enterprises to improve management efficiency and accuracy

3 Motivation

This paper addresses this pressing need by presenting a novel method designed to analyze and identify business processes ideal for RPA integration. This methodology is unique in that it can be applied to processes in numerous business domains. Furthermore, since it is tailored for business analysts [14] using CBRANK algorithm, there

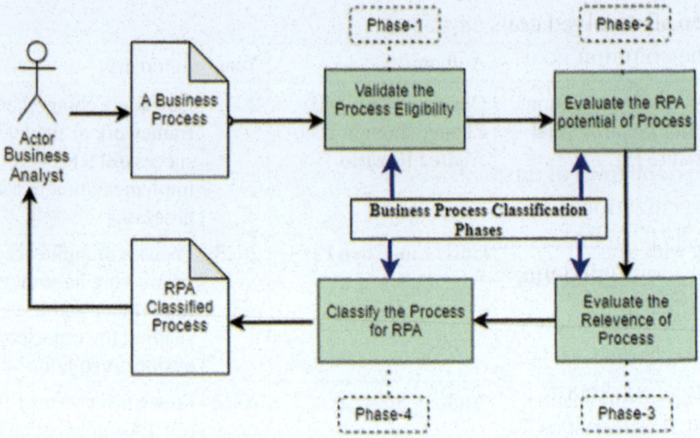

Fig. 1 Analytical framework for assessing business processes for RPA adoption

is no need for them to become experts in RPA. Through this paper, we aim to provide organizations with a practical tool to unlock the transformative potential of RPA and pave the way for more efficient, streamlined, and automated business processes.

4 Contribution

In our paper, Process Analysis for the goal of the Robotic Process Automation Adoption Framework is to create a fresh methodology that first studies business processes before determining if they can be automated [15] using RPA. The classification of Business Processes is based on the outcome of the proposed method. The suggested model's classification standards are not appropriate for RPA; moderately suited for RPA; less suitable for RPA; or highly suitable for RPA (Fig. 1).

5 Proposed Model (Framework for Robotic Process Automation Adoption)

5.1 Phase 1: To Validate the Process Eligibility

Phase-1 is the approach for determining if a process is suitable for RPA implementation [16], our proposed methodology assesses two key aspects:

- Property-1 MATURITY
- Property-2 STANDARDIZATION.

(1) Do all involved units anticipate the same service within the process and have the common goal?
(2) Does the relationship between the business process and the software applications have a clear definition and a plan?

To assess the level of process standardization (Property-2), it is essential to confirm that the organization is implementing a standardized procedure to achieve its business objectives [7]. To understand how the processes are carried out to achieve a certain goal, their thorough definition calls for four different viewpoints [17]:

a. The functional view depicts the processes that are taking place as well as the interdependencies, such as producer–consumer dependencies [10].
b. Information about the activities' control dependencies and sequencing is provided by the dynamic view.
c. The organizational perspective identifies the individuals who execute each process activity, their locations within the organization, and the channels of the information flow in term of arguments.
d. The informational view contains an explanation of the data that the process manipulates, such as artifacts and entity information.

It is required to validate that tasks are carried out uniformly throughout all divisions of the organization and also validate that all process views organizational view [18], informational view, functional view, dynamic view work according to specified criteria.

5.2 *Phase 2: Evaluate RPA Potential of a Business Process*

This step evaluates a business process's potential for automation using the RPA methodology. We have introduced a process' RPA potential as its ability to set up the robots to perform the specified tasks [19] which were formerly carried out by humans. There are two tasks in this step.

- At first, we need to check the potential weight of each process involved.
- In next step using the RPA potential of each individual action, we will determine the RPA potential weight of the entire business process.

5.2.1 Process Activity Weighting for RPA Potential

Our approach provides two criteria to assess a process activity's RPA automation potential. The activity must, at the very least, need a manual interface [20] with a piece of software in order to satisfy the first criteria (Property-3). Second, the business must adhere to well-established business standards [21] (Property-4). As a result, our method establishes the importance of an activity's.

- Property-3 manual communication with a software program;

- Property-4 preset business rules.

Potential of RPA process is indicated by POT(AP) of activity A within process P. In next step according to The Object Management Group activities are classified as Atomic or Non-atomic process activity [11, 22]. So, in order to weigh the RPA potential of a process activity, we separate the following.

Case-1

- **The activity is atomic**: In Fig. 2, we have described the complete procedure for weighing the RPA potential of individual activity. We have proposed a rank-based algorithm CBRANK, a set of questions generated to get the results.

  Question 1: Is manual communication with a software program included?
  Question 2: Is the action supported by well-stated business rules?

The result of activity A in a process P will be taken into consideration. If the response is negative to at least one, the result's RPA potential is zero (**Pot(Ap) = 0**), but if it is positive, the result's RPA potential [23] is one (**Pot(Ap) = 1**).

Case-2

The activity is compound: A process activity is made up of multiple other subactivities (compound activity); the percentage of (sub)activities is used to compute the RPA potential for each individual process activity.

(i) Software application requires a manual communication with a software application (Property-3).
(ii) Predefined business rules are also incorporated in it (Property-4): A compound activity A having n number of individual activities (A1.....An).

Our method applies the process illustrated as discussed in Fig. 3; we have denoted the percentage of subactivities as ***Perc*(*Property* 3)** The percentage of (sub)activities

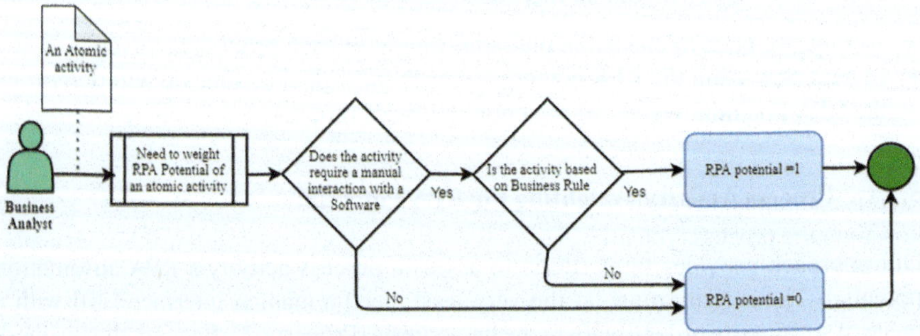

Fig. 2 Framework for RPA potential of an atomic activity weighting procedure

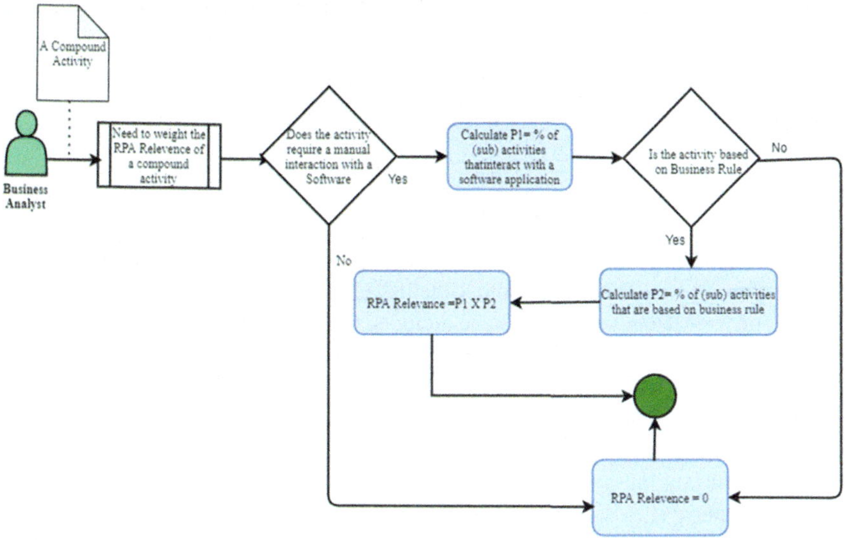

Fig. 3 Framework for RPA potential of compound activity weighting procedure

that engage with a piece of software will be calculated. In addition to (sub)activities [24] that communicate with software, it determines the Perc(Property 4) which denotes the proportion of (sub)activities that are based on pre-established business standards:

$$Pot(Ap) = Perc(Property\,3) \times Perc(Property\,4) \tag{1}$$

$$Pot(Pn) = \sum_{i=1}^{n} Pot(Api)$$

5.2.2 Weighting Each Process Activity's RPA Relevance

It is a procedure with a modest level of complexity and a high volume of transactions. Using the formula as the average daily number of transactions, the volume of transactions is computed. In addition to calculating the process complexity, take into account how long the process typically takes [25] to complete. A basic process takes a human only a few minutes to complete; on the other hand, a complex process takes hours. Consequently to assess a business process's RPA Relevance our approach evaluates its level of complexity (Property-6) and volume of transactions (Property-5). Two complementing tasks are used to do this study:

- **Task-1** It determines the weight of each process activity's RPA relevance to automate each process activity that is being used.
- **Task-2** Using the findings from the previous assignment, it evaluates the relevance of automating the entire business process using the RPA approach:

$$Rel(Pn) = \sum_{i=1}^{n} Rel(Api) \tag{2}$$

Therefore, a value between 0 and 1 inclusively represents the potential weight of the RPA process. In our study, we have included the Scale ***for Assigning a Relevance Weight*** (SARW) diagram in Fig. 4. The "Scale for Assigning a Relevance Weight Model" is a method or system for determining and assigning relevance weights to various elements, factors, or criteria in a given context. This model is used to quantify the importance or significance of each element in relation to a particular decision, evaluation, or ranking process.

Each property (i.e., the amount of transactions and level of complexity) is given a value on this scale, which ranges from Low to High. The business analyst will analyze the volume of transactions (i.e., Property-5) of a process activity A.

5.2.3 Average Number of Transactions Performed by the Activity A

According to our proposed method, the monthly transaction volumes of all process users must be determined, then this sum must be divided by 30 to determine the average number of transactions. On the basis of the significance scale modified from [8], in Fig. 4, the volume of transactions is regarded as LOW if this figure is fewer than 30. The number of transactions is regarded as MODERATE if it is between 30 and 145, and HIGH if it is higher.

Analysis of the degree of complexity of an activity *A*. If task A's average completion time is less than four minutes, between four and thirty minutes, or longer than that, the business analyst must react with LOW, MODERATE, or HIGH. The technique uses the SARW model to weight the RPA relevance of each activity A that comprises the process by choosing one of the five values (0, 0.25, 0.5, 0.75, or 1). This is done after the business analyst has analyzed both features ***P* (i.e., *Rel*(*Ap*))**.

5.3 *Phase-3 Weighting the RPA Relevance of a Business Process*

Our method makes use of the outcomes from the preceding stage to determine the weighted RPA relevance of a business process P made up of n activities, denoted as Rel(Pn) [12]. Rel(Pn) is therefore defined as the process activities' average RPA

Transactions Volume			
>145/day High	1	0.5	0
[30-145/day Moderate	0.25	0.75	0.25
<30/day Low	0	0.5	1
	<4min Low	[4-30] min Moderate	> 30 min High

Degree of Complexity

Fig. 4 Relevancy matrix for business process

relevance weight. Therefore, the RPA potential relevance of a business process is therefore a value between 0 and 1 inclusive.

5.4 *RPA Classification of a Business Process*

The proposed method's final step assigns an RPA classification to a business process based on the process's RPA relevance and potential (Steps 2 and 3), which are classified using the quadrant shown in Fig. 5. (REL (P) > 0.5) lends itself well to the RPA technique. It is categorized as A in the quadrant. A business process that is somewhat suitable for the RPA technique has both high RPA promise and low RPA relevance. It is categorized as B in the quadrant.

A business process that has both high RPA relevance and low RPA potential is less suited to the RPA strategy. It is categorized as a C in the quadrant. Finally, when a business process's RPA potential and relevance are limited, it is not appropriate to automate it using the RPA approach. To clarify the result let us consider the business processes Px and Py as an example. The RPA Potential of the first process is 0.75, meaning that the RPA approach can automate 75% of its tasks while the remaining 25% require manual interaction with software [26]. Px is typically executed 50 times per day on average, and its activities take an average of 3 min, so its RPA relevance is 0.25. As a result, Px is classified by RPA as having a moderate level of suitability for the RPA technique, or B, in the quadrant. Let's look at a process called Py that has an RPA potential of 0.60, meaning that 60% of its actions can be automated using RPA. Py's RPA relevance is 0.75 if it runs 50 times per day on average and takes 20 min to complete each task. Py's RPA classification in the quadrant is therefore A, which is very appropriate for RPA.

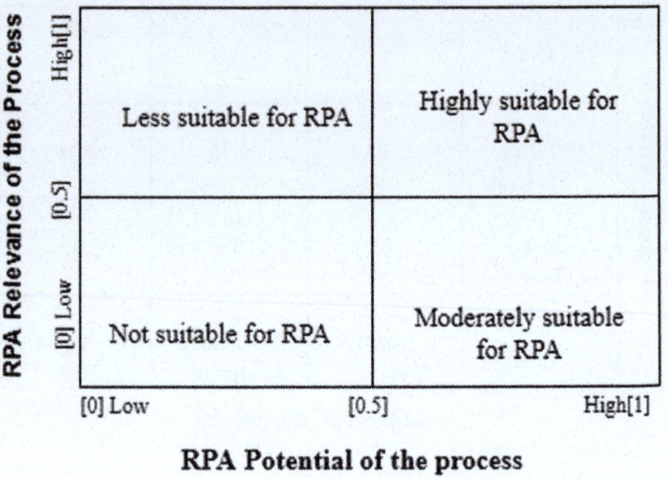

Fig. 5 Classification of business process based on relevance and potential of process

6 Conclusion and Future Work

In conclusion, "Navigating Automation: Process Analysis Framework for Robotic Process Automation Adoption" has provided valuable insights into the intricate landscape of Robotic Process Automation (RPA) adoption. This research has explored the complexities and challenges associated with integrating RPA into organizational processes and presented a structured framework to navigate this transformative journey effectively.

The framework outlined in this study emphasizes the significance of meticulous process selection. Not all processes are created equal, and the choice of which to automate is pivotal. Process complexity, scalability, and potential returns must be carefully considered. The framework offers a structured approach for organizations to identify processes that are best suited for automation, setting the stage for a successful RPA implementation.

In summary, our research contributes a valuable resource for organizations embarking on their RPA adoption journey. The framework presented here provides a structured approach to selecting, implementing, and managing RPA within the context of existing business processes. By adhering to the principles outlined in this framework, organizations can maximize the benefits of automation, drive operational efficiency, and remain competitive in an increasingly automated world. The path to successful RPA adoption may be challenging, but with the right guidance and tools, organizations can navigate automation with confidence and reap the rewards it offers. Future work focuses on applying the frame work in real case studies with some Business Process Automation area as proof of concept.

References

1. Luo H (2021) RPA and artificial intelligence in budget management based on multiperspective recognition based on network communication integration. https://www.researchgate.net/journal/Wireless-Communications-and-Mobile-Computing-1530-8677
2. Rodriguez JF, Silva LD, Gomes LF (2021) Robotic Process Automation in business process improvement: a systematic literature review. In: 2021 IEEE Latin American conference on computational intelligence (LA-CCI), pp 1–6. https://doi.org/10.1109/LA-CCI52669.2021.9543879
3. Bindiya AJ, Anitha N, Indumathi S (2023) Robotic Process Automation (RPA): a software bot for healthcare sector. In: 2023 international conference on intelligent and innovative technologies in computing, electrical and electronics (IITCEE), pp 685–689
4. Flechsig C et al (2021) Robotic process automation in purchasing and supply management: a multiple case study on potentials barriers and implementation. J Purch Supply Manag 100718
5. Zhang C, Issa H, Rozario A (2022) Robotic Process Automation: a review of organizational grey literature. Int J Inf Syst Project Manag. https://doi.org/10.12821/ijispm100101
6. Liu C, Li C (2023) Innovating with bots: strategies for successful RPA implementation. CC BY 4.0
7. Sobczak A (2022) Robotic process automation as a digital transformation tool for increasing organizational resilience in polish enterprises. Collegium Econ Anal Warsaw Sch Econ 162. Niepodległości Poland
8. Johnson CD, Bauer BC, Niederman F (2021) The automation of management and business science. https://doi.org/10.5465/amp.2017.0159
9. Kim Y-J, Hong J-W, You Y-Y (2021) A convergence model for environment analysis of SME using RPA
10. Liu F, Li J (2023) Application exploration of robot process automation in digital labor time management system. In: 3rd international conference on machine learning and big data analytics for IoT security and privacy procedia computer science, vol 228, pp 89–97. https://doi.org/10.1016/j.procs.2023.11.012
11. Fattah AM, Taha TM (2021) Robotic Process Automation (RPA) adoption and its effect on employee performance: a conceptual model. In: IEEE conference on business engineering and industrial applications (ICBEIA), pp 27–32. https://doi.org/10.1109/ICBEIA52286.2021.9473499
12. Flechsig C, Anslinger F, Lasch R (2022) Robotic process automation in purchasing and supply management: a multiple case study on potentials, barriers, and implementation. J Purch Supply Manage 28(1):100718. https://doi.org/10.1016/j.pursup.2021.100718
13. Sucahyo TLJL, Thaha RHM, Sumarsono MA (2021) Robotic Process Automation adoption in Indonesian state-owned enterprises. In: 2021 international conference on engineering, technologies, and applied sciences (ICETAS), pp 1–5. https://doi.org/10.1109/ICETAS52997.2021.9596921
14. Willcocks L, Lacity M, Craig A (2018) Robotic process automation at Xchanging (Paper 15-03). In: The outsourcing unit working research paper series; Davenport TH (2018) The rise of robotic process automation. Harv Bus Rev 96(4):104–114
15. Viale L, Zouari D (2020) Impact of digitalization on procurement: the case of robotic process. Supply Chain Forum: An Int J. https://doi.org/10.1080/16258312.2020.1776089
16. da Silva Costa DA, São Mamede H, da Silva MM (2022) Robotic Process Automation (RPA) adoption: a systematic literature review. Eng Manage Prod Serv 2:1–12
17. Impedovo A, Barracchia EP, Rizzo G (2023) Intelligent robotic process automation for supplier document management on e-procurement platforms. In: 2023, machine learning, optimization, and data science lecture notes in computer science, pp 156–166. https://doi.org/10.1007/978-3-031-25599-1_12
18. Hofmann P, Samp C, Urbach N (2020) Robotic process automation. Electron Markets 30(1):99–106. https://doi.org/10.1007/s12525-019-00365

19. Waiyanet P, Madonkha P (2022) A study on the implementation of robotic process automation (RPA) to decrease the time required for the documentation process: a case study of abc co., ltd. In: 2022 7th international conference on business and industrial research (ICBIR), pp 430–434. https://doi.org/10.1109/ICBIR54589.2022.9786521
20. Cabral TS (2018) Robotics and AI in the European union: opportunities and challenges. UNIO-EU Law J 4(2):135–146
21. Séguin S (2021) Minimizing the number of robots required for a robotic process automation (RPA) problem. Group for Research in Decision Analysis (GERAD)
22. Van den Bergh J, Mannaert H (2018) The impact of robotic process automation on the internal audit function. Int J Audit 22(3):388–404
23. Leshob A, Bourgouin A, Renard L (2018) Towards a process analysis approach to adopt robotic process automation. In: 2018 IEEE 15th international conference on e-business engineering (ICEBE), Xi'an, China, pp 46–53. https://doi.org/10.1109/ICEBE.2018.00018
24. Kedziora D, Penttinen E (2021) Governance models for robotic process automation: the case of Nordea Bank. J Inf Technol Teach Cases 11:2–61
25. Lacity M, Willcocks L, Craig A (2015) Robotic process automation: mature capabilities in the energy sector. In: The outsourcing unit working research paper series
26. Barrasa AR (2020) Robotic process automation in the financial industry: opportunities and challenges. In: 2020 15th Iberian conference on information systems and technologies (CISTI), pp 1–5. https://doi.org/10.23919/CISTI49556.2020.9140761

Unraveling COVID-19 Sentiments on Twitter: In-Depth Analysis with LSTM and BART

Fiza Khan, Fareen Ansari, Jeba Ansari, Janhvi Gupta, and Arushi Singh

Abstract The increasing ubiquity of Internet usage has made sentiment analysis an essential field of research in the rapidly developing field of natural language processing. This research makes a new contribution to sentiment analysis during the COVID-19 pandemic by looking at the emotional responses expressed in comments and tweets. Emotions are extracted from a large set of annotated tweets (March 23, 2020, to July 15, 2020) using content mining. A variety of methods, such as sentiment lexicon-based approaches, topic modeling, and keyword extraction, are used in content mining. Several content mining approaches are used to extract emotional feelings from the tweets after they have been pre-processed to remove noise and extraneous information. Bidirectional and Auto-Regressive Transformers (BART) and Long Short-Term Memory (LSTM) models increase the accuracy of emotion classification. Comparing models reveals COVID-19 Twitter emotions for informed outbreak control. Consistency is ensured by human annotators who classify tweets into emotion categories such as happiness, grief, fury, fear, disgust, and surprise. Annotation efficiency can be improved with automated sentiment tools.

Keywords Sentiment analysis · LSTM · BART · NLP

1 Introduction

In the current digital era, the Internet is a huge textual data source that includes thoughts, feelings, and opinions from a variety of sources, including news articles, social media, online forums, and customer evaluations. One important component of Natural Language Processing (NLP) that is devoted to extracting feelings from text data is sentiment analysis, sometimes referred to as opinion mining. This study makes use of linguistic approaches, deep learning architectures, and machine learning

F. Khan · F. Ansari · J. Ansari · J. Gupta · A. Singh (✉)
Department of Information Technology, KIET Group of Institutions, Delhi-NCR, UttarPradesh, Muradnagar, Ghaziabad 201206, India
e-mail: arushi.singh@kiet.edu

P. J. S. Gonçalves et al. (eds.), *Proceedings of Fifth International Conference on Computing, Communications, and Cyber-Security*, Lecture Notes in Networks and Systems 1128, https://doi.org/10.1007/978-981-97-7371-8_42

algorithms to identify the polarity—positive, negative, or neutral—of opinions found in the text. In the field of sentiment analysis and opinion mining (SAOM), written language (text) opinions and thoughts expressed by the public are automatically extracted from social media, blogs, reviews, news, and other online platforms. Its goal is to examine how people view various entities—such as goods, people, organizations, and services—based on their own experiences, viewpoints, feelings, and attitudes [1].

The COVID-19 pandemic has sparked significant interest in sentiment analysis of social media data, particularly on platforms like Twitter, due to its ability to capture public opinion, emotions, and responses. This study utilizes a dataset of 92,276 unique tweets obtained through the Twitter API. Despite the absence of sentiment labels, the analysis employs techniques such as topic modeling, sentiment lexicon-based analysis, and clustering to uncover underlying sentiment patterns. The dataset is characterized by its large volume, consistent hashtag (#covid19), lack of supervised labels, and real-time nature, reflecting diverse discourse including news updates, personal experiences, opinions, and discussions related to COVID-19. Leveraging this data provides valuable insights into evolving public sentiment and opinion dynamics surrounding the pandemic on social media, with opportunities for future refinement and integration of additional features to enhance analysis.

Models are now able to capture complex emotions, contextual variances, and cultural subtleties in a variety of textual datasets because of developments in machine learning and natural language processing. This paper plays a pivotal role in shaping business strategies, predicting market trends, enhancing user experiences, and fostering sentiment-aware applications. Sentiment Analysis, a cornerstone application, holds profound implications in marketing, customer feedback analysis, political opinion tracking, and social media monitoring. Its core aim is to computationally discern and quantify sentiments within textual data, enabling valuable insight extraction from unstructured text. The advent of sophisticated machine learning and deep learning architectures has revolutionized sentiment analysis, extracting nuanced sentiments and contextual understanding with unprecedented accuracy and efficiency. Long Short-Term Memory (LSTM) networks are particularly good at recognizing temporal connections and contextual information in textual data when used for sentiment analysis. Word-by-word processing of input sequences combined with built-in memory cells allows LSTMs to detect minute differences in sentiment, outperforming conventional methods. While using sophisticated natural language processing (NLP) methods, taking advantage of Bidirectional and Auto-Regressive Transformers' (BART) capabilities in particular. Pre-trained transformer-based model BART has shown impressive performance in a variety of natural language processing tasks, such as text generation and sequence-to-sequence modeling. Specifically, we use BART to classify sentiment on COVID-19 tweets in a zero-shot manner to identify complex emotions present in user-generated information. This study offers a thorough analysis of sentiment analysis, emphasizing its importance, approaches, uses, and difficulties. It explores the capabilities of two well-known approaches, LSTM networks and BART. An essential part of NLP, sentiment analysis shows itself to be a potent tool for gleaning information from textual input. The article highlights the use of sentiment analysis in decision-making through a discussion of various applications, including social media monitoring and business intelligence. The research helps to effectively utilize textual data for informed decision-making and actionable insights by expanding understanding and comparing techniques. The succinct synopsis recognizes the dynamic field of sentiment analysis research, with broad implications for higher education, business, and society.

2 Related Work

In 2023, N. Braig and A. Benz extended the research on sentiment analysis of COVID-19-related Twitter data, building upon the foundations outlined above. Their paper offers an updated viewpoint on the use of sentiment analysis to Twitter data during the epidemic, improving both the systematic literature evaluation and the investigation of particular areas of interest. The researchers delved into sophisticated algorithms, feature extraction techniques, and datasets, going beyond the realm of social and behavioral sciences. Braig and Benz improved the technique across four crucial stages: identification, screening, eligibility, and inclusion. They did this by adhering to the Preferred Reporting Items for Systematic Reviews and Meta-Analyses (PRISMA) criteria. A thorough evaluation of pertinent research was made easier by their thorough literature search, which was detailed over five databases. A more sophisticated and current knowledge of sentiment analysis applied to COVID-19-related twitter data is made possible by the ensuing improvements in technique and insights [2].

Amit Kumar Sharma and Gagandeep Kaur published a study in the Journal of Big Data in 2023 that described a hybrid feature extraction strategy-based deep learning model for consumer sentiment analysis. This article presents a fresh strategy that may be able to alleviate some of the shortcomings noted in previous studies. Significant limitations on model performance and dataset size have plagued earlier research. Restricted datasets, such as those mentioned in the table, have presented difficulties for sentiment analysis research, sometimes resulting in biased models and impeding generalizability. Furthermore, it can be difficult for models created using fewer datasets to fully represent the variety of language subtleties and contextual variables. By presenting a deep learning model that integrates sophisticated feature extraction methods with deep learning, the research by Kaur and Sharma improves this subject [6].

The study was suggested in 2023 by H. Rahman and J. Tariq that the sentiment analysis architecture makes use of a multi-tier method with three hierarchical models (M1, M2, and M3) that use SVM, Decision Trees, and Naive Bayes. The five sentiment classifications of positive, negative, neutral, very positive, and strongly negative text data are used to train these models. Using Python's NLTK package, stop word elimination and lemmatization are part of the pre-processing. After TF-IDF weighting, "TF-IDF vectorization" or "TF-IDF encoding" was carried out. The pre-processed data is divided into test, training, and development sets using a 10:70:20 ratio. The main objective is to improve sentiment analysis performance for machine learning models by analyzing succinct textual input. The literature survey's sentiment categorization is accurate, thanks to this approach [9].

Xu, Chang, and Jayne's systematic study highlights important shortcomings in earlier research on sentiment analysis based on social media. Interestingly, these research frequently use limited or specialized datasets, which compromise the robustness of the models in a variety of languages and contexts. These restrictions limit the real-world application by making it difficult to capture a range of verbal patterns and moods. Furthermore, a restricted concentration on performance indicators ignores useful consequences, such as trend identification or decision assistance. Furthermore, validity and reproducibility are hampered by reporting methodologies that lack sufficient transparency. In order to improve model generalizability and real-world applicability, future research should give priority to large, diverse datasets and transparent methodology in order to overcome these problems. By addressing these problems, the field will advance toward more significant solutions for sentiment analysis [11].

Due to their popularity and proven efficacy, SVM and Naive Bayes classifiers are frequently used in sentiment prediction and optimization procedures. On the other hand, hierarchical machine learning techniques perform mediocrely in classification settings. On the other hand, SVM and Multinomial Naive Bayes have demonstrated better accuracy and optimization capabilities. Deep convolutional and recursive neural network algorithms still struggle to capture word compositionality, despite the limited investigation of sentiment analysis utilizing neural network architectures. Convolutional neural networks are particularly good at extracting characteristics at the word or sentence level, such as stems and morphological tags; however, they have trouble extracting information at the character level for complicated words. Interestingly, J48, BFTree, and OneR classifiers have shown flexibility in a variety of classification applications, including textual emotion recognition and twitter text categorization, while not being widely used in sentiment prediction tasks [10]. Table 1a provides a comprehensive summary of various sentiment analysis studies focusing on COVID-19, presenting a diverse array of approaches and associated challenges. Each study endeavors to analyze sentiment regarding COVID-19, aiming to identify the most effective strategies while employing a range of methodologies with potential implications for future applications. The table offers an overview of research articles, serving as condensed summaries derived from different reviews. It serves as a valuable resource for individuals seeking insights into COVID-19 sentiment analysis and the complexities associated with it (Tables 1b, 1c).

3 Proposed Work

The research introduces two distinct sentiment analysis models tailored specifically for evaluating sentiments in COVID-19-related tweets. The first model employs a Deep Long Short-Term Memory (LSTM) neural network architecture, pre-trained on a meticulously curated dataset subjected to necessary pre-processing steps like

Table 1a Reviewed papers with comparable research themes (Part 1)

References	Journal/ conference	Year	Methods used	Future scope	Constraints
[4]	Procedia Computer Science	2016	Examines multiple sentiment analysis techniques, including rule-based, and lexicon-based methods with a focus on evaluating performance	Avenues for improving sentiment and semantic analysis with n-grams and utilizing social media reviews	Achieving precise review categorization due to language nuances and contextual complexities
[9]	Hum. Cent. Comput. Inf. Sci. 7	2017	Employed J48, BFTree, OneR, and Naive Bayes classifiers to optimize sentiment prediction with Weka	Deep neural networks can be employed with word embeddings and then extended to include convolutional neural networks	Restricted to extracting emoticons, foreign words, and extended words, and perform potential overlook of nuanced features
[10]	International Journal of Engineering and Technology	2017	Utilizes sentiment package, word cloud, and Bayes algorithm for sentiment and polarity analysis	Integration of big data techniques, exploration of advanced sentiment algorithms	Restricted analysis due to processing constraints, potential oversight of nuanced sentiments
[3]	Electronics	2020	Compares DNN, CNN, and RNN models using word embedding and TF-IDF	Explore hybrid approaches to enhance classification accuracy and reduce computational costs	Lack of comprehensive comparative analysis on computational time

tokenization, padding, and text cleaning. Optimization during training involves employing the Binary cross-entropy loss function and the Adam optimizer. Post-training evaluation measures the model's performance through average test loss and accuracy, facilitating the classification of attitudes in tweets as positive or negative.

The COVID-19 tweets dataset undergoes pre-processing steps such as normalization, stop word removal, and stemming/lemmatization to prepare it for analysis. Exploratory Data Analysis (EDA) involves generating word clouds, plotting geographical distributions, and analyzing tweet frequency per user. For unsupervised sentiment analysis, Word2Vec embeddings project tweets into a vector space,

Table 1b Reviewed papers with comparable research themes (Part 2)

References	Journal/ conference	Year	Methods used	Future scope	Constraints
[11]	International Journal of Scientific and Technology Research Volume 9	2020	Leverage ML, lexicon method, N-Gram, POS tagging, stemming for sentiment analysis accuracy boost	Explore diverse text data combinations to enhance prediction accuracy	Research beyond SVM and Naive Bayes to assess and enhance diverse sentiment analysis approaches
[8]	Social Network Analysis and Mining	2021	Features can be extracted using methods such as Bag of Words, N-grams, and TF-IDF, using confusion matrix	Improvements in deep learning architectures may increase the precision of sentiment and emotion recognition	Lexicon methods face nuances, deep learning demands extensive resources
[7]	Computers, Materials and Continua	2022	Multitier architecture for sentiment analysis using Naive Bayes, Decision Tree, and SVM algorithms	Negation handling, spam review detection, feature optimization, emoticons, POS tagging	Increased feature vector size, class imbalance, learning high-level textual features
[6]	Decision Analytics Journal	2022	Using the PRISMA framework outlines sentiment analysis objectives, identifies issues	Extend scope, global collaboration for diverse literature, tackle sentiment analysis challenges inclusively	Limited to recent studies, the exclusion of non-English publications may hinder understanding of sentiment analysis

followed by dimensionality reduction with PCA or t-SNE. DBSCAN clustering categorizes tweets based on density. Transfer learning is explored via XGBoost and self-attention networks. XGBoost trains gradient boosted decision trees on labeled datasets similar to COVID-19 tweets. Self-attention networks, using PyTorch, are trained with a data loader for inference on the COVID-19 dataset. Overall, preprocessing methods clean and structure text data, followed by EDA and modeling to analyze sentiment and explore transfer learning techniques (Fig. 1).

LSTM networks excel in sequential data processing, capturing long-term dependencies, and addressing the vanishing gradient problem, though they lack bidirectional context understanding. Conversely, Bidirectional and Auto-Regressive Transformers (BART) leverages pre-training on extensive corpora, auto-regressive

Table 1c Reviewed papers with comparable research themes (Part 3)

References	Journal/ conference	Year	Methods used	Future scope	Constraints
[2]	J Big Data 10	2023	Utilized hybrid feature vector, RRF, ARF to improve sentiment classification (LSTM)	Expanding consumer review summarization, evaluating with diverse datasets, exploring language-independent NLP techniques	Expanding consumer review summarization, evaluating with diverse datasets, exploring language-independent NLP techniques
[5]	IEEE access	2023	Sentiment analysis of Twitter data related to COVID-19 using PRISMA criteria	Social and behavioral science perspective to enhance understanding and application	Potential representation errors, English dataset bias in Twitter data may limit the accuracy

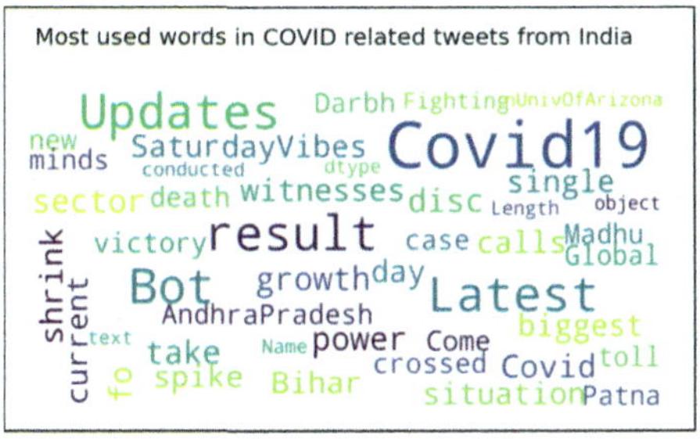

Fig. 1 Word cloud of Indian tweets

decoding during training, and bidirectional attention mechanisms, achieving state-of-the-art performance across various NLP tasks. BART outperforms LSTM in text summarization and machine translation due to its bidirectional context awareness and pre-training techniques. LSTM, however, is better suited for tasks with sequential input and offers interpretability.

Using pre-trained versions for accessibility, LSTM and BART both provide better sentiment analysis results on the dataset. A second model allows for zero-shot sentiment categorization by leveraging BART architecture. By pre-training on sentiment labels such as "happy" and "sad," this model produces outputs for sentiment analysis that are sophisticated and include confidence scores for different sentiment labels. Users can submit unique tweets or use sample ones to assess the models; the results are displayed in an easy-to-read manner. The BART model offers several sentiment labels and confidence ratings, enabling thorough analysis of sentiment in COVID-19-related tweets, whereas the LSTM model classifies sentiment as just positive

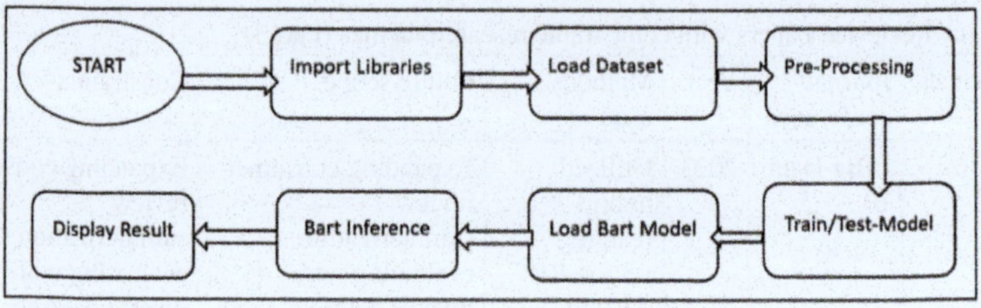

Fig. 2 Flowchart of applying BART and LSTM

or negative. This all-inclusive strategy supports an in-depth analysis of sentiment dynamics in COVID-19 conversations.

Figure 2 shows a sentiment classification process that makes use of BART and LSTM models for transfer learning. First, the code imports certain necessary libraries. Next, a pre-trained tweet dataset is ingested. The definition of pre-processing and inference modules—which include tokenization, padding, and dataset construction for data preparation—occurs in subsequent stages. After that, the LSTM model is defined and put through training and testing. The main program loop controls how the BART model is loaded and how inference is carried out. The final stage entails presenting the sentiment classification results.

4 Results and Discussion

4.1 Model Training

In Fig. 3, the model received 19,600 steps in total over its 8 epochs of training. The training loss consistently decreased, reaching extremely low values close to zero. The validation loss, however, showed fluctuations and did not follow a clear decreasing trend, peaking at 1.234672. Overfitting is evident in the model, since there is a large difference between the validation and training losses. Variations in validation loss point to training instability, which might be brought on by insufficient regularization or a rapid learning rate. Concerns regarding noise fitting and memorizing are raised by the almost minimal training loss. It is suggested that regularization be included (dropout, for example) and that learning concerns regarding dataset analysis for anomalies be adjusted. For improved performance, more research into sophisticated designs or hyperparameter fine-tuning is suggested.

Epoch	Step	Train Loss	Validation Loss
1	200	0.052577	0.081068
1	400	0.051915	0.046640
1	600	0.067061	0.061379
1	800	0.539269	0.763887
1	1000	0.020809	0.022876
1	1200	0.733645	0.086998
1	1400	0.051352	0.053301
1	1600	0.058821	0.085194
1	1800	0.073794	0.138465
...	...	...	...
8	19200	0.000000	0.000003
8	19400	0.000004	0.000003
8	19600	0.000000	0.000002

Fig. 3 Model training

```
Average Test Loss: 0.9906
Average Test Accuracy: 0.8576
```

Fig. 4 Model performance

4.2 Model Performance

As shown in Fig. 4, the model demonstrates a decent performance with an average test loss of 0.9986 and an accuracy of 0.8576. Additional insights could be gained through a more detailed examination of its performance on specific classes or by comparing it against alternative models.

The program allows users to select an option by entering a corresponding number. Based on their choice, the program prompts the user to input custom text or displays sentiment analysis results for sample tweets. After processing the input, the model presents sentiment predictions with confidence scores, offering valuable insights into the conveyed sentiment.

4.3 Result on Applying LSTM

Sentiment analysis of COVID-19-related tweets is done accurately by the model. For example, it accurately detects the negative emotion in a tweet regarding job losses and the risk posed by COVID-19.

In Fig. 5, adeptly discerns an affirmative attitude in an additional tweet that conveys contentment and familial security throughout challenging circumstances. The model is skilled at sentiment analysis. To sum up, the neural sentiment analysis model

```
Original input sentence: Many lost their jobs because of covid and it is highly dangerous

Pre-processed input sentence: ['many', 'lost', 'jobs', 'covid', 'highly', 'dangerous']

Sentiment: Negative

----------------------------------------------------------------------------------------

Original input sentence: I am happy that my family members are safe in this tough times

Pre-processed input sentence: ['happy', 'family', 'members', 'safe', 'tough', 'times']

Sentiment: Positive
```

Fig. 5 Result on applying LSTM

distinguishes between positive and negative opinions in COVID-19-related information by efficiently interpreting and assessing sentiments in textual data. It exhibits the capacity to identify both happy and unhappy feelings in such material.

5 Result on Applying BART

BART's sentiment analysis of COVID-19 tweets reveals nuanced emotion probabilities. In the negative tweet on job loss and danger, "warn" dominates at 51.29%, followed by "alert" at 26.02%. Conversely, the positive tweet on family safety is predominantly labeled "happy" at 75.24%. These probability distributions offer nuanced insights, enriching our understanding of sentiments beyond simple binary classifications. The program concludes as the user opts to exit, completing the comprehensive analysis of COVID-19 sentiments through BART (Fig. 6).

```
Entered input sentence: Many lost their jobs because of covid and it is highly dangerous

Sentiment of the tweet (Probability Distribution):
Labels Confidence Scores
   warn        51.290 %
  alert        26.023 %
    sad        12.588 %
 sorrow        4.8833 %
  angry        3.1200 %
  grief        1.2809 %
neutral        0.5886 %
  happy        0.2237 %
```

Fig. 6 Result on applying BART

6 Conclusion

This research study presents a unique method for sentiment analysis of COVID-19 tweets by utilizing cutting-edge techniques like BART and LSTM. Exhibiting an impressive average test performance of 87.7%, the LSTM model was trained only using COVID-19 tweet data. On the other hand, BART, a transformer model that has been trained beforehand, shows potential in zero-shot classification since it can correctly forecast the sentiment probability distributions of input words. We hope to improve the accuracy and depth of sentiment categorization by capturing the subtleties seen in social media conversation connected to COVID-19 by including LSTM and BART into our analytic pipeline. Governments, businesses, and public health organizations may get important information from real-time sentiment analysis of COVID-19 tweets to help them make educated decisions, modify their communication plans, evaluate the success of their policies, and comprehend the views of their constituents throughout the pandemic.

Acknowledgements We appreciate all of the support we received in concluding this project, especially from our guide, Ms. Arushi Singh, and the Department of Information Technology of the KIET Group of Institutions, Ghaziabad.

References

1. Ainapure BS, Pise RN, Reddy P, Appasani B, Srinivasulu A, Khan MS, Bizon N (2023) Sentiment analysis of covid-19 tweets using deep learning and lexicon based approaches. Sustainability 15(3)
2. Braig N, Benz A, Voth S, Breitenbach J, Buettner R (2023) Machine learning techniques for sentiment analysis of covid-19-related twitter data. IEEE Access 11:14778–14803
3. Dang C, Moreno-García MN, De La Prieta F (2020) Sentiment analysis based on deep learning: a comparative study. Electronics 9:483
4. Devika MD, Sunitha C, Ganesh A (2016) Sentiment analysis: a comparative study on different approaches. Procedia Comput Sci 87:44–49; In: Fourth international conference on recent trends in computer science engineering (ICRTCSE 2016)
5. Fang X, Zhan JZ (2015) Sentiment analysis using product review data. J Big Data
6. Kaur G, Sharma AK (2023) A deep learning-based model using hybrid feature extraction approach for consumer sentiment analysis. J Big Data
7. Kawade D, Oza K (2017) Sentiment analysis: machine learning approach. Int J Eng Technol 9:2183–2186
8. Pandya S, Mehta P (2020) A review on sentiment analysis methodologies, practices and applications
9. Rahman H, Tariq J, Masood A, Subahi A, Khalaf O, Alotaibi Y (2022) Multitier sentiment analysis of social media text using supervised machine learning. Comput Mater Continua 74:5527–5543
10. Singh J, Singh G, Singh R (2017) Optimization of sentiment analysis using machine learning classifiers. HCIS 7:12
11. Xu QA, Chang V, Jayne C (2022) A systematic review of social media-based sentiment analysis: emerging trends and challenges. Decis Anal J 3:100073

IoT in Health Care: Architectural Explorations, Pandemic Perspectives, and Future Horizon

Oroos Arshi, Aryan Chaudhary, Gunjan Chhabra, Keshav Kaushik, Rashmi Singh, and Neha Bhardwaj

Abstract The integration of IoT in health care is a game changer, transforming patient care, clinical workflows, and health management. Its significance stems from the ability to monitor patient vitals in real time, share data seamlessly, and make informed decisions. This work is an attempt to investigate the integration of Internet of Things (IoT) technologies within healthcare systems. It offers insights into its architectural underpinnings, pandemic applications, and future horizons. It explores the essential elements and system architectures specific to healthcare IoT, followed by an examination of current applications and their significance amidst the COVID-19 pandemic. Through a comprehensive review, this paper analyzes the technological landscape of IoT in health care during the pandemic, highlighting key innovations and their implications for patient care. It also provides a case study, including Philips Healthcare's Care Coordinator, to provide practical insights into IoT-enabled healthcare solutions and their effectiveness. However, the paper also highlights challenges such as privacy concerns and interoperability issues that hinder the widespread adoption of IoT in health care. By providing a multidimensional perspective, this paper contributes to a deeper understanding of IoT's role in reshaping healthcare

O. Arshi
School of Computer Science, University of Petroleum and Energy Studies, Dehradun, Uttarakhand, India

A. Chaudhary (✉)
BioTech Sphere Research, Kolkata, India
e-mail: aryanchaudhary500@gmail.com

G. Chhabra
Department of CSE, Graphic Era Hill University Dehradun, Dehradun, India

Graphic Era Deemed to Be University, Dehradun, Uttarakhand, India

K. Kaushik
Amity School of Engineering and Technology, Amity University, Punjab, India

R. Singh
Amity Institute of Applied Sciences, Amity University, Noida, Uttar Pradesh, India

N. Bhardwaj
Sharda School of Basic Sciences and Research, Sharda University, Greater Noida, India

P. J. S. Gonçalves et al. (eds.), *Proceedings of Fifth International Conference on Computing, Communications, and Cyber-Security*, Lecture Notes in Networks and Systems 1128, https://doi.org/10.1007/978-981-97-7371-8_43

delivery, emphasizing its potential to address critical challenges and shape the future of healthcare systems.

Keywords Internet of Things · IoT technology · Health Care · COVID-19 pandemic · Remote patient monitoring

1 Introduction

The healthcare sector has undergone a transformative shift with the advent of Internet of Things (IoT) technology, which has been integrated into medical practices since the early 2000s [1]. This innovation facilitates remote patient monitoring, thereby reducing the necessity for in-person visits and enhancing patient outcomes. However, the healthcare industry confronts formidable challenges, including escalating costs, the prevalence of chronic diseases, and aging populations [2]. Harnessing IoT for healthcare app development offers a promising solution to these challenges by potentially increasing access to affordable medical services, curbing hospital expenses, and enabling convenient medical examinations in patients' homes. Such advancements not only enhance the effectiveness of healthcare practitioners but also elevate the quality of patient care, particularly benefiting elderly patients [3]. In this paper, we delve into the intricacies of IoT system architecture, identify research challenges, and explore the potential benefits of IoT-enabled health care, particularly amidst the COVID-19 pandemic.

The imperative for IoT-enabled health care has become even more pronounced in the context of the COVID-19 crisis. With stringent social distancing measures and restrictions on non-essential travel in place, IoT-enabled remote patient monitoring has emerged as a crucial tool for both containing the spread of the virus and ensuring uninterrupted patient care [4]. Furthermore, IoT technologies facilitate the swift deployment of telemedicine solutions and remote consultations, empowering healthcare professionals to adapt swiftly to the dynamic pandemic landscape. By enabling real-time monitoring and analysis of patient's health statuses, IoT-enabled devices equip medical practitioners with invaluable insights, enabling them to detect issues promptly and intervene as necessary. However, previous research in this domain has exhibited certain shortcomings, including a lack of comprehensive exploration of IoT system architectures, insufficient attention to the unique challenges posed by the COVID-19 pandemic [5], and limited analysis of the potential benefits for healthcare practitioners and patients alike. Our study seeks to address these gaps by offering a detailed examination of IoT system layouts, delineating key research challenges, and elucidating the multifaceted benefits of IoT-enabled health care, particularly in the context of the ongoing pandemic. Through rigorous analysis and empirical evidence, we aim to underscore the transformative potential of IoT technology in revolutionizing healthcare delivery and improving patient outcomes. IoT integration has the potential to completely transform the healthcare sector by resolving major issues and enhancing patient care. Healthcare organizations can improve patient outcomes, maximize resource utilization, and improve access to medical services by utilizing Internet of Things applications, especially for elderly patients who need specialized care. The adoption of IoT-enabled technologies will surely play a crucial part in determining how health care is delivered in the future as the industry continues to develop. This paper provides a comprehensive analysis of IoT in health care, focusing on architectural exploration, current applications, and pandemic adaptations. Unlike existing studies, it delves deeply into the specific architecture required for healthcare settings and examines how IoT technologies are being utilized during the COVID-19 pandemic. It also discusses the practical applications of IoT in health care and the challenges associated with its implementation. Additionally, this paper explores the technological landscape and offers insights into future horizons. While its analysis is extensive, we acknowledge potential biases and the evolving nature of both IoT technologies and healthcare practices.

In the survey, the relevance of individual research categories is essential for understanding the diverse landscape of IoT applications in health care and identifying areas of focus for future research and development. By delineating specific research categories such as sensor technologies, communication protocols, data analytics, and security measures, the survey aims to provide a comprehensive overview of the multifaceted challenges and opportunities in healthcare IoT. Each research category addresses distinct aspects of IoT deployment, from data capture and transmission to storage, analysis, and security. Understanding the relevance of these categories allows researchers to prioritize efforts, allocate resources effectively, and drive innovation in key areas to address critical healthcare needs and enhance patient outcomes.

2 Literature Survey

There have been several IoT health tracking system solutions recently modeled. Damian et al. [1] developed indoor and outdoor information systems that utilize the Internet of Things, or IoT for short. The developed methodology made use of three-axis accelerometers which are magnetometers, thresholding, and PDR, which are decision trees. A design methodology (DM), which analyzes the design purpose from the points of view of the contracting the government, stakeholders, and possible users, is also necessary. Jie Wang et al. [2] applied Cloud WISE and BASN technology. WISE cloud makes it simpler to store and retrieve cloud data, while wearables help patients keep track of their health. In Baker et al. [3], healthcare IoT systems that can be used in typical situations are discussed. One of the most significant developments in technology is wearable technology. In the healthcare industry, technology like cloud storage is the greatest method for maintaining data security. Islam et al. [4] present the most recent network structures, applications, and market trends in IoT-enabled healthcare solutions. In addition, the study looked at several IoT security and privacy features, including security criteria, attack taxonomies, and threat models. Nazir et al. [5] remove obstacles related to time, place, and other factors, while also expanding its reach and efficacy, to improve healthcare delivery. The true nature of the Internet of Things (IoT) health revolution may allow customers to receive exceptional care at affordable costs.

3 Essential Elements of Healthcare IoT Architecture

Healthcare IoT architecture comprises several essential elements that collectively enable the seamless integration of IoT devices, data communication, storage, processing, and analysis. These elements play a crucial role in facilitating remote patient monitoring, real-time data analytics, and proactive healthcare delivery [4].

Below are the key components of a healthcare IoT architecture:

(a) Sensor Devices

Sensor devices are at the core of healthcare IoT architecture, capturing real-time data from patients, medical equipment, and environmental factors. These devices can include wearable sensors, medical monitoring devices, environmental sensors, and smart medical devices embedded with sensors for measuring vital signs, physiological parameters, and environmental conditions.

(b) Communication Protocols

Communication protocols govern the transmission of data between IoT devices, gateways, and backend systems.

Common communication protocols used in healthcare IoT include Bluetooth Low Energy (BLE), Zigbee, Wi-Fi, cellular (3G/4G/5G), and Message Queuing Telemetry Transport (MQTT). These protocols ensure reliable, secure, and efficient data transmission, with considerations for low power consumption, bandwidth efficiency, and interoperability [6].

(c) Data Storage and Processing Systems

Data storage and processing systems manage the storage, retrieval, and analysis of IoT-generated data. Edge computing platforms enable real-time data processing and analytics at the network edge, closer to where data is generated, reducing latency and bandwidth usage.

Cloud infrastructure provides scalable and secure storage solutions, enabling centralized data storage, advanced analytics, and machine learning algorithms for predictive analytics and decision support [7].

(d) Edge Computing

Edge computing refers to the processing and analysis of data at or near the source of data generation, rather than in centralized data centers.

Edge computing enables real-time data analytics, decision-making, and response, making it ideal for applications requiring low latency, such as remote patient monitoring and critical care. Edge computing platforms may include edge servers, gateways, and edge devices equipped with processing capabilities and local storage.

(e) Cloud Infrastructure

Cloud infrastructure provides scalable and on-demand computing resources, storage, and services for IoT applications. Cloud platforms offer secure data storage, backup, and disaster recovery solutions, ensuring data integrity and availability. Cloud-based analytics and machine learning tools enable advanced data processing, predictive analytics, and insight generation from large volumes of IoT data [8].

In summary, the essential elements of healthcare IoT architecture encompass sensor devices for data capture, communication protocols for data transmission, data storage and processing systems for data management and analysis, edge computing for real-time processing, and cloud infrastructure for scalable storage and advanced analytics. These elements work together to enable innovative IoT solutions for remote patient monitoring, personalized healthcare delivery, and improved clinical outcomes.

4 Essential Elements of Healthcare IoT Architecture

The Internet of Things (IoT) is a structured design that consists of sensors gathering environmental data, sending it via IoT gateways, and maintaining data on cloud servers. These components work together to process data, providing equipment intelligence and enabling informed decision-making. The IoT's revolutionary potential is demonstrated through its comprehensive architecture, ensuring a connected and intelligent world. Figure 1 illustrates the architecture of the Internet of Things.

(a) Sensors

The Internet of Things (IoT) is a network of sensors that collect and transmit data related to various aspects of the environment, including temperature, weather, and the human body, via communication devices.

Fig. 1 Internet of things architecture

(b) IoT Gateways

Using this, we can connect to the devices or objects we require for online interaction. The central server and the gadgets can communicate with each other through gateways. The gateway collects information gathered by the sensors and the infrastructure, which is often a cloud server, receives it after that [6].

(c) Cloud Server

Data is stored and managed on both on-premises and cloud-based servers, providing machinery intelligence through precise procedures and diagnostics on cloud servers before being delivered to users' devices [6].

5 Essential Elements of Healthcare IoT Architecture

In health care, the Internet of Things (IoT) emerges as a revolutionary force, transforming the environment via a dynamic system architecture depicted in Fig. 2. The healthcare industry is revolutionizing patient care and monitoring with a sophisticated IoT architecture. This system uses hardware and software components to collect sensor signals, display them on devices, and utilize connectivity technologies like Bluetooth and Wi-Fi for easy data collection.

(a) IoT Product Infrastructure

It is comprised of the hardware or software components that collect sensor signals and display them on a particular device. A heart rate monitor or a component of IoT software [6].

(b) Sensors

The Internet of Things (IoT) medical industry uses a range of sensor devices to ascertain the patient's current condition, including arterial blood pressure detection devices, temperature sensors, pulse-oximeter sensors, and EKG sensors.

Fig. 2 System architecture of IoT system

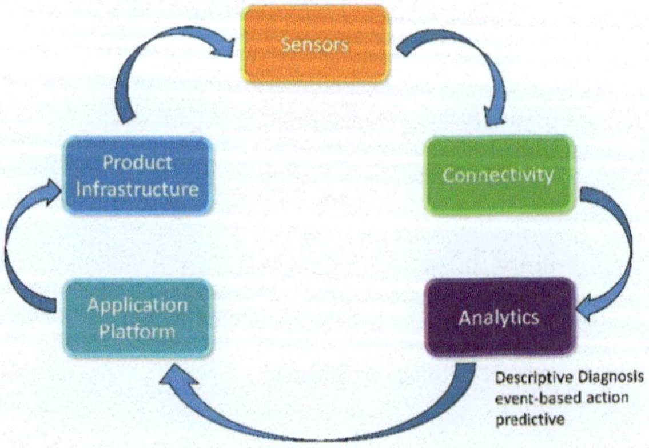

(c) Connectivity

By improving connections (using Internet connections like Bluetooth, Wi-Fi, etc.), IoT systems make it easier to read data from devices or sensors [6].

(d) Analytics

The patient's health is enhanced as a result of the healthcare system's analysis of the data, which is done by examining data from sensors and associates to ascertain the patient's health indicators. The descriptive, diagnostic, event-based action and predictive phases of analytics are the four phases.

(e) Application Platform

Mobile applications facilitate the transfer of specialized health data between patient devices and medical personnel [6, 7].

6 Application of IoT in Health Care

Since IoT has a wide range of applications across many industries, its growth is exciting for everyone. There are many uses in health care. Applications for the Internet of Things in health care include the following, which are notable:

- reducing the wait time in emergency rooms
- keeping tabs on hospital employees, inventory, and patients
- improved medication management
- ensuring that essential hardware is available.

IoT has also brought forth many wearables and technologies that have improved the quality of life for patients. These gadgets are listed below in Fig. 3.

(a) Hearables: Hearables are cutting-edge hearing aids that have dramatically changed how persons with hearing problems interact with the outside world. Modern wearables can connect to your smartphone via Bluetooth, making this possible. It is possible to apply extra layers of equalization, and filtration effects to real-world sounds. Doppler Labs serves perhaps the best illustration [8].
(b) Ingestible Sensors: The miracle of modern science is ingestible sensors. The sensors, which are roughly the size of a pill, monitor the medicine in our bodies and notify us if they detect any deviations. These sensors could save the lives of diabetics by reducing their symptoms and alerting them to impending serious health issues. Proteus Digital Health is an example of one such company.
(c) Moodables: Thync and Halo Neurosciences are working on moldable technology, head-mounted gadgets that enhance mood by delivering weak brain currents. Despite appearing like science fiction, this technology is close to reality, providing a way to feel wonderful all day. Their progress is impressive.

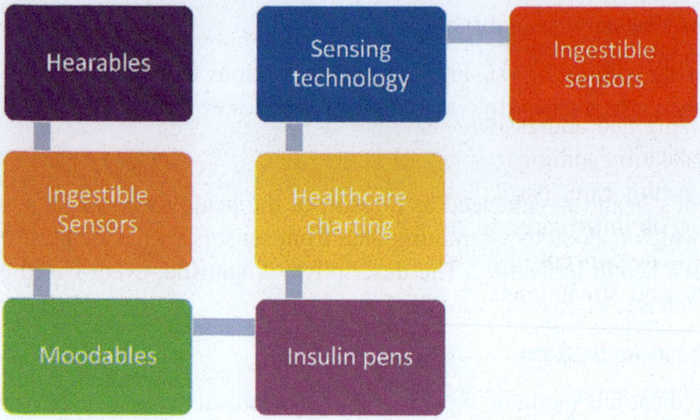

Fig. 3 Wearable technologies using Internet of Things

(d) Sensing technology: The development of computer vision technology has made it now possible for AI to simulate the experience of sight and the inferences made from it. Computer vision is used by drones like Skydio to recognize and avoid objects. Thanks to this technology, those with vision problems may now navigate [8].
(e) Healthcare charting: IoT gadgets like Audemix greatly cut down on the amount of manual work a doctor must do when documenting patients. Data about a patient is captured by voice activation. Reviewing the patient's data is made simple, thanks to this. By around 15 h per week, it lessens the workload for doctors.
(f) Insulin pens: These devices are used for real-time blood glucose monitoring and data sharing via a customized mobile app. With the aid of these devices, diabetic patients may keep an eye on the amount of glucose in their blood and even email their findings to their doctor and other qualified medical experts.
(g) Ingestible sensors: Medical sensors help patients take their prescribed medications by using a tiny, ingestible medical sensor that delivers a faint signal to a patient's portable receiver, which subsequently transmits information to a specialized smartphone app.

7 IoT-Enabled Health Care During the COVID-19 Pandemic

The Internet of Things (IoT) has significantly impacted health care, provided personalized care, and reduced wait times. During the COVID-19 pandemic, IoT technology enabled the use of various devices for monitoring patients, enhancing inpatient treatment, and improving patient comfort. IoT has been used in imaging, healthcare operations, pharmaceutical dispensing, patient tracking, laboratory tests, and medication administration. It has also enabled data analysis to find solutions to problems, enabling better patient care and informed decision-making. IoT has also helped in recognizing and addressing potential health issues, such as asthma attacks and timely medication administration [9]. The Internet of Things (IoT) has significantly impacted health care, particularly in managing COVID-19 patients. IoT devices monitor and optimize care, store and analyze patient data, and contribute to effective treatments. The concept of a Smart Hospital offers personalized services, reduced wait times, and streamlined administrative processes. IoT also monitors hygiene conditions, predicts patient arrivals, and provides real-time alerts. IoT is essential to healthcare crisis management, as seen in Fig. 4, especially in the context of the COVID-19 pandemic.

IoT has revolutionized numerous aspects of patient care and public health response during the COVID-19 epidemic, playing a crucial role in healthcare crisis management. IoT technologies minimize physical contact and lower the risk of viral transmission by enabling healthcare personnel to monitor vital signs and symptoms remotely through remote patient monitoring. Furthermore, IoT makes early epidemic detection easier by tracking environmental factors and facilitating real-time data processing. IoT technologies improve resource management by optimizing the distribution of hospital beds, ventilators, and medical supplies according to demand. IoT-enabled technologies, which track people's movements and aid in enforcing social separation measures, support contact tracing activities.

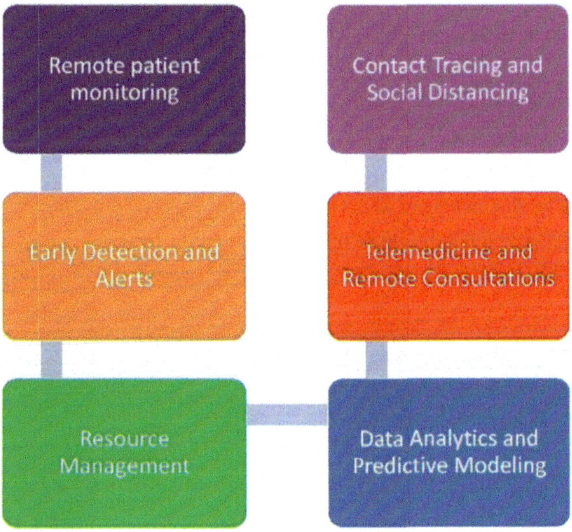

Fig. 4 IoT essential healthcare crisis management during COVID-19 pandemic

8 Challenges of IoT in Health Care

IoT implementation in health care comes with a number of considerations that must be carefully considered. First and foremost, considering the sensitive nature of patient data and the susceptibility of IoT devices to cyberattacks, it is imperative to have strong data security and privacy protections [10]. Second, interoperability issues arise when combining several devices with different communication protocols, impeding smooth data aggregation and sharing. Third, to extract valuable insights and guarantee precision and applicability for medical decision-making, handling the enormous volumes of data produced by IoT devices necessitates the use of advanced analytics tools and decision support systems. To overcome these obstacles, stakeholders must work together to create standardized procedures, improve data security, and successfully manage regulatory obligations. In the end, this will maximize the advantages of IoT in enhancing patient outcomes and healthcare delivery [6].

The following are some of the difficulties in implementing IoT in health care as shown in Fig. 5.

(a) Data security and privacy: The Internet of Things (IoT) poses significant challenges to data security and privacy in the healthcare sector [28]. Many IoT devices lack data security laws, making them vulnerable to hackers who can steal patient information, fabricate identification documents, and use patient names for false insurance claims.
(b) Integration of multiple devices and protocol: The integration of multiple devices in the healthcare sector is hindered by disagreements between manufacturers over communication standards and protocols, causing data aggregation and scalability challenges of the Internet of Things (IoT) [29].
(c) Accuracy and a wealth of Information: Data aggregation is challenging due to multiple communication protocols and requirements. IoT gadgets collect

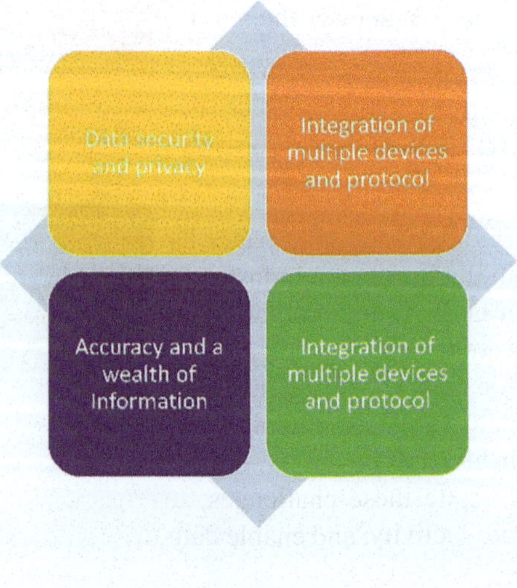

Fig. 5 Challenges in implementing IoT in health care

vast amounts of data, providing valuable insights but making medical decisions harder, raising concerns about patients' safety. This issue is growing as more devices are connected [7].

(d) Integration of multiple devices and protocol: The integration of multiple devices in the healthcare sector is hindered by disagreements between manufacturers over communication standards and protocols, causing data aggregation and scalability challenges of the Internet of Things (IoT) [6].

9 Case Study: Philips Healthcare's Care Coordinator

Philips Healthcare's eCareCoordinator is a remote patient monitoring (RPM) platform designed to improve the management of patients with chronic conditions such as congestive heart failure (CHF) and chronic obstructive pulmonary disease (COPD). By leveraging IoT technologies, eCareCoordinator enables healthcare providers to remotely monitor patients' vital signs and symptoms in real time, facilitating proactive interventions to prevent exacerbations and hospitalizations.

The eCareCoordinator platform consists of wearable sensors, home monitoring devices, and a secure cloud-based platform for data aggregation and analysis. Patients with chronic conditions are equipped with wearable devices that continuously monitor vital signs such as heart rate, blood pressure, and oxygen saturation levels. These devices transmit data wirelessly to the eCareCoordinator platform, where it is processed and analyzed in real time [9].

Healthcare providers access the eCareCoordinator platform via a secure web portal or mobile application, allowing them to monitor patients' health status remotely and intervene promptly when necessary. The platform uses advanced analytics algorithms to detect early warning signs of deterioration and alerts healthcare providers to take appropriate action. Additionally, eCareCoordinator provides patients with educational resources, medication reminders, and self-management tools to empower them to take an active role in their health care.

10 Conclusion and Future Directions

In conclusion, this paper has provided an in-depth exploration of IoT technology in the healthcare industry, addressing both its potential benefits and the challenges it faces. From facilitating remote patient monitoring to enabling telemedicine innovation and supporting population health management, IoT holds immense promise for revolutionizing healthcare delivery. However, challenges such as data security and regulatory compliance must be addressed through collaborative efforts among healthcare providers, device manufacturers, regulators, and technology specialists. Despite these challenges, IoT has the potential to improve patient care, enhance productivity, and enable data-driven decision-making in health care. Looking ahead, future research and development efforts should focus on further integrating IoT into healthcare systems, refining data analytics capabilities, and ensuring interoperability and scalability of IoT solutions. By leveraging the transformative power of IoT, the healthcare industry can move toward patient-centered, data-driven, and technologically enabled care delivery paradigms, ultimately leading to better patient outcomes and a healthier, more connected world.

References

1. Dziak D, Jachimczyk B, Kulesza WJ (2017) IoT-based information system for healthcare application: design methodology approach. Appl Sci 7(6):596
2. Wan J et al. (2018) Wearable IoT enabled real-time health monitoring system/ EURASIP J Wirel Commun Networking 1:1–10
3. Baker SB, Xiang W, Atkinson I (2017) Internet of things for smart healthcare: technologies, challenges, and opportunities. IEEE Access 5:26521–26544
4. Islam MdM, Rahaman A, Islam MdR (2020) Development of smart healthcare monitoring system in IoT environment. SN Comput Sci 1:1–11
5. Nazir S et al. (2019) Internet of things for healthcare using effects of mobile computing: a systematic literature review. Wirel Commun Mobile Comput
6. Joshi K, Kumar A, Anandaram H, Kumar A, Aggarwal A, Chhabra G (2023) Incorporating AI and IoT for smart poultry monitoring and health detection. 2023 Second international conference on augmented intelligence and sustainable systems (ICAISS), Trichy, India, pp 76–81, https://doi.org/10.1109/ICAISS58487.2023.10250564
7. Sharma, KK, Pahuja V, Chhabra G (2023) Employing convolutional neural network for IoT healthcare attack detection in intensive care unit. 2023 6th international conference on contemporary computing and informatics (IC3I), Gautam Buddha Nagar, India, pp 798–803, https://doi.org/10.1109/IC3I59117.2023.10398040
8. Vimal V et al. (2022) Enhance software-defined network security with IoT for strengthen the encryption of information access control. Comput Intell Neurosci, art. no. 4437507
9. Tanwar S, Wierzchon ST, Singh PK, Ganzha M, Epiphaniou G (Eds) (2023) Proceedings of fourth international conference on computing, communications, and cyber-security: IC4S 2022. Springer
10. Valsalan P, Baomar TAB, Baabood AHO (2020) IoT based health monitoring system. J Critical Rev 7(4):739–743

Constructing a Hierarchical Model Based on the Blockchain in Health Care

Ravendra Singh, Anupam Singh, and Raj Pal Singh

Abstract Building a hierarchical conceptual structure for smart health care, identifying the effects of distributed ledger technology on clever medical treatment, and then developing a system of applications for smart health care via blockchain that utilizes the theory of involvement of stakeholders are the goals of the research. However, besides the essential characteristics and the connections between different features and characteristics, such a hierarchy conceptual framework should also consider the involvement of several stakeholders. The study offers decision-making including analytical lab (DEMATEL) to control the intricate relationships between various features and attributes. Finally, a hierarchical conceptual framework is created by splitting the hierarchy employing Interpretive Structural Modeling (ISM). The results show that the top-level organization, medical records leadership, and general doctor management are among the system's core reasons. A specific application of blockchain in the field of healthcare automation requires a decentralized agreement that depends on how health information is handled within system constraints. The secret to updating a system is application optimization. Environmental control, health insurance, and internal and external regulation all contribute to the system's development and the successful protection of stakeholder interests. The financing layer, data layer, as well as stakeholder layer must form the basis of a blockchain-based application structure for smart health care.

R. Singh · R. P. Singh (✉)
Department of Computer Science and Engineering, Pranveer Singh Institute of Technology, Kanpur, India
e-mail: raj.1august@gmail.com

A. Singh
Department of Computer Science and Engineering, Graphic Era Hill University, Dehradun, Uttarakhand, India

Department of Computer Science and Engineering, Graphic Era Deemed to be University, Dehradun, Uttarakhand, India

A. Singh
e-mail: anupamsing.cse@geu.ac.in

P. J. S. Gonçalves et al. (eds.), *Proceedings of Fifth International Conference on Computing, Communications, and Cyber-Security*, Lecture Notes in Networks and Systems 1128, https://doi.org/10.1007/978-981-97-7371-8_44

Keywords Blockchain-based application · Blockchain · Internet of Things (IoT) · Decentralized · Security protocol · E-healthcare

1 Introduction

The advancement of global healthcare innovation and the intelligent healthcare system have been aided by the emergence of new technologies like big data and the Web of Everything. Innovative health care is an innovative healthcare framework that includes electronic medical records, care archives, as well as medical Internet of Things as its core components to create medical and healthcare services with the best management. Even though the field of smart health care is now developing rapidly, problems with network security and data security still exist. As the fourth industrial revolution's primary technology, blockchain is characterized by decentralization, secrecy, tamper obstruction, and transparency [1]. The potential for enhancing customer-focused smart healthcare organizations and addressing long-standing industry pain points with safety of data, privacy protection, and information sharing exists with the integration of blockchain-based technologies with smart health care. It might also establish a chain of multiparty wellness alliances comprising corporations, governments, and private citizens to further the commercialization of smart health care. Right now, the entire industry is interested in blockchain. Basic blockchain techniques from the bottom up [2, 3], managing bitcoin keys [4, 5], analyzing blockchain signings over the long time for authentication [6], and other findings are also part of current study. However, blockchain-related studies regarding the field of cognitive health care is still in their early stages. Despite the substantial research that has been done on the use of blockchain all through the field of connected medical care in the preceding literature, the establishment of a system for imaginative health care according to blockchain technology remains unclear and lacks a systematic examination. Our research's conceptual framework and use system may offer the smart healthcare sector useful direction on how to apply blockchain technology, in addition to giving academics revolutionary theoretical sources related to the development powered by blockchain intelligent health care.

2 Literature Review

To ensure that the medical industry gradually accumulates data, intelligent medicine realizes the relationship between patients, medical staff, healthcare facilities, and medical devices. This is done by creating a health archive close to a medical data platform and utilizing the most cutting-edge Internet of Things technological advances. Communicating medical data is crucial to raising the bar of care and improving the medical system's cognitive abilities [7]. Since the sharing of patient data between institutions is still an ongoing project, the blockchain provides the perfect current remedy for this problem [8]. New technology called blockchain makes it possible for vast networks of outsiders to exchange transactional and decentralized data. A decentralized network of data with many independent nodes is referred to as a blockchain [9]. Its features include timestamps, programmability, decentralization, and tamper-proofing [10]. The primary objective of the ongoing study is to develop an independent computing platform by fusing the implementation of virtual currency with information technology breakthroughs. For example, etheric blockchain systems can be used to create a smartphone app called Medical Records, which combines huge quantities of data with a shared ledger to allow users to collaborate with health-related data. Blockchain technology can also be used in tandem with OPAL to create a verification procedure for medical transactions [11]. However, this is a dearth of thorough study on the widespread application of blockchain relies innovations in the arena of cognitive medicine.

The project will establish a stakeholder-based, theory-based, customer-centered cryptocurrency intelligent medical application system to track its development path. We develop an application design based on ten characteristics and 22 criteria after classifying and evaluating each system stakeholder.

2.1 There is Both External and Internal Oversight

Internal supervision frequently refers to ongoing oversight of supply chains, equipment, and pharmaceuticals by medical facilities. Furthermore, control over the medical supply chain, control over the medical device, and oversight over the medical waste treatment process comprise the three primary categories of external regulation. The executive committee serves as the body in charge of regulation. A chain of custody is a mechanism that involves the transfer of goods or services from suppliers to clients. It consists of individuals, teams, assignments, information, and assets. Ensuring the safety of fragile items during the transportation process is its main objective. The movement of commodities and goods by the company depends on supply chain management. Monitoring the source of raw materials, the manufacturing process, and the delivery of completed goods are essential elements of managing the supply chain for medicines [12]. If supply networks are run by centralized systems, they are susceptible to fraud, theft, and manipulation [13]. In the medical field, effective supply chain management is crucial since reliable supply has a direct impact on the results for patients and health. Using blockchain technology is one way to enhance the security, integrity, and utility of a trustworthy supply chain [14]. Blockchain technology significantly reduces the amount of time needed and the chance of human error by enabling continuous, transparent, and safe supply chain transaction monitoring [15]. Furthermore, the application of blockchain technology can result in safer medical supplies and equipment. Each medical gadget needs to have a unique device identification saved. Firmware upgrades should be distributed and tracked via smart contracts. Invariance can also be used in blockchain-based surveillance systems to stop medical equipment from being stolen, lost, or deliberately altered [12].

2.2 *Electronic Medical Record*

Electronic healthcare record management is a subset of healthcare record administration. Maintaining an up-to-date patient history is made easier with the use of electronic medical records, which enable Internet-based electronic access to conventional paper medical information [16]. It is necessary to let peers know about this very personal information. Sensitive data accessibility, confidentiality, or privacy cannot be guaranteed by current electronic health records, and patient data is still dispersed across multiple medical facility systems for diagnosis and treatment [16, 17]. Treatment optimization (2.3). Two requirements for optimizing treatment are telemedicine and tailored therapy. The number of people receiving health care has increased dramatically in many countries, which makes it harder for patients to get timely treatment from doctors or nurses [10, 18,19]. A blockchain's information contract can be utilized to enable security assessment and supervision of healthcare sensors because it is visible, unchangeable, and anonymous [14].

By way of illustration, Medical Chain has developed an intelligent agreement that allows patients to permit doctors to remotely evaluate their medical conditions and offer advice or contrary opinions [15]. Additionally, the material can be categorized by drug development and study institutes after user authorization to perform pharmaceutically targeted research and development, and the data can be safely transmitted based on an asymmetrical encryption [9].

3 Method

The complexity and fuzziness of the data can be captured using the fuzzy method. DEMATEL makes it easier to negotiate the intricate connections between certain standards. ISM can investigate causal connections between many parameters. The research uses DEMATEL, an ISM methodology, and the idea of fuzzy sets to get meaningful results. In order to account for the internal relationships within the factor that influences the system, it carefully examines ten variables: external regulation (A1), health information leadership (A2), opinions optimizing (A3), medical leadership as a career (A4), health coverage (A5), internal oversight (A6), cost reduction (A7), the highest level of planning (A8), social frameworks (A9), and controlling the environment (A10). To standardize the calculation of the vector and the mixed matrix, the next step is to produce the mix intrinsic contribution matrix using the power source mixed DEMATEL method. Then, to identify the crucial factors influencing smart health care, we estimate the centrality and accidentality of blockchain.

3.1 Unstable DEMATEL

Fuzzy mathematics with DEMATEL is combined in a process known as fuzzy DEMATEL. The triangular fuzzier number (TFN) transforms our language preferences into a computable form. The DEMATEL method is a practical and effective way to find influential elements. In the original DEMATEL technique, the complex interactions between elements are often represented by exact numerical values, which may fail to adequately capture the true context of the situation. In order to overcome the shortcomings of the conventional DEMATEL technique, the fuzzy-mathematical technique is suggested in the study along with a definition of TFN. This method produces more reliable evaluation results that provide a more solid basis for managerial decision-making.

Step 1: With the research problem in mind, we select F1, F2, and Fn to create the system influencing system.

Step 2: Expert scoring, which is represented as a matrix, is used to establish the link between two factors. The degree of influence among components is divided into five categories using a scale we developed to rate semantic knowledge.

Step 3: An immediate impact matrix among the elements and a triangle-shaped fuzzy matrix are created using the Cripps Scoring (CFCS) method to convert fuzzy data.

The CFCS method's precise operation is as follows:

Initial triangular fuzzy number standardization

$$\begin{aligned} xa_{1_{ij}}^k &= \frac{a_{1_{ij}}^k - \min a_{1_{ij}}^k}{\Delta_{\min}^{\max}} \\ xa_{2_{ij}}^k &= \frac{a_{2_{ij}}^k - \min a_{1_{ij}}^k}{\Delta_{\min}^{\max}} \\ xa_{3_{ij}}^k &= \frac{a_{3_{ij}}^k - \min a_{1_{ij}}^k}{\Delta_{\min}^{\max}} \end{aligned} \tag{1}$$

In the second step, we normalize the right and left sides:

Semantic translation variable language and respective TFN values are shown in Table 1.

$$\begin{aligned} xls_{ij}^k &= \frac{xa_{2_{ij}}^k}{\left(1 + xa_{2_{ij}}^k - xa_{1_{ij}}^k\right)} \\ xrs_{ij}^k &= \frac{xa_{3_{ij}}^k}{\left(1 + xa_3^k - xa_{2_{ij}}^k\right)} \end{aligned} \tag{2}$$

Table 1 Semantic translation

Variable language	TFN
No effect, or N	(0, 0.2, 0)
Very low impact (VL)	(0.4, 0, 0.2)
L for minimal effect	(0.6, 0.2, 0.4)
High impact H	(0.6, 0.8, 0.4)
Very high impact, or VH	(1, 0.8, 1)

Third, we determine each expert's score's clear value following blurring:

$$x_{ij}^{k} = \frac{\left[xls_{ij}^{k}\left(1 - xls_{ij}^{k}\right) + xrs_{ij}^{k}xrs_{ij}^{k}\right]}{\left[1 - xls_{ij}^{k} + xrs_{ij}^{k}\right]}$$

$$z_{ij}^{k} = \min a_{1_{ij}}^{k} + x_{ij}^{k} \times \Delta_{\min}^{\max} \tag{3}$$

Fourth, we figure up the typical clarity value:

$$z_{ij}^{k} = \frac{\left(z_{ij}^{1} + z_{ij}^{1} + \cdots + z_{ij}^{k}\right)}{n} \tag{4}$$

Step 4: We normalize the Z-matrix of direct influence:

$$\lambda = \frac{1}{\max_{1 \le i \le n} \sum_{j=1}^{n} z_{ij}}, \quad \mathrm{G} = \lambda Z \tag{5}$$

Step 5: Using T (= G + Gn + G2) or T (= G(E − 1) − 1) as a foundation, e is the constant matrix, and the composite impact matrix T is taken.

Step 6: The composite impact matrix's values, which are created by summing each row or each column, show how much influence and how much being under influence there is Formulas to calculate e include

$$D_i = \sum_{j=1}^{n} t_{ij}, \tag{6}$$

$$R_i = \sum_{j=1}^{n} t_{ij} \tag{7}$$

The function of system components is represented by the significance, which is the outcome of the influence degree and impacted degree. The difference between the strength of an influence and the strength of an impact is one of causation.

4 Results

This paper presents a description of ten elements influencing blockchain-based smart medical care determined by an examination and review of the literature. Seven people with at least 10 years of expertise in the healthcare industry were carefully selected for interviews in November of 2020. As seen in Table 2, the effect degree, affected degree, proximity to amenities, and causation may all be determined using Eqs. (6) through (7). As shown in Fig. 1, a causal network is produced. Figure 1 depicts the numbers graphically. The significance and causality of the items in Fig. 1 are clearly visible. Groups the 10 danger indicators into cause sets or consequence sets depending on how severe the underlying cause is: A2 management of health records, A4 doctor administration, A3 therapy improvement, A5 charitable organizations, A7 cost savings, and A6 internal control are a few of the seven components. Their primary motivators are A8 design at the graduate level, A2 health-related data management, and A4 doctor management. The top three variables, namely A2, A4, and A8, have successive influence degrees of 2.3122, 2.4441, and 2.4071. This shows that these three factors influence other variables the most. Table 3 displays the accessible matrix.

As can be seen in the above table, variables A5 and A10 reflect the intersection of both the accessibility setting and the public set. At this point, elements A5 and A10 become first-order impact factors. We exclude all the columns and rows in modules

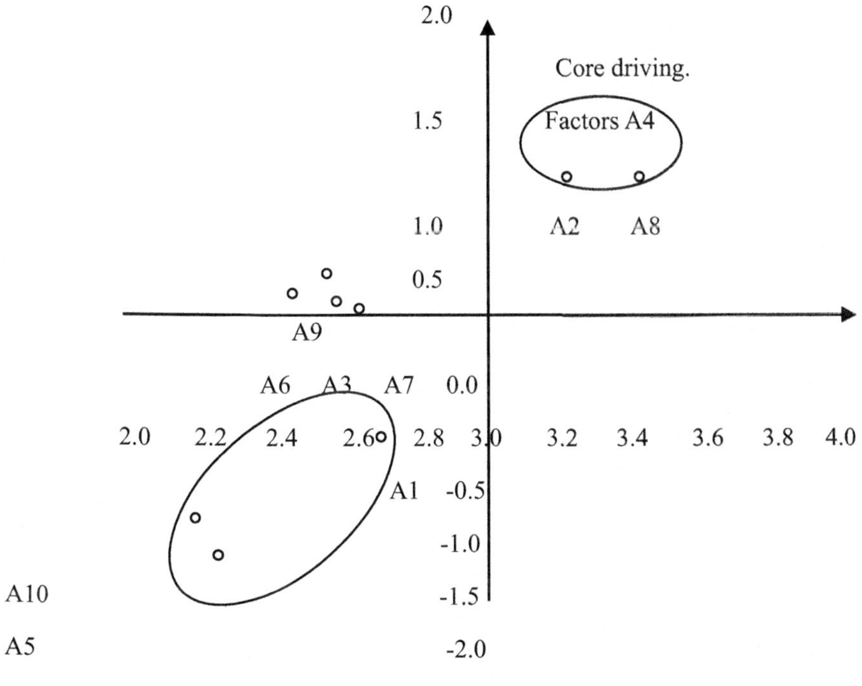

Fig. 1 Causation DEMATEL diagram

M that are connected to the variable. A5 and A10 in order to produce a more intricate breakdown matrix. The variables are determined at all levels using an analogy. After multiple division steps, three first-level nodes N1 5, 10; six second-level columnar N2 1, 6; three third-level networking N3 3, 9, 7; and three fourth-level network N4 4, 2, 8 of all the levels are eventually generated. The ISM, whose is depicted in Figure, was established by the study using the approach stated above.

The ISM analysis indicates that the first areas where cryptocurrencies have an impact on intelligent health care are the top-level design (A8), managing medical records (A2), and healthcare administration (A4). The challenge is to determine the optimum technique to monitor and control these elements as soon as they start to change.

In summary, in the end, several factors may influence how blockchain technology is used to create intelligent health care. A system integration paradigm for developing sustainable, intelligent medicine.

As seen in Table 2, the effect degree, affected degree, proximity to amenities, and causation may all be determined using Eqs. (6) through (7).

As shown in Fig. 1, a causal network is produced. Figure 1 depicts the numbers in Table 2 graphically. The significance and causality of the items in Fig. 1 are clearly visible.

Table 3 groups the 10 danger indicators into cause sets or consequence sets depending on how severe the underlying cause is. A2 management of health records, A4 doctor administration, A3 therapy improvement, A5 charitable organizations, A7 cost savings, and A6 internal control are a few of the seven components. Their primary motivators are A8 design at the graduate level, A2 health-related data management, and A4 doctor management.

Table 3 groups the 10 danger indicators into cause sets or consequence sets depending on how severe the underlying cause is. A2 management of health records, A4 doctor administration, A3 therapy improvement, A5 charitable organizations, A7 cost savings, and A6 internal control are a few of the seven components. Their primary

Table 2 Comprehensive impact matrix analysis

Factor	Influence Degree	Affected degree	Centrality	Causality
A1	0.297	1.0786	2.7643	−0.3719
A2	2.3211	1.7454	3.8440	1.4104
A3	1.2345	1.432	2.7634	0.8106
A4	2.341	1.8273	3.1653	1.7245
A5	0.8976	2.6453	2.5344	−1.6848
A6	1.423	1.8374	2.3464	0.5182
A7	1.8769	1.8972	2.8123	0.6001
A8	2.0386	1.1397	3.87364	1.8075
A9	1.2345	1.0873	2.1234	0.1418
A10	0.9876	1.1911	2.0159	−1.1763

Table 3 Reachable matrix

M	A1	A2	A3	A4	A5	A6	A7	A8	A9	A10
A1	1	0	0	0	1	1	0	0	0	1
A2	0	1	1	1	1	1	0	1	1	1
A3	1	0	1	0	0	1	0	0	0	0
A4	1	1	1	1	1	0	0	1	1	1
A5	1	0	0	0	1	0	1	0	0	0
A6	1	0	0	0	1	1	1	0	0	0
A7	0	0	0	0	0	0	0	0	0	0
A8	1	1	1	1	1	0	1	1	1	1
A9	0	0	0	0	1	0	0	0	1	1
A10	0	0	0	0	0	0	0	0	0	1

motivators are A8 design at the graduate level, A2 health-related data management, and A4 doctor management.

A structure built around mathematics was necessary because the influence technique, impact system, and scope of action vary depending on the topics covered by the blockchain.

5 Discussions

Examining the blockchain-based intelligent medical advancement system is the aim of this research. The subject of cognitive health care has not made considerable use of blockchain technology, and it lacks a comprehensive hierarchical framework. This study not only provided hierarchical architecture but also some recommendations for the smart healthcare sector. The smart acceptance, a sort of intellectual protocol intended to disseminate, authenticate, or convey an arrangement throughout an information format, is used in all system interactions. As a result, the system's overall compass is the top-level design. A blockchain-based smart medical system cannot be built or improved unless the top-level architecture is improved. An intelligent healthcare system based on blockchain is predicated on the comprehensive administration of medical records. Electronic medical records are essential in the healthcare industry and contain extremely private information that must be constantly exchanged across peers [8]. To enable continuous monitoring of patients and medical operations, as well as to keep an encrypted track of who triggered these actions, a smart contractual system sends messages to both patients and community physicians [20]. By using this data, the community doctor lowers resident care costs by determining if patients need additional care at the institution before initiating the resident's first diagnosis and course of treatment. Intelligent agreement can also be used to create community health initiatives utilizing tenancy health data, protecting the health of tenants. The cost savings cover both the price of discovering and studying new pharmaceuticals as well as the costs associated with interactions between physicians and patients. A critical first step toward improving the quality of service and intelligence of the medical system is the sharing of healthcare information on a blockchain, which also helps patients control their prescriptions and expenses [7]. Multisite studies can reduce trial costs for drug development and research businesses, and multisite medical trials can reduce management costs because of a system for managing data built on smart contract blockchain technology [11]. They represent the final stage of connecting the blockchain to the sophisticated healthcare system. The two elements of health insurance are source control and enhancing risk management capabilities.

6 Conclusions

However, the current state of the intelligent health sector makes it unsafe for institutions to share patient data with one another, and safeguarding privacy has become a bottleneck [21]. Current research in the field of smart healthcare focuses on discrete information technologies and the growing role of the Internet of all that in health care [22]. While others [20] focus on using a particular blockchain characteristic to better medical care, there are examinations that combine digital currencies with smart medical care as well. These studies usually use individuals and healthcare providers as research participants. Few research considers multi-agent cooperative development, and none of the studies propose a blockchain-based software framework for intelligent medical development. Finally, China provides the great majority of the data that is contained in this study. Subsequent research endeavors could explore a multifaceted and intricate structure for blockchain-driven intelligent medical governance, considering an extensive array of technical and cultural viewpoints.

References

1. Mirazand MH, Donald DC (2018) Application of blockchain in booking and registration systems of securities exchanges. In: Proceedings of the 2018 international conference on computing, electronics & communications engineering (iCCECE). IEEE, Southend, UK, pp 35–40
2. Andolfatto D (2018) Blockchain: what it is, what it does, and why you probably don't need one. Fed Reserve Bank St. Louis Rev 100(2):87–95
3. Appelbaum D, Smith SS (2018) Blockchain basics and hands-on guidance: taking the next step toward implementation and adoption. CPA J 88(6):28–37
4. Pal O, Alam B, Thakur V, Singh S (2019) Key management for blockchain technology. In: ICTExpress
5. Hyla T, Pejaś J (2020) Long-term verification of signatures based on a blockchain. Comput Electr Eng 81:106523

6. Dagher GG, Mohler J, Milojkovic M, Marella PB (2018) Ancile: privacy-preserving framework for access control and interoperability of electronic health records using blockchain technology. Sustain Cities Soc 39:283–297
7. Yue X, Wang H, Jin D, Li M, Jiang W (2016) Healthcare data gateways: found healthcare intelligence on blockchain with novel privacy risk control. J Med Syst 40(10):218
8. Gong T, Huang H, Li P, Zhang K, Jiang H (2015) A medical healthcare system for privacy protection based on IoT. In: Proceedings of the 2015 seventh international symposium on parallel architectures, algorithms and programming (PAAP). IEEE, Nanjing, China, pp 217–222
9. Underwood S (2016) Blockchain beyond bitcoin. Commun ACM 59(11):15–17
10. Yuanand Y, Wang FY (2016) Blockchain: the state of theart and future trends. Acta Automatica Sinica 42(4):481–494
11. Azaria A, Ekblaw A, Vieira T, Lippman A (2016) Medrec: using blockchain for medical data access and permission management. In: Proceedings of the international conference on open & BigData, Vienna, Austria
12. Boulos MNK, Wilson JT, Clauson KA (2018) Geospatial blockchain: promises, challenges, and scenarios in health and healthcare. Int J Health Geogr 17:25
13. Azzi R, Chamoun RK, Sokhn M (2019) The power of a blockchain-based supply chain. Comput Ind Eng 135:582–592
14. Rawat A., Suryavanshi RS, Nagar V, Yadav D (2023) Development of blockchain-enabled e-healthcare appointment system. In: International conference on trends in computational and cognitive engineering, Singapore. Springer Nature, pp. 359–369
15. Jamil F, Hang L, Kim K, Kim D (2019) A novel medical blockchain model for drug supply chain integrity management in a smart hospital. Electronics 8(5):505
16. Guo R, Shi H, Zhao Q, Zheng D (2018) Secure attribute-based signature scheme with multiple authorities for block-chain in electronic health records systems. IEEE Access 6:11676–11686
17. Oliveira MT, Carrara GR, Fernandesetal NC (2019) Towardsa performance evaluation of private blockchain frameworks using a realistic workload. In: Proceedings of the 2019 22nd conference on innovation in clouds, internet and networks and workshops (ICIN). IEEE, Paris, France, pp 180–187
18. Zheng Z, Xie S, Dai HN, Chen X, Wang H (2018) Blockchain challenges and opportunities: a survey. Int J Web Grid Serv 14(4):352–375
19. Thomason J, Ahmad M, Bronderetal P (2018) Block-chain—powering and empowering the poor in developing countries. In: Transforming climate finance and green investment with Blockchains Academic Press, Cambridge, MA, USA
20. Sylim P, Liu F, Marcelo A, Fontelo P (2018) Blockchain technology for detecting falsified and substandard drugs in distribution: pharmaceutical supply chain intervention. JMIR Res Protoc 7(9)
21. Bocek T, Rodrigues BB, Strasser T, Stiller B (2017) Blockchains everywhere-a use-case of blockchains in thepharmasupply-chain. In: Proceedings of the 2017IFIP/IEEE symposium on integrated network and service management (IM). IEEE, Lisbon, Portugal, pp 772–777
22. Kumar G, Saha R, Rai MK, Thomas R, Kim T-H (2019) Proof-of-work consensus approach in block chain technology for cloud and fog computing using maximization-factorization statistics. IEEE Internet of Things J 6(4):6835–6842
23. Hathaliya JJ, Tanwar S, Tyagi S, Kumar N (2019) Securingelectronics healthcare records in healthcare 4.0: a biometric-based approach. Comput Electr Eng 76:398–410
24. Kosba A, Miller A, Shi E, Wen Z, Papamanthou C (2016) Hawk: the blockchain model of cryptography and privacy-preserving smart contracts. In: Proceedings of the 2016 IEEE symposium on security and privacy (SP). IEEE, SanJose, CA, USA, pp 839–858
25. Gramoli V (2016) On the danger of private blockchains. In: Proceedings of the workshop on distributed cryptocurrencies and consensus ledgers (DCCL'16), Chicago, IL, USA

Securing Digital Integrity: Proposed Comprehensive Framework for Deepfake Detection and Blockchain Validation

Anant Jain, Adamya Gaur, Gauranshi Gupta, Shubhangi Mishra, Rahul Johari, and Deo Prakash Vidyarthi

Abstract This paper proposes a comprehensive solution to combat the growing threat of Deepfake technology, employing Convolutional Neural Networks (CNNs) and Blockchain. CNNs analyze video frames for anomalies indicative of Deepfake manipulation, while Blockchain ensures content integrity through timestamping and IPFS hashes. Leveraging transfer learning with EfficientNet-B1 architecture and dropout layers to prevent overfitting, the CNN model attains a 98.53% accuracy on testing data. Smart contracts are utilized to store social media content and AI verification results. The methodology integrates private storage on social media platforms, AI-based CNN verification, Blockchain timestamping, smart contract verification, and continuous improvement. By synergizing AI and Blockchain technologies, our approach aims to bolster defenses against deceptive content proliferation, enhancing the reliability of online media ecosystems.

Keywords Blockchain · IPFS · Convolutional neural network · Deepfake

A. Jain · A. Gaur · G. Gupta · S. Mishra · R. Johari (✉)
SWINGER: Security, Wireless, IoT Network Group of Engineering and Research, University School of Automation and Robotics (USAR), Guru Gobind Singh Indraprastha University, Delhi, India
e-mail: rahul@ipu.ac.in

A. Jain
e-mail: anant.07019011921@ipu.ac.in

A. Gaur
e-mail: adamya.04719011921@ipu.ac.in

G. Gupta
e-mail: gauranshi.00719011921@ipu.ac.in

S. Mishra
e-mail: shubhangi.09919011721@ipu.ac.in

D. P. Vidyarthi
School of Computer and Systems Sciences, Parallel and Distributed System Lab JNU, Delhi, India
e-mail: dpv@mail.jnu.ac.in

P. J. S. Gonçalves et al. (eds.), *Proceedings of Fifth International Conference on Computing, Communications, and Cyber-Security*, Lecture Notes in Networks and Systems 1128, https://doi.org/10.1007/978-981-97-7371-8_45

1 Introduction

Deepfake is a potent Artificial Intelligence-powered threat wherein Artificial Intelligence produces realistic fake content, often in the form of video or audio, impersonating real individuals. This technology enables malicious actors to manipulate and spread deceptive content, posing severe risks to individuals and society. Deepfakes can deceive, defame, and incite misinformation, contributing to identity theft, fraud, and political manipulation.

AI algorithms, particularly Convolutional Neural Networks (CNNs), play a pivotal role in detecting fake content by analyzing video frames. CNNs excel at identifying anomalies within individual frames but detection of Deepfake gets harder when the input images are of low quality, as the distinguishing feature visibility reduces. The dataset used in the experiment consists of low-quality videos, which further aids in creating a more robust model. During analysis, the network identifies patterns, textures, and facial features unique to the subject. Temporal dependencies were also considered as CNNs evaluated the consistency of these features across frames. By learning from extensive datasets, the algorithm distinguished between genuine and manipulated content, uncovering discrepancies in pixel-level details and facial expression. This approach empowers AI to expose inconsistencies indicative of Deepfake manipulation, contributing to the battle against deceptive digital content.

Blockchain is a decentralized and tamper-resistant ledger technology that records transactions across a network of computers. It utilizes hashing techniques to ensure data integrity, immutability, and transparency. Each block in the chain contains a timestamp and a hash of the previous block, forming a secure and chronological sequence of records. In the context of combating fake content, Blockchain serves as a second layer of validation. Timestamps on the Blockchain provide an immutable record of when the original video was verified. By comparing timestamps between the original and suspected fake videos, inconsistencies can be detected, revealing potential manipulations. The use of InterPlanetary File System (IPFS) hashes on the Blockchain ensures content integrity. If the IPFS hash of the suspected fake video differs from the hash recorded for the original on the Blockchain, it indicates alterations. This dual validation, through timestamps and IPFS hashes, enhances content verification, reinforcing the authenticity of original content and detecting potential fakes.

2 Problem Statement

The rise of Deepfake technology poses a critical threat to individuals and society, allowing malicious actors to create malicious intent. This technology exposes individuals and society to severe risks, including deception, defamation, and the spread of misinformation, contributing to identity theft, fraud, and political manipulation.

This research paper addresses the need for robust solutions to combat Deepfake by proposing a comprehensive approach that integrates advanced AI algorithms, specifically Convolutional Neural Networks (CNNs), with the tamper-resistant capabilities of Blockchain technology.

This research focuses on leveraging CNNs to analyze video frames, identifying patterns, textures, and facial features unique to subjects. Temporal dependencies are considered, enabling the algorithm to evaluate consistency across frames and expose inconsistencies indicative of Deepfake manipulation. The research also explored the integration of Blockchain as a second layer of validation, utilizing timestamps and IPFS hashes to ensure content integrity. Through this research aimed to contribute to the ongoing battle against deceptive digital content and fortify the resilience of online media ecosystems.

3 Literature Survey

In [1], the author proposes a Deepfake detection system to combat manipulated videos on platforms like YouTube, mitigating risks of political manipulation and social unrest. Utilizing Blockchain for transparency, multiple content providers collaborate, each employing their deep learning model for detection. Integrated voting combines results for objectivity, with user involvement enhancing accuracy through collective intelligence. The system fosters cooperation among companies with competing interests, establishing a consortium contract environment to combat Deepfake threats effectively.

In [2], the challenges posed by Artificial Intelligence-Generated Content (AIGC) are addressed in terms of service latency, security, and trustworthiness. It introduces a novel Blockchain-empowered framework for managing the lifecycle of edge AIGC products. The proposed Proof-of-AIGC protocol safeguards ownership. Additionally, a multi-weight subjective logic-based reputation scheme assists AIGC producers in selecting trustworthy edge service providers. Numerical results validate the effectiveness of the approach, and the survey concludes by highlighting avenues for future research in the intersection of AIGC, Blockchain, and edge networks.

The paper [3] reviews the proliferation of Deepfake technology in creating convincingly fake videos and images, often targeting public figures. It discusses notable research contributions and popular Deepfake tools, presenting two taxonomies for categorization. The analysis compares these models and tools based on algorithms, datasets, and accuracy. Despite advancements, challenges encountered include difficulties in detecting low-quality Deepfake, time-consuming implementation of detection techniques in real-world settings, and training Deepfake models on limited data for specific characters.

In [4], the author discusses that misinformation poses a significant threat across society, economy, and politics, fueled by the rapid dissemination facilitated by AI technologies like deep learning. To combat this, a novel approach integrates Blockchain and AI. Blockchain ensures content authenticity, while AI models detect fake news. Challenges include integrating these technologies seamlessly and ensuring transparency. This synergy marks a promising avenue, albeit with ongoing research to refine implementation and address potential limitations.

In [5], the author explores that Deepfake technology presents a critical challenge in discerning between genuine and fraudulent digital content online. Using AI, machine learning, and deep learning, Deepfake produces hyper-realistic yet deceitful material. Addressing this issue requires innovative approaches to reliably distinguish between authentic and fake multimedia. This paper reviews Deepfake technology, its associated threats, challenges, and future directions, emphasizing the need for robust detection methods. Critically, it highlights key techniques for securing digital integrity, crucial amidst the evolving landscape of fake content dissemination.

In [6], the author offers a comprehensive examination of social media data privacy policies and introduces a trust index model leveraging Blockchain technology to combat fake news and Deepfake. Privacy breaches on social media have become a pressing concern, especially highlighted during the COVID-19 infodemic. The proposed model aims to restore trust and reliability by providing transparent and secure information dissemination. Challenges include integrating Blockchain seamlessly into social media platforms and ensuring user adoption. This model represents a crucial step toward fostering a more trustworthy online environment amidst rampant misinformation.

In [7], the surge in tempered electronic content facilitated by advanced artificial intelligence, particularly Deepfake, is addressed. The study highlights the inadequacy of current solutions in tracking digital media's history and provenance. It explores the potential of Blockchain technology, including innovations like Smart Contracts and Hyperledger Fabric, to combat the Deepfake problem. The paper proposes a Blockchain-dependent framework for preserving data integrity and preventing Deepfake. The study also delves into the current state of Deepfake video detection research, covering the generating process, detection algorithms, and benchmarks.

In [8], prevailing Deepfake literature predominantly emphasizes AI-assisted detection methods, overlooking the crucial aspect of content 'authentication'. This paper diverges by exploring Blockchain integration for tamper-proof evidence of genuine content, addressing a significant gap in current research. Despite scattered efforts in this direction, a comprehensive guide for the effective adoption of Blockchain against Deepfake is absent. The paper provides insights into potential use cases and solutions, offering a holistic guide to leverage Blockchain effectively in navigating the challenges posed by Deepfake technology.

The paper [9] explores the influence of digital deception, like Deepfake and misinformation, via the Internet and social media. It suggests leveraging distributed ledger technologies, specifically Blockchain, to combat such deception. By offering transparent and immutable records in a peer-to-peer framework, DLTs enhance data authenticity and traceability. The survey delves into DLT applications, tackles associated challenges, and suggests future research directions to fortify resilience against cyber threats.

The paper [10] addresses the proliferation of Deepfake, driven by advances in AI and deep learning, posing threats to reality and trust. Emphasizing the need for proof of authenticity, the paper critiques current solutions for lacking robust history tracking. Proposing a solution, it introduces an Ethereum smart contract framework using IPFS hashes to trace digital content back to its origin, applicable to various digital forms. The approach prioritizes traceability to trusted sources for content authenticity. The open-sourced smart contract code on GitHub encourages collaboration to combat challenges posed by fake digital content.

4 Methodology Adopted

4.1 Flowchart

Figure 1 shows the architecture of AI and Blockchain-based model for Deepfake detection.

4.2 Simulation Environment

The proposed smart contract for the Blockchain was simulated and executed using the Remix IDE, a powerful development tool. Remix IDE facilitated the compilation, deployment, and testing of the smart contract through its user-friendly interface. Google Colab was employed to simulate the process of Deepfake detection. This cloud-based platform allowed for the execution of the Deepfake detection algorithm collaboratively and efficiently. The combination of Remix IDE for Blockchain simulation and Google Colab for Deepfake detection provided a comprehensive testing environment for the proposed solution.

4.3 Screenshot of Blockchain Simulation in Remix IDE

Figure 2 highlights three primary functions within the smart contract: timestampVideo(), responsible for storing the video hash and AI verification result on the Blockchain, along with several read functions such as getAIVerification() which takes videoId as input and returns true or false that AI verification for a video is done or not and getVideoDetails() that also takes videoId as input and returns video hash as output, enabling retrieval of results from the Blockchain.

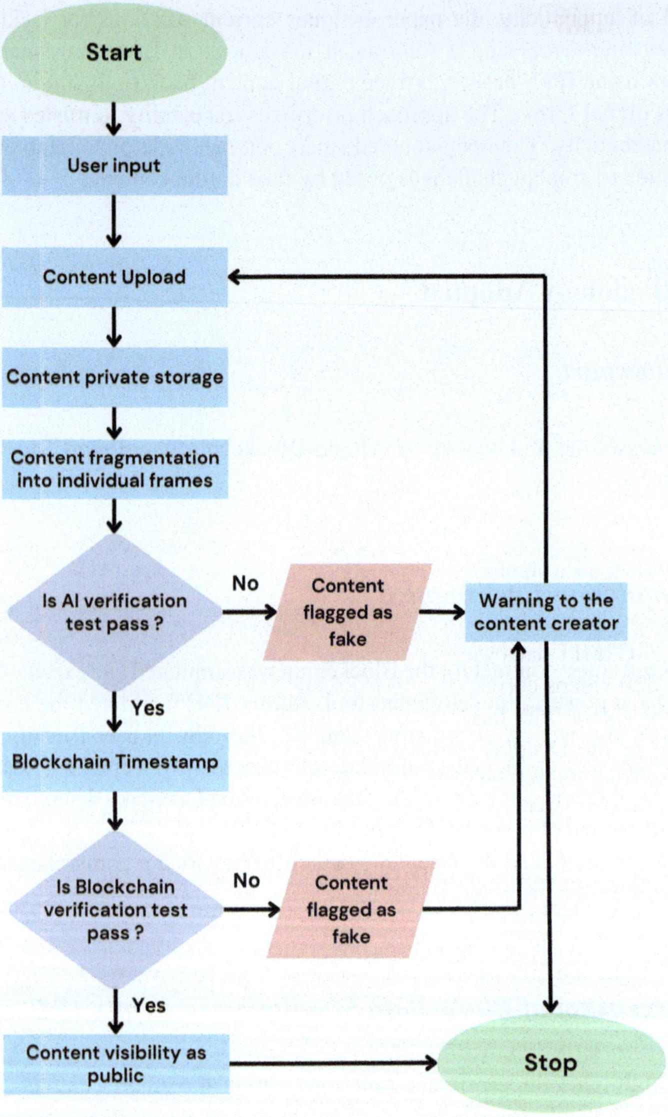

Fig. 1 Flow of the Deepfake content detection system

4.4 Dataset and Pre-processing

Data used for the training of a model plays a crucial role in its performance. Data from DeepfakeTIMIT [11] and VidTIMIT [12] were used for training the model. It contains 10 face-swapped videos of each individual and no audio content. For creating the DeepfakeTIMIT dataset, 16 similar-looking pairs of people were selected manually

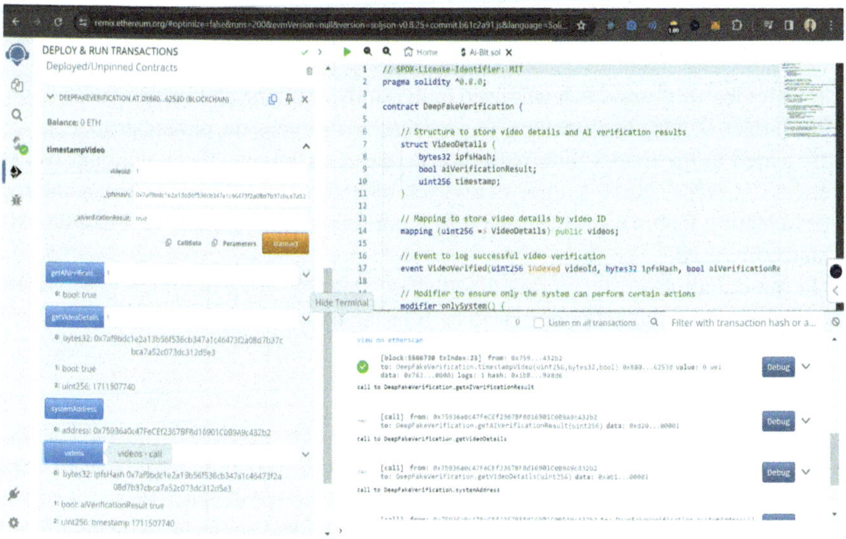

Fig. 2 Implementation of solidity code in Remix IDE

from the VidTIMIT database. As there were 10 videos per person in the VidTIMIT database, 320 videos corresponding to each version were generated which resulted in 620 total videos with faces swapped. Different methodologies were used for the generation of the Deepfake videos. Further, frames were extracted out from the videos, enabling the model to perform a frame-by-frame analysis of the Deepfake content. Figure 3 displays the real and fake frames of 5 such individuals. To create a robust model, the frames were further processed using different techniques in the training data. This included random horizontal and vertical shifting and zoom in and out up to 10%. Shearing and horizontal flipping of images were also done. All the tensor values were standardized. The data was further divided into 80% training, 10% validation, and 10% testing sets.

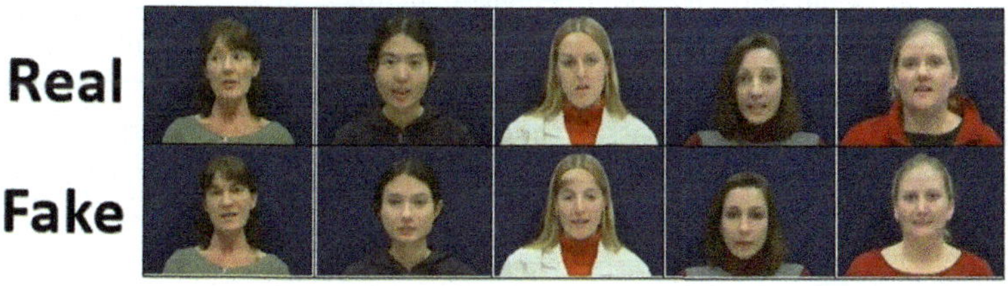

Fig. 3 Real and fake images of individuals [11, 12]

5 Results

A transfer learning approach was used in the study, utilizing and further customizing EfficientNet-B1 [13] architecture. To prevent overfitting by the model, dropout layers were added to the model. Internally, the model is based on a Convolutional Neural Network (CNN), where a filter convolves over the image, leading to an internal representation that is useful for extracting important features. Figure 4 displays the architecture used.

In the output layer, the sigmoid activation function is used, which provides the likelihood of a frame being real or fake. It is calculated as

$$S(x) = \frac{1}{1 + e^{-x}}$$

where e is Euler's number and x is the input value. Using the aggregate of these scores and threshold values, the final authenticity score is calculated. Five fold cross-validation has been conducted on the model, to check for overfitting shown in Fig. 5. The proposed model achieved a 98.53% accuracy score on the testing set. Figure 6 displays the accuracy versus epoch and loss versus epoch line plots.

6 Conclusion and Future Work

A smart contract was successfully developed and tested to store timestamped videos, fetch AI test results, and retrieve video details. The smart contract, implemented on Remix IDE, demonstrated robust functionality. Additionally, the Deepfake detection system, utilizing CNN on Google Colab, yielded an impressive accuracy of 98.53%. This integrated solution showcases the efficacy of combining Blockchain and AI in combating Deepfake, offering a reliable framework for authenticating and securing digital content.

In future work, the research will expand its focus beyond visual content to include audio content, as the current model primarily analyzes image frames from videos. Additionally, the study will be extended to develop a fully integrated Deepfake detection system, combining Blockchain and AI technologies seamlessly. This integrated system will encompass all necessary components to detect and authenticate both visual and audio content, ensuring a comprehensive approach to combat Deepfake.

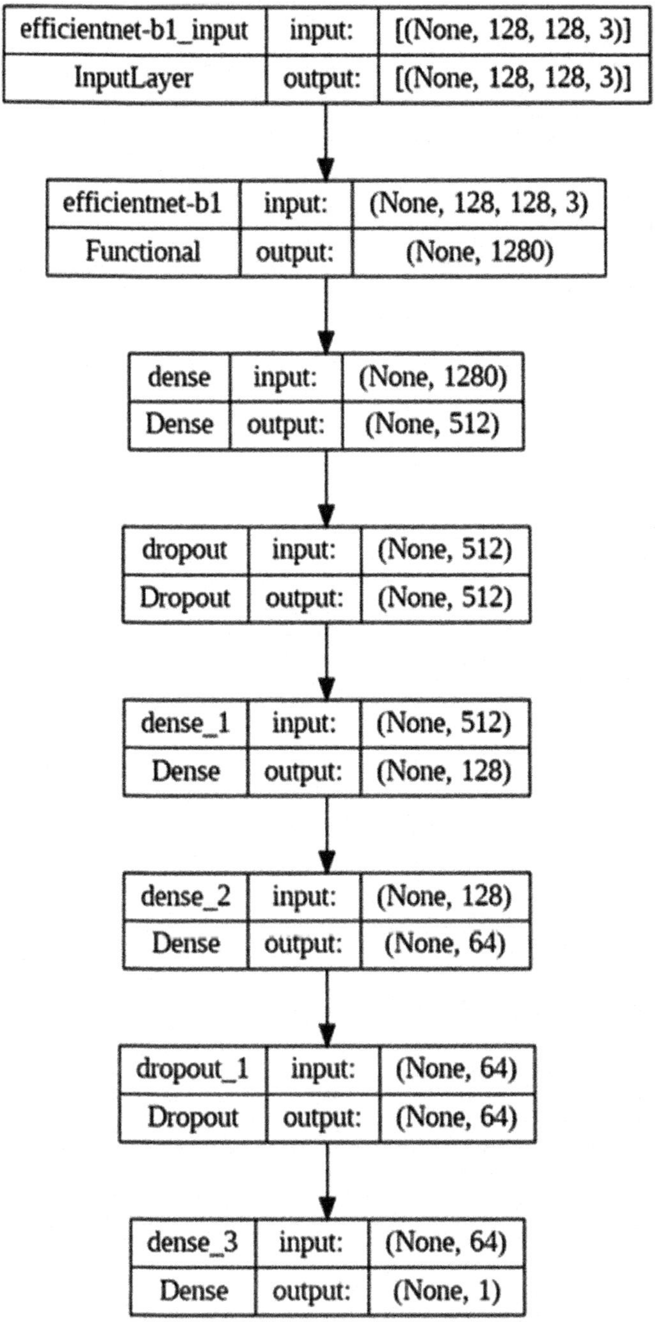

Fig. 4 Deep learning model architecture

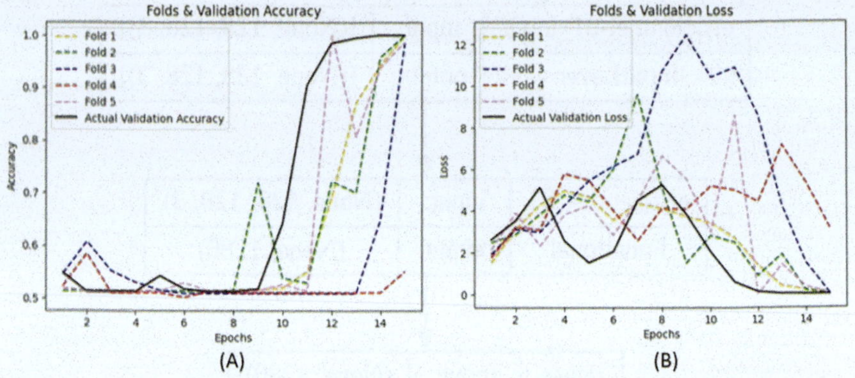

Fig. 5 fivefold cross-validation **a** Accuracy versus Epoch curves and **b** Loss veruss Epoch curves for training and validation

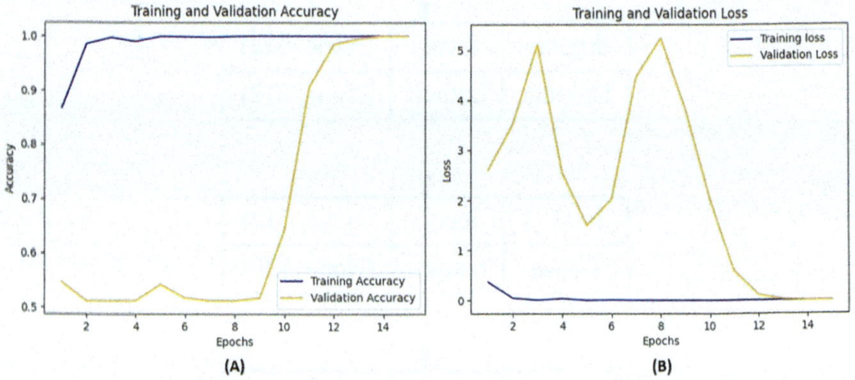

Fig. 6 **a** Accuracy versus Epoch curves and **b** Loss versus Epoch curves for training and validation

References

1. Choi N, Kim H (2023) DDS: deepfake detection system through collective intelligence and deep-learning model in blockchain environment. Appl Sci 13(4):2122
2. Liu Y, Du H, Niyato D, Kang J, Xiong Z, Miao C-Y, Jamalipour A (2023) BlockChain-empowered lifecycle management for AI-generated content (AIGC) products in edge networks. arXiv preprint arXiv:2303.02836
3. Mukta, MdSH, Ahmad J, Azam Khan Raiaan M, Islam S, Azam S, Ali ME, Jonkman M (2023) An investigation of the effectiveness of Deepfake models and tools. J Sensor Actuator Netw 12(4):61
4. Seneviratne O (2022) Blockchain for social good: Combating misinformation on the web with AI and blockchain. In: Proceedings of the 14th ACM web science conference 2022, pp 435–442
5. Sharma M, Kaur M (2021) A review of Deepfake technology: an emerging AI threat. Soft Comput Secur Appl Proc ICSCS 2022:605–619
6. Jing TW, Murugesan RK (2021) Protecting data privacy and prevent fake news and deepfake in social media via blockchain technology. In: Advances in cyber security: second international

conference, ACeS 2020, Penang, Malaysia, December 8–9, 2020, Revised Selected Papers 2, pp. 674–684. Springer Singapore

7. Rashid MdM, Lee S-H, Kwon K-R (2021) BlockChain technology for combating Deepfake and protect video/image integrity 24(8):1044–1058
8. Yazdinejad A, Parizi RM, Srivastava G, Dehghantanha A (2020) Making sense of BlockChain for AI Deepfake technology. In: 2020 IEEE Globecom workshops (GC Wkshps, pp 1–6. IEEE)
9. Fraga-Lamas P, Fernandez-Carames TM (2020) Fake news, disinformation, and Deepfake: leveraging distributed ledger technologies and BlockChain to combat digital deception and counterfeit reality. IT profes sional 22(2):53–59
10. Hasan HR, Salah K (2019) Combating deepfake videos using blockchain and smart contracts. IEEE Access 7:41596–41606
11. Korshunov P, Marcel S (2018) Deepfake: a new threat to face recognition? assessment and detection. arXiv preprint arXiv:1812.08685 (2018). https://zenodo.org/records/4068245
12. Sanderson C, Lovell BC (2009) Multi-region probabilistic histograms for robust and scalable identity inference. Advances in Biometrics: Third International Conference, ICB 2009, Alghero, Italy, June 2–5, 2009. Proceedings 3. Springer, Berlin Heidelberg, pp 199–208
13. Tan M, Le Q (2019) Efficientnet: Rethinking model scaling for convolutional neural networks. In: International conference on machine learning, pp 6105–6114. PMLR

Author Index

P. J. S. Gonçalves et al. (eds.), *Proceedings of Fifth International Conference on Computing, Communications, and Cyber-Security*, Lecture Notes in Networks and Systems 1128, https://doi.org/10.1007/978-981-97-7371-8

SPRINGER NATURE

GPSR Compliance

The European Union's (EU) General Product Safety Regulation (GPSR) is a set of rules that requires consumer products to be safe and our obligations to ensure this.

If you have any concerns about our products, you can contact us on ProductSafety@springernature.com

In case Publisher is established outside the EU, the EU authorized representative is:

Springer Nature Customer Service Center GmbH
Europaplatz 3
69115 Heidelberg, Germany

The manufacturer's authorised representative in the EU is Springer Nature Customer Service Centre GmbH, Europaplatz 3, 69115 Heidelberg, Germany. If you have any concerns regarding our products, please contact ProductSafety@springernature.com

Printed and bound by CPI Group (UK) Ltd, Croydon, CR0 4YY
01/10/2026
02225901-0011